*Immunopharmacology
of the
Respiratory System*

THE HANDBOOK OF IMMUNOPHARMACOLOGY

Series Editor: Clive Page
King's College London, UK

Titles in this series

Cells and Mediators

Immunopharmacology of Eosinophils
(edited by H. Smith and R. Cook)

The Immunopharmacology of Mast
Cells and Basophils
(edited by J.C. Foreman)

Lipid Mediators
(edited by F. Cunningham)

Immunopharmacology of Neutrophils
(edited by P.G. Hellewell and
T.J. Williams)

Immunopharmacology of
Macrophages and other
Antigen-Presenting Cells
(edited by
C.A.F.M. Bruijnzeel-Koomen and
E.C.M. Hoefsmit)

Adhesion Molecules
(edited by C.D. Wegner)

Immunopharmacology of Lymphocytes
(edited by M. Rola-Pleszczynski)

Immunopharmacology of Platelets
(edited by M. Joseph)

Immunopharmacology of Free Radical
Species
(edited by D. Blake and
P.G. Winyard)

Cytokines
(edited by A. Mire-Sluis, forthcoming)

Systems

Immunopharmacology of the
Gastrointestinal System
(edited by J.L. Wallace)

Immunopharmacology of Joints
and Connective Tissue
(edited by M.E. Davies and J. Dingle)

Immunopharmacology of the Heart
(edited by M.J. Curtis)

Immunopharmacology of Epithelial
Barriers
(edited by R. Goldie)

Immunopharmacology of the
Renal System
(edited by C. Tetta)

Immunopharmacology of the
Microcirculation
(edited by S. Brain)

Immunopharmacology of the
Respiratory System
(edited by S.T. Holgate)

The Kinin System
(edited S. Farmer, forthcoming)

Drugs

Immunotherapy for Immune-
related Diseases
(edited by W.J. Metzger,
forthcoming)

Immunopharmacology of AIDS
(forthcoming)

Immunosuppressive Drugs
(forthcoming)

Glucocorticosteroids
(forthcoming)

Angiogenesis
(forthcoming)

Phosphodiesterase Inhibitors
(edited by G. Dent, K. Rabe
and C. Schudt, forthcoming)

Immunopharmacology
of the
Respiratory System

edited by

Stephen T. Holgate
University Medicine
Southampton General Hospital
Southampton, UK

ACADEMIC PRESS
Harcourt Brace and Company, Publishers
London San Diego New York
Boston Sydney Tokyo Toronto

ACADEMIC PRESS LIMITED
24/28 Oval Road
London NW1 7DX

United States Edition published by
ACADEMIC PRESS INC.
San Diego, CA 92101

This book is printed on acid-free paper

A catalogue record for this book
is available from the British Library

ISBN 0-12-352325-7

Typeset by Mathematical Composition Setters Ltd, Salisbury, Wiltshire
Printed and bound in the United Kingdom
Transferred to Digital Printing, 2011

Contents

4. Mast Cells and Basophils: Their Role in Initiating and Maintaining Inflammatory Responses 53

Peter Bradding, Andrew F. Walls *and* Martin K. Church

5. Eosinophils: Effector Leukocytes of Allergic Inflammatory Responses 85

Kaiser G. Lim *and* Peter F. Weller

6. Cytokine Regulation of Chronic Inflammation in Asthma 101

P.J. Barnes, K.F. Chung and I. Adcock

7. Neural Networks in the Lung 123

Stephanie A. Shore, Craig M. Lilly, Benjamin Gaston and Jeffrey M. Drazen

8. The Microvasculature as a Participant in Inflammation 147

Jeffrey J. Bowden *and* Donald M. McDonald

9. Regulation of Airway Smooth Muscle 169

Charles Twort

10. *The Airway Epithelium: The Origin and Target of Inflammatory Airways Disease and Injury* 187

Clive Robinson

11. *Transition Between Inflammation and Fibrosis in the Lung* 209

A. Wangoo *and* R.J. Shaw

12. The Cell Biology of the Resolution of Inflammation 227

C. Haslett

Glossary 245

Key to Illustrations 255

Index 261

The color plates are located between pp 208–209.

Contributors

I. Adcock
Department of Thoracic Medicine,
National Heart and Lung Institute,
Dovehouse Street,
London SW3 6LY,
UK

P.J. Barnes
Department of Thoracic Medicine,
National Heart & Lung Institute,
Dovehouse Street,
London SW3 6LY,
UK

Jeffrey J. Bowden
Department of Medicine,
Flinders Medical Centre,
Bedford Park, SA 5042
Australia

Peter Bradding
Department of Respiratory Medicine,
The Glenfield Hospital,
Groby Road
Leicester LE3 9QP
UK

K.F. Chung
Department of Thoracic Medicine,
National Heart & Lung Institute,
Dovehouse Street,
London SW3 6LY,
UK

Martin K. Church
Immunopharmacology Group,
Southampton General Hospital,
Southampton SO16 6YD,
UK

C.J. Corrigan
Department of Allergy & Clinical Immunology,
National Heart & Lung Institute,
Dovehouse Street,
London SW3 6LY,
UK

Jeffrey M. Drazen
Respiratory Division,
Brigham and Women's Hospital,
75 Francis Street,
Boston,
MA 02115,
USA

Benjamin Gaston
Department of Pediatrics,
Naval Medical Center,
San Diego,
CA,
USA

Chris Haslett
Respiratory Medicine Unit,
Department of Medicine,
The University of Edinburgh,
Royal Infirmary,
Lauriston Place,
Edinburgh EH3 9YW,
UK

Patrick G. Holt
Division of Cell Biology,
Institute for Child Health Research,
PO Box 855,
West Perth,
Western Australia,
Australia 6872

D.M. Kemeny
Department of Allergy & Allied Respiratory
Disorders,
Division of Medicine,
United Medical and Dental Schools of Guy's and St
Thomas's Hospital,
St Thomas St,
London SE1 9RT,
UK

Craig M. Lilly
Department of Medicine,
Brigham and Women's Hospital,
and
Department of Medicine,
Harvard Medical School,
Boston,
MA,
USA

Kaiser G. Lim
Department of Medicine,
Harvard Medical School,
Beth Israel Hospital
Dana 617,
330 Brookline Avenue,
Boston,
MA 02215,
USA

Donald M. McDonald
Cardiovascular Research Institute,
University of California,
San Francisco,
CA 94143-0130,
USA

Clive Robinson
Department of Pharmacology & Clinical Pharmacology,
St George's Hospital Medical School,
Cranmer Terrace,
London SW17 0RE,
UK

R.J. Shaw
Department of Respiratory Medicine,
St Mary's Hospital Medical School,
London W2 1PG,
UK

Stephanie A. Shore
Physiology Program,
Harvard School of Public Health,
Boston,
MA,
USA

Charles Twort
Division of Medicine,
St Thomas' Hospital,
Lambeth Palace Road,
London SE1 7EH,
UK

Andrew F. Walls
Immunopharmacology Group,
Southampton General Hospital,
Southampton SO16 6YD
UK

A. Wangoo
Department of Respiratory Medicine,
St Mary's Hospital Medical School,
London W2 1PG,
UK

Peter F. Weller
Beth Israel Hospital,
Dana 617,
330 Brookline Avenue,
Boston,
MA 02215,
USA

Series Preface

The consequences of diseases involving the immune system such as AIDS, and chronic inflammatory diseases such as bronchial asthma, rheumatoid arthritis and atherosclerosis, now account for a considerable economic burden to governments worldwide. In response to this, there has been a massive research effort investigating the basic mechanisms underlying such diseases, and a tremendous drive to identify novel therapeutic applications for the prevention and treatment of such diseases. Despite this effort, however, much of it within the pharmaceutical industries, this area of medical research has not gained the prominence of cardiovascular pharmacology or neuropharmacology. Over the last decade there has been a plethora of research papers and publications on immunology, but comparatively little written about the implications of such research for drug development. There is also no focal information source for pharmacologists with an interest in diseases affecting the immune system or the inflammatory response to consult, whether as a teaching aid or as a research reference. The main impetus behind the creation of this series was to provide such a source by commissioning a comprehensive collection of volumes on all aspects of immunopharmacology. It has been a deliberate policy to seek editors for each volume who are not only active in their respective areas of expertise, but who also have a distinctly *pharmacological* bias to their research. My hope is that *The Handbook of Immunopharmacology* will become indispensable to researchers and teachers for many years to come, with volumes being regularly updated.

The series follows three main themes, each theme represented by volumes on individual component topics.

The first covers each of the major cell types and classes of inflammatory mediators. The second covers each of the major organ systems and the diseases involving the immune and inflammatory responses that can affect them. The series will thus include clinical aspects along with basic science. The third covers different classes of drugs that are currently being used to treat inflammatory disease or diseases involving the immune system, as well as novel classes of drugs under development for the treatment of such diseases.

To enhance the usefulness of the series as a reference and teaching aid, a standardized artwork policy has been adopted. A particular cell type, for instance, is represented identically throughout the series. An appendix of these standard drawings is published in each volume. Likewise, a standardized system of abbreviations of terms has been implemented and will be developed by the editors involved in individual volumes as the series grows. A glossary of abbreviated terms is also published in each volume. This should facilitate cross-referencing between volumes. In time, it is hoped that the glossary will be regarded as a source of standard terms.

While the series has been developed to be an integrated whole, each volume is complete in itself and may be used as an authoritative review of its designated topic.

I am extremely grateful to the officers of Academic Press, and in particular to Dr Carey Chapman, for their vision in agreeing to collaborate on such a venture, and greatly hope that the series does indeed prove to be invaluable to the medical and scientific community.

C.P. Page

Preface

The *Handbook of Immunopharmacology* provides an opportunity of bringing together often disparate areas of science and clinical medicine towards a comon theme centred on the molecular and cellular basis of disease. In this volume we have concentrated our efforts on the lung – a major interface between the environment and internal milieux. In inviting contributors to this volume I wanted to focus attention specifically on the mechanisms of asthma since, over the last decade, our understanding of this disease has greatly increased and opportunities for novel therapeutic interventions have been revealed.

Since many of the principles of immunopharmacology cut across specific diseases, particular attention was paid to include aspects of disease pathogenesis that are common to many lung disorders. Two examples of this are the chapters on mechanisms of fibrosis by Dr R. Shaw and a focus on the resolution of inflammatory responses by Professor C. Haslett. To obtain maximum value from this volume it is important to recognize that cell–cell interactions are of paramount importance in inflammatory response and to read each chapter within the context of the others.

As we begin to unravel the complexities of such lung diseases as asthma, we are reminded of the importance of understanding the basic mechanisms behind the origins of this and related disease as well as its progression. It is my hope that this volume provides the reader with an up-to-date view of a rapidly evolving branch of medical science.

Finally, I would like to thank all our contributors for the time and enormous effort they put in to provide each of the contributions which together has produced a most exciting and timely volume in this series.

S.T. Holgate

1. Macrophage and Dendritic Cell Populations in the Respiratory Tract

Patrick G. Holt

1 Introduction

Epithelial surfaces in the respiratory tract are continuously exposed to a wide array of antigens from the environment, some of which (notably those of microbial origin) are intrinsically pathogenic and accordingly require rapid "neutralization" via both inert and adaptive immunoinflammatory mechanisms, and others which are essentially inert and ideally should be eliminated without stimulating active host responses. Discrimination between these different classes of antigen requires fine control, particularly of local adaptive immune responses. Populations of dendritic cells (DCs) and mononuclear phagocytic cells (macrophages; Macs) are at the focal point of the control mechanisms which maintain local immunological homeostasis in the lung and airways. Recent findings relating to the immunobiology of these important cells, in healthy and inflamed tissues, are reviewed below.

1.1 DISTRIBUTION AND PHENOTYPE

1.1.1 Dendritic Cells

The presence of DCs in lung tissues was first recognized in the context of the disease histiocytosis X, but subsequent observations from a number of laboratories (reviewed in Holt et al., 1990a) indicate that large populations of these cells are present in both the lung parenchyma and in the airway mucosa of healthy experimental animals and humans.

The identification of these cells has relied almost exclusively upon immunostaining for surface class II major histocompatibility complex (MHC) Ia glycoprotein antigen, expression of which is generally acknowledged as a constitutive property of DCs from all tissues of normal adult animals. In the steady state, DC populations in the respiratory tract additionally display a wide range of surface markers, many of which are function associated. The typical profile reported for the mouse includes T200, NLDC145 and F480 (Sertl et al., 1987; Breel et al., 1988; Pollard and Lipscomb, 1990). In the normal rat, lung (in particular airway) DCs express high levels of CD4, low levels of the common β-chain of CD11a/18 and the markers defined by the monoclonal antibodies (mAbs) Ox41 and Ox43, and also express intercellular adhesion molecule-1 (ICAM-1; Holt and Schon-Hegrad, 1987; Holt et al., 1988; Schon-Hegrad et al., 1991; Xia et al., 1991; Gong et al., 1992). They also stain with mAb Ox62 (Nelson et al., 1994) which defines a putative integrin common to DCs and dendritic epithelial T cells, but are themselves CD3⁻ (Nelson et al., 1994), and a significant subset stains positively for the marker ED1

Immunopharmacology of the Respiratory System
ISBN 0-12-352325-7

(Holt and Schon-Hegrad, 1987; Holt et al., 1988) which is shared by DCs and Macs, but unlike the latter they do not stain for ED2 (Holt et al., 1992).

Human respiratory tract DCs express Ia in conjunction with CD1a or CD1c (e.g. Soler et al., 1989), T-200 (Sertl et al., 1987; Casolaro et al., 1988), RFD1 (Munro et al., 1987; Nicod et al., 1990), and in addition are stained by a variety of mAbs against surface molecules associated with trafficking and cell–cell interaction. The latter include both $\beta 1$ and $\beta 2$ integrins, and the adhesins leucocyte function-associated antigen-3 (LFA-3) and ICAM-1 (Nicod and El Habre, 1992).

Unlike their counterparts in skin, the epidermal Langerhans cell (LC) population, respiratory tract DC populations display limited expression of intracellular Birbeck granules (Bg). These are found only occasionally in sections of normal lung tissue in mouse and human lung and airway tissue samples (Hanau et al., 1985; Sertl et al., 1987; Soler et al., 1989) and are virtually absent in rat (P.G. Holt, unpublished). They are found in increased frequency in airway tissues of human smokers (Soler et al., 1989), which suggests that Bg expression may require local stimulation for initial induction (Hanau et al., 1985).

The surface phenotype of respiratory tract DCs alters markedly with inflammation. The most notable changes include upregulation of expression of Ia, CD1, and the β-chain of CD11a/18 (Soler et al., 1989; Schon-Hegrad et al., 1991; Kradin et al., 1991; McWilliam et al., 1994; Nelson et al., 1994). It is additionally evident that the surface phenotype of rodent DCs is variable at different levels of the respiratory tract (Holt et al., 1992; Holt and Schon-Hegrad, 1987) and at least some of the variation may be attributable to differences in the level of stimulation from inhaled irritants at different sites (Nelson et al., 1994). Consistent with this suggestion, administration of exogenous pro-inflammatory cytokines, particularly interferon (IFNγ), markedly upregulates baseline Ia expression (Kradin et al., 1991; Nelson et al., 1994). Similar conclusions follow from the results of comparative studies on lung DCs from smokers versus non-smokers (Soler et al., 1989), and from observations on changes in the phenotype of nasal mucosal DCs in rhinitis patients during the pollen season (Fokkens et al., 1989).

A small population of DCs has also been observed on the luminal surface of the alveolar spaces (Casolaro et al., 1988; Soler et al., 1989; Havenith et al., 1992), but the presence of these cells in bronchoalveolar lavage (BAL) fluids appears restricted to situations of local inflammatory stimulation (further discussion below).

Quantitative studies on the distribution of respiratory tract DC populations, at least those present within the epithelium of the conducting airways, has been facilitated by the development in our laboratory of techniques for sectioning isolated airway segments in a tangential plane parallel to the long axis of the airway (Holt et al., 1989,

1990b). This provides an en face view of DCs within the airway epithelium, and produces a pattern of immunostaining which is essentially equivalent to that achieved with isolated epidermal sheets (Fig. 1.1, top). Importantly, this procedure has resolved the vexed issue of the "heterogeneity" of Ia staining within the airway epithelium which is seen in conventional transverse or longitudinal sections (Fig. 1.1, middle) and demonstrates that in normal tissues close to 100% of Ia staining can be accounted for by cells with the classical dendriform morphology of DCs, both in humans (Holt et al., 1989) and experimental animals (Holt et al., 1990a).

Employing this technique, we (Schon-Hegrad et al., 1991) and others (Gong et al., 1992) have shown that airway intraepithelial DC density in adult animals is inversely related to airway diameter, a distribution pattern consistent with the notion that stimulation from inhaled irritants/antigens provides the principal signal(s) for maintenance of the density and surface phenotype of the airway DC population.

The Ia$^+$ airway DC network appears to develop almost entirely postnatally (Fig. 1.1, middle versus bottom) from Ia$^-$ precursors, recruitment of which commences in late fetal life (Nelson et al., 1994). It is of interest to note that a small but significant subset of Ia$^-$ DCs (accounting for up to 25% of the overall population) persist in the airway epithelium into adulthood (Nelson et al., 1994).

The overall density of the principal Ia$^+$ DC population in young adult animals varies from 500–800 mm^{-2} tracheal epithelium to <100 mm^{-2} in the small airways (Schon-Hegrad et al., 1991; Gong et al., 1992); however, in older animals (>1 year old) these differences narrow markedly, presumably reflecting net inflammatory "history" (unpublished). The density of DCs within the epithelium of small bronchioles in humans (sectioned as shown in Holt et al., 1990a) is in the order of 500–600 mm^{-2}. In parenchymal lung tissue, DCs are typically found in interstitial tissues at the junction of adjoining alveoli ("interseptal junctional zones"; Holt et al., 1993); no reliable methods are available for their accurate quantitation at these sites.

1.1.2 Macrophages

Macs are ubiquitous throughout the respiratory tract, and discrete populations can be discerned in the airway mucosa, the lung parenchyma ("interstitial" macrophages; IMs), the luminal surface of the alveoli (pulmonary alveolar macrophages; PAMs) and the conducting airways, and in the vascular bed ("intravascular" macrophages; IVMs). The latter represent a stable "marginated" population, intimately associated with the endothelial basement membrane; they are most common in ruminants (Winkler, 1989) but also occur in humans.

The largest Mac population in the lung is the PAMs. The distribution of these cells is not random, as examination of lung tissues rapidly fixed by perfusion with

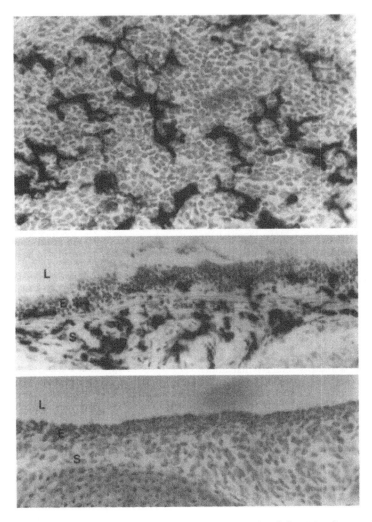

Figure 1.1 Intraepithelial dendritic cells in rat airway wall. Top panel: adult rat trachea sectioned through the airway epithelium parallel to the basement membrane stained by immunoperoxidase for Ia; note darkly stained dendritic cells. Middle and bottom panels: longitudinal frozen sections of adult (middle) and 8-day-old (bottom) rat trachea stained as in top. Note Ia⁺ dendritic cells in and below airway epithelium of adult, compared to preweanling. L, airway lumen; E, airway epithelium; S, submucosa. Reproduced with permission from Holt et al. (1990c).

glutaraldehyde through the vascular bed has revealed that in both humans (Parra *et al.*, 1986) and rats. (Holt *et al.*, 1993) 90% occur in intimate association with the alveolar epithelial surface at interseptal junctions. Similar fixation techniques have revealed a comparable (albeit less dense) population in intimate association with ciliated airway epithelial cells (Brain *et al.*, 1984).

In the steady state, the surface marker profile of the vast majority of the population is superficially consistent with that of mature tissue macrophages. However, closer examination reveals additional expression of markers which are generally seen only in tissue sections from central lymphoid organs, such as the murine marker NLDC145 which stains interdigitating cells (IDCs; Bilyk

and Holt, 1991). In contrast, in both experimental models of lung inflammation and in a variety of immunoinflammatory diseases, the spectrum of surface marker expression alters markedly to reflect a large contribution from recruited monocytes, in particular high level expression of macrophage-1 (Mac-1) and F4/80 in the mouse and OKM1 in humans. Typical examples of these changes in humans are in atopic asthma (Hance *et al.*, 1985) and in sarcoidosis (Mattoli *et al.*, 1991) which both demonstrate major lung Mac population "shifts" towards the monocytic end of the spectrum; the RFD1/RFD7 markers have proven of particular value in defining these changes in the latter disease (Poulter, 1990).

1.2 ORIGIN AND TURNOVER

1.2.1 Dendritic Cells

Recent studies from our laboratory have provided the first quantitative data on the population dynamics of lung DC populations. These studies have been performed in the rat, employing a radiation chimera model employing congenic animals expressing allelic variants of CD45 which are discernible with appropriate mAbs. In this system, animals from one congenic strain are X-irradiated to ablate bone marrow, and reconstituted with marrow from their congenic partners; sequential tracheal epithelial tissue samples were taken and host and donor DCs identified in the same sections employing dual colour immunostaining for Ia plus CD45. Plotting the waxing and waning respectively of donor and host DCs against time provides an accurate estimate of population half-life, which is in the order of 2–3 days (Holt *et al.*, 1994; Fig. 1.2). This figure is remarkably short, and compares with approximately 7 days for DCs in the lung parenchyma and 2–3 weeks for epidermal LCs (see Fig. 1.3) and their counterparts in heart and kidneys (Fossum, 1989).

Ongoing studies in our laboratory indicate that under stimulation from local inflammation, the turnover time of airway DCs may be further accelerated. This is suggested by the finding that the density (and Ia expression) of these cells increases in chronic inflammation, and is further reinforced by findings from models of acute

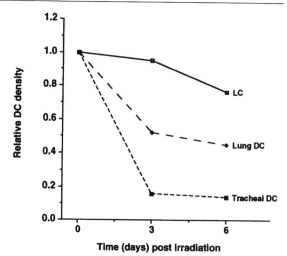

Figure 1.3 Comparative rates of decline in the density of epidermal LCs, lung parenchymal DCs, and tracheal epithelial DCs, after whole body X-irradiation of rats. Data were initially derived as mean number of DCs per unit surface area (for tracheal epithelium and epidermis) or per microscope field (for lung) using groups of three to four animals per time point and normalized to respective day 0 control figures for presentation. Reproduced with permission from Holt *et al.* (1994).

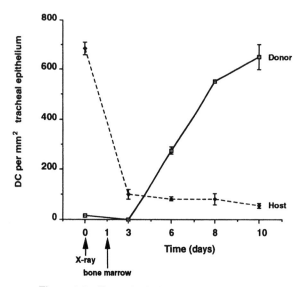

Figure 1.2 X-ray depletion and bone marrow reconstitution of tracheal intraepithelial DCs in rats transplanted with congenic marrow. Recipient (host) and donor DCs were identified by dual colour immunostaining. Data shown are means ± SD derived from three to five animals per time point. Reproduced with permission from Holt *et al.* (1994).

inflammation. Notably inhalation of a solution containing bacterial lipopolysaccharide (LPS; Schon-Hegrad *et al.*, 1991) or heat-killed *Moraxella catarrhalis* organisms (McWilliam *et al.*, 1994) recruits large numbers of DCs into the airway epithelium (Fig. 1.4b), with kinetics which closely parallel those of incoming neutrophils (Fig. 1.4a). Unlike the neutrophils which progress through the epithelium into the airway lumen, the recruited DCs remain within the epithelium for the duration of the acute phase of the inflammatory response, and then migrate selectively (over the ensuing 48 h) to the regional lymph nodes, presumably conveying samples of antigens acquired during their sojourn within the epithelium (McWilliam *et al.*, 1994).

DC infiltrates, albeit of smaller numbers, have also been observed in BAL fluids from rodents in a model of subacute peripheral lung inflammation (Havenith *et al.*, 1992), and in chronic inflammation in human smokers (Soler *et al.*, 1989).

1.2.2 Macrophages

The origin of lung Mac populations, particularly PAMs, has been the subject of considerable debate. In particular, the relative contribution of bone marrow-derived monocytes versus resident lung tissue Mac precursors to renewal of the PAM population has been questioned. While it is clear that monocyte recruitment from the blood plays a major role in population renewal during inflammation (Blussé van Oud Alblas *et al.*, 1983), it is

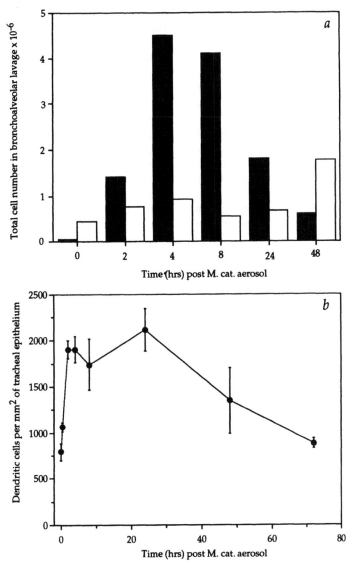

Figure 1.4 Cellular inflammatory response in the lung and airways after inhalation of whole bacteria. Adult PVG rats were exposed to aerosolized *Moraxella catarrhalis*. At various times after the aerosol exposure, bronchoalveolar lavage was performed using medium containing 0.35% (w/v) lignocaine. Cytospin preparations of the cells obtained were stained with Leishman's stain and differential counts performed. Tracheas were removed at each time point and frozen sections prepared and immunostained for Ia. (a) Recovery of neutrophils (filled bars) and macrophages (open bars) from BAL fluids. Cytospin preparations contained little or no Ia⁺ cells. (b) Fluctuations in the density of Ia⁺ DCs in the tracheal epithelium during the inflammatory response. The data shown are derived from three to five animals per time point. Reproduced with permission from McWilliam *et al.* (1994).

evident that the lung itself contains a reservoir of Mac precursors which can respond to monocyte colony-stimulating factor (M-CSF), potentially giving rise to locally derived monocytes (Stanley *et al.*, 1978). This conclusion is supported by a variety of studies from animal models demonstrating local monocyte division during both the steady state and during inflammation, and particularly studies showing that steady-state population size is unaffected by bone marrow ablation (Sawyer *et al.*, 1982; Sawyer, 1986a,b; Oghiso and Yamada, 1992).

Kinetic studies indicate that the normal half-life of PAMs is in excess of 21 days, but this figure may be accelerated during inflammation. There is no direct evidence on the half-life of other lung Mac populations, but our (unpublished) observations on rat suggest that airway mucosal Macs are long-lived with turnover times in excess of 14 days.

1.3 FUNCTIONS

1.3.1 Dendritic Cells

In tissues outside the respiratory tract, DCs are recognized as the most potent antigen-presenting cells (APCs) for presentation of inductive signals to naive T cells (Steinman, 1991).

The initial indication that lung DCs perform similar functions came from studies in our laboratory, on cells isolated from collagenase digests of lung tissue slices (Holt *et al.*, 1985). These experiments demonstrated that the most potent APC present in the resulting mixed cell population was non-phagocytic, non-adherent, surface IgG-FcR negative, Ia positive, and of ultra-low density on Percoll gradients, features characteristic of DCs. Follow-up studies produced immunohistochemical data to support this conclusion, and moreover demonstrated active uptake and processing of inhaled protein antigen by these cells *in vivo*, in particular the population present in the airway epithelium (Holt *et al.*, 1988). It was also noted that the capacity of these DCs to express APC activity *in vitro* was inversely related to the concentration of contaminating lung macrophages in the cell preparations (Holt *et al.*, 1988; further discussion below).

Earlier reports on epidermal LCs indicated that when freshly isolated, these cells expressed only a small proportion of their potential APC activity, and required a period of "maturation" *in vitro* in the presence of granulocyte–macrophage colony-stimulating factor (GM-CSF) [± tumour necrosis factor α (TNFα) and interleukin-1 (IL-1)] before they were able to efficiently activate T cells. We have recently demonstrated that the same situation applies to airway epithelial and lung parenchymal DC populations, as their capacity to present soluble protein antigens (acquired *in vivo*) to primed T cells, or to present "self" peptides to naive allogeneic T cells in a primary mixed lymphocyte reaction (MLR), upregulates on a log scale if they are pre-exposed to GM-CSF (Holt *et al.*, 1993). These findings support the notion that *in vivo*, lung DCs function as "sentinel" cells, acquiring and processing antigen, but do not develop the capacity to efficiently present these antigens to the immune system until they are given a "maturation" signal in the form of GM-CSF (Holt *et al.*, 1993); on the basis of a variety of studies involving comparisons between DCs during the peripheral tissue and lymph node phases of their life cycle, it is evident that this "maturation" process normally does not occur until they migrate to regional lymph nodes (Steinman, 1991).

The issue of whether airway and lung DCs may "mature" *in situ* in the presence of locally generated GM-CSF, for example during acute or chronic inflammation, remains to be resolved, but recent data (discussed below) indicate that an efficient control mechanism exists to downregulate this process.

1.3.2 Macrophages

Lung Mac populations play a variety of important roles in the maintenance of local homeostasis in the respiratory tract. As well as a primary function in the uptake and degradation of opsonized and unopsonized particulates and soluble material, particularly of microbial origin, lung Macs (analogous to their counterparts in other tissues) secrete a wide array of cytokines and mediators which regulate the function of both bone marrow-derived and mesenchymal cells. It is further evident that the spectrum of biologically active molecules they secrete varies with their state of "activation" and maturation, during their progress through the monocyte–mature Mac stages of their life cycle. A prime example is IL-1, which is secreted at high levels during the monocyte

Table 1.1 Principal secretory products from pulmonary alveolar macrophages

Enzymes
Elastase
Collagenase
Cathepsin
Lysozyme
β-Glucuronidase
Acid hydrolases
Angiotensin converting enzyme

Interleukins
IL-Iα and β
IL-IRa
IL-6, 8, 12
TNFα
MIP-1 and −2
IFNα, β, γ
TGFα, β
CSF(s)
Platelet-derived growth factor
Basic fibroblast growth factor

Other proteins
Antiproteases
Fibronectin
Complement components
Coagulation factors
Transferrin

Biologically active lipids
Thromboxane A_2
Prostaglandins (E_2, D_2, $F_{2\alpha}$)
Lipo-oxygenase metabolites
Leukotriene B_4, C_4, D_4
Platelet-activating factor

Reactive oxygen/nitrogen intermediates
Superoxide, H_2O_2, OH^-
NO, nitrites, nitrates (not human?)

Adapted from Sibille and Reynolds (1990).

stage, but not once the cell develops into a mature PAM (Wewers *et al.*, 1984).

Some of the most important mediators secreted by PAMs are summarized in Table 1.1.

In addition to the above, it has been recognized for some time that PAMs are capable of elaborating powerful "suppressive" signals which can limit the proliferative potential of T cells, and it has been suggested that this function is central to maintenance of local immunological homeostasis in the peripheral lung (Holt, 1986, 1993). It is additionally evident that considerable heterogeneity exists within the PAM population with respect to capacity to suppress T cell activation. In particular, purification of the "monocyte-like" subpopulation of PAMs either by size, buoyant density or expression of appropriate surface markers (Warner *et al.*, 1981; Holt *et al.*, 1982; Murphy and Herscowitz, 1984; Shellito and Kaltreider, 1984), unmasks underlying capacity for accessory cell activity, and it is accordingly likely that the net effect of an individual PAM population on local T cell activation will depend upon the balance between "competing" monocyte versus mature Mac subpopulations. This may underlie the findings that PAMs from asthmatics (Aubas *et al.*, 1984) and sarcoidosis patients (Lem *et al.*, 1985; Venet *et al.*, 1985; Poulter, 1990; Nagai *et al.*, 1991), syndromes characterized by a phenotypic "shift" in lung Mac populations towards monocytes (presumably via enhanced recruitment from blood), demonstrate significantly less suppressive activity than normal controls. It should be emphasized that demonstration of the full potential of this suppressive mechanism in humans requires the use of T cell activation

systems in which the stimulus is a nominal antigen, as opposed to polyclonal mitogens such as anti-CD3 or phytohaemagglutinin (PHA), or even MLR (Holt, 1986); providing antigenic stimulation is employed, PAM dose–response curves for human T cell suppression closely resemble those seen in murine systems (Upham *et al.*, 1994).

The available evidence suggests that the underlying mechanism for PAM-mediated T cell suppression is multifactorial, and includes both contact-dependent (Rich *et al.*, 1991) and contact-independent elements. The latter include prostaglandins (McCombs *et al.*, 1984; Rich *et al.*, 1987), interleukin-1 receptor autogonist (IL-1Ra; Kern *et al.*, 1988; Moore *et al.*, 1992), transforming growth factor β (TGFβ; Roth and Golub, 1993; Lipscomb *et al.*, 1994), nitric oxide (NO; Kawabe *et al.*, 1992; Bilyk and Holt, 1994; Strickland *et al.*, 1994; Upham *et al.*, 1994) and vitamin D3 metabolites (Mason *et al.*, 1984). NO appears to be the dominant mechanism in rodents, but different pathway(s) are employed in humans.

Recent studies from our laboratory additionally indicate that the transition from the monocyte to mature PAM "T-suppressive" phenotype is time dependent and cytokine sensitive. Following recruitment into the lung via a sterile inflammatory stimulus, murine monocytes require $\geqslant 5$ days for full maturation (Bilyk and Holt, 1994). In addition, the presence of cytokines such as GM-CSF (especially if free TNFα is also available) inhibits the expression of suppressive activity (Fig. 1.5). If mature PAMs are cultured continuously in the presence of GM-CSF, their T cell-interactive phenotype reverts to that of

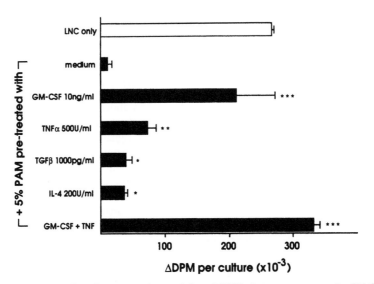

Figure 1.5 Modulation of the T cell-suppressive activity of PAMs by pre-exposure to GM-CSF ± TNF. Data shown are the proliferative responses of lymph node cells (LNC) to mitogen in the presence of PAMs at a concentration equivalent to 5% of the bold cells in culture; the PAMs were pretreated for 48 h with medium alone, or medium supplemented with the cytokines shown. Untreated PAMs exhibited suppressive activity comparable to the medium-treated group. Reproduced with permission from Bilyk and Holt (1993).

monocytes, providing a "window" for T cell activation which may be required during inflammation for effective mobilization of local adaptive immune defences. Significantly, if GM-CSF exposure is continued beyond 48 h, the PAMs revert to their original suppressive phenotype, probably as a result of down-modulation of GM-CSF receptors (Bilyk and Holt, 1994); this would effectively "close the window" for T cell activation, permitting re-establishment of local immunological homeostasis.

Ongoing studies in our laboratory additionally indicate that this "suppressive" mechanism is considerably more sophisticated than hitherto believed, and may indeed exist specifically to limit the local amplification (as opposed to expression) of secondary T cell immunity in the peripheral lung. It should firstly be remembered that PAMs do not constitutively express T cell suppressive activity, but do so only after direct interaction with preactivated T cells (Wamer et al., 1981), or more precisely with IFNγ released by activated T cells (Albina et al., 1991). We have recently performed a comprehensive kinetic analysis of T cells undergoing mitogen stimulation in the presence of PAMs, and find that despite being "locked" into G0/G1 in the cell cycle and being unable to proliferate, the T cells undergo normal Ca^{2+} flux, transient T cell receptor (TCR) down-modulation, IL-2 and IFNγ secretion, and IL-2R expression, i.e. they are unable to successfully "read-out" the IL-2 signal (Strickland et al., 1994). The latter is reversible, as if the PAMs are removed, the T cells then proliferate. It is noteworthy that while human PAMs employ NO-independent mechanisms for T cell suppression, the characteristic phenotype displayed by "PAM-suppressed" human T cells appears identical to that in rodents (Upham et al., 1994).

Thus, this mechanism appears to permit individual T cells to express their primary function of antigen recognition and cytokine release, but limits their responses to a "single-hit", thus minimizing the potential risk of local "bystander" tissue damage via a locally expanding T cell response.

Recent studies from our laboratory (Thepen et al., 1991, 1992a,b; Holt et al., 1993; Strickland et al., 1993) have supplied direct proof that, at least in rodents, this mechanism is also operative in vivo. It is now possible to create transiently "PAM-deficient" mice or rats via the "macrophage suicide technique" involving intratracheal inoculation of liposomes containing the MO cytotoxic drug dichloromethylene diphosphonate (DMDP); phagocytosis of the liposome results in PAM death over the ensuing 12–24 h. Primary immunity to antigens not previously encountered is unaffected by loss of PAMs (Thepen et al., 1991). However, aerosol challenge of primed animals results in markedly increased systemic and local T cell dependent immune responses, the latter characterized by the migration of large numbers of activated T cells into the lung, and accompanying T_H2-dependent IgE plasma cell responses in the lymph nodes

draining the respiratory tract (Thepen et al., 1992b). In addition, the cloning frequency of lung-derived T cells, which is very low in the steady state, improves markedly after PAM elimination (Strickland et al., 1993).

2. Macrophage–Dendritic Cell Interactions

The initial indication that the function(s) of lung DCs may be regulated by local Macs came from our initial studies on the isolation of peripheral lung DCs from the rat. These experiments identified the most potent APC population in collagenase digest of parenchymal lung tissue as DCs, but further demonstrated that expression of their full APC potential required initial elimination of contaminating Macs (Holt et al., 1985). Similar results were obtained with airway epithelial DCs (Holt et al., 1988).

More recent studies from our lab demonstrate that the capacity of lung and airway DCs to upregulate their APC functions and to acquire potent "presentation" capacity in response to GM-CSF in vitro, is inhibited by soluble signal(s) from the dominant local Mac population (PAMs), which include NO and TNFα (Holt et al., 1993). This mechanism also appears operative in vivo, as PAM elimination by the "macrophage suicide technique", involving intratracheal administration of DMDP in liposomes, results in the rapid upregulation of the APC activity of resident lung DCs to approximate that achieved in vitro by co-culture with GM-CSF (Holt et al., 1993; Fig. 1.6).

3. Macrophage–Dendritic Cell Regulation of T Cell Function(s) in Inflamed Lung and Airway Tissues

The review above argues that the efficiency of T cell activation in the lung and airways is determined by cross-regulating signals derived from both Macs and DCs. Moreover, many of these signals are likely to change both qualitatively and quantitatively during inflammation, and in addition the source of the signals may change. In particular, it is now clear that in chronic inflammatory diseases such as asthma, GM-CSF is produced in high levels within the airway epithelium, by the epithelial cells themselves (e.g. Burgess et al., 1977) and also by infiltrating bone marrow-derived cells.

Additionally, it is clear that a consistent feature of diseases such as asthma is the recruitment of monocytes into the alveolar spaces and the airway mucosa (Beasley et al., 1989; Poston et al., 1992). Monocytes are potent APCs for reactivation of primed T cells, and these cells additionally secrete cytokines such as GM-CSF and IL-1.

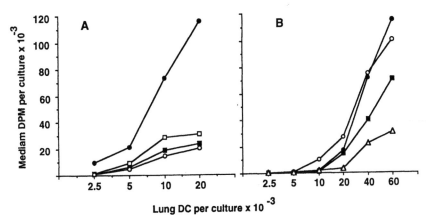

Figure 1.6 Antigen presentation by lung DCs in a primary MLR: effect of prior depletion of alveolar macrophages in DC donors. Semi-purified lung DCs from groups of control or treated BN rats ($n = 3$) were titrated into cultures of WAG lymph node cells, and resulting DNA synthesis determined as incorporation of [³H]thymidine at the 120 h time point. Panel A: lung DC pools were from untouched controls (■), and animals intratracheally inoculated 48 h previously with either phosphate-buffered saline (PBS) (○) or liposomes containing PBS (□) or DMDP (●). Panel B: MLR stimulatory activity of lung DCs prepared from BN rats at varying periods after intratracheal administration of DMDP liposomes: zero time control (▲); 24 h post-administration (●); 48 h (○); 72 h (■). Reproduced with permission from Holt *et al.* (1993).

TGFβ and TNFα secretion by resident lung Macs is also a feature of lung and airway inflammation, in a variety of models (Becker *et al.*, 1989; Khalil *et al.*, 1989; Kips *et al.*, 1992).

Recent studies from our laboratory have attempted to develop integrated *in vivo* and *in vitro* murine systems to model this dynamic process, and the results of these experiments (Bilyk and Holt, 1993, 1994) suggest the following scenario. In the steady state, local T cell activation in the lung and airways is limited by: (i) the direct inhibitory effect of resident Macs on T cell proliferation, (ii) the intrinsic "functional immaturity" of resident DCs, which is able to endocytose and process inhaled antigen, but are inefficient in presenting antigen to T cells; and (iii) the inhibitory effects of resident Macs upon GM-CSF-induced DC "maturation" *in situ*. In response to inflammatory stimuli: (i) monocytes (active APCs) are recruited and dilute the resident Mac population; and (ii) a variety of cells secrete cytokines, in particular GM-CSF (epithelial cells and monocytes), IL-1 (monocytes), TGFβ and TNFα (activated resident Macs). We have shown that GM-CSF plus TNFα "deprogramme" the T cell inhibitory activity of resident Macs (especially PAMs; Bilyk and Holt, 1993), and by virtue of their capacity to suppress NO production, inhibit the capacity of PAMs to block the inductive effect of GM-CSF upon DC maturation (Holt *et al.*, 1993).

The latter is unlikely to occur in humans by this precise mechanism, as human PAMs produce little or no NO,

however it is feasible that human PAMs may modulate DC suppression by other pathways – this suggestion appears reasonable on the basis of our recent finding that human PAMs modulate individual events in the T cell activation process in a fashion which appears identical to that achieved via NO by their murine counterparts (Upham *et al.*, 1994), indicating that alternative mechanisms are available in humans. The combination of GM-CSF, TNF and IL-1 represents the most potent stimulus known for promotion of DC "maturation" into the APC phenotype (Steinman, 1991), and it is thus feasible that in the absence of a countermanding signal from PAMs, resident lung DCs can potentially respond to such locally produced cytokine mixtures and thus develop the capacity to function as efficient stimulators for local T cell activation.

Results from the murine model suggest that this potential "window" for local T cell activation which may be opened during inflammation appears to be self-limiting, as PAMs continuously exposed to GM-CSF *in vitro* spontaneously reacquire the T cell suppressive phenotype by 72 h (Bilyk and Holt, 1994), presumably as a result of down-modulation of GM-CSF receptors which occurs as a result of prolonged exposure to ligand; TNFa is ineffective in its own right in modulating PAM suppressive activity, and can only synergize with effective GM-CSF signals (Bilyk and Holt, 1994), analogous to the situation described for the effects of this cytokine combination on DC maturation (Steinman, 1991).

It has also been reported that airway epithelial cells, in particular those from humans, produce NO during chronic inflammation. While this mediator has been proposed to exert a variety of effects in asthma including those directly involved with development of the symptoms of airways obstruction, our results in the animal models suggest that this function may also contribute to eventual limitation of T cell-mediated airways inflammation via suppression of GM-CSF-induced DC maturation.

4. References

Albina, J.E., Abate, J.A. and Henry, W.L. (1991). Nitric oxide production is required for murine resident peritoneal macrophages to suppress mitogen-stimulated T cell proliferation. Role of IFN-gamma in the induction of the nitric oxide-synthesizing pathway. J. Immunol. 147, 144–148.

Aubas, P., Cosso, B., Godard, P., Michel, F.B. and Clot, J. (1984). Decreased suppressor cell activity of alveolar macrophages in bronchial asthma. Am. Rev. Respir. Dis. 130, 875–878.

Beasley, R., Roche, W.R., Roberts, J.A. and Holgate, S.T. (1989). Cellular events in the bronchi of mild asthmatics after bronchial provocation. Am. Rev. Respir. Dis. 139, 806–817.

Becker, S., Devlin, R.B. and Haskill, J.S. (1989). Differential production of tumour necrosis factor, macrophage colony-stimulating factor, and interleukin 1 by human alveolar macrophages. J. Leuk. Biol. 45, 353–361.

Bilyk, N. and Holt, P.G. (1991). The surface phenotypic characterization of lung macrophages in C3H/HeJ mice. Immunology 74, 645–651.

Bilyk, N. and Holt, P.G. (1993). Inhibition of the immunosuppressive activity of resident pulmonary alveolar macrophages by granulocyte/macrophage colony-stimulating factor. J. Exp. Med. 177, 1773–1777.

Bilyk, N. and Holt, P.G. (1994). Cytokines modulate the immunosuppressive phenotype of pulmonary alveolar macrophages via regulation of nitric oxide production (submitted).

Blussé van Oud Alblas, A., Mattie, H. and van Furth, R. (1983). A quantitative evaluation of pulmonary macrophage kinetics. Cell Tiss. Kinet. 16, 211–219.

Brain, J.D., Gehr, P. and Kavet, R.I. (1984). Airway macrophages. The importance of the fixation method. Am. Rev. Respir. Dis. 129, 823–826.

Breel, M., Van der Ende, M., Sminia, T. and Kraal, G. (1988). Subpopulations of lymphoid and non-lymphoid cells in bronchus-associated lymphoid tissue (BALT) of the mouse. Immunology 63, 657–662.

Burgess, A.W., Camakaris, J. and Metcalf, D. (1977). Purification and properties of colony-stimulating factor from mouse lung-conditioned medium. J. Biol. Chem. 252, 1998–2003.

Casolaro, M.A., Bernaudin, J.F., Saltini, C., Ferrans, V.J. and Crystal, RG. (1988). Accumulation of Langerhans' cells on the epithelial surface of the lower respiratory tract in normal subjects in association with cigarette smoking. Am. Rev. Respir. Dis. 137, 406–411.

Fokkens, W.J., Vroom, T.M., Rijntjes, E. and Mulder, P.G. (1989). Fluctuation of the number of CD-1(T6)-positive dendritic cells, presumably Langerhans cells, in the nasal mucosa of patients with an isolated grass-pollen allergy before, during, and after the grass-pollen season. J. Allergy Clin. Immunol. 84, 39–43.

Fossum, S. (1989). The life history of dendritic leukocytes. In "The Cell Kinetics of the Inflammatory Reaction" (ed. O.H. Iversen), pp. 101–124. Springer Verlag, Berlin.

Gong, J.L., McCarthy, K.M., Telford, J., Tamatani, T., Miyasaka, M. and Schneeberger, E.E. (1992). Intraepithelial airway dendritic cells: a distinct subset of pulmonary dendritic cells obtained by microdissection. J. Exp. Med. 175, 797–807.

Hanau, D., Fabre, M., Lepoittevin, J.P., Stampf, J.L., Grosshans, E. and Benezra, C. (1985). ATPase and morphologic changes induced by UVB on Langerhans cells in guinea pigs. J. Invest. Dermatol. 85, 135–138.

Hance, A.J., Douches, S., Winchester, R.J., Ferrans, J. and Crystal, R.J. (1985). Characterisation of mononuclear phagocyte subpopulations in the human lung by using monoclonal antibodies: changes in alveolar macrophage phenotype associated with pulmonary sarcoidosis. J. Immunol. 134, 284–292.

Havenith, C.E., Breedijk, A.J. and Hoefsmit, E.C. (1992). Effect of Bacillus Calmette-Guerin inoculation on numbers of dendritic cells in bronchoalveolar lavages of rats. Immunobiology 184, 336–347.

Holt, P.G. (1986). Downregulation of immune responses in the lower respiratory tract: the role of alveolar macrophages. Clin. Exp. Immunol. 63, 261–270.

Holt, P.G. (1993). Regulation of antigen-presenting function(s) in lung and airway tissues. Eur. Respir. J. 6, 120–129.

Holt, P.G. and Schon-Hegrad, M.A. (1987). Localization of T cells, macrophages and dendritic cells in rat respiratory tract tissue: implications for immune function studies. Immunology 62, 349–356.

Holt, P.G., Warner, L.A. and Papadimitriou, J.M. (1982). Alveolar macrophages. VII. Functional heterogeneity within macrophage populations from rat lung. Aust. J. Exp. Biol. Med. Sci. 60, 607–618.

Holt, P.G., Degebrodt, A., O'Leary, C., Krska, K. and Plozza, T. (1985). T cell activation by antigen-presenting cells from lung tissue digests: suppression by endogenous macrophages. Clin. Exp. Immunol. 62, 586–593.

Holt, P.G., Schon-Hegrad, M.A. and Oliver, J. (1988). MHC class II antigen-bearing dendritic cells in pulmonary tissues of the rat. Regulation of antigen presentation activity by endogenous macrophage populations. J. Exp. Med. 167, 262–274.

Holt, P.G., Schon-Hegrad, M.A., Phillips, M.J. and McMenamin, P.G. (1989). Ia-positive dendritic cells form a tightly meshed network within the human airway epithelium. Clin. Exp. Allergy 19, 597–601.

Holt, P.G., Schon-Hegrad, M.A. and McMenamin, P.G. (1990a). Dendritic cells in the respiratory tract. Int. Rev. Immunol. 6, 139–149.

Holt, P.G., Schon-Hegrad, M.A., Oliver, J., Holt, B.J. and McMenamin, P.G. (1990b). A contiguous network of dendritic antigen-presenting cells within the respiratory epithelium. Int. Arch. Allergy Appl. Immunol. 91, 155–159.

Holt, P.G., McMenamin, C. and Nelson, D. (1990c). Primary sensitisation to inhalant allergens during infancy. Paediatr Allergy Immunol. 1, 3–13.

Holt, P.G., Oliver, J., McMenamin, C. and Schon-Hegrad, M.A. (1992). Studies on the surface phenotype and functions of dendritic cells in parenchymal lung tissue of the rat. Immunology 75, 582–587.

Holt, P.G., Oliver, J., Bilyk, N., McMenamin, C., McMenamin, P.G., Kraal, G. and Thepen, T. (1993). Downregulation of the antigen presenting cell function(s) of pulmonary dendritic cells in vivo by resident alveolar macrophages. J. Exp. Med. 177, 397–407.

Holt, P.G., Haining, S., Nelson, D.J. and Sedgwick, J.D. (1994). Origin and steady-state of class II MHC-bearing dendritic cells in the epithelium of the conducting airways. J. Immunol. 153, 256–261.

Kawabe, T., Isobe, K.I., Hasegawa, Y., Nakashima, 1. and Shimokata, K. (1992). Immunosuppressive activity induced by nitric oxide in culture supernatant of activated rat alveolar macrophages. Immunology 76, 72–78.

Kerm, J.A., Lamb, R.J., Reed, J.C., Elias, J.A. and Daniele, R.P. (1988). Interleukin-1-beta gene expression in human monocytes and alveolar macrophages from normal subjects and patients with sarcoidosis. Am. Rev. Respir. Dis. 137, 1180–1184.

Khalil, N., Bareznay, O., Sporn, M. and Greenberg, A.H. (1989). Macrophage production of transforming growth factor beta and fibroblast collagen synthesis in chronic pulmonary inflammation. J. Exp. Med. 170, 727–737.

Kips, J.C., Tavernier, J. and Pauwels, R.A. (1992). Tumor necrosis factor (TNF) causes bronchial hyperresponsiveness in rats. Am. Rev. Respir. Dis. 145, 332–336.

Kradin, R.L., McCarthy, K.M., Xia, W.J., Lazarus, D. and Schneeberger, E.E. (1991). Accessory cells of the lung. I. Interferon-gamma increases Ia$^+$ dendritic cells in the lung without augmenting their accessory activities. Am. J. Respir. Cell. Mol. Biol. 4, 210–218.

Lem, V.M., Lipscomb, M.F., Weissler, J.C., Nunez, G., Ball, E.J., Stastny, P. and Toews, G.B. (1985). Bronchoalveolar cells from sarcoid patients demonstrate enhanced antigen presentation. J. Immunol. 135, 1766–1771.

Lipscomb, M.F., Pollard, A.M. and Yates, J.L. (1994). A role for TGF-beta in the suppression by murine bronchoalveolar cells of lung dendritic cell initiated immune responses. Reg. Immunol. 5, 151–157.

Mason, R.S., Frankel, T., Chan, Y.L., Lissner, D. and Posen, S. (1984). Vitamin D conversion by sarcoid lymph node homogenate. Ann. Intern. Med. 100, 59–61.

Mattoli, S., Mattoso, V.L., Soloperto, M., Allegra, L. and Fasoli, A. (1991). Cellular and biochemical characteristics of bronchoalveolar lavage fluid in symptomatic nonallergic asthma. J. Allergy Clin. Immunol. 87, 794–802.

McCombs, C.C., Michalski J.P. and Light, R.W. (1984). Specificity of immunosuppression by human alveolar cells. Exp. Lung Res. 7, 211–222.

McWilliam, A.S., Nelson, D., Thomas, J.A. and Holt, P.G. (1994). Rapid Dendritic Cell recruitment is a hallmark of the acute inflammatory response at mucosal surfaces. J. Exp. Med. 179, 1331–1336.

Moore, S.A., Streiter, R.M., Rolfe, M.W., Standiford, T.J., Burdick, M.D. and Kunkel, S.L. (1992). Expression and regulation of human alveolar macrophage-derived

interleukin-1 receptor antagonist. Am. J. Respir. Cell. Mol. Biol. 6, 569–575.

Munro, C.S., Campbell, D.A., Du, B.R., Mitchell, D.N., Cole, P.J. and Poulter, L.W. (1987). Monoclonal antibodies distinguish macrophages and epithelioid cells in sarcoidosis and leprosy. Clin. Exp. Immunol. 68, 282–287.

Murphy, M.A. and Herscowitz, H.B. (1984). Heterogeneity among alveolar macrophages in humoral and cell-mediated immune responses: separation of functional subpopulations by density gradient centrifugation on percoll. J. Leuk. Biol. 35, 39–54.

Nagai S., Aung, H., Takeuchi, M., Kusume, K. and Izumi, T. (1991). IL-1 and IL-1 inhibitory activity in the culture supernatants of alveolar macrophages from patients with interstitial lung diseases. Chest 99, 674–680.

Nelson, D.J., McMenamin, C., McWilliam, A.S., Brenan, M. and Holt, P.G. (1994). Development of the airway intraepithelial Dendritic Cell network in the rat from class II MHC (Ia) negative precursors: differential regulation of Ia expression at different levels of the respiratory tract. J. Exp. Med. 179, 203–212.

Nicod, L.P. and El Habre, F. (1992). Adhesion molecules on human lung dendritic cells and their role for T cell activation. Am. J. Respir. Cell. Mol. Biol. 7, 207–213.

Nicod, L.P., Galve de Rochemonteix, B. and Dayer, J.M. (1990). Dissociation between allogeneic T cell stimulation and interleukin-1 or tumor necrosis factor production by human lung dendritic cells. Am. J. Respir. Cell. Mol. Biol. 2, 515–522.

Oghiso, Y. and Yamada, Y. (1992). Heterogeneity of the radiosensitivity and origins of tissue macrophage colony-forming cells. J. Radiat. Res. 33, 334–341.

Parra, S.C., Burnette, R., Price, H.P. and Takaro, T. (1986). Zonal distribution of alveolar macrophages, type II pneumonocytes, and alveolar septal connective tissue gaps in adult human lungs. Am. Rev. Respir. Dis. 133, 908–912.

Pollard, A.M. and Lipscomb, M.F. (1990). Characterization of murine lung dendritic cells: similarities to Langerhans cells and thymic dendritic cells. J. Exp. Med. 172, 159–167.

Poston, R.N., Chanez, P., Lacoste, J.Y., Litchfield, T., Lee, T.H. and Bousquet, J. (1992). Immunohistochemical characterisation of the cellular infiltration in asthmatic bronchi. Am. Rev. Respir. Dis. 145, 918–921.

Poulter, L.W. (1990). Changes in lung macrophages during disease. FEMS Micro. Immunol. 64, 327–332.

Rich, R.A., Tweardy, D.J., Fujiwara, H. and Ellner, J.J. (1987). Spectrum of immunoregulatory functions and properties of human alveolar macrophages. Am. Rev. Respir. Dis. 136, 258–265.

Rich, R.A., Cooper, C., Toossi, Z., Leonard, M.L., Stucky, R.M., Wiblin, R.T. and Ellner, J.J. (1991). Requirement for cell-to-cell contact for the immunosuppressive activity of human alveolar macrophages. Am. J. Respir. Cell. Mol. Biol. 4, 287–294.

Roth, M.D. and Golub, S.H. (1993). Human pulmonary macrophages utilize prostaglandins and transforming growth factor beta1 to suppress lymphocyte activation. J. Leuk. Biol. 53, 366–371.

Sawyer, R.T. (1986a). The ontogeny of pulmonary alveolar macrophages in parabiotic mice. J. Leuk. Biol. 40, 347–354.

Sawyer, R.T. (1986b). The significance of local resident

pulmonary alveolar macrophage proliferation to population renewal. J. Leuk. Biol. 39, 77–87.

Sawyer, R.T., Strausbauch, P.H. and Volkman, A. (1982). Resident macrophage proliferation in mice depleted of blood monocytes by strontium–89. Lab. Invest. 46, 165–170.

Schon-Hegrad, M.A., Oliver, J., McMenamin, P.G. and Holt, P.G. (1991). Studies on the density, distribution, and surface phenotype of intraepithelial class II major histocompatibility complex antigen (Ia)-bearing dendritic cells (DC) in the conducting airways. J. Exp. Med. 173, 1345–1356.

Sertl, K., Takemura, T., Tschachler, E., Ferrans, V.J., Kaliner, M.A. and Shevach, E.M. (1987). Dendritic cells in cutaneous, lymph node and pulmonary lesions of sarcoidosis. Scand. J. Immunol. 25, 461–467.

Shellito, J. and Kaltreider, H.B. (1984). Heterogeneity of immunological function among subfractions of normal rat alveolar macrophages. Am. Rev. Respir. Dis. 129, 747–753.

Sibille, Y. and Reynolds, H.Y. (1990). Macrophages and polymorphonuclear netutrophils in lung defense and injury. Am. Rev. Respir. Dis. 114, 471–501.

Soler, P., Moreau, A., Basset, F. and Hance, A.J. (1989). Cigarette smoking-induced changes in the number and differentiated state of pulmonary dendritic cells/Langerhans cells. Am. Rev. Respir. Dis. 139, 1112–1117.

Stanley, E.R., Chen, B.D. and Lin, H.-S. (1978). Induction of macrophage production and proliferation by a purified colony-stimulating factor. Nature 274, 168–170.

Steinman, R.M. (1991). The dendritic cell system and its role in immunogenicity. Annu. Rev. Immunol,9, 271–296.

Strickland, D.H., Thepen, T., Kees, U.R., Kraal, G. and Holt, P.G. (1993). Regulation of T cell function in lung tissue by pulmonary alveolar macrophages. Immunology 80, 266–272.

Strickland, D.H., Kees, U.R. and Holt, P.G. (1994). Suppression of T cell activation by Pulmonary Alveolar Macrophages: dissociation of effects on TcR, IL-2R expression, and proliferation. Eur. Resp. J. 7, 2124–2130.

Thepen, T., McMenamin, C., Oliver, J., Kraal, G. and Holt, P.G. (1991). Regulation of immune response to inhaled antigen by alveolar macrophages: differential effects of in vivo alveolar macrophage elimination on the induction of tolerance vs. immunity. Eur. J. Immunol. 21, 2845–2850.

Thepen, T., Hoeben, K., Breve, J. and Kraal, G. (1992a). Alveolar macrophages down-regulate local pulmonary immune responses against intratracheally administered T cell-dependent, but not T cell-independent antigens. Immunology 76, 60–64.

Thepen, T., McMenamin, C., Girn, B., Kraal, G. and Holt, P.G. (1992b). Regulation of IgE production in presensitised animals: In vivo elimination of alveolar macrophages preferentially increases IgE responses to inhaled allergen. Clin. Exp. Allergy 22, 1107–1114.

Upham, J.W., Strickland, D.H., Bilyk, N. and Holt, P.G. (1995). Alveolar macrophages from humans and rodents selectively inhibit T cell proliferation but permit activation and cytokine secretion. Immunology, 84, 142–147.

Venet, A., Hance, A.J., Saltini, C., Robinson, B.W.S. and Crystal, R.G. (1985). Enhanced alveolar macrophage-mediated antigen-induced T lymphocyte proliferation in sarcoidosis. J. Clin. Invest. 75, 293–301.

Warner, L.A., Holt, P.G. and Mayrhofer, G. (1981). Alveolar macrophages. VI. Regulation of alveolar macrophage-mediated suppression of lymphocyte proliferation by a putative T cell. Immunology 42, 137–147.

Wewers, M.D., Rennard, S.I., Hance, A.J., Bitterman, P.B. and Crystal, R.G. (1984). Normal human alveolar macrophages obtained by lavage have a limited capacity to release interleukin-1. J. Clin. Invest. 74, 2208–2218.

Winkler, G.C. (1989). Review of the significance of pulmonary intravascular macrophages with respect to animal species and age. Exp. Cell Biol. 57, 281–286.

Xia, W.J., Schneeberger, E.E., McCarthy, K. and Kradin, R.L. (1991). Accessory cells of the lung. II. Ia+ pulmonary dendritic cells display cell surface antigen heterogeneity. Am. J. Respir. Cell. Mol. Biol. 5, 276–283.

2. The Role of T Lymphocytes in Mucosal Protection and Injury

C.J. Corrigan

1. Introduction

T lymphocytes play a fundamental role in the initiation
and regulation of inflammatory responses. Through their
specific antigen receptors, they are capable of recognizing
invading foreign antigens and initiating appropriate
immune responses, which may be characterized
predominantly by "cell-mediated" reactions, in which
effector immune cells play a major role, or "humoral"
reactions, in which antibody responses are more prominent. Although immature B lymphocytes express surface
IgM with low binding affinity for foreign antigens, these
cells are largely dependent on T lymphocytes for their
subsequent activation and proliferation, and for antibody
affinity maturation. There is now abundant evidence that
T lymphocytes orchestrate both the initiation and the
propagation of immune responses largely through the
secretion of cytokines, and that the particular combinations of cytokines secreted during the course of these
inflammatory responses is responsible for the type of

inflammatory reaction with ensues, including whether
the reaction is predominantly cellular or humoral in
nature. It is the purpose of this chapter to describe the
fundamental properties of T lymphocytes and the factors
which govern their secretion of cytokines in various
inflammatory conditions. The mechanism of recruitment
of T lymphocytes to sites of inflammation, particularly
mucosal surfaces, will be discussed. Finally, it will be
shown how the secretion of various patterns of cytokines
can influence the course and nature of the ensuing
inflammatory response. In some cases these responses
serve to protect the host against infection and ensure subsequent healing and repair. In other cases, however,
these responses may be chronic and result in long-term
damage to host tissues.

2. T Lymphocyte-derived Cytokines

T lymphocytes secrete an array of protein mediators
involved in cell growth, inflammation, immunity,

differentiation and repair. Over the last 15 years, these mediators have been given various names. "Lymphokines" were originally defined as cell free soluble factors generated by sensitized T lymphocytes in response to specific antigen (Dumonde *et al.*, 1969). The terms "cytokine" and "interleukin" served to broaden this definition to include factors originating from many different cell types (Aarden *et al.*, 1979). Cytokines were defined initially on the basis of their activities, but the cloning of the genes for these products has greatly facilitated their classification, and cytokine expression both *in vitro* and *in vivo* can now be studied at the gene, mRNA or protein level. Identification of cytokine genes has also allowed examination of factors regulating the production of different cytokines. Most recently, a new family of chemotactic cytokines or "chemokines" are being identified. These chemokines are secreted by a wide variety of cell types. Two particularly striking functional features of cytokines are their extensive pleiotropy and their redundancy; most cytokines have multiple functions, and any one function can generally be mediated by more than one cytokine (Paul, 1989). Chemokines are less pleiotropic than cytokines because they are not potent inducers of other cytokines and exhibit more specialized functions in

inflammation and repair. In addition, their activities are generally limited to a narrower spectrum of specific target cells.

In contrast to hormones, which are carried by the blood stream throughout the entire body, cytokines, with certain exceptions, tend to mediate localized effects. Studies with administration of recombinant cytokines *in vivo* have shown that their half-life in the circulation is brief, typically less than 30 min, substantiating the notion that recirculation is not usually a fundamental aspect of cytokine physiology. Cytokines are highly potent mediators and exert their actions by binding to high affinity receptors (K_d typically 10^{-11} M) expressed at low numbers (typically 100–1000 receptors per cell) on different cell types. These receptors generally consist of two or more chains, each containing single membrane-spanning domains. Frequently, one, and sometimes both, of these receptor chains belongs to a family of cytokine receptors characterized by a degree of structural homology in their extracellular domains, including conservation of key cysteine residues and the expression of a common pentameric amino acid motif adjacent to the cell membrane. Most cytokines are glycoproteins, although glycosylation is generally not required for

Table 2.1 Summary of properties and alternative cell sources of T lymphocyte-derived cytokines

Cytokine	Cell sources	Activities
IL-1α/β	Many cell types Macrophages Keratinocytes Endothelial cells Some T and B cells Glial cells Epithelial cells	Increased proliferation of T and B cells Fibroblasts: proliferation, PGE_2 synthesis Macrophages: PGE_2, chemokine secretion, cytotoxicity Endothelium: PGE_2, synthesis adhesion molecule expression Participates in septic shock CNS: fever, sleep, anorexia Bone and cartilage resorption
IL-2	Activated T cells Activated B cells *in vitro*	Activates CD4/CD8 T cells, B cells, macrophages and NK cells Enhances cytolytic activity of LAK and NK cells

IL-3	T cells
	Thymic epithelium
	Mast cells
	Keratinocytes
		Enhances survival, proliferation and differentiation of multipotential stem
		   cells and committed precursors of the granulocyte, macrophage, erythroid,
		   eosinophil, neutrophil, megakaryocyte, mast cell and basophil lineages
		Enhances effector functions of myeloid end cells such as phagocytosis and
		   cytotoxicity

IL-4	Subpopulations of CD4
	   and CD8 T cells
	Mast cells
	Basophils
	CD4⁻ CD8⁻T cells in
	   spleen and thymus
	B cells
		Regulates many phases of B cell development, including viability and MHC
		   class II expression, FcεRII (CD23) receptor expression.
		Promotes antibody synthesis and Ig class switching by B cells (IgG4, IgE)
		Growth factor for helper and cytolytic T cells
		Promotes differentiation of T_H2-type T cells but impedes that of T_H1-type
		   cells
		Enhances MHC class II expression, tumoricidal activity and cytokine
		   synthesis by monocytes
		Promotes growth of mast cells
		Enhances growth of myeloid and erythroid pregenitors in combination with
		   GM-CSF, IL-6 and erythropoietin

Table 2.1 *(Continued)*

Cytokine	Cell sources	Activities
IL-5	T cells Mast cells Eosinophils	Enhances differentiation and proliferation of more mature eosinophil progenitors Principal regulator of mature eosinophil numbers Growth factor for murine B cells; effect on human B cells not seen in all systems
IL-6	T cells Monocytes Macrophages Fibroblasts Hepatocytes Endothelial cells	B cells: cofactor for growth and Ig section T cells: growth cofactor Hepatocytes: acute phase proteins Osteoclasts: bone resorption Megakaryocyte maturation
IL-9	T cells	Growth factor for some T cell clones Supports erythroid colony formation with erythropoietin Enhances IL-3-induced growth of bone marrow-derived mast cells in mouse, not human
IL-10	T cells Monocytes Macrophages Keratinocytes Activated B cells	Inhibits cytokine secretion by T_H1- and T_H2-type T cell clones Impairs accessory function of macrophages and dendritic cells Inhibits cytokine secretion by macrophages ($TNF\alpha$, IL-1, IL-6, IL-8, GM-CSF) and microbicidal activity Activates B cells (proliferation and secretion of IgM, IgG and IgA) Induces proliferation of thymocytes
IL-13	T cells	Induces B cell growth and differentiation Inhibits cytokine secretion by macrophages
IFNγ	T cells NK cells	Activates tumoricidal and microbicidal activities of monocytes and macrophages Enhances MHC class I and II molecule expression on monocytes, macrophages, epithelium, endothelium and connective tissue cells Inhibits induction and proliferation of T_H2-type T cells Regulates antibody synthesis by direct effects on B cells; inhibits IgE synthesis Activates NK cells and cytotoxic T cells
GM-CSF	T cells Mast cells Eosinophils Monocytes Fibroblasts Endothelial cells Neutrophils	Enhances survival, proliferation and effector activity of progenitors of granulocytes, monocytes, eosinophils, but not mast cells
TNFα	T cells NK cells LAK cells Macrophages Astrocytes Endothelial cells Mast cells	Enhances T cell proliferation Enhances B cell proliferation Enhances adhesion molecule expression on endothelium Neutrophils: activation, adhesion Fibroblasts: proliferation, cytokine secretion CNS: fever Promotes bone resorption
TNFβ	T cells (especially T_H1-type CD4 T cells and cytotoxic CD8 T cells)	Enhances B cell proliferation Enhances fibroblast proliferation Cytotoxic for a wide variety of tumour cell lines Accounts for some of the cytolytic activity of T cells, NK cells and LAK cells

biological function, since non-glycosylated forms generally retain all biological activities. T cell-derived cytokine genes exist as single copies in the haploid genome and generally show a related genomic organization consisting of four exons and introns (Arai *et al.*, 1990). Several T cell-derived cytokine genes [including interleukin-3 (IL-3), IL-4, IL-5, granulocyte–macrophage colony-stimulating factor (GM-CSF), monocyte colony-stimulating factor (M-CSF) and its receptor] are clustered on the long arm of human chromosome 5 (Arai *et al.*, 1990).

An exhaustive list of the functions of the various cytokines and chemokines is beyond the scope of this chapter, although a list of the principal T lymphocyte-derived cytokines, their known cellular sources (other than T lymphocytes) and principal activities are listed in Table 2.1. The principal members of the chemokine family of inflammatory cytokines are listed in Table 2.2.

Table 2.2   Summary of the properties and cell sources of some chemokines

Name	Cell sources	Inducers	Activities
α-Chemokines			
IL-8	Monocytes Neutrophils Fibroblasts Endothelium T cells	Endotoxin Mitogens Viruses IL-1, TNF, IL-3 (IFNγ co-stimulatory)	Chemotaxis: neutrophils T cells, basophils Basophil degranulation Neutrophil activation
MGSA/Gro	Monocytes Fibroblasts Endothelium	Endotoxin IL-1, TNF	Chemotaxis: neutrophils Neutrophil activation
PF4	Platelets	Platelet activators	Chemotaxis: fibroblasts Induces fibroblast growth
β-TG/NAP-2	Monocytes Platelets	Platelet activators	Chemotaxis: neutrophils Neutrophil activation
CTAP-III/β-TG	Monocytes	Platelet activators	Chemotaxis: fibroblasts Neutrophil activation
γIP-10	Monocytes Fibroblasts Endothelium	Endotoxin IFNγ	Chemotaxis: monocytes, activated T cells, DTH reactions
ENA-78	Epithelium	IL-1, TNF	Chemotaxis: neutrophils Neutrophil activation
β-Chemokines			
MCAF/MCP-1	Monocytes Fibroblasts Endothelium	Endotoxin Mitogens IL-1, TNF, PDGF, IFNγ	Chemotaxis: monocytes, basophils Basophil degranulation
RANTES	T cells Platelets	Mitogens	Chemotaxis: monocytes, memory T cells, eosinophils, basophils Basophil degranulation
LD-78 (MIP-1α)	T cells Monocytes	Mitogens Endotoxin TNF	Chemotaxis: monocytes, activated CD8 T cells, eosinophils
ACT-2 (MIP-1β)	T cells Monocytes	Mitogens TNF, IL-2	Chemotaxis: monocytes, activated CD4 T cells
I-309	T cells	Mitogens	Chemotaxis: monocytes
MCP-3	Tcells	Mitogens	Chemotaxis: eosinophils, basophils

MGSA, melanoma growth simulatory activity; Gro, growth-related oncogene; PF$_4$, platelet factor 4; β-TG, β-thromboglobulin; NAP-2, neutrophil-activating peptide-2; CTAP-III, connective tissue-activating peptide; MCAF, monocyte chemoattractant and activating factor; MIP-1α, macrophage inflammatory protein-1α; MIP-1β, macrophage inflammatory protein-1β; MCP-1 monocyte chemotactic protein-1.

These chemokines have been further subdivided into two subsets, α and β. The α-chemokine genes are clustered on chromosome 4 and share a characteristic amino acid sequence such that the first two of their four cysteine groups are separated by one amino acid (C-X-C). Genes encoding the β-chemokines are located on chromosome 17. These proteins have no intervening amino acid between the first two cysteine residues (C-C).

3. Functional Heterogeneity of T Lymphocytes in Human Immune Responses

In recent years it has become clear that the nature of the immune response initiated by CD4 T lymphocytes is at least partly dependent on the "selection" or preferential activation of particular subsets of CD4 T lymphocytes which secrete defined patterns of cytokines (Table 2.3). These patterns of cytokine release result in the initiation and propagation of distinct immune effector mechanisms. Initial studies of mouse CD4 T lymphocyte clones revealed that these could be divided into two basic functional subsets termed T_H1 and T_H2. T_H1 T lymphocytes were characterized by the predominant secretion of IL-2, interferon γ (IFNγ) and tumour necrosis factor β (TNFβ), while T_H2 cells characteristically predominantly secreted IL-4, IL-5, IL-6 and IL-10. Other cytokines, such as TNFα, IL-3 and GM-CSF, were produced by both T_H1 and T_H2 cell subsets (Mosmann et al., 1986; Mosmann and Moore, 1991). These differing patterns of cytokine secretion by CD4 T lymphocytes result in distinct effector functions (Mosmann and Coffman, 1989). Broadly speaking, T_H1 cells participate in delayed type hypersensitivity (DTH) reactions (Cher and Mosmann, 1987), but also provide help for B lymphocyte immunoglobulin synthesis under certain circumstances. T_H2 cells, on the other hand, by their

pattern of secretion of B lymphocyte co-stimulatory cytokines, enhance the synthesis of all immunoglobulins, including IgE, in immune responses. In the centre of this spectrum, T lymphocyte clones secreting cytokines characteristic of both T_H1 and T_H2 cells, termed T_H0 cells, were also described (Firestein et al., 1989). It is still not clear whether these T_H0 cells represent distinct functional subsets or precursor cells in the process of differentiating into T_H1 or T_H2 cells.

Initial studies on mitogen-stimulated or alloreactive T lymphocyte clones raised from the peripheral blood of normal donors suggested that only a few CD4 T lymphocyte clones showed clear-cut T_H1 or T_H2 phenotypes, while the majority resembled T_H0 cells (Paliard et al., 1988). A different picture emerged, however, when human T lymphocyte clones specific for particular antigens were raised, particularly when these antigens were implicated in prototype immunological mechanisms such as DTH to mycobacterial antigens or nickel on the one hand, and IgE-mediated responses to helminth allergens on the other. For example, study of a large series of T lymphocyte clones raised from the peripheral blood of healthy individuals specific for purified protein derivative (PPD) of *Mycobacterium tuberculosis* or the excretory/secretory antigens of *Toxocara canis* (TES) revealed that the majority of the PPD-specific clones showed a T_H1 cytokine profile, with excess secretion of IL-2 and IFNγ, whereas the TES-specific clones showed a T_H2 profile, with excess IL-4 and IL-5 secretion (Del Prete et al., 1991a). Similarly, many T lymphocyte clones specific for intracellular bacteria such as *Borrelia burgdorferi* showed a T_H1 pattern of cytokine secretion (Yssel et al., 1991). In patients with atopic dermatitis, T lymphocyte clones specific for house dust mite (*Dermatophagoides pteronyssinus*) allergen raised from the peripheral blood secreted IL-4 and not IFNγ, whereas all clones specific for *Candida albicans* or tetanus toxoid raised from the same donors secreted an excess of IFNγ and relatively little IL-4 (Parronchi et al., 1991). CD8 T lymphocytes also appear to exhibit functional heterogeneity in terms of their cytokine secretion profile. For example, CD8 T cell clones specific for *Mycobacterium leprae* antigens showed patterns of cytokine secretion consistent with both T_H1 and T_H2 phenotypes (Salgame et al., 1991).

There is some evidence for mutual cross-inhibition of proliferation and cytokine secretion by T_H1 and T_H2 T lymphocytes, which is mediated by particular cytokines. For example, exogenous IL-4 favours the growth and proliferation of T_H2 T lymphocytes, whereas T_H2 clones are exquisitely sensitive to inhibition by IFNγ. IL-2 is a growth factor for both T_H1 and T_H2 cells. Whereas in mice, IL-10 significantly inhibits the proliferation and cytokine secretion of T_H1 cells (Fiorentino et al., 1989), human IL-10 significantly inhibits proliferation of and cytokine secretion by both T_H1 and T_H2 clones raised in response to specific antigen or lectin, and is itself a

Table 2.3 Cytokine profiles of subsets of human T lymphocytes

Cytokine	$T_H\rho$	T_H0	T_H1	T_H2
IL-2	+	+	+	−
IFNγ	−	+ +	+ +	−
TNF-β	−	−	+ +	−
IL-4	−	+ +	−	+ +
IL-5	−	+ +	−	+ +
IL-9	−		−	+ +
IL-10	−	+ +	−	±
TNF-α	−		+ +	+
GM-CSF	−	+	+ +	+
IL-3	−	+ +	+ +	+ +

−, little or no secretion; ±, secretion by some clones; +, moderate secretion; + +, marked secretion.

product of both T_H1 and T_H2 clones (Del Prete *et al.*, 1993).

Human T_H1 and T_H2 cells also differ in the nature of their help for immunoglobulin (Table 2.4) synthesis by autologous B lymphocytes and in their cytolytic potential (Table 2.4). In the presence of specific antigen, T_H2 clones were able to induce the synthesis of IgM, IgG, IgA and IgE by autologous B lymphocytes (Del Prete *et al.*, 1991b). Under the same conditions, T_H1 clones induced synthesis of IgM, IgG and IgA, but not IgE, with a peak response at a T cell: B cell ratio of 1 : 1. At higher T cell: B cell ratios, Ig synthesis was reduced, possibly reflecting the cytolytic activity of the T_H1 clones against autologous B lymphocytes at higher T cell: B cell ratios (Del Prete *et al.*, 1991b). This inhibition of Ig synthesis by T_H1 cells may represent an important mechanism whereby the production of antibodies other than IgE in immune responses is self regulated. In contrast, the failure of such an intrinsic regulatory mechanism in the case of T_H2 cells may explain, at least partly, why IgE antibody responses may persist despite the cessation of antigen exposure.

3.1 MECHANISMS OF DIFFERENTIATION OF HUMAN T_H1 AND T_H2 CD4 T LYMPHOCYTES

The experimental data discussed above suggest that T_H1 and T_H2 CD4 T lymphocyte clones exist in humans, and that one factor determining their profiles of cytokine secretion is their antigen specificity. The observation, however, that T lymphocyte clones specific for any given antigen, such as tetanus toxoid, can exhibit T_H1, T_H2 and T_H0 phenotypes (Parronchi *et al.*, 1991) suggests that the cytokine secretion profile of individual T lymphocytes is not irrevocably "set" in a T_H1 or T_H2 pattern by the criterion of antigen specificity alone. Indeed, it is hard to envisage how this could occur, since T lymphocyte antigen specificity is acquired at random during differentiation in the thymus, and therefore predetermination of a particular cytokine secretion profile according to antigen specificity would have to invoke some mech-

anism whereby such a profile is "imprinted" on a T lymphocyte before it has encountered its specific antigen, which is not impossible but seems unlikely.

In order to address this problem, it has been suggested that, at least in mice, T_H1 and T_H2 cells may represent memory cells that have matured into different functional phenotypes in the face of repeated stimulation by specific antigen. This hypothesis invokes the putative existence of antigen-naive precursor T lymphocytes (T_Hp) which secrete principally IL-2 and develop into early memory T_H0 effector cells after first encounter with specific antigen (Swain *et al.*, 1990). These cells then terminally differentiate into T_H1 or T_H2 cells following repetitive antigen stimulation.

The concept that T_H0, T_H1 and T_H2 effector cells differentiate from a common pool of precursor cells raises the question of which factors influence the differentiation of a T lymphocyte to the T_H1 or T_H2 phenotype. There is some evidence that exogenous cytokines may play a role in this differentiation process. For example, early addition of IL-4 to uncloned peripheral blood mononuclear cells stimulated with PPD shifted the subsequent differentiation of PPD-specific T lymphocytes from a T_H1 towards a T_H0 or T_H2 phenotype. Conversely, in similar cultures, early addition of both IFNγ and anti-IL-4 antibody induced many allergen- or TES-specific T lymphocyte clones to differentiate into T_H0 and T_H1, instead of T_H2 clones (Maggi *et al.*, 1992). These data suggest that the presence or absence of exogenous IL-4 or IFNγ at the time of antigen stimulation of resting T cells may influence their subsequent development into T_H1 or T_H2 clones, regardless of any pre-existing functional bias. In addition, the cytokines IFNα and IL-12 [both activators of natural killer (NK) cells which induce IFNγ synthesis] were also found to promote the differentiation of allergen- or TES-specific T lymphocyte clones towards a T_H0 or T_H1 phenotype instead of the usual T_H2 phenotype, whereas neutralization of endogenous IL-12 by specific antibody promoted the differentiation of PPD-specific T lymphocytes towards a T_H0 or T_H2 phenotype instead of the typical T_H1 phenotype (Romagnani, 1992). These observations have led to the hypothesis (Romagnani, 1992) that infection of cells such as macrophages with viruses and intracellular bacteria may favour a T_H1 response through local release of IFNα and IL-12, which might in turn activate NK cells resulting in local release of IFNγ. It is less clear what could favour the differentiation of T_H2 effector T lymphocytes from precursors *in vivo*, although it is possible to speculate that allergens or helminth-derived antigens, in contrast to intracellular parasites, might invoke relatively little IL-12 release from monocytes and therefore low local concentrations of IFNγ. This, coupled with the presence of local IL-4 released from IgE-sensitized mast cells in patients with atopic disease or helminthic infections, might favour the local differentiation of T_H2 effector T lymphocytes.

Table 2.4 Helper function for Ig synthesis and cytolytic activity of human CD4 T lymphocyte subsets

	T_H1	T_H2	T_H0
Cytolytic activity	+ + +	±	+ +
Helper function for:			
IgE synthesis	–	+ + +	±
Synthesis of other Ig classes			
Low T/B cell ratios	+ +	+ +	+ +
High T/B cell ratios	–	+ + +	±

–, absent; ±, detectable; + +, moderate; + + +, marked.

3.2 POSSIBLE ROLE OF T$_H$1 AND T$_H$2 T LYMPHOCYTES IN MUCOSAL PROTECTION AND INJURY

The existence of CD4 and CD8 T lymphocytes which secrete defined combinations of cytokines in response to various antigenie stimuli, and a consideration of the properties of these cytokines as summarized in Table 2.1 above, allows the view that particular patterns of cytokine release during inflammatory responses play an important role in determining the type of immune effector mechanism that a foreign antigen may elicit. Thus, T$_H$1 cells, through their predominant secretion of IL-2 and IFNγ, might be expected to promote the development of cytotoxic T lymphocytes, enhance the bactericidal activity of macrophages and stimulate the activity of Nk and lymphocyte-activated killer (LAK) cells. These effector mechanisms would be particularly effective for the elimination of viruses and other intracellular pathogens. On the other hand, the secretion of cytokines such as IL-4, IL-6 and IL-10 by T$_H$2 T lymphocytes would be expected to inhibit local activation of macrophages and dendritic cells but would favour the synthesis and release of specific antibody by activated B lymphocytes. Such humoral responses have a well-defined role in the elimination of invading extracellular microorganisms such as bacteria and their secreted products. Cytokines derived from T$_H$2 T lymphocytes are also implicated in the eosinophilia (IL-3, IL-5 and GM-CSF) and IgE response (IL-4) which accompany helminthic parasitic infections, although whether such responses play an important role in elimination of the parasite is in many cases in doubt.

In addition to their possible role in eliminating certain infections, there is also some evidence that secretion of T$_H$2 cytokines may be detrimental to the host. In several infectious diseases of humans, a T$_H$1 pattern of cytokine secretion is associated with resistance to infection, whereas a T$_H$2 pattern may be associated with progressive, uncontrolled infection. The T$_H$2 cytokines IL-4, IL-10 and IL-13 (Minty et al., 1993), although enhancing antibody production, may suppress cell-mediated immune responses. For example, IL-4 blocks the IL-2 dependent proliferation of T$_H$1 T cells through down-regulation of IL-2 receptors (Martinez et al., 1990), abrogates the antimicrobial activity of monocytes activated by IFNγ (Lehn et al., 1989), and blocks macrophage nitric oxide production, which is essential for the killing of intracellular parasites (Liew and Cox, 1991). IL-10 inhibits antigen-specific T lymphocyte responses by interfering with antigen presentation by monocytes and macrophages and by down-regulation of their major histocompatibility complex (MHC) class II molecule expression (De Waal Malefyt et al., 1991b). IL-10 also inhibits the release from macrophages of cytokines with antimicrobial functional properties (De Waal Malefyt

et al., 1991a). A good example of how T$_H$2 cytokines might impair host responses to infection is provided by leprosy, which presents as a clinical and immunological spectrum of disease where, on the one hand patients with tuberculoid leprosy exemplify the resistant response restricting growth of the pathogen, whereas on the other hand patients with lepromatous leprosy suffer from uncontrolled proliferation of the *Mycobacterium leprae* organism despite demonstrating a marked humoral response to the pathogen. These clinical patterns of disease are associated with distinct cytokine patterns in skin lesions of patients with leprosy (Yamamura et al., 1992), with elevated concentrations of mRNA encoding IL-2 and IFNα in tuberculoid lesions and elevated concentrations of mRNA encoding IL-4, IL-5 and IL-10 in lepromatous lesions. Spontaneous "conversion" of the disease from the lepromatous to the tuberculoid form was associated with an alteration of the corresponding cytokine pattern observed in skin lesions. These differences in cytokine profiles were confirmed in both CD4 and CD8 T lymphocytes at the clonal level (Salgame et al., 1991). In addition to providing further evidence for the existence of T$_H$1 and T$_H$2 CD4 and CD8 T lymphocytes, these studies also implicate T$_H$2 cytokines in suppressing cell-mediated immunity to such an extent that infection is allowed to proceed unabated.

Some of the most convincing evidence for a role for T$_H$1 and T$_H$2 cytokines in orchestrating distinct inflammatory responses *in vivo* has come from studies of various immunopathological disorders. In studies employing *in situ* hybridization, the cutaneous inflammatory responses to challenge with allergen in atopic subjects and tuberculin in non-atopic subjects were compared (Kay et al., 1991; Tsicopoulos et al., 1992). Both types of response (late-phase allergic and DTH) were associated with an influx of activated CD4 T lymphocytes, but whereas mRNA molecules encoding IL-2 and IFNα were abundant within the tuberculin reactions, very little mRNA encoding these cytokines was observed in the late phase allergic reactions. Conversely, mRNA encoding IL-4 and IL-5 was abundant in the late phase allergic but not the tuberculin reactions. In effect, the profiles of cytokine secretion in the allergic and tuberculin reactions closely paralleled those of T$_H$2 and T$_H$1 CD4 T lymphocytes, respectively. Furthermore, the relative numbers and types of granulocytes infiltrating these reactions reflected these different patterns of cytokine release (Gaga et al., 1991). These observations provide direct evidence in support of the hypothesis that activated T lymphocytes, through their patterns of cytokine secretion, regulate the types of granulocyte which participate in inflammatory reactions. Furthermore they demonstrate that T$_H$1 and T$_H$2 CD4 T lymphocyte responses can be detected in humans under physiological conditions, and that the antigen specificity of the T lymphocytes might be one factor which determines which type of response is initiated. In another study employing *in situ* hybridization

(Robinson *et al.*, 1992) it was shown that significantly elevated percentages of cells in the bronchoalveolar lavage (BAL) fluid of asthmatics expressed mRNA encoding IL-2, IL-3, IL-4, IL-5 and GM-CSF but not IFNγ in mild atopic asthmatics compared with non-atopic normal controls. Allergen challenge of sensitized atopic asthmatics was associated with increased numbers of activated T lymphocytes and eosinophils and increased expression of mRNA encoding IL-5 and GM-CSF in the bronchial mucosa (Bentley *et al.*, 1993). Taken together, these studies have provided a considerable body of evidence for the existence of T_H1 and T_H2 *patterns* of cytokine secretion, although it cannot be ascertained from such studies whether or not these cytokines originate from the same cells. More direct evidence that these cytokines may indeed have arisen from the same cells has been obtained from other studies showing that following mitogen stimulation, the majority of T lymphocytes derived from the conjunctival infiltrates of patients with vernal conjunctivitis developed into T_H2 clones (Maggi *et al.*, 1991). Similarly, high proportions of T_H2 clones were obtained from the skin lesions of patients with atopic dermatitis (Van der Heijden *et al.*, 1991; Ramb-Lindhauer *et al.*, 1991). A proportion of these skin-derived T lymphocyte clones were specific for house dust mite (Van der Heijden *et al.*, 1991) or grass pollen (Ramb-Lindhauer *et al.*, 1991) allergens.

In summary, these experimental data provide convincing evidence for specific patterns of cytokine release by both CD4 and CD8 T lymphocyte clones in the course of immune responses to invading microorganisms and in various immunopathological diseases. Whilst responses to invading microorganisms are generally protective, there is some evidence that T_H2 cytokines may be instrumental in producing host susceptibility to infections with intracellular microorganisms. Finally, the cytokine secretion pattern of T lymphocytes, particularly CD4 T lymphocytes, is strongly reflected in their functional capacities, particularly in the expression of cytolytic activity and help for Ig synthesis.

4. The Role of Cytotoxic T Lymphocytes in Mucosal Protection Against Invading Microorganisms

Cytotoxic T lymphocytes are important for the protection of mucosal surfaces against invading microorganisms, particularly organisms which replicate intracellularly. Most cytotoxic T lymphocytes (CTLs) express CD8, although cytocidal CD4$^+$ T lymphocytes have been described (Fleischer and Wagner, 1986). Antigen-specific CTLs recognize foreign antigen in association with major histocompatibility molecules on the surface of infected target cells through the α/β T lymphocyte antigen receptor. In addition, some CTLs recognize and lyse infected target cells via γ/δ T lymphocyte antigen receptors in a non-MHC-restricted fashion (Allison and Havran, 1991). Virtually all cells are susceptible to infection by viruses and some cells, particularly phagocytic cells, by bacteria and parasites. Pathogen-derived peptides are presented to CD8$^+$ CTLs in conjunction with MHC class I molecules.

Whereas virus-neutralizing antibody can prevent viral infections, only cell-mediated antiviral responses can eliminate established viral infections (Mackenzie *et al.*, 1989). It is not clear exactly how such infections are halted by T lymphocytes (Martz and Gamble, 1992). Direct CTL-mediated lysis of infected cells, or secretion of inhibitory cytokines such as IFNγ and TNFα may be involved. The fact that adoptive transfer of cloned, influenza virus-specific CTLs can protect mice against lethal doses of virus against which the cloned CTLs were generated, but not against a non-cross-reacting influenza virus strain mixed and administered with the immunizing virus, suggests a vital role for antigen-specific direct effector cell contact (Lukacher *et al.*, 1984). MHC class I-restricted CD8$^+$ CTL also appear to be important for resistance to microbial pathogens and parasites such as malarial parasites.

Lysis of infected target cells depends on conjugate formation with antigen-specific CTLs (Zagury *et al.*, 1975) mediated by the T lymphocyte antigen receptor and class I MHC molecules of the target cells and accessory molecules such as leucocyte function-associated antigen-1 (LFA-1) and intercellular adhesion molecule-1 (ICAM-1). The killing of infected target cells by CTL may be divided, for the purposes of analysis, into several stages, although these stages are not necessarily sequential, and may not all be utilized in every case.

1. "Productive" conjugate formation between the antigen-specific CTL and the infected target cells.
2. Recognition of specific antigen on the surface of the infected target cell by the CTL antigen receptor.
3. Transmembrane triggering of the CTL following engagement of the antigen receptor.
4. Induction of nuclear events in the target cell including apoptosis.
5. Activation of cytoplasmic processes in the CTL, including granule exocytosis and perforin secretion.

CTLs make contact with most surrounding cells when co-cultured with these cells *in vitro*. Although a predominance of non-lethal engagements is observed even with antigen-bearing target cells (Sitkovsky, 1988), only antigen-bearing target cells are able to trigger killing by the CTLs after cross-linking of their antigen receptors. A critical determinant for CTL activation by antigenic peptides appears to be the occupancy of localized regions of the CTL antigen receptor complex, which comprises the T cell antigen receptor and the CD8 molecule over a contiguous region of the cell surface, since a large number of small artificial antigen-presenting particles were not able to mimic triggering of CTLs by large particles (Mescher, 1992). CD8 molecules are considered important in this process both as co-receptors and adhesion molecules by enhancing the contact interactions of CTL with target cells which may express low surface numbers of MHC class I molecules or CTLs which have low-affinity antigen receptors for the antigenic peptide (O'Rourke and Mescher, 1992). Accessory adhesion molecules, including LFA-1 and ICAM-1, are also involved.

Following contact of the CTL with the target cell, intracellular signals generated within the CTL appear to be important regulators of CTL triggering. For example, one such important second messenger is the cyclic AMP-dependent protein kinase A: cytotoxicity, granule exocytosis and IFNγ secretion were inhibited by pretreatment of CTLs with anti-sense mRNA oligonucleotides encoding the Cα catalytic subunit of protein kinase A (Sugiyama et al., 1992).

CTL contact with target cells also induces changes in the target cells themselves, particularly fragmentation of DNA which is characteristic of the apoptotic process of cell death. This has long been considered to be an indispensable event in CTL-induced target cell death, although evidence that contradicts this view is accumulating. For example, addition of inhibitors of DNA topoisomerases I and II inhibited DNA fragmentation in target cells in a dose-dependent manner without affecting CTL-mediated cytotoxicity (Nishioka and Welsh, 1992). Other CTL granule proteins, such as the polyadenylate-binding protein PIA-1, may also be involved in a target cell DNA fragmentation (Tian et al., 1991). Thus, CTLs may have specialized "weapons" for particular target cells, and DNA fragmentation may not be indispensable.

Another well-recognized mechanism for target cell killing is the compromising of the integrity of the target cell plasma membrane by release of the protein perforin which is secreted by CTL and Nk cells. This protein spans the membrane of the target cells and forms pores in it (Ortaldo et al., 1992). Perforin secretion has been detected in infiltrating CD8 T lymphocytes during the course of viral infections, autoimmune reactions, transplant rejection and tumour rejection. Perforin release occurs in the context of a polarized secretory event in the CTL triggered by its surface receptors resulting in the generation of second messengers, cytoplasmic polarization and the translocation of the bulk of the CTL secretory granules towards the synapse-like junctional cleft formed between the CTL and the target cell. After granule exocytosis, the high concentration of granule components in this cleft enables them to diffuse into the target cell cytoplasm if plasma membrane permeabilization is effected by perforin. One of several granule enzymes implicated in this process is the serine protease granzyme A, which causes breakdown of target cell DNA when secreted along with perforin (Shiver et al., 1992).

While there are still questions to be answered regarding the mechanisms of CTL-mediated cytotoxicity at mucosal surfaces, the current consensus is that multiple mechanisms, including perforin secretion, granule enzyme secretion and initiation of apoptosis, may account for CTL-mediated cytotoxicity (Smyth et al., 1992).

5. Suppression of Immune Responses by T Lymphocytes at Mucosal Surfaces

Mucosal surfaces are exposed to a wide variety of foreign antigens, and not all of these induce an immune response. The mechanisms by which this selectivity is brought about are obviously very complex, and may depend on many factors other than the functioning of immune cells. For example, it may be that owing to (as yet unrecognized) variability in the structure of mucosal surfaces, antigen is allowed access to immune cells in some individuals but not in others. It is, nevertheless, appropriate to consider the mechanisms by which T lymphocytes, which normally initiate immune reactions to foreign antigens, may be rendered unresponsive to such antigens. This should enable a full understanding of what goes wrong when inappropriate immune responses, as for example in autoimmune diseases, are initiated.

Three principal mechanisms have been proposed to explain immunological tolerance: clonal deletion, clonal anergy and T lymphocyte-mediated immunosuppression. Clonal deletion refers to the physical elimination of self-reactive T lymphocytes, and is believed to occur principally in the thymus during T lymphocyte development. Clonal anergy, which is reversible, refers to the induction of cellular unresponsiveness when T lymphocytes fail to receive co-stimulatory signals necessary for cellular activation, and is believed to occur primarily in the periphery. Both of these processes involve ligation of the T lymphocyte antigen receptor, which ultimately results in cell death or inactivation. In contrast, immunosuppression is mediated by antigen-specific suppressor T lymphocytes, which inhibit the immune response initiated by other cells. Of these mechanisms, the latter two are more relevant to the possible therapy of human disease since they are subject to manipulation. They are considered in more detail below.

5.1 T LYMPHOCYTE CLONAL ANERGY

T lymphocyte antigen receptor (TCR) signalling in isolation is not sufficient to activate T lymphocytes, since purified T lymphocytes do not secrete IL-2 in response to mitogenic lectins or anti-TCR antibodies in the absence of viable accessory cells, despite the fact that under these conditions TCR-associated second messengers are generated (Schwartz, 1992). Furthermore, antigen-presenting B lymphocytes differ markedly in their capacity to stimulate cytokine production by T lymphocytes under conditions where they present the same numbers of peptide–MHC complexes to the antigen receptors on responsive T lymphocytes (Jenkins et al., 1990). These observations suggest that accessory cells do more than present antigen to T lymphocytes; they must provide other critical "co-stimulatory" signals.

Experiments investigating co-stimulation have been based on the premise that TCR-mediated and co-stimulatory signals could be provided by ligands present on different surfaces. For example, the TCR-mediated signal could be provided to the T lymphocyte by anti-TCR antibody immobilized on a plastic surface and the co-stimulatory signal by any accessory cell of interest. This approach has been used (Mueller et al., 1989) to identify several features of the co-stimulatory signal.

1. It is delivered by T lymphocyte/accessory cell contact, not by soluble factors such as cytokines.
2. It does not require MHC compatibility between the responding T lymphocyte and the co-stimulating accessory cell.
3. It is provided most efficiently by dendritic cells, less effectively by macrophages and activated B lymphocytes, poorly by resting B lymphocytes, and not at all by T lymphocytes.
4. It is not mediated by an increase in intracellular calcium concentrations or by inositol phospholipid metabolism.

Of the molecules studied to date, only the biology of the CD28–B7 receptor–ligand pair meets these criteria. CD28 is a dimeric cell surface glycoprotein expressed by all human CD4 T lymphocytes and approximately 50% of CD8 T lymphocytes, those with cytotoxic functions. Cytolytic T lymphocyte-associated antigen (CTLA)-4 is a cell surface protein closely related to CD28. Like CD28, CTLA-4 binds B7, but is expressed on T lymphocytes only following activation (Linsley et al., 1992).

Evidence that CD28 is a co-stimulatory molecule for T lymphocytes has been provided by the finding that non-stimulatory anti-CD28 antibody Fab fragments completely inhibit the provision of co-stimulation to T lymphocytes by activated B lymphocytes (Harding et al., 1992). CD28 cross-linking in the presence of phorbol ester results in IL-2 production by T lymphocytes which is resistant to inhibition by cyclosporin A (CSA; June et al., 1990), suggesting that CD28 signals by a biochemical pathway distinct from that used by the T lymphocyte antigen receptor.

The CD28 ligand B7 (also known as BB1) is a heavily glycosylated membrane glycoprotein. As might be expected for a co-stimulatory ligand, B7 is constitutively expressed on dendritic cells, is induced on activated B lymphocytes and monocytes, and is not expressed on resting T lymphocytes. Expression of B7 on B lymphocytes is induced by cross-linking of their surface Ig or MHC class II molecules (Schwartz, 1992; Nabavi et al., 1992), as may occur during presentation of antigen to T lymphocytes. Anti-B7 antibodies or a fusion protein comprised of the extracellular domain of CTLA-4 fused to the human IgG1 Fc region (CTLA-4-Ig) blocked the proliferation of T lymphocytes in response to antigen presented by a variety of B7-expressing accessory cells (Jenkins et al., 1991).

The critical importance of the co-stimulatory signal transduced by CD28, and the possible therapeutic potential for manipulating CD28/B7 interactions, have been illustrated by studies in animals. Blocking of this interaction in vivo prevented rejection of xenogeneic pancreatic islets (Lenschow et al., 1992). The recipients of these xenogeneic grafts retained their grafts even when the blocking agent (CTLA-4-Ig) was no longer administered, suggesting that blocking of CD28–B7 interactions at the time of novel antigenic exposure may result in immunological tolerance. Similarly, in T lymphocyte clones, TCR signalling in the absence of accessory cell derived co-stimulatory ligands in vitro results in the induction of a long-lasting state of unresponsiveness to antigenic stimulation (Tan et al., 1992), suggesting an important role for the CD28–B7 interaction in preventing the induction of immune unresponsiveness.

What is the relevance of these interactions between T lymphocytes and antigen-presenting cells with regard to the induction and suppression of immune responses at mucosal surfaces? These interactions might be important for regulating the immune response to soluble antigens, more particularly the selective induction of T_H1 and T_H2 responses, and also suggest a critical role for antigen-presenting cells in determining whether or not an immune response is initiated, and, if so, the nature of this response. For example, at sites where the dominant antigen-presenting cells are dendritic cells, which constitutively express B7, both T_H1 and T_H2 responses might be initiated. On the other hand, soluble antigen present at low concentrations in the periphery, such as at a mucosal surface, might be preferentially taken up and presented by resting B lymphocytes expressing antigen-specific surface immunoglobulin. When B7-negative resting B lymphocytes serve as antigen-presenting cells, however, T_H1 T lymphocytes fail to become activated and furthermore are rendered unresponsive (Gilbert and Weigle, 1992). In contrast, it is difficult to induce unresponsiveness in T_H2 T lymphocytes (Williams et al., 1992), which can be activated, at least in mice, by resting B lymphocytes (Gajewski et al., 1991), suggesting that the CD28–B7 interaction is not essential for activation of T_H2 T lymphocytes. IL-10 secreted by activated T_H2 T lymphocytes might then inhibit macrophage co-stimulation. This, coupled with the lack of secretion of IFNγ, which is required for the induction of B7 expression on resting macrophages, might further down-regulate T_H1 responses in the presence of low concentrations of antigen.

5.2 T LYMPHOCYTE-MEDIATED IMMUNOSUPPRESSION

As referred to above, immunosuppression may be mediated by suppressor T lymphocytes, which inhibit the immune responses mediated by other cells in an antigen-specific fashion (Dorf and Benacerraf, 1984; Dorf et al., 1992). Suppressor T lymphocytes express conventional antigen receptor α and β genes, and the molecular basis for their recognition of at least some (and probably most) antigens appears to be similar to that of other T lymphocytes. There are several mechanisms by which suppressor T lymphocytes are thought to inhibit immune responses in an antigen-specific fashion.

1. By secretion of soluble antigen-specific suppressor factors. There is some evidence that T lymphocyte antigen receptor α and/or β genes encode molecules associated with these factors (Fairchild et al., 1993). These factors can be bound to and eluted from antigen affinity columns, and inhibit antigen-specific immune reactions in vivo. If these factors resemble "soluble" T lymphocyte antigen receptor molecules, however, it is not clear how such complexes could recognize intact antigens, particularly in the absence of MHC class I and class II molecules. Furthermore, solubilized $\alpha\beta$ T lymphocyte antigen receptor molecules show extremely low binding affinities for peptide–MHC complexes (Jorgensen et al., 1992). Finally, it is not clear exactly how such factors could suppress an immune response.
2. By antigen-specific killing of effector T lymphocytes, B lymphocytes or antigen-presenting cells. Specific suppression could be achieved by killing of T lymphocytes bearing clonally distributed antigen receptors, through recognition of idiotypic antigen receptor–peptide–MHC complexes. Antigen-specific T lymphocyte killing of B lymphocytes results in specific suppression of antibody responses (Shinohara et al., 1991), while immunosuppression by T lymphocyte lysis of antigen-presenting cells has also been demonstrated (Rock et al., 1992).
3. By release of inhibitory cytokines following recognition of antigen–MHC complexes on antigen-presenting cells. Cytokines which have been implicated in inducing immunosuppression in this manner include IFNγ (Street and Mosmann 1991), TGFβ, (Miller et al., 1992) and IL-10.

6. Accumulation of T Lymphocytes at Mucosal Sites

T lymphocyte migration serves to bring the entire range of T lymphocyte receptor specificities in contact with antigen. It also serves to disseminate the products of an immune response, effector cells and memory cells, to other regions of the body, resulting in systemic immunity.

Lymphocyte homing is not random. The phenotype of T lymphocytes, the nature of their antigen receptors ($\alpha\beta$, $\gamma\delta$) and the state of differentiation or activation of the cells affects their distribution and migration. The vast majority of T lymphocytes migrate through the body by passing from the peripheral blood to secondary lymphoid tissues via a specialized endothelium lining the "high" endothelial venules. T lymphocytes are returned to the circulation via the efferent lymphatic ducts and the thoracic duct. T lymphocyte migration to the spleen is exceptional, in that the spleen does not contain lymphatic ducts, so that migrating cells that extravasate into the spleen are returned directly to the bloodstream. Small numbers of T lymphocytes also leave the blood in the peripheral tissues by crossing "flat" endothelium, and such traffic is increased markedly in the presence of inflammation. T lymphocytes entering tissues eventually accumulate in the afferent lymphatics, which serve to drain cells and antigen to local lymph nodes.

A recent conceptual development has been that adhesion and extravasation of T lymphocytes, and leucocytes in general, occurs through several sequential steps (Butcher, 1991; Lawrence and Springer, 1991; Shimizu et al., 1992).

1. A primary interaction between a selectin and its sugar-bearing receptor results in adhesion which is unstable under shear force, resulting in characteristic "rolling" of the leucocyte along the vessel wall. Specificity of interactions may be observed even at this stage. For example, T lymphocytes bearing the cutaneous lymphocyte-associated antigen (CLA) selectively bind to E-selectin expressed on the endothelium of inflamed skin (Picker et al., 1991a).
2. An activation step, delivered to leucocytes by certain cytokines or cell surface interactions which leads to a conformational change in their β_1- or β_2-integrins, causing a change in their binding avidity and a strong adhesion which is stable under shear force. For example, the integrin $\alpha_4\beta_7$ mediates preferential homing of T lymphocytes to the gut mucosa (Hu et al., 1992).
3. A transendothelial migration step following firm adhesion.

One fundamental factor regulating the migration of T lymphocytes into mucosal surfaces is their naive/memory status. Controversy still surrounds the nature and identity of naive and memory T lymphocytes. At present, it is uncertain whether long-lived memory cells actually exist, since there is good evidence that a major component (and possibly all) of immunological memory depends on continuous antigenic stimulation (Gray, 1992). The most widely accepted model for T lymphocyte memory (Mackay, 1993) holds that naive T lymphocytes (CD45RA$^+$) transform to effector/activated cells following antigen stimulation with the appropriate co-stimulatory signals. These cells then transform to memory cells (CD45RO$^+$), in the course of which they become smaller and lose some of the markers associated with acute activation. The question whether CD45RO$^+$ memory T lymphocytes can revert to CD45RA$^+$ cells is at present a contentious subject (Michie *et al.*, 1992; Bell, 1992). Whatever the case, CD45RA$^+$ and CD45RO$^+$ T lymphocytes are phenotypically and functionally distinct subsets, which migrate and localize in different ways.

Naive T lymphocytes exported from the thymus immediately begin to recirculate through lymphoid tissue, which provides the appropriate environment for their stimulation by antigen (Mackay *et al.*, 1990). The phenotype of naive T lymphocytes is consistent with their lymphoid tissue homing. Nearly all naive T lymphocytes express L-selectin (CD62L), which mediates the selectin interaction for lymph node homing, whereas a large proportion of memory T lymphocytes are L-selectin negative. Binding mediated by L-selectin is favoured by conditions of low flow shear. The lectin moiety of L-selectin binds weakly to oligosaccharide-containing molecules expressed on lymphoid endothelial cells, including CD34 and glycosylation-dependent cell adhesion molecule-1 (GlyCam-1) and the mucosal vascular addressin MadCam-1 expressed by high endothelial venules of mucosal surfaces (Berg *et al.*, 1993). Another ligand for L-selectin is the carbohydrate expressed on high endothelial venules of peripheral lymph nodes, which is recognized by the monoclonal antibody (mAb) Meca-79 (Berg *et al.*, 1991), which also recognizes the carbohydrate moieties of MadCam-1 and CD34. Firm binding and transmigration of naive T lymphocytes under non-inflamed conditions may be partly dependent on integrin-dependent interactions through, for example, $\alpha_4\beta_1$, $\alpha_5\beta_1$ and LFA-1 (Shimizu *et al.*, 1992).

In contrast to naive T lymphocytes, memory T lymphocytes migrate non-randomly in a partially tissue-specific fashion (Salmi *et al.*, 1992), suggesting that the mucosal environment in which T lymphocyte activation occurs may "imprint" a pattern of selectivity on the activated cell. The selective migration of a subset of T lymphocytes through the gut is probably the best characterized example of tissue-specific homing (Mackay, 1992). Gut-tropic cells in humans express the $\alpha_4\beta_7$ integrin, which is a counter-receptor for the mucosal vascular addressin MadCam-1 expressed on high endothelial venules of Peyer's patches and mesenteric lymph nodes, and on the endothelium of the gut mucosa (Briskin *et al.*, 1993). $\alpha_4\beta_7$-positive T lymphocytes in humans are of the memory phenotype. The T lymphocytes that localize to and migrate through the skin are almost exclusively of the memory phenotype, but unlike gut memory cells, skin memory T lymphocytes express elevated $\alpha_4\beta_1$ integrin and CLA (Pieker *et al.*, 1991a, Mackay *et al.*, 1992b), a ligand for E-selectin expressed on the endothelium of inflamed skin.

It is predominantly memory T lymphocytes which localize to inflammatory lesions, especially during the early phases of inflammatory responses. Simple phenotypic analysis of cells harvested from a variety of pathological lesions confirmed that they were CD45RO$^+$ (Shimizu *et al.*, 1992). In addition to T lymphocytes, local high endothelial venules may also be affected by the presence of an inflammatory response. For example, antigen-challenged lymph node vascular endothelial cells up-regulate adhesion molecules such as vascular cell adhesion molecule-1 (VCAM-1), and show markedly increased T lymphocyte traffic (Mackay *et al.*, 1992a). In addition, certain endothelial adhesion molecules, for example E-selectin and VCAM-1, selectively bind to memory T lymphocytes (Shimizu *et al.*, 1991). These molecules are not expressed by endothelial cells of non-inflamed tissue, and expression in inflammatory conditions results at least partly from exposure of the endothelium to cytokines such as TNFα and IL-1. Cytokines and chemokines undoubtedly play an important role in T lymphocyte recruitment to inflammatory sites. For example, MIP-1β preferentially augments the binding of CD8 T lymphocytes to VCAM-1 (Tanaka *et al.*, 1993). MIP-1β is immobilized on the endothelial surface by binding to proteoglycans such as CD44, thus facilitating local concentration of the chemokine at inflammatory sites.

More stable T lymphocyte binding to endothelium at inflammatory sites is largely mediated by integrin adhesion receptors such as VLA-4 and LFA-1. Binding of memory T lymphocytes through these receptors in inflammatory conditions does not result from increased expression of these molecules by T lymphocytes, but appears to be related to conformational changes in their structure. For example, LFA-1 can be functionally activated *in vitro* by cross-linking of CD3 molecules on the surface of T lymphocytes (Dustin and Springer, 1989). VLA-4 recognition of VCAM-1 plays a prominent role in the binding of T lymphocytes to cytokine-activated endothelial cells (Oppenheimer-Marks *et al.*, 1991). There are other adhesion receptors that could play a role in mediating T lymphocyte interactions with endothelium at sites of inflammation, including vascular adhesion protein-1 (VAP-1) which is expressed in inflamed lymph nodes (Salmi *et al.*, 1993). CD2 binding to its counter receptors CD58 (LFA-3), CD59 or CD48, all of which are expressed on endothelium, may also facilitate T lymphocyte binding.

As mentioned above, T lymphocyte binding to endothelium and transmigration of the T lymphocytes across the endothelium are separable events, since not all adherent T lymphocytes transmigrate through an endothelial cell layer *in vivo* or *in vitro*. Blocking experiments with mAbs have demonstrated that transendothelial migration of T lymphocytes bound to endothelial cells involves LFA-1 and ICAM-1, but not VLA-4, VCAM-1, LFA-3 or E-selectin (Oppenheimer-Marks *et al.*, 1991). These observations emphasize the fact that a number of adhesion receptors that mediate the initial adhesive events between T lymphocytes and endothelial cells do not necessarily play a major role during subsequent transendothelial migration.

In summary, transendothelial migration is an intrinsic property of distinct T lymphocyte subpopulations which exhibit specific phenotypic characteristics. The activities of a variety of adhesion receptors play a central role in the selective recruitment of these T lymphocyte populations into inflammatory sites, and thus in the pathogenesis of chronic inflammatory diseases. The process of T lymphocyte migration involves combinatorial steps, such that the type of T lymphocyte which binds to endothelium in a particular tissue depends not only on the specificity of the initial selectin interaction, but also on the actions of locally secreted cytokines and the types of integrins and their receptors which are expressed. Small differences in the expression of certain adhesion molecules by two cell types (for example naive and memory T lymphocytes) may be amplified through these different steps, resulting in markedly different efficiencies of endothelial binding and transmigration.

7. T Lymphocytes and the Regulation of Immunoglobulin Synthesis

Primary antibody responses are mounted by antibodies of the IgM class. Since these antibodies generally express germline-determined variable region genes that have not yet been modified by somatic mutation, they bind antigen only with low affinity. This is partially compensated for by their 10 antigen binding sites, which enable avid binding to multimeric antigens. IgM antibodies can also destroy or opsonize target cells through their very efficient fixation of complement. Nevertheless, the low affinity of IgM antibody, its pentameric structure which limits its diffusion capacity and its short half-life *in vivo* together suggest that IgM antibodies are probably less well adapted than IgG antibodies to protect the host against repeated infections by pathogens.

Antibodies of the IgG, IgA and IgE classes are made later than IgM during a primary immune response, but account for most of the antibody that is made during a memory response. Although isotype switching and affinity maturation are independent processes, they usually occur simultaneously (Fish *et al.*, 1989), so that the increased affinity of bivalent antibodies of isotypes other than IgM is rapidly enhanced. IgG antibody is the predominant isotype in plasma and lymph, while IgA antibody predominates at mucosal surfaces and in secretions from these surfaces.

IgG antibodies have a long half-life, ensuring high concentrations in the serum. In humans IgG_1 and IgG_3 fix complement, while IgG_2 has some ability in this regard. $Fc\gamma RI$ receptors (on monocytes, macrophages and neutrophils) bind to IgG_1 and IgG_3 most avidly and to IgG_2 to a lesser degree. $Fc\gamma RII$ (on macrophages, monocytes, neutrophils and B lymphocytes) and $Fc\gamma RIII$ (on monocytes, macrophages, NK cells, neutrophils and some T lymphocytes) selectively bind, with low affinity, to IgG_1 and IgG_3. IgG antibodies also bind to placental Fc receptors through which they can transported into the fetal circulation. Different antigenic stimuli result in the predominant production of different IgG subclasses. Soluble protein antigens evoke predominant synthesis of IgG_2 in humans, although IgG_4 is more predominant after repeated immunization. Viruses tend to provoke IgG_1 and IgG_3 responses, whereas nematode parasites preferentially stimulate IgG_4 synthesis. The different functions of these IgG antibody subclasses, and their preferential production in infections do not necessarily indicate that the various subclasses are essential for protection of the host against the particular pathogens. Most patients with isolated IgG_2 deficiency are immunologically normal, although some show an increased susceptibility to infection with encapsulated bacteria. In such cases, it is not clear whether this susceptibility to infection is associated with a particular lack of IgG_2 or a generalized lack of antibodies specific for the relevant bacterial epitopes, regardless of isotype. For similar reasons, it is difficult to interpret whether the association between low serum IgG_1 and IgG_3 concentrations and chronic lung infections represents a particular requirement for these complement-fixing antibodies for host defence against pulmonary pathogens.

As with the IgG subclasses, it is easier to point out properties of IgA antibodies that should make them particularly well suited to protect against host infection than to demonstrate that these properties are essential for the well-being of the host. The association of one or two IgA dimers with a molecule of secretory component facilitates transport of this isotype into secretions at mucosal surfaces. The oligomeric nature of IgA enhances its ability to interact with viruses and bacteria that may invade mucosal surfaces, while the lack of barriers to diffusion in secretions obviates the effect of its high molecular weight. IgA lacks the ability to fix complement by the classical pathway, but is the most effective isotype at fixing complement by the alternative pathway. IgA antibodies, like IgG antibodies, bind well to Fc receptors on granulocytes. The affinity of IgA antibodies for mucus may reinforce the mucus barrier to pathogen penetration at mucosal surfaces. Despite these apparent advantages, most IgA-deficient individuals have no obvious deficiency in immune function, although this deficiency occurs with a frequency of approximately 0.1% in the general population. A minority of IgA-deficient individuals do suffer from respiratory infections, malabsorption and autoimmune disorders at increased frequency. In most IgA-deficient individuals, IgM antibodies associate with secretory component and thereby replace IgA as the predominant secretory isotype (Bkrandtzaeg *et al.*, 1968). In at least some of the IgA-deficient patients who have frequent infections, this substitution fails to occur, leading to a more global immune defect (Bkrandtzaeg *et al.*, 1968).

IgE antibodies have the unique ability to bind with high affinity to Fc$_\epsilon$RI receptors on mast cells and basophils, and induce degranulation and cytokine secretion by these cells when they are cross-linked by antigen. In addition to its central role in the pathogenesis of atopic disorders, it has been speculated that IgE-mediated mast cell degranulation may have a role in the expulsion of gut and respiratory tract parasites. IgE may also have a role in eosinophil killing of parasites by IgE-mediated antibody-dependent cell-mediated cytotoxicity. Despite these observations, it is unclear whether IgE plays an indispensable role in protection against disease.

Cytokines play an important role in the control of switching of antibody isotypes in humans away from IgM during the evolution of the immune response. Cytokines which play a role in this process may be classified as follows.

1. Those cytokines which are responsible for the specificity of the isotype switch event in B lymphocytes. They include IL-4 and TGFβ, and have in common the ability to induce the transcription of an immature (germline) form of RNA encoding the heavy chain isotype to which switching will be induced.

2. Those cytokines which are generally permissive or inhibitory for antibody production by B lymphocytes and may influence the secretion of a number of different antibody isotypes, depending on which of the specific stimuli are present. These cytokines include IL-2, IL-5, IL-6 and IFNγ (but this is not an exhaustive list). Permissive stimuli other than cytokines, for example the co-stimulatory signals delivered by T lymphocytes, may also be included in this category.

The control of switching to IgE synthesis in human B lymphocytes has been the subject of many studies. IL-4 is a critical stimulus for IgE switching in human B lymphocytes cultured with Epstein–Barr virus, anti-CD40 antibody and activated T lymphocytes (Gascan *et al.*, 1991). No other cytokine has been able to substitute for IL-4 in any of these systems. The inducement of a large increase in steady-state germline ϵ-chain RNA expression in B lymphocytes stimulated with IL-4 and endotoxin (Gauchat *et al.*, 1990) also provides evidence for IL-4 promotion of switching to IgE. In addition to IgE, IL-4 also stimulates switching of human B lymphocytes to synthesis of IgG4, although IgG4 is detectable earlier, consistent with the interpretation that the switch from IgM to IgE typically involves an initial switch to an IgG4 and then a second switch to IgE (Gascan *et al.*, 1991).

The cytokines IL-2, IL-5 and IL-6 enhance IL-4-induced secretion of IgE by human B lymphocytes (Vercelli *et al.*, 1989; Jabura *et al.*, 1990), and probably act as "permissive" stimuli through their non-specific activation of B cells, since there is no evidence that they can induce IgE switching. Several cytokines also inhibit IgE synthesis by human B lymphocytes *in vitro*. Both IFNγ and IFNα specifically inhibit IgE secretion by B lymphocytes cultured with Epstein–Barr virus (Thyphronitis *et al.*, 1989), although neither of these cytokines inhibits IL-4-dependent induction of increased steady-state concentrations of germline ϵ-chain RNA in B lymphocytes (Gauchat *et al.*, 1990). IL-12 has been reported to reduce IgE synthesis by B lymphocytes stimulated with IL-4 and cortisol (Kiniwa *et al.*, 1992).

In the case of the other immunoglobulin classes, much less information is available as regards regulation of their synthesis in humans. TGFβ has been shown selectively to stimulate IgA class switching by B lymphocytes in several *in vitro* systems (van Vlasselaer *et al.*, 1992). Little is known, however, about the control of the synthesis of human IgG and its various subclasses. There is no evidence that IFNγ, IL-4 or TNFβ regulate switching to any of the IgG subclasses. Although IL-4 enhances IgG$_4$ secretion in B lymphocytes cultured with anti-CD40 antibodies (Gascan *et al.*, 1991), it has not been shown to induce expression of the germline form of γ4 RNA.

In summary, it is clear that cytokines play a critical role in the regulation of both the amount and the class of immunoglobulins secreted by B lymphocytes at mucosal surfaces. Conversely, B lymphocytes themselves may influence local cytokine synthesis by T lymphocytes, as exemplified by their differential capacity to present soluble antigens to T$_H$1 and T$_H$2 T lymphocytes (described previously).

8. The Role of T Lymphocytes in Recruitment of Granulocytes to Mucosal Surfaces

8.1 NEUTROPHILS

By analogy with T lymphocytes, the migration of neutrophils from the peripheral circulation into mucosal surfaces has been conceptualized as consisting of three separate steps (Butcher, 1991).

1. Selectins mediate the initial weak tethering of neutrophils to the endothelial wall. This results in characteristic "rolling" of the neutrophils along the endothelium under conditions of heightened shear force. L-selectin (LECAM-1) is constitutively expressed on neutrophils (Lasky, 1991), whereas E-selectin (ELAM-1) expression is induced on endothelial cells by exposure to endotoxin, TNFα or

IL-1 (Lasky, 1991). P-selectin (GMP-140, PADGEM) is rapidly expressed on endothelial cells after release from intracellular storage granules (Weibal–Palade bodies) on exposure to a variety of inflammatory mediators such as complement, thrombin and histamine. The selectins are important molecules for the promotion of neutrophil adhesion. Their ligands are sialylated derivatives of the Lewis X and Lewis A oligosaccharides which are found on several molecules including L-selectin itself (Picker *et al.*, 1991b).

2. β_2-Integrins expressed on the neutrophil surface mediate their firm adhesion to endothelium, by binding to immunoglobulin-like molecules such as ICAM-1 (CD54) on the endothelium. Of the β_2-integrins, CD11b/CD18 (Mac-1) is expressed predominantly on neutrophils, although CD11a/CD18 (LFA-1) and CD11c/CD18 (p150,95) are also expressed on all granulocytes (Kishimoto *et al.*, 1989). As with T lymphocytes, the local generation of cytokines and chemokines may bring about activation of neutrophils within the vascular compartment, resulting in up-regulation of surface expression of β_2-integrins and shedding of surface L-selectin. In addition, certain cytokines, such as IL-1, may up-regulate the constitutive expression of endothelial ICAM-1. The importance of β_2-integrins for extravasation of granulocytes has been shown in animal models, where prior administration of anti-CD18 antibodies attenuates the neutrophil-dependent increase in pulmonary vascular permeability, haemorrhage and neutrophil accumulation in the lung which accompanies haemorrhagic shock (Vedder *et al.*, 1988).

3. Finally, β_2-integrin interactions with ICAM-1 and other adhesion molecules of the immunoglobulin supergene family mediate firm adhesion of neutrophils and diapedesis beyond the vascular compartment. This process may also be dependent on the local release of cytokines and chemokines. For example, the α-chemokines IL-8 and NAP-2 (a proteolytic cleavage product of connective tissue activating peptide III) are potent chemotactic and activating factors for neutrophils (Baggiolini *et al.*, 1989; Walz and Baggiolini, 1989).

8.2 EOSINOPHILS

The mucosal inflammatory responses which accompany asthma, allergic inflammation and helminthic infestations are characterized by a marked and specific infiltration of eosinophils into the relevant mucosal surfaces. Of the cytokines secreted by activated T lymphocytes, IL-3, IL-5 and GM-CSF promote maturation, activation and prolonged survival of the eosinophil (Lopez *et al.*, 1986; Rothenberg *et al.*, 1988, 1989). IL-5 is unique in that, unlike IL-3 and GM-CSF, it acts specifically on

eosinophils in terms of activation, hyperadhesion and terminal differentiation of the eosinophil precursor (Lopez *et al.*, 1988; Walsh *et al.*, 1990). IL-5 may be the most important cytokine for eosinophil differentiation since it is released principally by T lymphocytes, and the eosinophilia associated with parasitic infections is T lymphocyte dependent (Basten and Beeson, 1970). This hypothesis is further supported by the observation that transgenic mice constitutively expressing the IL-5 gene show a marked, specific expansion of blood and tissue eosinophils (Dent *et al.*, 1990).

One fundamental problem with asthma pathogenesis is the mechanism by which eosinophils preferentially accumulate in the inflamed mucosa. Local expression of eosinophil-specific cytokines such as IL-5 may partly account for this phenomenon by selectively enhancing eosinophil differentiation and survival (Hamid *et al.*, 1991; Robinson *et al.*, 1992). Chemoattractants may also play a role. Cytokines such as IL-5, IL-3 and GM-CSF have been shown to prime eosinophils for an enhanced chemotactic response to other chemoattractants including chemokines such as IL-8 (Warringa *et al.*, 1991; Sehmi *et al.*, 1992). The T lymphocyte-derived cytokines, lymphocyte chemoattractant factor (LCF) and IL-2 are also relatively potent eosinophil chemoattractants (Rand *et al.*, 1991a,b). The LCF protein exerts its activity by binding to CD4 molecules, so that in addition to affecting lymphocytes it also specifically acts on eosinophils compared with other granulocytes, since only eosinophil granulocytes express CD4 (Rand *et al.*, 1991a). An exciting recent observation has been that the chemokines RANTES, MCP-3 and, to a lesser extent, MIP-1α are powerful and selective chemoattractants for eosinophils and basophils *in vitro* (Kameyoshi *et al.*, 1992; Baggiolini and Datinden, 1994). An investigation of whether these chemokines are released at mucosal surfaces involved with asthmatic and allergic inflammation is now urgently required.

Another revealing field of study has been the role of adhesion molecules in selective eosinophil migration. As with neutrophils, eosinophils can bind to endothelium using L-selectin, P-selectin and E-selectin, with no apparent differences between these cells and neutrophils. As with neutrophils, eosinophils exhibit surface shedding of L-selectin on activation, which may facilitate endothelial transmigration (Smith *et al.*, 1992). Eosinophil β_2-integrins such as LFA-1 and Mac-1, as with neutrophils, probably mediate firm adhesion of these cells to endothelium and subsequent transmigration by binding to molecules such as ICAM-1 expressed on the endothelium (Bochner *et al.*, 1991; Kyan-Aung *et al.*, 1991). The expression of LFA-1 and Mac-1 on eosinophils can be up-regulated by inflammatory mediators (Hartnell *et al.*, 1990). Eosinophils appear to be unique, however, in that IL-3 and IL-5 up-regulate eosinophil, but not neutrophil, adhesion to unstimulated endothelial cells (Walsh *et al.*, 1990). Furthermore, eosinophils, but not neutrophils, express the β_1-integrin VLA-4, which is a ligand for the adhesion molecule VCAM-1 expressed on the surface of stimulated endothelial cells (Walsh *et al.*, 1991). Eosinophils may also use VLA-4 for binding to tissue fibronectin (Elices *et al.*, 1990), which prolongs their survival *in vitro*, possibly by inducing autocrine secretion of IL-3 (Anwar *et al.*, 1993). The expression of VCAM-1 on endothelial cells is increased by exposure to IL-4, which enhanced VLA-4/VCAM-1-dependent adherence of eosinophils, but not neutrophils, to endothelium (Schleimer *et al.*, 1992). In summary, these mechanisms offer several possible explanations for the selective accumulation of eosinophils observed in asthmatic and allergic inflammation. The important role of T lymphocytes, which are clearly an important source of IL-3, IL-4, IL-5 and GM-CSF, and may be an important source of chemokines such as RANTES, MCP-3 and MIP-1α, is self-evident.

9. Summary

T lymphocytes clearly play a vital role in immune mechanisms which lead both to damage and to protection of mucosal surfaces. Through their specific antigen receptors, they are responsible for the initiation and orchestration of immune responses. It is clear that the particular patterns of cytokines secreted by activated T lymphocytes determine the nature of the ensuing inflammatory response, and in particular, whether cell-mediated reactions or humoral reactions predominate. T lymphocytes also play a role in the genesis of immunological tolerance, which may abrogate unwanted inflammatory responses in mucosal surfaces exposed to a wide variety of foreign antigens. Finally, T lymphocytes play a major role in the recruitment of particular granulocytes to inflammatory reactions at mucosal surfaces. These granulocytes may be responsible for the repulsion of invading microorganisms, or alternatively for chronic tissue damage. A more intimate knowledge of the functions of T lymphocytes in both health and disease will surely create a wider scope for therapeutic intervention.

10. References

Aarden, L.A. Brummer, T.K., Cerottini, J-C., Dayer, J-M., de Weck, A.L. *et al.* (1979). Revised nomenclature for antigen-non-specific T cell proliferation and helper factors. J. Immunol. 123, 2928–2929.

Allison, J.P. and Havran, W.L. (1991). The immunobiology of T cells with invariant $\gamma\delta$ antigen receptors. Annu. Rev. Immunol. 9, 679–705.

Anwar, A.R.E., Moqbel, R., Walsh, G.M., Kay, A.B. and Wardlaw, A.J. (1993). Adhesion to fibronectin prolongs eosinophil survival. J. Exp. Med. 177, 819–824.

Arai, K. Lee, F. Miyajima, A. Miyatake, S., Arai, N. and Yokota, T. (1990). Cytokines: co-ordinators of immune and inflammatory responses. Annu. Rev. Biochem. 59, 783.

Baggiolini, M., Walz, A. and Kunkel, S.L. (1989). Neutrophil-activating peptide-1/interleukin 8, a novel cytokine that activates neutrophils. J. Clin. Invest. 84, 1045–1049.

Baggioloni, M. and Datinden, C.A. (1994). CC chemokines in allergic inflammation. Immunol. Today 15, 127–133.

Basten, A. and Beeson, P.B. (1970). Mechanism of eosinophilia II: role of the lymphocyte. J. Exp. Med. 131, 1288–1309.

Bell, E.B. (1992). Function of CD4 T cell subsets *in vivo*: expression of CD45R isoforms. Semin. Immunol. 4, 43–50.

Bentley, A.M., Menq, Q., Robinson, D.S., Hamid, Q, Kay, A.B. and Durham, S.R. (1993). Increases in activated T lymphocytes, eosinophils and cytokine messenger RNA for IL-5 and GM-CSF in bronchial biopsies after allergen inhalation challenge in atopic asthmatics. Am. J. Respir. Cell Mol. Biol. 8, 35–42.

Berg, E.L. Robinson, M.K., Warnock, R.A. and Butcher, E.C. (1991). The human peripheral lymph node vascular addressin is a ligand for LECAM-1, the peripheral lymph node homing receptor. J. Cell Biol. 114, 343–349.

Berg, E.L., McEvoy, L.M., Berlin, C., Bargatze, R.F. and Butcher, E.C. (1993). L-selectin mediates lymphocyte rolling on MADCAM-1. Nature 366, 695–698.

Bkrandtzaeg, P., Fjellanger, I. and Geruldsen, S.T. (1968). Immunoglobulin M: local synthesis and selective secretion in patients with IgA deficiency. Science 160, 789–791.

Briskin, M.J., McEvoy, L.M. and Butcher, E.C. (1993). MADCAM-1 has homology to immunoglobulin and mucin-like adhesion receptors and to IgA. Nature 363, 461–464.

Bochner, B.S., Luscinskas, F.W., Gimbrone, M.A.J., Newman, W., Sterbinsky, S.A., Derse-Anthony, C.P., Klunk, D. and Schleimer, R.P. (1991). Adhesion of human basophils, eosinophils and neutrophils to IL-1 activated human vascular endothelial cells: contributions of endothelial cell adhesion molecules. J. Exp. Med. 173, 1553–1557.

Butcher, E.C. (1991). Leukocyte-endothelial cell recognition three (or more) steps to specificity and diversity. Cell 67, 1033–1036.

Cher, D.J. and Mosmann, T.R. (1987). Two types of murine helper T cell clone II: delayed-type hypersensitivity is mediated by Th 1 clones. J. Immunol. 138, 3688–3694.

De Waal-Malefyt, R., Abrams, J., Bennett, B., Figdor, C.G. and De Vries, J.E. (1991a). Interleukin-10 (IL-10) inhibits cytokine synthesis by human monocytes: an autoregulatory role of IL-10 produced by monocytes. J. Exp. Med. 174, 1209–1220.

De Waal-Malefyt, R., Haanen, J., Spits, H., Roncarto, M.G., Velde, A., Figdor, C., Johnson, K., Kastelein, R., Yssel, H. and De Vries, J.E. (1991b). Interleukin-10 (IL-10) and viral IL-10 strongly reduce antigen-specific human T cell proliferation by diminishing the antigen presenting capacity of monocytes via down-regulation of class II major histocompatibility complex expression. J. Exp. Med. 174, 915–924.

Del Prete, G.F., De Carli, M., Mastromauro, C., Biagiotti, R., Macchia, D., Falagiani, P., Ricci, M. and Romagnani, S. (1991a). Puriiled protein derivative of *Mycobacterium tuberculosis* and excretory–secretory antigen(s) of *Toxocara canis* expand *in vitro* human T cells with stable and opposite (type 1 T helper or type 2 T helper) profile of cytokine production. J. Clin. Invest. 88, 346–350.

Del Prete, G.F., De Carli, M., Ricci, M. and Romagnani, S. (1991b). Helper activity for immunoglobulin synthesis of Th1 and Th2 human T cell clones. The help of Th1 clones is limited by their cytolytic capacity. J. Exp. Med. 174, 809–813.

Del Prete, G.F., De Carli, M., Almerigogna, F., Giudizi, M.G., Biagiotti, R. and Romagnani, S. (1993). Human IL-10 is produced by both type 1 helper (Th1) and type 2 helper (Th2) T cell clones and inhibits their antigen-specific proliferation and cytokine production. J. Immunol. 150, 353–360.

Dent, L.A., Strath, M., Mellor, A.L. and Sanderson, C.J. (1990). Eosinophilia in transgenic mice expressing interleukin-5. J. Exp. Med. 172, 1425–1431.

Dorf, M.E. and Benacerraf, B. (1984). Suppressor cells and immunoregulation. Annu. Rev. Immunol. 2, 127–158.

Dorf, M.E., Kuchroo, V.K. and Collins, M. (1992). Suppressor T cells: some answers but more questions. Immunol. Today 13, 241–243.

Dumonde, D.C., Wolstencraft, R.A., Panayi, G.S., Matthew, M., Morley, J. and Howson, W.T. (1969). "Lymphokines": non-antibody mediators of cellular immunity generated by lymphocyte activation. Nature 224, 38–42.

Dustin, M.L. and Springer, T.A. (1989). T cell receptor cross-linking transiently stimulates adhesiveness through LFA-1. Nature 341, 619.

Elices, M.J., Osborn, L., Takada, Y., Crouse, C., Luhowskyj, S., Hemler, M.E. and Lobb, R.R. (1990). VCAM-1 on activated endothelium interacts with the leucocyte integrin VLA-4 at a site distinct from the VLA-4/fibronectin binding site. Cell 60, 577–584.

Fairchild, R.L., Palmer, E. and Moorhead, J.W. (1993). Production of DNP-specific/class I MHC-restricted suppressor molecules is linked to the expression of T cell receptor α- and β-chain genes. J. Immunol. 150, 67–77.

Fiorentino, D.F., Bond, M.W. and Mossmann, T.R. (1989). Two types of mouse T helper cells IV. Th2 clones secrete a factor that inhibits cytokine production by Th1 clones. J. Exp. Med. 170, 2081–2095.

Firestein, G.S. Roeder, W.D., Laxer, J.A., Towsend, K.S., Weaver, C.T. Hum, J.T. et al. (1989). A new murine CD4+ T cell subset with an unrestricted cytokine profile. J. Immunol. 143, 518–525.

Fish, S., Zenowich, E., Fleming, M. and Manser, T. (1989). Molecular analysis of original antigenic sin. I. Clonal selection, somatic mutation and isotype switching during a memory B cell response. J. Exp. Med. 170, 1191–1209.

Fleischer, B. and Wagner, H. (1986). Significance of T4 or T8 phenotype of human cytotoxic T-lymphocyte clones. Curr. Top. Microbiol. Immunol. 126, 101–109.

Gaga, M., Frew, A.J., Varney, V.A. and Kay, A.B. (1991). Eosinophil activation and T-lymphocyte infiltration in allergen-induced late phase skin reactions and classical delayed-type hypersensitivity. J. Immunol. 147, 816–822.

Gajewski, T.F., Pinnas, M., Wong, T. and Fitch, F.W. (1991). Murine Th1 and Th2 clones proliferate optimally in response to distinct antigen-presenting cell populations. J. Immunol. 146, 1750–1758.

Gascan, H., Gauchat, J-F., Aversa, G., Van Vlasselaer, P. and de Vries, J.E. (1991). Anti-CD40 monoclonal antibodies or CD4+ T cell clones induce IgG4 and IgE switching in purified human B cells via different signalling pathways. J. Immunol. 147, 8–13.

Gauchat, J-F., Lebman, D.A., Coffman, R.L., Gascan, H. and de Vries, J.E. (1990). Structure and expression of germline E transcripts in human B cells induced by IL-4 to switch to IgE production. J. Exp. Med. 172, 463–473.

Gilbert, K.M. and Weigle, W.O. (1992). B cell presentation of a tolerogenic signal to Th clones. Cell Immunol. 139, 58–71.

Gray, D. (1992). The dynamics of immunological memory. Semin. Immunol. 4, 29–34.

Hamid, Q., Azzawi, M., Ying, S., Moqbel, R., Wardlaw, A.J., Corrigan, C.J., Bradley, B., Durham, S.R., Collins, J.V., Jeffery, P.K., Quint, D.J. and Kay, A.B. (1991). Expression of mRNA for IL-5 in mucosal bronchial biopsies from asthma. J. Clin. Invest. 87, 1541–1546.

Harding, F.A., McArthur, J.G., Gross, J.A., Raulet, D.M. and Allison, J.P. (1992). CD28-mediated signalling co-stimulates murine T cells and prevents induction of anergy in T cell clones. Nature 356, 607–609.

Hartnell, A., Moqbel, R., Walsh, G.M., Bradley, B. and Kay, A.B. (1990). Fcγ and CD11/CD18 receptor expression on normal density and low density eosinophils. Immunology 69, 264–270.

Hu, M.C., Crowe, D.T., Weissman, I.L. and Holzmann, B. (1992). Cloning and expression of mouse integrin beta p (beta 7): a functional role in Peyer's patch-specific lymphocyte homing. Proc. Natl. Acad. Sci. USA 899, 1924–1928.

Jabura, H.H., Fu, S.M., Geha, R.S. and Vercelli, D. (1990). CD40 and IgE: synergism between anti-CD40 monoclonal antibody and IL-4 in the induction of IgE synthesis by highly purified human B cells. J. Exp. Med. 172, 1861–1864.

Jenkins, M.K., Burrell, E. and Ashwell, J.D. (1990). Antigen presentation by resting B cells. Effectiveness at inducing T cell proliferation is determined by co-stimulatory signals, not T cell receptor occupancy. J. Immunol. 144, 1585–1590.

Jenkins, M.K., Taylor, P.S., Norton, S.D. and Urdhal, K.B. (1991). CD28 delivers a co-stimulatory signal involved in antigen-specific IL-2 production by human T cells. J. Immunol. 147, 2461–2466.

Jorgensen, J.L., Reay, P.A., Ehrich, E.W. and Davis, M.M. (1992). Molecular components of T cell recognition. Annu. Rev. Immunol. 10, 835–873.

June, C.H., Ledbetter, J.A., Linsley, P.S. and Thompson, C.B. (1990). Role of the CD28 receptor in T cell activation. Immunol. Today 11, 211–216.

Kameyoshi, Y., Dorschner, A., Mallet, A.I., Christophers, E. and Schroder, J.M. (1992). Cytokine RANTES released by thrombin-stimulated platelets is a potent attractant for human eosinophils. J. Exp. Med. 176, 587–592.

Kay, A.B., Ying, S., Varney, V., Gaga, M., Durham, S.R., Moqbel, R., Wardlaw, A.J. and Hamid, Q. (1991). Messenger RNA expression of the cytokine gene cluster IL-3, IL-4, IL-5 and GM-CSF in allergen-induced late phase cutaneous reactions in atopic subjects. J. Exp. Med. 173, 775–778.

Kiniwa, M., Gately, M., Galler, V., Chizzonite, R., Fargeas, C. and Delespesse, G. (1992). Recombinant IL-12 suppresses the synthesis of IgE by IL-4 stimulated human lymphocytes. J. Clin. Invest. 90, 262–266.

Kishimoto, T.K., Larson, R.S., Corbi, A.L., Dustin, M.L., Staunton, D.E. and Springer, T.A. (1989). The leukocyte integrins: LFA-1, Mac-1 and p150,95. Adv. Immunol. 46, 149–182.

Kyan-Aung, U., Haskard, D.O., Poston, R.N., Thornhill, M.H. and Lee, T.H. (1991). Endothelial leukocyte adhesion molecule 1 and intercellular adhesion molecule 1 mediate adhesion of eosinophils to endothelial cells *in vitro* and are expressed by endothelium in allergic cutaneous inflammation *in vitro*. J. Immunol. 146, 521–528.

Lasky, L.A. (1991). Lectin cell adhesion molecules (LEC-CAMs): a new family of cell adhesion proteins involved with inflammation. J. Cell. Biochem. 45, 139–146.

Lawrence, M.B. and Springer, T.A. (1991). Leukocytes roll on a selectin at physiologic flow rates: distinction from and pre-requisite for adhesion through integrins. Cell 65, 859–873.

Lehn, M, Weiser, W.Y., Engelharn, S.J., Gillis, S. and Remold, H.G. (1989). IL-4 inhibits H_2O_2 production and anti-leishmanial capacity of human cultured monocytes medicated by IFNγ. J. Immunol. 143, 3020–3024.

Lenschow, D.J., Zeng, Y., Thistlewaite, J.R., Montag, A., Brady, W., Gibson, M.L., Linsley, P.S. and Bluestone, J.A. (1992). Long-term survival of xenogeneic pancreatic islet grafts induced by CTLA-4-Ig. Science 257, 789–792.

Liew, F.Y. and Cox, F.E. (1991). Non-specific defence mech-anism: the role of nitric oxide. Immunol. Today 12, A17–A21.

Linsley, P.S., Greene, J.L., Tan, P., Bradshaw, J., Ledbetter, J.A., Anasetti, C. and Damle, N.K. (1992). Co-expression and functional cooperation of CTLA-4 and CD28 on activated T lymphocytes. J. Exp. Med. 176, 1595–1604.

Lopez, A.F., Williamson, D.J., Gamble, J.R., Begley, C.G., Harian, J.M., Klebanoff, S.J., Waltersdorph, A., Wong, G., Clark, S.C. and Vadas, M.A. (1986). Recombinant human granulocyte-macrophage colony stimulating factor stimulates *in vitro* mature human eosinophil and neutrophil function, surface receptor expression and survival. J. Clin. Invest. 78, 1220–1228.

Lopez, A.F., Sanderson, C.J., Gamble, J.R., Campbell, H.D., Young, I.G. and Vadas, M.A. (1988). Recombinant human interleukin-5 is a selective activator of human eosinophil function. J. Exp. Med. 167, 219–224.

Lukacher, A.E., Braciale, V.L. and Braciale, T.J. (1984). *In vivo* effector function of influenza virus-specific cytotoxic T-lymphocyte clones is highly specific. J. Exp. Med. 160, 814–826.

Mackay, C.R. (1992). Migration pathways and immunologic memory among T lymphocytes. Semin. Immunol. 4, 51–58.

Mackay, C.R. (1993). Immunological memory. Adv. Immunol. 53, 217–265.

Mackay, C.R., Marston, W.L. and Dudler, L. (1990). Naive and memory T cells show distinct pathways of lymphocyte recirculation. J. Exp. Med. 171, 801–817.

Mackay, C.R., Marston, W. and Dudler, L. (1992a). Altered patterns of T cell migration through lymph nodes and skin following antigen challenge. Eur. J. Immunol. 22, 2209–2210.

Mackay, C.R., Marston, W.L., Dudler, L., Spertini, O., Tedder, T.F. and Hein, W.R. (1992b). Tissue-specific migration pathways by phenotypically distinct subpopulations of memory T cells. Eur. J. Immunol. 22, 887–895.

Mackenzie, C.D., Taylor, P.M. and Askonas, B.A. (1989). Rapid recovery of lung histology correlates with clearance of influenza virus by specific CD8[+] cytotoxic cells. Immunology 67, 375–381.

Maggi, E., Biswas, P., Del Prete, G., Parronchi, P., Macchia, D., Simonelli, C., Emmi, L., De Carli, M., Tiri, A., Ricci, M. and Romagnani, S. (1991). Accumulation of Th2-like helper T cells in the conjunctiva of patients with vernal conjunctivitis. J. Immunol. 146, 1169–1174.

Maggi, E., Parronchi, P., Manetti, R., Simonelli, C., Piccinni, M.P., Rugiu, F.S., De Carli, M., Ricci, M. and Romagnani, S. (1992). Reciprocal regulatory role of IFNγ and IL-4 on the *in vitro* development of human Th1 and Th2 clones. J. Immunol. 148, 2142–2147.

Martinez, O.M., Gibbons, R.S., Gavoroy, M.R. and Aronson, F.R. (1990). IL-4 inhibits IL-2 receptor expression and IL-2-dependent proliferation of human T cells. J. Immunol. 144, 2211–2215.

Martz, E. and Gamble, J.R. (1992). How do CTL control virus infections? Evidence for pre-lytic halt of herpes simplex. Viral Immunol. 5, 81–91.

Mescher, M.F. (1992). Surface contact requirements for acti-vation of cytotoxic T lymphocytes. J. Immunol. 149, 2402–2405.

Michie, C.A., McLean, A., Alcock, G. and Beverley, P.C.L. (1992). Lifespan of human lymphocyte subsets defined by CD45 isoforms. Nature 360, 264–265.

Miller, A., Lider, O., Roberts, A.B., Sporn, M.B. and Weiner, H.L. (1992). Suppressor T cells generated by oral tolerisation to myelin basic protein suppress both *in vitro* and *in vivo* immune responses by the release of transforming growth factor-β after antigen-specific triggering. Proc. Natl. Acad. Sci. USA 89, 421–425.

Minty, A., Chalon, P., Derocq, J-M., Dumont, X., Guillemot, J-C., Kaghad, M., Labit, C., Lepatois, P., Liazun, P., Miloux, B. *et al.* (1993). Interleukin-13 is a new human lymphokine regulating inflammatory and immune responses. Nature 362, 248–250.

Mosmann, T.R. and Coffman, R.L. (1989). Th1 and Th2 cells: different patterns of lymphokine secretion lead to different functional properties. Annu. Rev. Immunol. 7, 145–173.

Mosmann, T.R. and Moore, K.W. (1991). The role of IL-10 in cross-regulation of Th1 and Th2 responses. Immunoparasitol. Today 12, 49–53.

Mosmann, T.R., Cherwinski, H., Bond, M.W., Gedlin, M.A. and Coffman, R.L. (1986). Two types of murine helper T cell clones. I. Definition according to profiles of lymphokine activities and secreted proteins. J. Immunol. 136, 2348–2357.

Mueller, D.L., Jenkins, M.K. and Schwartz, R.H. (1989). Clonal expansion versus functional clonal inactivation: a co-stimulatory signalling pathway determines the outcome of T cell antigen receptor occupancy. Annu. Rev. Immunol. 7, 445–480.

Nabavi, N., Freeman, G.J., Gault, A., Godfrey, D., Nadler, L.M. and Glimcher, L.H. (1992). Signalling through the MHC class II cytoplasmic domain is required for antigen presentation and induces B7 expression. Nature 360, 266–268.

Nishioka, W.K. and Welsh, R.M. (1992). Inhibition of cytotoxic T-lymphocyte-induced target cell DNA frag-mentation, but not lysis, by inhibitors of DNA topoisomerases I and II. J. Exp. Med. 175, 23–27.

Oppenheimer-Marks, N., Davis, L.S., Brogue, D.T., Ramberg, J. and Lipsky, P.E. (1991). Differential utilization of ICAM-1 and VCAM-1 during the adhesion and transendothelial migration of human T lymphocytes. J. Immunol. 147, 2913.

O'Rourke, A.M. and Mescher, M.F. (1992). Cytotoxic T-lymphocyte activation involves a cascade of signalling and adhesion events. Nature 358, 253–255.

Ortaldo, J.R., Winkler-Pickett, R.T., Nagashima, K., Yagita, H. and Okumura, K. (1992). Direct evidence for release of pore-forming protein during NK cellular lysis. J. Leucocyte Biol. 52, 483–488.

Paliard, X, de Waal Malefijt, R., Yssel, H., Blanchard, D., Chretien, L., Abrams, J. et al. (1988). Simultaneous production of IL-2, IL-4 and interferon-gamma by activated human CD4$^+$ and CD8$^+$ T cell clones. J. Immunol. 141, 849–855.

Parronchi, P, Macchia, D., Piccinni, M.P., Biswas, P., Simonelli, C., Maggi, E., Ricci, M., Ansari, A.A. and Romagnani, S. (1991). Allergen and bacterial antigen-specific T-cell clones established from atopic donors show a different profile of cytokine production. Proc. Natl. Acad. Sci. USA 88, 4538–4542.

Paul, W.E. (1989). Pletotropy and redundancy: T cell-derived lymphokines in the immune response. Cell 57, 521.

Picker, L.J., Kishimoto, T.K., Smith, C.W., Warnock, R.A. and Butcher, E.C. (1991a). ELAM-1 is an adhesion molecule for skin homing T cells. Nature 349, 796–799.

Picker, L.J., Warnock, R.A., Burns, A., Doerschuk, C.M., Berg, E.L. and Butcher, E.C. (1991b). The neutrophil selectin LECAM-1 presents carbohydrate ligands to the vascular selectins ELAM-1 and GMP-140. Cell 66, 921–933.

Ramb-Lindhauer, C., Feldmann, A., Rotte, M. and Neumann, C. (1991). Characterization of grass pollen reactive T cell lines derived from lesional atopic skin. Arch. Dermatol. Res. 283, 71–76.

Rand, T.H., Cruikshank, W.W., Center, D.M., Weller, P.F. (1991a). CD4-mediated stimulation of human eosinophils: lymphocyte chemoattractant factor and other CD4-binding ligands elicit eosinophil migration. J. Exp. Med. 173, 1521–1528.

Rand, T.H., Silberstein, D.S., Kornfeld, H., Weller, P.F. (1991b). Human eosinophils express functional IL-2 receptors. J. Clin. Invest. 88, 825–832.

Robinson, D.S., Hamid, Q., Ying, S., Tsicopoulos, A., Barkans, J., Bentley, A.M., Corrigan, C., Durham, S.R. and Kay, A.B. (1992). Evidence for predominant "Th2-type" bronchoalveolar lavage T-lymphocyte population in atopic asthma. N. Engl. J. Med. 326, 298–304.

Rock, K.L., Rothstein, L., Fleishacker, C. and Gamble, S. (1992). Inhibition of class I and class II MHC-restricted antigen presentation by cytotoxic T lymphocytes specific for an exogenous antigen. J. Immunol. 148, 3028–3033.

Romagnani, S. (1992). Induction of Th1 and Th2 response: a key role for the "natural" immune response? Immunol. Today 13, 379–381.

Rothenberg, M.E., Owen, W.F., Silberstein, D.S., Woods, J., Soberman, R.J., Austen, K.F. and Stevens, R.L. (1988). Human eosinophils have prolonged survival, enhanced functional properties and become hypodense when exposed to human interleukin-3. J. Clin. Invest. 81, 1986–1992.

Rothenberg, M.E., Petersen, J., Stevens, R.L., Silberstein, D.S., McKenzie, D.T., Austen, K.F. and Owen, W.F.

(1989). IL-5 dependent conversion of normodense human eosinophils to the hypodense phenotype uses 3T3 fibroblasts for enhanced viability, accelerated hypodensity and sustained antibody-dependent cytotoxicity. J. Immunol. 143, 2311–2316.

Salgame, P., Abrams, J.S., Clayberger, C., Goldstein, H., Convit, J., Modlin, D.L. and Bloom, B.R. (1991). Differing cytokine profiles of functional subsets of human CD4 and CD8 T cell clones. Science 254, 279–282.

Salmi, M., Granfors, K., Leirsalo-Repo, M., Hamalainen, M., MacDermott, R., Leino, R., Havia, T. and Jalkanen, S. (1992). Selective endothelial binding of interleukin-2-dependent human T cell lines derived from different tissues. Proc. Natl. Acad. Sci. USA 89, 11436–11440.

Salmi, M., Kalimo, K. and Jalkanen, S. (1993). Induction and function of vascular adhesion protein-1 at sites of inflammation. J. Exp. Med. 178, 2255.

Schleimer, R.P., Sterbinsky, S.A., Kaiser, J., Bickel, C.A., Klunk, D.A., Tomoika, K., Newman, W., Luscinskas, F.W., Gimbrone, M.A., McIntryre, B.W. and Bochner, B.S. (1992). IL-4 induces adherence of human eosinophils and basophils but not neutrophils to endothelium: association with expression of VCAM-1. J. Immunol. 148, 1086–1092.

Schwartz, R.H. (1992). Co-stimulation of T lymphocytes: the role of CD28, CTLA-4 and B7/BB1 in IL-2 production and immunotherapy. Cell 71, 1065–1068.

Sehmi, R., Wardlaw, A.J., Cromwell, O., Kurthara, K., Waltmann, P. and Kay. A.B. (1992). IL-5 selectively enhances the chemotactic response of eosinophils obtained from normal but not eosinophilic subjects. Blood 79, 2952–2959.

Shimizu, Y., Shaw, S., Graber, N., Gopal, T.V., Horgan, K.J., Vant, S.G. and Newman, W. (1991). Activation-independent binding of human memory T cells to adhesion molecule ELAM-1. Nature 349, 799–802.

Shimizu, Y., Newman, W., Tanaka, Y. and Shaw, S. (1992). Lymphocyte interactions with endothelial cells. Immunol. Today 13, 106–112.

Shinohara, N., Huang, Y-Y. and Maroyama, A. (1991). Specific suppression of antibody responses by soluble protein-specific, class II restricted cytotoxic T lymphocyte clones. Eur. J. Immunol. 21, 23–27.

Shiver, J.W., Su, L. and Henkart, P.A. (1992). Cytotoxicity with target DNA breakdown by rat basophilic leukaemia cells expressing both cytolysin and granzyme A. Cell 71, 315–322.

Sitkovsky, M.V. (1988). Mechanistic, functional and immuno-pharmacological implications of biochemical studies of antigen receptor-triggered cytolytic T-lymphocyte activation. Immunol. Rev. 103, 127–160.

Smith, J.B., Kunjummen, R.D., Kishimoto, T.K. and Anderson, D.C. (1992). Expression and regulation of L-selectin on eosinophils from human adults and neonates. Pediatr. Res. 32, 465–471.

Smyth, M.J. Norisha, Y. and Ortaldo, J.R. (1992). Multiple cytolytic mechanisms displayed by activated human peripheral blood T cell subsets. J. Immunol. 148, 55–62.

Street, N.E., and Mosmann, T.R. (1991). Functional diversity of T lymphocytes due to secretion of different cytokine patterns. FASEB J. 5, 171–177.

Sugiyama, H., Chen, P., Hunter, M., Taffs, R. and Sitkovsky, M. (1992). The dual role of the cAMP-dependent protein kinase Ca subunit in T cell receptor-triggered T-lymphocyte effector functions. J. Biol. Chem. 267, 25256–25263.

Swain, S.L., Weinberg, A.D. and English, M. (1990). CD4+ T cell subsets: lymphokine secretion of memory cells and effector cells which develop from precursors *in vitro*. J. Immunol. 144, 1788–1798.

Tan, P., Anasetti, C., Hansen, J.A., Melrose, J., Brunvand, M., Bradshaw, J., Ledbetter, J.A. and Linsley, P.S. (1992). Induction of alloantigen-specific hyperresponsiveness in human T lymphocytes by blocking interaction of CD28 with its natural ligand B7/BB1. J. Exp. Med. 177, 165–173.

Tanaka, Y., Adams, D.H., Hubscher, S., Hirano, H., Siebenlist, S. and Shaw, S. (1993). Adhesion induced by proteoglycan-immobilized cytokine MIP-1β. Nature 361, 79–82.

Thyphronitis, G., Toskos, G.C., June, C.H., Levine, A D. and Finkelman, F.D. (1989). IgE secretion by Epstein-Barr virus infected purified human B lymphocytes is stimulated by IL-4 and suppressed by IFNγ. Proc. Natl. Acad. Sci. USA 86, 5580–5584.

Tian, Q., Streuli, M., Saito, H., Schlossman, J.F. and Anderson, P. (1991). A polyadenylate binding protein localised to the granules of cytolytic lymphocytes induces DNA fragmentation in target cells. Cell 67, 629–639.

Tsicopoulos, A., Hamid, Q., Varney, V., Ying, S., Moqbel, R., Durham, S.R. and Kay, A.B. (1992). Preferential messenger RNA expression of Th1-type cells (IFNγ+, IL-2+) in classical delayed-type (tuberculin) hypersensitivity reactions in human skin. J. Immunol. 148, 2058–2061.

Van der Heijden, F.L., Wierenga, E.A., Bos, J.D. and Kapsenberg, M.L. (1991). High frequency of IL-4 producing CD4+ allergen-specific T lymphocytes in atopic dermatitis leisonal skin. J. Invest. Dermatol. 97, 389–394.

Van Vlasselaer, P., Punnonen, J. and de Vries, J.E. (1992). Transforming growth factor-β directs IgA switching in human B cells. J. Immunol. 148, 2062–2067.

Vedder, V.B., Winn, R.K., Rice, C.L., Chi, E.Y, Arfors, K.E. and Harlan, J.M. (1988). A monoclonal antibody to the adherence promoting leucocyte glycoprotein CD18 reduces organ injury and improves survival from hemorrhagic shock and resuscitation in rabbits. J. Clin. Invest. 81, 939–945.

Vercelli, D., Jabura, H.H., Arai, K-I., Yokota, T. and Geha, R.S. (1989). Endogenous IL-6 plays an obligatory role in IL-4-dependent human IgE synthesis. Eur. J. Immunol. 19, 1419–1424.

Walsh, G.M., Hartnell, A., Wardlaw, A.J., Kurihara, K., Sanderson, C.J. and Kay, A.B. (1990). IL-5 enhances the *in vitro* adhesion of human eosinophils, but not neutrophils, in a leucocyte integrin (CD11/18)-dependent manner. Immunology 71, 258–265.

Walsh, G.M., Hartnell, A., Mermod, J-J., Kay, A.B. and Wardlaw, A.J. (1991). Human eosinophil, but not neutrophil adherence for IL-1 stimulated HUVEC is α4β1 (VLA-4) dependent. J. Immunol. 146, 3419–3423.

Walz, A. and Baggiolini, M. (1989). Novel cleavage product of B-thromboglobulin formed in cultures of stimulated mononuclear cells activates human neutrophils. Biochem. Biophys. Res. Commun. 159, 969–975.

Warringa, R.A., Koenderman, L., Kok, P.T., Kreukniet, J. and Bruijnzeel, P.L. (1991). Modulation and induction of eosinophil chemotaxis by granulocyte/macrophage colony stimulating factor and IL-3. Blood 77, 2694–2700.

Williams, M.E., Shea, C.M., Lichtman, A.H. and Abbas, A.K. (1992). Antigen receptor-mediated anergy in resting lymphocytes and T cell clones. J. Immunol. 149, 1921–1926.

Yamamura, M., Wagn, X-H., Ohmen, J.D., Uyemura, K., Rea, T.H, Bloom, B.R. and Modlin, R.L. (1992). Cytokine patterns of immunologically mediated tissue damage. J. Immunol. 149, 1470–1475.

Yssel, H, Shanafelt, M.C., Soderberg, C., Schneider, P.V. Anzola, J. and Peitz, G. (1991). *Borrelia burgdorferi* activates a T helper type 1-like T cell subset in lyme arthritis. J. Exp. Med. 174, 593–601.

Zagury, D., Bernard, J., Thierness, N., Feldman, M. and Berke, G. (1975). Isolation and characterisation of individual functionally reactive cytotoxic T-lymphocytes. Conjugation, killing and recycling at the single cell level. Eur. J. Immunol. 5, 812–818.

3. The Regulation of Immunoglobulin E Synthesis

D.M. Kemeny

1. Introduction

Largely because it is present at such low concentrations, immunoglobulin E (IgE) was the last of the five different classes of immunoglobulin to be identified (Ishizaka and Ishizaka, 1967; Johansson and Bennich, 1967; Bennich et al., 1968). Since its discovery there have been many theories about how IgE is regulated (Katz, 1980; Geha, 1987; Ishizaka, 1988) but it was not until the advent of molecular immunology that the factors which control IgE production were properly defined (Coffman and Carty, 1986; Snapper et al., 1988). All B cells contain a single gene for each of the immunoglobulin isotypes, yet the amount of IgE produced and the frequency of IgE antibody responses to common environmental antigens differs greatly between individuals. Clearly there are genetic as well as environmental factors which determine whether or not an individual will make an IgE response and it is unclear whether the absence of IgE responses in normal individuals is due to the failure of the immune system to respond to small quantities of antigen, to an inability to generate appropriate T cell help or to an increased capacity to activate IgE suppressor T cells.

The failure of non-atopic humans to mount an IgE response to grass pollen, for example, does not appear to be due to an inability of the immune system to respond to the small quantities that are inhaled (ng year^{-1}), (Platts-Mills, 1987) although it is possible that less is required to sensitize the atopic individual so that IgE responses can be established before IgE suppressor mechanisms are activated. Nearly all humans make IgG antibodies to grass pollen and house dust mite (Kemeny *et al.*, 1989) and most have T cells which proliferate in response to these allergens. Furthermore, most people possess the ability to mount an IgE response if immunized under suitable conditions. For example, novice beekeepers produce IgE antibody to bee venom when first stung (Aalberse *et al.*, 1983); a significant proportion of normal babies produce low levels of IgE antibody to foods (Hattevig *et al.*, 1987); and most non-atopic adults immunized with grass pollen together with alum mount a pollen-specific IgE antibody response (Marsh *et al.*, 1972). In non-atopic individuals, IgE responses tend to be short lived, suggesting that the IgE response is limited by the presence of active IgE suppressive mechanisms. In this chapter I will describe the process of IgE synthesis by the B cell, the effect of different cytokines on IgE synthesis and the contribution of different immune and accessory cells to this process.

2. Characteristics of the Human IgE Response

The link between production of IgE antibodies to environmental allergens, allergic sensitization and allergic disease such as asthma, rhinitis, conjunctivitis and eczema is well established. The allergic immune response has, however, proved to be very resistant to manipulation either by vaccine immunotherapy (Committe on Safety of Medicines, 1986) or by treatment with interferon (IFNα; Boguniewicz *et al.*, 1990). IgE antibodies to grass pollen continue to be synthesized throughout the year despite the absence of antigen, although levels are boosted each summer, and bee venom-specific IgE antibodies can persist for over 20 years following the last recorded sting (Harries *et al.*, 1984; Urbanek *et al.*, 1986; Devey *et al.*, 1989). Interestingly, year-long immunotherapy with ragweed pollen vaccine has been shown to suppress the seasonal rise in ragweed-specific IgE (Yunginger and Gleich, 1973) indicating that at least part of the IgE response is susceptible to therapeutic manipulation.

Small amounts of IgE (0.4–80 IU ml^{-1}) are detected in the sera of almost all non-allergic adults. There are rare individuals who have undetectable levels of IgE and they appear healthy which suggests that IgE is not essential. In certain parasitic diseases IgE levels are greatly elevated, for example, infection with metazoan helminths such as *Echinococcus granulomatus* is characterized by increased serum IgE levels (Dessaint *et al.*, 1975), although protozoan infections do not affect IgE levels (Rademecker *et al.*, 1974). Other immunoglobulin isotypes remain unaffected in patients with helminth infections indicating that the effect of the parasite is selective for the IgE isotype. Patients with atopic diseases such as conjunctivitis, rhinitis, asthma and atopic dermatitis have elevated IgE levels.

As stated above, year-long, high levels of serum IgE antibody to grass pollen are maintained without constant exposure to allergen in hay fever patients. Since the average half-life of IgE in the serum is 2.3 days in humans (Waldmann, 1969; Table 3.1) IgE antibodies must be formed continuously. Indeed, if the rate of clearance is taken into account this would translate into a normal range of 10–1000 IU ml^{-1} (0.024–2.4 μg ml^{-1}) with IgE levels in the sera of some atopic dermatitis patients as high as 50 000 IU ml^{-1} (1.2 mg ml^{-1}). Similarly, as much of the IgE is secreted via the various mucosae, and so continually removed, the daily production of this immunoglobulin can, in some patients, be comparable to that of other immunoglobulins. The half-life of rat and mouse IgE is even shorter at less than 0.5 days.

3. The Structure of IgE

Like other immunoglobulins, IgE is comprised of two pairs of heavy and light chains (Fig. 3.1) joined by disulphide bridges. Following digestion with the enzyme papain, IgE is split into a fragment containing the antibody combining site (Fab) and a second fragment (Fc). IgE has four constant heavy chain domains, one more than IgG, IgD and IgA. The second heavy chain domain of IgE substitutes for the hinge region. The light chains exist in two distinct forms, $\varkappa$ and λ, although the two light chains in any given IgE molecule are normally the same. The sequence of amino acids for both heavy and light chains is remarkably conserved along most of their length. Recent studies have indicated that, in contrast to IgG$_1$, IgE is bent (Zheng *et al.*, 1992).

Table 3.1 The serum half-life of human and rodent immunoglobulin in days. Data from various sources: Cremer *et al.* (1973), Bazin *et al.* (1974), Tada *et al.* (1975).

	IgA	IgM	IgD	IgG1·	IgG2	IgG3	IgG4	IgE
Human	5.8	5.1	2.8	21	21	7	21	2.3

	IgA	IgM	IgG	IgG1	IgG2a	IgG2b	IgG2c	IgE
Rat	ND	2.6	1.6	13	5	15	3	0.5

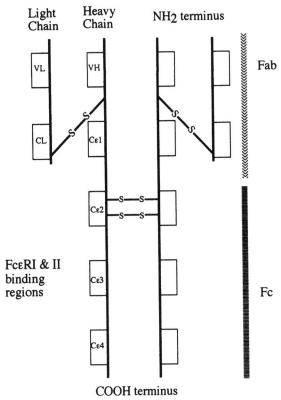

Figure 3.1 The structure of IgE showing the substitution of Cε2 for the hinge region and the location of the FcεRI and II receptor sites.

3.1 THE IGE ANTIBODY COMBINING SITE

At the amino terminal end of the molecule are variable regions which comprise the antibody binding site and confer specificity on the IgE molecule. Although both heavy and light chain variable regions may contribute to the antigen binding site this is not always so, and some antibodies, for example egg lyosyme (Smith-Gill *et al.*, 1987), can possess all of their antigen binding activity in the heavy or light chain variable region. Furthermore, variability is not evenly distributed along the variable region; some regions show exceptional variability (hypervariable) and it is these regions which are directly involved in the formation of the antibody combining site. These regions have also been called complementarity-determining regions, because their structure complements that of the antigen, and are named CDR1, CDR2 and CDR3. Adjacent to the hypervariable regions are highly conserved framework regions which dictate the overall structure of the combining site. On x chains there are four, and on λ chains there are six, framework regions. The framework regions are not entirely constant and do show some variation.

3.2 GENERATION OF IGE ANTIBODY DIVERSITY

The heavy chain variable region (V_H) is composed of J, D and V exon-encoded proteins whereas the light chain variable region (V_L) only contains J and V exon products. In order to be expressed in a functional form the respective exons, J_H, D_H and J_H or J_L and V_L, must be assembled in their germline configuration by site-specific recombination. It is during this recombination that antibody diversity is generated. We know that somatic mutation takes place because many antibody specificities can be generated from single V gene. These mutations occur during the proliferation and differentiation of B cells as the result of errors in DNA replication around the V(D)J region which are not corrected (deleted). Further diversity is generated by recombination of different variable region gene products. It can also be generated by joining different heavy and light chain variable gene products. Thus antibodies will be variants of the original germline transcript. It is likely that, in the germinal centre, antigen selectively binds to the highest affinity antibody-bearing B cells which take up more antigen and so express more antigen peptide to specific T cells. Interaction with these T cells protects the B cell from apoptosis and so increases the likelihood that high affinity IgE antibody-secreting B cells will survive.

3.3 THE IGE CONSTANT REGION

The constant region of immunoglobulins is divided into into three (IgD, IgA, IgG) or four (IgM, IgE) heavy chain domains which are named (ε1, ε2, ε3 etc.) according to the immunoglobulin class concerned. It is in this part of the molecule that the structures that are responsible for the effector functions of the immunoglobulin are located. In the case of IgA and IgM it is here that the J chain, which forms dimeric IgA and pentameric IgM, and a secretory component, which protects these molecules from enzymic digestion, bind. For IgG, this is where C1q binds, although the flexibility of the hinge region dictates whether it does so, as for IgG$_1$ and IgG$_3$, or not, as with IgG$_2$ and IgG$_4$.

4. *IgE and its Receptors*

The unique feature of IgE is its ability to bind to high affinity receptors, Fc$_\varepsilon$RI, on mast cells and basophils (Ravetch and Kinet, 1991; Metzger, 1992). IgE has been cloned and the Fc$_\varepsilon$RI binding site localized to the Cε2′–Cε3′ region (Gould *et al.*, 1987; Helm *et al.*, 1988). This fragment was shown to block sensitization of human skin mast cells *in vivo* (Geha *et al.*, 1985). The affinity of binding of IgE to these receptors is far greater than for other classes of immunoglobulin to their respective receptors. In addition to the high affinity receptor Fc$_\varepsilon$RI there is the so-called low affinity IgE receptor, Fc$_\varepsilon$RII (CD23), which exists in two forms, Fc$_\varepsilon$RIIa and Fc$_\varepsilon$RIIb. The binding affinity of IgE to CD23 is similar to that of IgG to its receptor. The region of IgE which binds to both the Fc$_\varepsilon$RI and Fc$_\varepsilon$RII has been mapped to the Cε3 domain (Beavil *et al.*, 1992). As well as activating mast cells and basophils when IgE–Fc$_\varepsilon$RI is cross-linked by antigen, IgE–Fc$_\varepsilon$RII activates eosinophils to kill schistosome parasites (Joseph *et al.*, 1983). Furthermore, Fc$_\varepsilon$RI on Langerhans cells in the skin (Rieger *et al.*, 1992) and Fc$_\varepsilon$RII on B cells and macrophages facilitate antigen uptake as described above (Mudde *et al.*, 1990).

5.   *Molecular Organization of IgE*

During an immune response newly activated B cells undergo a process of somatic mutation in which a large number of variable genes are expressed and those B cells that make high affinity antibody are selected. Following activation of the B cell, the IgE heavy chain or μ gene is transcribed. In general, once activated, a B cell only expresses a single heavy chain gene. What leads it to express the IgE heavy chain is discussed below.

5.1   ORGANIZATION OF IMMUNOGLOBULIN GENES

The loci for human and mouse immunoglobulin heavy chain constant region genes are similar (Fig. 3.2; Flanagan and Rabbits, 1982). Each starts with a cluster of variable region genes [V(D)J] followed by the Cμ gene and Cδ genes. These are followed in humans by Cγ3, Cγ1, a pseudo-C$\varphi\varepsilon$, gene followed by Cα1, a second pseudo-C$\varphi\gamma$ gene, Cγ2, Cγ4, Cε and Cα2. It is likely that the heavy chain γ, ε and α genes in the human constant region arose as the result of gene duplication. Thus there are two types of IgA (IgA$_1$ and IgA$_2$) and two complement fixing (IgG$_1$ and IgG$_3$) and non-complement fixing (IgG$_2$ and IgG$_4$) forms of IgG. In mice such gene

duplication does not appear to have occurred and the Cμ and Cδ genes are followed by Cγ3, Cγ1, Cγ2b, Cγ2a, Cε and Cα.

5.2   EXPRESSION OF IMMUNOGLOBULIN GENES

Synthesis of IgE requires rearrangements within the immunoglobulin gene complex to produce productive mRNA transcripts encoding both the variable and constant regions of the molecule. The variable regions are encoded by multiple germline DNA elements that are assembled into complete V(D)J regions by recombinase enzymes (Schatz *et al.*, 1989). B cells retain their antigen specificity, as determined by the variable region, throughout their life. Isotype switching occurs during B cell differentiation so that a different heavy chain region is produced in association with the same V(D)J segment. This process involves a DNA recombination event which deletes regions between the V(D)J segments and the genes encoding different heavy chain regions (C$_H$ genes). Genes are organized along chromosomes as exons (the code for the protein concerned) and are separated by introns (intervening stretches of DNA). Synthesis of proteins involves transcription (copying) of the relevant gene, post-transcriptional modification of the transcribed gene, translation (assembly of the protein according to the RNA message), and post-translational modification such as glycosylation (the addition of sugar groups to the polypeptide backbone). Initiation of transcription requires activation of a promoter which is a segment of DNA upstream of the 5$'$ end of the gene and initiates transcription. During transcription the DNA unwinds and copies (transfer RNA) are made. The initial transcript can be modified in a number of ways (Fig. 3.3) by capping at the 5$'$ end and by polyadenylation [addition of a poly(A) tail] at the 3$'$ end. Introns are spliced out and exons joined to form the mature transcript. For immunoglobulins it appears that exons as well as introns may be spliced out in the form of a loop or switch circle. Recombination occurs in highly repetitive regions of

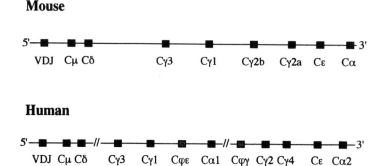

Mouse

VDJ   Cμ Cδ Cγ3   Cγ1   Cγ2b   Cγ2a   Cε Cα

Human

VDJ   Cμ Cδ Cγ3   Cγ1   C$\varphi\varepsilon$ Cα1   C$\varphi\gamma$ Cγ2 Cγ4   Cε Cα2

Figure 3.2   Human and mouse immunoglobulin heavy chain genes.

Gene

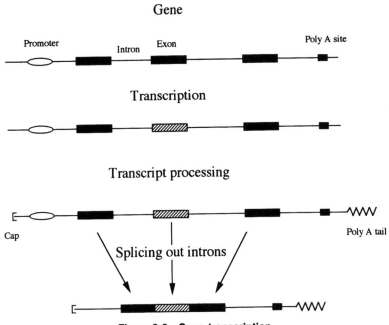

Figure 3.3 Gene transcription.

DNA located immediately 5′ of each C_H gene at switch (S) regions, with the exception of Cδ. Isotype switching is preceded by transcription of a germline transcript of the C_H gene to which switching will be directed. This transcript is of a region upstream of the C_H gene and cannot be translated into mature protein (Ichiki *et al.*, 1992). The excised DNA forms a loop of extrachromosomal,

circular DNA, or a switch circle (Fig. 3.4; Harriman *et al.*, 1993). In the case of switching to IgE, recombination occurs in B cells which express surface IgM and IgD by deletion of the regions between the Sμ and Sε switch regions adjacent to the IgM (Cμ) and IgE (Cε) heavy chain region genes. Experiments in mice in which the switch from IgG_1 to IgE is blocked have shown that it is

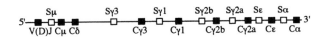

Mouse immunoglobulin heavy chain genes

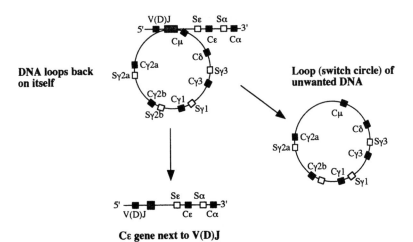

Figure 3.4 A possible scheme for IgE transcription in which the genes for other immunoglobulins are looped out to leave the IgE transcript next to the variable gene transcript.

possible to switch directly from IgM to IgE although switching via IgG₁ also occurs (Siebenkotten *et al.*, 1992; Mandler *et al.*, 1993; Jung *et al.*, 1994). Similar evidence exists for a direct μ to ε switch in human B cells (Brinkmann *et al.*, 1992; van der Stoep *et al.*, 1994) although from the analysis of switch circles it also appears that there is μ to γ to ε and μ to α to γ to ε switching (Zhang *et al.*, 1994b). Furthermore, it is possible to identify at least four alternatively spliced forms of IgE in human B cells (Zhang *et al.*, 1994a) which explains some of the disparity between the observed and predicted molecular weight of IgE (Zhang *et al.*, 1992). The functional significance of these different forms of IgE is presently unknown. Other switch factors include interleukin-10 (IL-10) for IgG₁ and IgG₃ (Briere *et al.*, 1994) and transforming growth factor β (TGFβ) for IgA (van Vlasselaer *et al.*, 1992).

6. *Control of IgE Synthesis by Cytokines*

The process of B cell immunoglobulin class switching is controlled by number of cytokines of which only two have been shown to selectively promote switching to IgE, IL-4 (Coffman *et al.*, 1986; Snapper and Paul, 1987) and more recently IL-13 (Punnonen *et al.*, 1993b; Fig. 3.5). There is a requirement for T–B cell contact (discussed below; Del Prete *et al.*, 1988; Parronchi *et al.*, 1990). In addition to cytokines that promote expression of IgE and immunoglobulin synthesis, a number of cytokines have been identified which inhibit IgE production – most notably IFNγ and IFNα (Snapper and Paul, 1987) and IL-8 (Kimata *et al.*, 1992). Other cytokines, such as IL-12, may also be important, but these act independently to influence the generation of IL-4-producing T cells rather that the B cell itself (Romagnani, 1992).

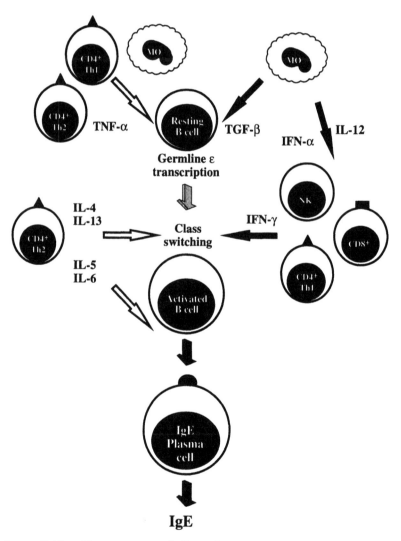

Figure 3.5 IgE class switching. The open arrows indicate those stimuli that support IgE synthesis and the grey arrows those that inhibit it directly or indirectly.

6.1 ISOTYPE-SPECIFIC ENHANCERS OF IGE SYNTHESIS

IL-4 is the most important factor for determining switching to IgE. In the presence of increasing concentrations of IL-4 Snapper et al. (1988) showed that B cells stimulated with lipopolysaccharide (LPS) were switched from IgM and IgG3 to IgG1 and, at still higher concentrations of IL-4, to IgE production. Interestingly, the levels of IgE produced were of the same order of magnitude for all isotypes. IL-4 alone is sufficient to initiate synthesis of a 1.7–1.9 kb non-productive germline Cε transcript, containing a 2 kb region upstream of the ε S region, and the four exons encoding the heavy chain domains of IgE (Gauchat et al., 1990). Transcription of this product precedes IgE responses in vivo (Thyphronitis et al., 1993). IL-4 alone is not sufficient, however, to induce synthesis of the mature IgE transcript. A second signal to the B cell is required which can be provided in a number of different ways which are described below.

IL-13 was discovered through its sequence similarity with IL-4 (Minty et al., 1993). It is the human homologue of mouse P600 which is produced by T$_H$2 cells (Defrance et al., 1994) but has no effect on murine B cells (Zurawski and de Vries, 1994). IL-13 is able to induce IgE synthesis from purified human B cells and appears able to act in the absence of IL-4 (Punnonen et al., 1993b). Although the IL-4 receptor does not bind IL-13, it appears that IL-4 and IL-13 receptors share a common subunit that is important for signal transduction (Zurawski and de Vries, 1994); furthermore, an IL-4 mutant protein inhibits both IL-4- and IL-13-induced IgG4 and IgE synthesis (Aversa et al., 1993a). IL-13 exerts similar effects on monocytes to IL-4 but, unlike IL-4, IL-13 was reported not to affect T cells (Punnonen and de Vries, 1994). However it has recently been shown that IL-13 activates the recently described IL-4-dependent transcription factor NF-IL-4 (Kohler et al., 1994) and clearly there are questions about IL-13/T cell interactions that remain unresolved.

6.2 SOURCE OF IL-4 AND IL-13

IL-4 and IL-13 are derived from a number of cells. T cells are the major source of these cytokines and this is addressed in greater detail below. However a number of other cells make IL-4, such as mast cells (Plaut et al., 1989) and basophils (Brunner et al., 1993). These also express CD23 and CD40-L, can trigger B cells via CD21 and CD40, respectively (see below), and serve as a co-stimulus for IgE production. Indeed, mast cell and basophil lines (Gauchat et al., 1993) and IL-3-stimulated splenic and bone marrow non-T, non-B cells (Ledermann et al., 1992) are capable of switching B cells to make IgE. IL-4-producing mast cells have been identified in the lung (Bradding et al., 1992), in the nasal mucosae (Bradding et al., 1993), and in parasite-infested animals

there is a significant increase in IgE receptor-positive non-T, non-B cells in the spleen from 0.53% to 3.8% 8–9 weeks after infection with Schistosoma mansoni. These cells, although present in smaller numbers than T cells (c. 70%), secrete similar amounts of IL-4 (Williams et al., 1993). The extent to which non-T cell-derived IL-4 is important in IgE synthesis in vivo is unclear.

Confirmation of the importance of IL-4 in IgE production in vivo came from the work of Finkelman and colleagues who showed that anti-IL-4, and anti-IL-4 receptor (IL-4R), antibody inhibited ongoing IgE synthesis in mice infected with the nematode Nippostrongylus braziliensis and in mice immunized with the polyclonal B cell activator, goat anti-mouse IgD (Finkelman et al., 1988, 1991). IL-4 appeared to be critical during the early phase (first 4 days) of the IgE response (Finkelman et al., 1988). However CD4$^+$ T cells were still required after this time indicating that cell–cell contact or other cytokines were required (Finkelman et al., 1989). To determine whether the continued presence of IL-4 was required for IgE production, Finkelman and colleagues showed that the majority of the IgE response in animals sensitized with TNP-KLH or with the nematode Heligmosomoides polygyrus, which induces a chronic IgE response, was inhibited with anti-IL-4 (Urban et al., 1991). This fits our understanding of B cell immunobiology as B cells do not live for long and when they differentiate into IgE-producing cells must continually switch to IgE for which they require IL-4. The definitive proof that IL-4 is required for IgE production comes from IL-4 knockout mice (Kuhn et al., 1991) which are completely unable to produce IgE.

6.3 NON-ISOTYPE-SPECIFIC ENHANCERS OF IGE

Although IL-4 and IL-13 are the only currently known IgE switch factors, a number of other cytokines have been shown to modulate IgE synthesis. These include other T cell-derived cytokines such as IL-5 and IL-6 (Pene et al., 1988a; Vercelli et al., 1989) which promote the subsequent growth and development of IgE-switched B cells into mature plasma cells (Pene et al., 1988b; Vercelli et al., 1989a; Purkerson and Isakson, 1992), although there is evidence that it is not essential for IgE in vivo (van Ommen et al., 1994). IL-6 may also be a switch factor for mouse IgG1. Tumour necrosis factor α (TNFα), which is produced by both T$_H$1 and T$_H$2 CD4$^+$ T cells, enhances germline IgE transcription induced by IL-4 (Gauchat et al., 1992a). IL-4 may inhibit IgM (Snapper and Paul, 1987; Snapper et al., 1988) as IL-4 inhibited lipopolysaccharide (LPS)-stimulated mouse IgM production. Furthermore, treatment of rheumatoid patients with IL-4 reduced the production of IgM rheumatoid factor (Hidaka et al., 1992). IL-9 is reported to potentiate IL-4-induced IgE and IgG1 release from LPS-primed murine B lymphocytes (Petit-Frere et al., 1993).

IL-3 promotes IgE from B cells induced by non-T cells (Le Gros *et al.*, 1990; Ledermann *et al.*, 1992).

6.4   INHIBITORS OF IGE SYNTHESIS

The observation that IFNγ antagonized the IL-4-induced IgE switch in murine B cells was made concurrent with the discovery that IL-4 was able to switch B cells to produce IgE (Snapper and Paul 1987; Snapper *et al.*, 1988) and was subsequently confirmed with human peripheral blood B cells (Pene *et al.*, 1988a). It is still unclear how IFNγ and IFNα suppress IgE production but it does not appear to be at the level of ε germline transcription (Gauchat *et al.*, 1992b). IL-8 is also reported to inhibit IL-4-induced IgE and IgG₄ production (Kimata *et al.*, 1992) although this remains to be confirmed. IL-8 is made by a number of cells including epithelial cells. It is also made by monocytic cell lines (Friedland *et al.*, 1992) and by alveolar macrophages (Dentener *et al.*, 1993). IFNα is produced by monocytes and macrophages which may moderate IgE production. The role of IFNγ and IFNα in antagonizing IgE was confirmed by Finkelman and colleagues who showed that pretreatment of mice with IFNγ or IFNα blocked the anti-IgD-induced IgE response. Although clinically effective, these cytokines have failed to suppress IgE production in patients with atopic dermatitis (Boguniewicz *et al.*, 1990).

The mechanism for these regulatory effects is not completely understood. IFNγ and IFNα are unable to inhibit IL-4 and anti-CD40-induced IgE production or IL-4-induced germline ε chain transcription (Gauchat *et al.*, 1990, 1992a,b; Gascan *et al.*, 1991; Rousset *et al.*, 1991). TGFβ and TNFα act directly on B cells to block or enhance ε transcription, respectively (de Vries *et al.*, 1991). Epstein–Barr vivus (EBV)/IL-4-induced IgE synthesis is reported to be refractory to the effects of TGFβ (Wu *et al.*, 1992).

7.   *Control of IgE Synthesis by Cell Surface Molecules*

In addition to secreted cytokines there are a number of other stimuli, delivered by interaction between the B cells and other immune cells (principally the T cell), which are required for IgE synthesis (Fig. 3.6). These include cognate T cell–B cell interactions involving the TcR/CD3 complex and major histocompatibility complex (MHC) class II-associated antigens (Vercelli *et al.*, 1989), and non-cognate interactions via accessory molecules on T cells (Parronchi *et al.*, 1990) although there is no evidence that this interaction exerts an isotype-specific effect. In mice bacterial LPS serves as a co-stimulus for IL-4 (Coffman *et al.*, 1986), and infection with EBV (Thyphronitis *et al.*, 1989) induces high levels of IgE in B cells cultured with IL-4. CD58 (LFA-3)–CD2 interaction stimulates IgE production of B cells cultured with IL-4 (Diaz-Sanchez *et al.*, 1994). Membrane TNFα on CD4⁺ T cell can also provide a co-stimulatory stimulus for IL-4 and B cells to produce IgE and IgG₄ (Aversa *et al.*, 1993b).

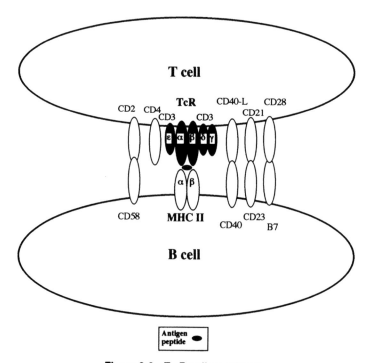

Figure 3.6   T–B cell co-stimuli.

It is also possible to induce *in vitro* IgE synthesis in the absence of T cells, most commonly using ligation of the CD40 antigen on B cells by anti-CD40 monoclonal antibodies (mABs; Jabara *et al.*, 1990; Brinkmann *et al.*, 1992; Armitage *et al.*, 1993) or by treatment with hydrocortisone (Jabara *et al.*, 1991), which presumably activates the same intracellular pathway as these other cell surface stimuli. The counterstructure for CD40 on T cells has been cloned and is termed CD40L (Spriggs *et al.*, 1992). Transfection of cells with CD40L stimulated proliferation of B cells and, in the presence of IL-4, induced IgE synthesis (Spriggs *et al.*, 1992). Defective expression of CD40L causes X-linked immunodeficiency with hyper-IgM production (Korthauer *et al.*, 1993). There are also cell surface interactions that specifically regulate IgE. Stimulation of CD21, for example with anti-CD21 antibodies, promotes the rescue of germinal centre B cells from apoptosis (Bonnefoy *et al.*, 1993; Henchoz *et al.*, 1994). Soluble IgE FcεRII (sCD23) can trigger CD21 on B cells (Aubry *et al.*, 1992). Cross-linking of CD23 antigen by its natural ligand (IgE) or by anti-CD23 antibody prevents B lymphocyte proliferation and differentiation (Luo *et al.*, 1991); this has recently been reviewed (Sutton and Gould, 1993).

8. CD4+ T Cell Production of IgE Regulatory Cytokines

8.1 MURINE CD4+ T CELL SUBSETS

T cells are the principal source of IgE regulatory cytokines. CD4+ T cells can be subdivided on the basis of the cytokines they secrete. These subsets have been termed T helper 1 (T_H1) and T helper 2 (T_H2) (Mosmann *et al.*, 1986; Mosmann, 1992) on the basis of the cytokines that they produce. T_H1 cells make IFNγ, IL-2 and TNFβ and effect cell-mediated immunity. In the mouse, T_H1 cells support IgG$_{2a}$ antibody production. T_H2 cells make IL-4, IL-5, IL-6 and IL-10, support IgE responses and cause eosinophilia. Some cytokines, e.g. IL-3, TNFα and granulocyte–macrophage colony stimulating factor (GM-CSF), are secreted in similar amounts by both subsets (Table 3.2). In addition to T_H1 and T_H2 cells there are other subtypes such as T_H0, which exhibit an unrestricted cytokine profile (Firestein *et al.*, 1989), and there are probably others, such as a non-cytolytic vesicular stomatitis virus-specific CD4+ T cell clone which produces IL-6, TNFα and TNFβ but not IL-2, IL-4 or IFNγ (Cao *et al.*, 1993).

8.2 HUMAN CD4+ T CELL SUBSETS

Available evidence indicates that the cytokine profile of human T cell clones resembles that seen in the mouse although there are some notable exceptions. Most, but

Table 3.2 The comparative amounts of cytokines produced by mouse CD4+ and human CD8+ T cell clones. The data in this table were compiled from various sources: Mosmann *et al.* (1986), Mosmann and Moore (1991), Salgame *et al.* (1991), Yamamura *et al.* (1991).

	CD4+ cell clones			CD8+ T cell clones	
	T_H1	T_H2	T_H0	Type 1	Type 2
GM-CSF	+ +	+	+ +	+	+ +
IL-3	+ +	+ +	+ +		
TNFα	+ +	+	+ +	+ +	+
IL-2	+ +	−	+ +	+ +	+
IFNγ	+ +	−	+ +	+ +	+
TNFβ	+ +	−	+ +		
IL-4	−	+ +	+ +	−	+ +
IL-5	−	+ +	+ +	−	+ +
IL-6	−	+ +	+ +	+ +	−
IL-10	−	+ +	+ +		
P600 (IL-13)	−	+ +	+ +		

+ +, strong producer; +, moderate producer; −, non producer.

not all, T cell clones prepared from atopic donors were T_H2-like (Parronchi *et al.*, 1991), while the majority derived from non-atopic individuals were closer to T_H1 (Wierenga *et al.*, 1990a,b; Del Prete *et al.*, 1991; Parronchi *et al.*, 1991). Some human T_H1 clones, however, make appreciable amounts of IL-6 (Wierenga *et al.*, 1991). Allergen-specific peripheral human T cell clones commonly have a T_H0 profile as do some of those from the skin of atopic dermatitis patients (Van Reijsen *et al.*, 1992). The extent to which the conditions under which cells are cloned dictates their cytokine profile is unclear but addition of cytokines can certainly influence the type of clones produced. Indeed if clones are prepared without prior bulk culture there is less evidence for bias to one subtype or another, suggesting that the presence of specific cells present in peripheral blood influences the cytokine profile of the T cell clones produced (Fig. 3.7).

8.3 CD4+ T CELL DIFFERENTIATION

The mechanism whereby one subset is activated in preference to the other is not completely understood but appears to involve IL-4 and IFNγ (Fig. 3.8; Swain, 1991; Swain *et al.*, 1991; Chatelain *et al.*, 1992). In mice, IL-4 supports the generation of T_H2 cells (Swain *et al.*, 1990) and suppresses T_H1 cell development, while IFNγ suppresses the development of T_H2 cells (Gajewski *et al.*, 1989a). TGFβ induces a cell population distinct from T_H1 and T_H2 which predominantly secretes IL-2 and expresses a memory phenotype (Swain *et al.*, 1991). CD4+ T cells, activated in the presence of IL-6, are reported to have a T_H0 cytokine profile (Croft and Swain, 1991). Similar effects of IL-4 and IFNγ are seen in the rat

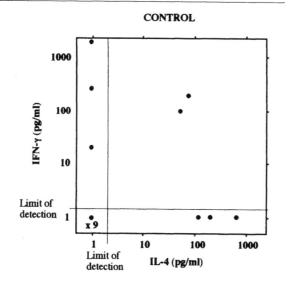

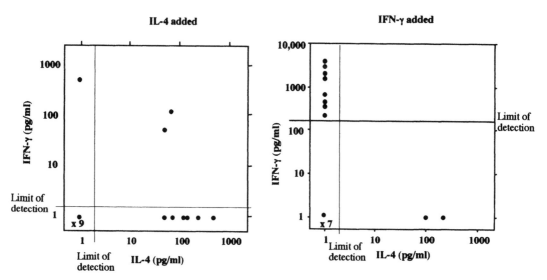

Figure 3.7   The effect of IL-4 and IFNγ on the generation of *Dermatophagoides pteronyssinus*-specific CD4$^+$ T$_H$1/T$_H$2/T$_H$0 T cell clones generated at limiting dilution without prior bulk culture.

(Diaz-Sanchez *et al.*, 1993b; Noble *et al.*, 1993b,c) where IL-4 enhanced production of mRNA for IL4, IL-5 and IL-10.

In addition, different types of antigen-presenting cell may influence T$_H$1/T$_H$2 CD4$^+$ T cell subset development (Chang *et al.*, 1990). Macrophages are reported to favour T$_H$1 cells by producing cytokines such as IFNα and IL-12 (Hsieh *et al.*, 1993; Manetti *et al.*, 1993) while B cells appear to favour T$_H$2 cell growth. IL-10 may also be involved in T$_H$1/T$_H$2 cross-regulation (Mosmann and Moore, 1991) by inhibiting the accessory cell function of monocytes (Punnonen *et al.*, 1993a), in particular their production of IL-12 (D'Andrea *et al.*, 1992). Romagnani and his colleagues reported that IL-4 enhanced the formation of T$_H$2 clones (Romagnani, 1992) and inhibited the formation of T$_H$1 clones while IFNγ and IL-12

(Maggi *et al.*, 1992; Hsieh *et al.*, 1993; Manetti *et al.*, 1993) supported the differentiation of T$_H$1 T cells but inhibited the formation of T$_H$2 cells. The number of cytolytic clones generated was increased by IFNγ and decreased by IL-4 (Parronchi *et al.*, 1992).

8.4   GROWTH AND SURVIVAL OF CD4$^+$ T CELL SUBSETS

As well as regulating the differentiation of CD4$^+$ T cells, some cytokines can inhibit the proliferation and cyto-toxic capability of T$_H$1 and T$_H$2 cells (Sher *et al.*, 1992). T$_H$2 but not T$_H$1 cells express IL-1β receptors and can accept IL-1β as a second signal for clonal expansion (Greenbaum *et al.*, 1988; Lichtman *et al.*, 1988; Munoz *et al.*, 1990a). TGFβ is reported to inhibit the growth of

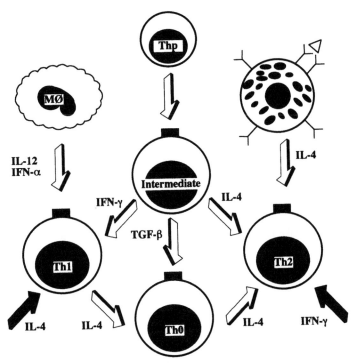

Figure 3.8 CD4 differentiation. The open arrows indicate those stimuli that support and the grey arrows those that inhibit the growth and differentiation of T cell subtypes.

T_H1-like tumour (CSA1M)-specific CD4$^+$ T cells (Li *et al.*, 1993). Down-regulation of host cell-mediated immunity to some parasites can, at least in part, be attributed to the action of IL-4, IL-10 and TGFβ (Sher *et al.*, 1992). Fitch and his group have shown that IFNγ inhibits proliferation of T_H2 clones (Gajewski and Fitch, 1990). This effect is only seen in those clones which do not make IFNγ and is not seen with T_H1 or T_H0 clones (Greenbaum *et al.*, 1988; Lichtman *et al.*, 1988). We have observed that T_H2-like cells generated by culture of splenic rat T cells with IL-4 and anti-IFNγ, proliferated in response to recombinant rat IL-4 (Noble *et al.*, 1993b), but was inhibited by IFNγ.

Overstimulation of the T cell receptor complex with high doses of aqueous protein antigens (Burstein and Abbas, 1993) or anti-CD3 (Gajewski *et al.*, 1989b) induces T cell tolerance in which T_H1 cells are inhibited, but T_H2 cells are not. This does not appear to be due to gross differences in CD3 expression but rather to differences in TCR-associated signal transduction pathways for lymphokine gene expression. Inositol phosphates, for example, were readily detected in T_H1 but not T_H2 clones activated via the TCR complex (Gajewski *et al.*, 1990). Anergy, mediated by high doses of IL-2, can be induced in T_H1 but not T_H2 clones (Schwartz, 1990; Williams and Unanue, 1990; Kang *et al.*, 1992). Unlike IFNγ, which only affected non-IFNγ-producing cells, IL-2-induced anergy was only seen in IL-2-producing CD4$^+$ (T_H1 and T_H0; Williams and Unanue, 1990;

Williams *et al.*, 1990) and CD8$^+$ T cell clones (Otten and Germain, 1991). Indeed, faced with an anergic stimulus T_H0 clones are reported to lose the capacity to respond to IL-2 and to exhibit a T_H2-like cytokine profile (Jenkins *et al.*, 1990). The resistance of T_H2 cells to IL-2-induced anergy may explain their participation in long-lived IgE antibody responses.

T_H1 clones are also reported to be more sensitive to inhibitors of some second messenger pathways such as cholera toxin, 8-bromoadenosine 3′5′-cyclic monophosphate, and cyclosporin A than are T_H2 clones (Munoz *et al.*, 1990b). Indeed IL-1-induced c-*jun* gene transcription and mRNA expression proceeds by a pathway which is dependent on protein tyrosine kinase activity. This mechanism of signal transmission was independent of protein kinase C (PKC) whereas c-*fos* mRNA expression was linked to the 80 kDa IL-1 receptor (IL-1R; Munoz *et al.*, 1992). To date little is known about differential signal transduction in T_H1 and T_H2 cells except that T_H1 cells do not appear as responsive to IL-1 as T_H2 cells (Greenbaum *et al.*, 1988; Lichtman *et al.*, 1988; Munoz *et al.*, 1990a). The response of T_H1 and T_H2 cells to cytotoxic drugs also differs. Ricin has been shown to inhibit T_H1 cytokine production (Diaz-Sanchez *et al.*, 1993b; Kemeny *et al.*, 1994) in rats *in vivo*. Cyclosporin A is reported to inhibit mouse T_H1 rather than T_H2 clones (Munoz *et al.*, 1990b). T cells cultured with IL-2 and IL-4 are claimed to be resistant to glucocorticoid-induced apoptosis

(Zubiaga *et al.*, 1992) and to have a reduction in glucocorticoid receptor binding affinity (Kam *et al.*, 1993).

9. CD8⁺ T cells

9.1 SUPPRESSION OF IGE BY CD8⁺ T CELLS

In addition to CD4⁺ T cells, CD8⁺ T cells also appear to play an important part in IgE regulation. Our attention was drawn to immunoregulatory CD8⁺ T cells while investigating the mechanism of castor bean sensitization. Virtually all people exposed to castor bean dust, regardless of their genetic background, make IgE antibodies to castor bean proteins (Thorpe *et al.*, 1988). Rats immunized with castor bean extract became sensitized to castor bean proteins (Thorpe *et al.*, 1989). The substance responsible is ricin, a toxic lectin which, when injected together with a bystander antigen, potentiates the IgE, but not the IgG, antibody response to bystander antigen (Diaz-Sanchez and Kemeny, 1990; Hellman, 1994). The IgE response thus induced can be boosted by repeated injection of ricin and antigen and is long-lived (Diaz-Sanchez and Kemeny, 1991; Fig. 3.9).

In these animals there was a profound (>50%) reduction in the number of CD8⁺ T cells in the spleen. These appear to have been targeted because they bore an increased number of ricin-binding glycoproteins on their surface 12–24 h after immunization (Diaz-Sanchez *et al.*, 1993a). When adoptively transferred to naive, syngeneic recipients, early-activated CD8⁺ T cells suppressed the IgE response by up to 95% (Diaz-Sanchez *et al.*, 1993a; Fig. 3.10). Sedgwick and Holt (1985) have shown that IgE regulatory CD8⁺ T cells are generated following inhalation of ovalbumin. Such cells bear the γδ form of the T cell receptor (McMenamin and Holt, 1994) and are discussed in more detail in Chapter 1.

The fact that CD8⁺ T cells can regulate immune responses to antigens normally processed via the MHC class I pathway poses some interesting questions. Are these cells antigen specific? Holt's work indicates that they are although the early activated CD8⁺ T cells that we described did not appear to be, possibly because of the large number of cells transferred (1×10^8). Are these cells MHC class I restricted? IFNγ secretion by these cells appears to be partially inhibited by anti-MHC class I (McMenamin and Holt, 1993). We have recently found that proliferation of ovalbumin-specific CD8⁺ T cell lines generated following parenteral immunization are inhibited by anti-MHC class I antibodies in a dose-dependent manner. In the mouse, ovalbumin-specific CD8⁺ MHC class I restricted T cell clones have been produced either by immunization with a monomethoxypolyethylene glycol–ovalbumin conjugate or with

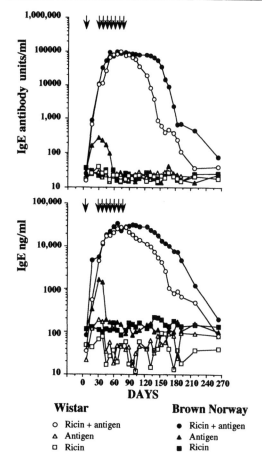

Figure 3.9 The persistent IgE response induced by repeated immunization of Brown Norway and Wistar rats with ricin and antigen.

trypsin-digested ovalbumin fragments (Chen *et al.*, 1992; Ishioka *et al.*, 1989). Ovalbumin-specific CD8⁺ T cell clones derived using the former regime suppressed *in vitro* production of IgG and IgE antibody but were not cytotoxic (Chen *et al.*, 1992), while those produced using the latter strategy were cytotoxic (Ishioka *et al.*, 1989).

9.2 CD8⁺ T CELL DEPLETION, A PARADOX

From these data it might be expected that removal of CD8⁺ T cells would enhance IgE production and favour T_H2-type immune responses. However, this appears to depend on the stage at which depletion is carried out. Thus, if CD8⁺ T cells are depleted in rats (using the mAb OX-8) in which an IgE response had been induced, then the usual decline in serum IgE levels is prevented (Holmes *et al.*, 1994). This supports the view that CD8⁺ T cells suppress IgE. However, if depletion of CD8⁺ T cells is performed prior to immunization, then the IgE response is not enhanced. Moreover, the IgE response is

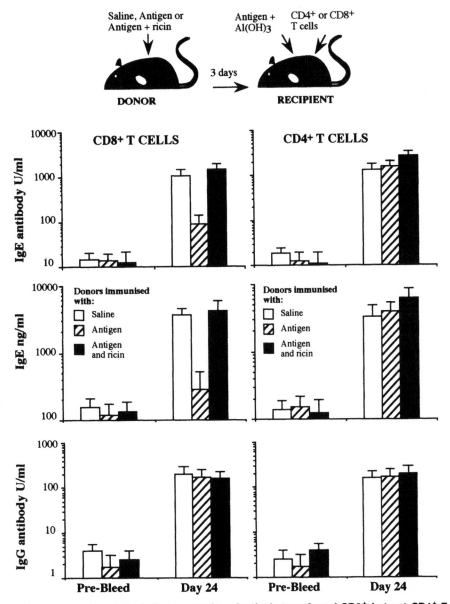

Figure 3.10 **Suppression of the IgE response by adoptively transferred CD8⁺ but not CD4⁺ T cells.**

inhibited if animals are immunized with ricin and a bystander antigen (Kemeny *et al.*, 1992). This suggests that CD8⁺ T cells may actually contribute to the IgE response. Indeed, some CD8⁺ T cell clones that inhibit cytotoxic CD4⁺ T cells secrete large amounts of IL-4 (Salgame *et al.*, 1991) and CD8⁺ T cells are capable of differentiating into CD4⁻CD8⁻ cells that secrete IL-4, IL-5 and IL-6 and provide B cell help (Erard *et al.*, 1993).

Discrepancies between the effects of CD8⁺ T cell depletion on the early and later stages of the immune response are not uncommon. In the mercuric chloride-induced model of autoimmune disease (Mathieson *et al.*, 1991),

depletion of CD8⁺ T cells did not prevent disease but rendered animals more susceptible to rechallenge. Depletion of CD8⁺ T cells had no apparent effect on *Plasmodium chabaudi* infection, although it increased recurring parasitaemia (Podoba and Stevenson, 1991). Furthermore, depletion of CD8⁺ T cells alone was insufficient to prevent reinfection with *Toxoplasma gondii* despite the finding that *in vitro* killing of mastocytoma cells infected with *T. gondii* was completely dependent on the presence of CD8⁺ T cells (Subauste *et al.*, 1991; Gazzinelli *et al.*, 1992). Thus, depletion of CD8⁺ T cells appears to do more than simply remove cells that suppress IgE or mediate cytotoxicity.

9.3 CD8$^+$ T CELL REGULATION OF T$_H$1 AND T$_H$2 CYTOKINE PRODUCTION

CD8$^+$ T cells might regulate IgE production: by suppressing IgE synthesis through the inhibitory effect of IFNγ on B cells; and/or by altering the differentiation and function of T$_H$2-like CD4$^+$ T cells. There is *in vivo* evidence for both of these mechanisms. Depletion of early-activated CD8$^+$ T cells in rats immunized with ricin and bystander antigen reduces the amount of IFNγ produced by CD4$^+$ and CD8$^+$ T cells (Kemeny et al., 1993). This treatment also results in a transient 10-fold increase in the capacity of splenocytes to produce the T$_H$2-type cytokines IL-4, IL-5 and IL-10 (Noble et al., 1993c). Furthermore, adoptive transfer of early-activated IgE-specific CD8$^+$ T suppressor cells enhances the production of IFNγ by splenocytes more than 1000-fold, although it is yet not clear if IFNγ is the product of the adoptively transferred CD8$^+$ donor cells or host CD4$^+$ or CD8$^+$ T cells (Kemeny et al., 1994).

The influence of polyclonal CD8$^+$ T cells on CD4$^+$ T cell differentiation can also be observed *in vitro*. Splenic T cells were cultured for 4 days with concanavalin A (Con A) and various cytokines (Noble et al., 1993b) before isolating CD4$^+$ T cells by positive selection and determining the cytokines released by these cells after restimulation. It was observed that prior removal of CD8$^+$ T cells enhanced the capacity of T$_H$2 CD4$^+$ T cells to produce IL-4 and IL-5, and increased the ability of these cells to proliferate in response to recombinant rat IL-4 (Noble et al., 1993a). However, removal of CD8$^+$ T cells also increased the capacity of T$_H$1 CD4$^+$ T cells to produce IFNγ.

These data suggest either that CD8$^+$ T cells inhibit CD4$^+$ T cell development in an unrestricted manner, or that there were different populations of CD8$^+$ T cells that exerted differential effects on T$_H$1 and T$_H$2 CD4$^+$ T cell development. Clearly, this will not be resolved until more CD8$^+$ T cells have been cloned and the effects of these clones on CD4$^+$ T cell development determined. The fact that immunoregulatory CD8$^+$ T cells appear to be active early in the immune response places these cells in an ideal position to influence the direction that the response will take. However, how CD8$^+$ T cells are recruited and activated, how they interact with accessory cells and recognize antigen, and how they mediate their effects, have yet to be determined.

9.4 CD8$^+$ T CELL SUBSETS

The strongest evidence for the existence of different subsets of CD8$^+$ T cells comes from the finding that CD8$^+$ T cell clones specific for *Mycobacterium leprae*, derived from the skin biopsies of patients with leprosy, could be divided into two subtypes according to their cytokine secretion profile (Salgame et al., 1991; Table 3.2). Type 1 CD8$^+$ T cell clones were derived from healed lesions of patients with tuberculoid leprosy, were cytotoxic, secreted IFNγ but not IL-4, and were restricted to MHC class I. Type 2 CD8$^+$ T cell clones were derived from patients with lepromatous leprosy, suppressed the killing of *M. leprae* by *M. leprae*-specific CD4$^+$ T cell clones by secretion of IL-4, secreted IFNγ, IL-4, IL-5 and IL-10, but not IL-6, and were restricted to MHC class II. Furthermore, production of IFNγ by mouse CD8$^+$ T cells has also been shown to be confined to a CD44 high subset (Rocken et al., 1992).

Control of cytokine release by CD8$^+$ T cells shares some features with CD4$^+$ T cells, although there is no direct correlation between CD8$^+$ subsets and CD4$^+$ T$_H$1 and T$_H$2 cell subsets. Mouse CD8$^+$ spleen cells cultured with IL-2, IL-4 and anti-CD3 antibodies produce large amounts of IL-4 (Seder et al., 1992). Rat CD8$^+$ T cells cultured with Con A, IL-2, IL-4 and anti-IFNγ generate increased amounts of IL-4 and IL-5 mRNA when restimulated with phorbol myristate acetate (PMA) and ionomycin (Noble, MacAry and Kemeny, 1995). However, in contrast to CD4$^+$ T$_H$2 cells, most alloreactive and antigen-specific CD8$^+$ T cell clones also produce large amounts of IFNγ (Rocken et al., 1992). A similar split between IL-4 and IL-5 on the one hand and IL-6 on the other has been described for *Mycobacterium leprae*-specific human type 1 and type 2 CD8$^+$ T cell clones (Salgame et al., 1991; Yamamura et al., 1991).

10. Conclusions

During the past few years there has been an explosion in our understanding of the processes that regulate IgE synthesis. We now know that B cells can switch directly or sequentially to IgE, that IL-4 and IL-13 are switch factors for IgE, and that TGFβ, IFNα and IFNγ inhibit IgE while TNFα and IL-5 enhance IgE synthesis. Many of the interactions between B cells and T cells such as CD58/CD2, CD40/CD40L, CD28/B7 and CD21/CD23 further serve to regulate the IgE response and many different cells contribute to the B cell cytokine microenvironment including mast cells and basophils, monocytes, macrophages, and dendritic cells, although the major stimulus appears to be delivered by the T cell. Although conveniently ordered into IgE helper (T$_H$2) and IgE suppressor (T$_H$1) cells, subtypes of CD8$^+$ T cells also appear to be important either by exerting direct effects on B cells or by influencing the direction that the CD4$^+$ T cell response takes (Kemeny et al., 1994). During the next few years we can expect that our understanding of the process of immune regulation will increase. I believe that the application of this knowledge to the IgE regulatory network will lead to the discovery of new immunomodulatory compounds and novel vaccines to control the IgE response in allergic disease.

11. References

Aalberse, R.C., Van-Der-Gaag, R. and Van-Leeuwen, J. (1983). Serologic aspects of IgG4 antibodies. I. Prolonged immunization results in an IgG4-restricted response. J. Immunol. 130, 722–726.

Armitage, R.J., Macduff, B.M., Spriggs, M.K. and Fanslow, W.C. (1993). Human B cell proliferation and Ig secretion induced by recombinant CD40 ligand are modulated by soluble cytokines. J. Immunol. 150, 3671–3680.

Aubry, J.P., Pochon, S., Graber, P., Jansen, K.U. and Bonnefoy, J.Y. (1992). CD21 is a ligand for CD23 and regulates IgE production. Nature 358, 505–507.

Aversa, G., Punnonen, J., Cocks, B.G., de Waal Malefyt, R., Vega, F., Jr., Zurawski, S.M., Zurawski, G. and de Vries, J.E. (1993a). An interleukin 4 (IL-4) mutant protein inhibits both IL-4 or IL-13-induced human immunoglobulin G4 (IgG4) and IgE synthesis and B cell proliferation: support for a common component shared by IL-4 and IL-13 receptors. J. Exp. Med. 178, 2213–2218.

Aversa, G., Punnonen, J. and de Vries, J.E. (1993b). The 26-kD transmembrane form of tumor necrosis factor alpha on activated CD4+ T cell clones provides a costimulatory signal for human B cell activation. J. Exp. Med. 177, 1575–1585.

Bazin, H., Beckers, A. and Querinjean, P. (1974). Three classes and four(sub)-classes of rat immunoglobulins: IgM, IgA, IgE and IgG1, IgG2a, IgG2b, IgG2c. Eur. J. Immunol. 4, 1974.

Beavil, A.J., Edmeades, R.L., Gould, H.J. and Sutton, B.J. (1992). α-Helical coiled-coil stalks in the low-affinity receptor for IgE (Fc epsilon RII/CD23) and related C-type lectins. Proc. Natl. Acad. Sci. USA 89, 753–757.

Bennich, H.H., Ishizaka, K., Johansson, S.G., Rowe, D.S., Stanworth, D.R. and Terry, W.D. (1968). Immunoglobulin E: a new class of human immunoglobulin. Immunology 15, 323–324.

Boguniewicz, M., Jaffe, H.S., Izu, A., Sullivan, M.J., York, D., Geha, R.S. and Leung, D.Y. (1990). Recombinant gamma interferon in treatment of patients with atopic dermatitis and elevated IgE levels. Am. J. Med. 88, 365–370.

Bonnefoy, J.Y., Henchoz, S., Hardie, D., Holder, M.J. and Gordon, J. (1993). A subset of anti-CD21 antibodies promote the rescue of germinal center B cells from apoptosis. Eur. J. Immunol. 23, 969–972.

Bradding, P., Feather, I.H., Howarth, P.H., Mueller, R., Roberts, J.A., Britten, K., Bews-JPA, Hunt, T.C., Okayama, Y., Heusser, C.H., Bullock, G.R., Church, M.K. and Holgate, S.T. (1992). Interleukin 4 is localized to and released by human mast cells. J. Exp. Med. 176, 1381–1386.

Bradding, P., Feather, I.H., Wilson, S., Bardin, P.G., Heusser, C.H., Holgate, S.T. and Howarth, P.H. (1993). Immunolocalization of cytokines in the nasal mucosa of normal and perennial rhinitic subjects. The mast cell as a source of IL-4, IL-5, and IL-6 in human allergic mucosal inflammation. J. Immunol. 151, 3853–3865.

Briere, F., Servet-Delprat, C., Bridon, J.M., Saint-Remy, J.M. and Bancherau, J. (1994). Human interleukin 10 induces naive surface immunoglobulin D+ (sIgD+) B cells to secrete IgG1 and IgG3. J. Exp. Med. 179, 757–762.

Brinkmann, V., Muller, S. and Heusser, C.H. (1992). T cell dependent differentiation of human B cells: direct switch from IgM to IgE, and sequential switch from IgM via IgG to IgA production. Mol. Immunol. 29, 1159–1164.

Brunner, T., Heusser, C.H. and Dahinden, C.A. (1993). Human peripheral blood basophils primed by interleukin 3 (IL-3) produce IL-4 in response to immunoglobulin E receptor stimulation. J. Exp. Med. 177, 605–611.

Burstein, H.J. and Abbas, A.K. (1993). In vivo role of interleukin 4 in T cell tolerance induced by aqueous protein antigen. J. Exp. Med. 177, 457–463.

Cao, B.N., Huneycutt, B.S., Gapud, C.P., Arceci, R.J. and Reiss, C.S. (1993). Lymphokine expression profile of resting and stimulated CD4(+) CTL clones specific for the glycoprotein of vesicular stomatitis virus. Cell. Immunol. 146, 147–156.

Chang, T.L., Shea, C.M., Urioste, S., Thompson, R.C., Boom, W.H. and Abbas, A.K. (1990). Heterogeneity of helper/inducer T lymphocytes: III. Responses of IL-2- and IL-4-Producing (TH1 and TH2) clones to antigens presented by different accessory cells. J. Immunol. 145, 2803–2808.

Chatelain, R., Varkila, K. and Coffman, R. L. (1992). IL-4 induces a TH2 response in Leishmania major-infected mice. J. Immunol. 148, 1182–1187.

Chen, Y., Takata, M., Maiti, P.K., Rector, E.S. and Sehon, A.H. (1992). Characterization of suppressor T cell clones derived from a mouse tolerized with conjugates of ovalbumin and monomethoxypolyethylene glycol. Cell. Immunol. 142, 16–27.

Coffman, R.L. and Carty, J. (1986). At T cell activity that enhances polyclonal IgE production and its inhibition by interferon-gamma. J. Immunol. 136, 949–954.

Coffman, R.L., Ohara, J., Bond, M.W. et. al. (1986). B cell stimulatory factor-1 enhances the IgE response of lipopolysaccharide-activated B cells. J. Immunol. 136, 4538–4541.

Committee on Safety of Medicines (1986). CSM update: desensitizing vaccines. Br. Med. J. 293, 948.

Cremer, N.E., Taylor, D., O, N,, Lenette, E.H. and Hagens, S.J. (1973). IgM production in rats infected with monkey leukemia virus. J. Natl. Cancer Inst. 51, 905.

Croft, M. and Swain, S.L. (1991). B cell response to T helper cell subsets. II. Both the stage of T cell differentiation and the cytokines secreted determine the extent and nature of helper activity. J. Immunol. 147, 3679–3689.

D'Andrea, A., Rengaraju, M., Valiante, N.M., Chehimi, J., Kubin, M., Aste, M., Chan, S. H., Kobayashi, M., Young, D., Nickbarg, E. et al. (1992). Production of natural killer cell stimulatory factor (interleukin 12) by peripheral blood mononuclear cells. J. Exp. Med. 176, 1387–1398.

Defrance, T., Carayon, P., Billian, G., Guillemot, J.C., Minty, A., Caput, D. and Ferrara, P. (1994). Interleukin 13 is a B cell stimulating factor. J. Exp. Med. 179, 135–143.

Del Prete, G., Maggi, E., Parronchi, P., Chretien, I., Tiri, A., Macchia, D., Ricci, M., Bancherau, J., De-Vries, J. and Romagnani, S. (1988). IL-4 is an essential factor for the IgE synthesis induced in vitro by human T cell clones and their supernatants. J. Immunol. 140, 4193–4198.

Del Prete, G.F., De Carli, M., Mastromauro, C., Biagiotti, R., Macchia, D., P., F., Ricci, M. and Romagnani, S. (1991). Purified protein derivative of Mycobacterium tuberculosis and excretory-secretory antigen(s) of Toxocara canis expand in vitro human T cells with stable and opposite (type 1 T helper or type 2 T helper) profile of cytokine production. J. Clin. Invest. 88, 346–350.

Dentener, M.A., Bazil, U, Von Asmuth, E.J., Ceska, M. and Buurman, W.A. (1993). Involvement of CD14 in lipopolysaccharide-induced tumor necrosis factor-alpha, IL-6 and IL-8 release by human monocytes and alveolar macrophages. J. Immunol. 150, 2885–2891.

Dessaint, J.P., Bout, D., Wattre, P. and Capron, A. (1975). Quantitative determination of specific IgE antibodies to Echinococcus granulosus and IgE levels in sera from patients with hydatid disease. Immunology 29, 813.

Devey, M.E., Lee, S.R., Richards, D. and Kemeny, D.M. (1989). Serial studies on the functional affinity and heterogeneity of antibodies of different IgG subclasses to phospholipase A-2 produced in response to bee-venom immunotherapy. J. Allergy Clin. Immunol. 84, 326–330.

De Vries, J.E., Gauchat, J.F., Aversa, G.G., Punnonen, J., Gascan, H. and Yssel, H. (1991). Regulation of IgE synthesis by cytokines. Curr. Opin. Immunol. 3, 851–858.

Diaz-Sanchez, D. and Kemeny, D.M. (1990). The sensitivity of rat CD8$^+$ and CD4$^+$ T cells to ricin *in vivo* and *in vitro* and their relationship to IgE regulation. Immunology 69, 71–77.

Diaz-Sanchez, D. and Kemeny, D.M. (1991). Generation of a long-lived IgE response in high and low responder strains of rat by co-administration of ricin and antigen. Immunology 72, 297–303.

Diaz-Sanchez, D., Lee, T.H. and Kemeny, D.M. (1993a). Ricin enhances IgE responses by inhibiting a subpopulation of early-activated CD8$^+$ T cells. Immunology 78, 226–236.

Diaz-Sanchez, D., Noble, A., Staynov, D.Z., Lee, T.H. and Kemeny, D.M. (1993b). Elimination of IgE regulatory CD8$^+$ T cells *in vivo* differentially modulates the ability of splenocytes to produce IL-4 and IFNγ but not IL-2. Immunology 78, 513–519.

Diaz-Sanchez, D., Chegini, S., Zhang, K. and Saxon, A. (1994). CD58 (LFA-3) stimulation provides a signal for human isotype switching and IgE production distinct from CD40. J. Immunol. 153, 10–20.

Erard, F., Wild, M.-T., Garcia-Sanz, J.A. and Le Gros, G. (1993). Switch of CD8$^+$ T cells to non-cytotoxic CD8$^-$ CD4$^-$ cells that make T$_H$2 cytokines and help B cells. Science 260, 1802–1805.

Finkelman, F.D., Katona, I.M., Urban, J.F., Holmes, J., Ohara, J., Tung, A.S., Sample, J.V. and Paul, W.E. (1988). IL-4 is required to generate and sustain *in vivo* IgE responses. J. Immunol. 141, 2335–2341.

Finkelman, F.D., Holmes, J., Urban, J., Jr., Paul, W.E. and Katona, I.M. (1989). T help requirements for the generation of an *in vivo* IgE response: A late acting form of T cell help other than IL-4 is required for IgE but not for IgG1 production. J. Immunol. 142, 403–408.

Finkelman, F.D., Urban, J.F.J.R., Beckmann, M.P., Schooley, K.A., Holmes, J.M. and Katona, I.M. (1991). Regulation of murine *in vivo* IgG and IgE responses by a monoclonal anti-IL-4 receptor antibody. Int. Immunol. 3, 599–607.

Firestein, G.S., Roeder, W.D., Laxer, J.A., Townsend, K.S., Weaver, C.T., Hom, J.T., Linton, J., Torbett, B.E. and Glasebrook, A.L. (1989). A new murine CD4($^+$) T cell subset with an unrestricted cytokine profile. J. Immunol. 143, 518–525.

Flanagan, J.G. and Rabbits, T.H. (1982). Arrangement of human immunoglobulin heavy chain constant region genes implies evolutionary duplication of a segment containing gamma, epsilon and alpha genes. Nature (London) 300(5894), 709–713.

Friedland, J.S., Remick, D.G., Shattock, R. and Griffin, G.E. (1992). Secretion of interleukin-8 following phagocytosis of *Mycobacterium tuberculosis* by human monocyte cell lines. Eur. J. Immunol. 22, 1373–1378.

Gajewski, T.F. and Fitch, F.W. (1990). Anti-proliferative effect of IFN-gamma in immune regulation. IV. Murine CTL clones produce IL-3 and GM-CSF, the activity of which is masked by the inhibitory action of secreted IFN- gamma. J. Immunol. 144, 548–556.

Gajewski, T.F., Joyce, J. and Fitch, F.W. (1989a). Antiproliferative effect of IFN-gamma in immune regulation. III. Differential selection of TH1 and TH2 murine helper T lymphocyte clones using recombinant IL-2 and recombinant IFN-gamma. J. Immunol. 143, 15–22.

Gajewski, T.F., Schell, S.R., Nau, G. and Fitch, F.W. (1989b). Regulation of T cell activation: Differences among T cell subsets. Immunol. Rev. 111(79–110).

Gajewski, T.F., Schell, S.R. and Fitch, F.W. (1990). Evidence implicating utilization of different T cell receptor-associated signalling pathways by T(H)1 and T(H)2 clones. J. Immunol. 144811), 4110–4120.

Gascan, H., Gauchat, J.F., Roncarolo, M.G., Yssel, H., Spits, H. and De Vries, J.E. (1991). Human B cell clones can be induced to proliferate and to switch to IgE and IgG4 synthesis by interleukin 4 and a signal provided by activated CD4($^+$) T cell clones. J. Exp. Med. 173, 747–750.

Gauchat, J.F., Lebman, D.A., Coffman, R.L., Gascan, H. and De-Vries, J.E. (1990). Structure and expression of germline epsilon transcripts in human B cells induced by interleukin 4 to switch to IgE production. J. Exp. Med. 172, 463–473.

Gauchat, J.F., Aversa, G., Gascan, H. and De-Vries, J.E. (1992a). Modulation of IL-4 induced germline epsilon RNA synthesis in human B cells by tumor necrosis factor-alpha, anti-CD40 monoclonal antibodies or transforming growth factor-beta correlates with levels of IgE production. Int. Immunol. 4, 397–406.

Gauchat, J.F., Gascan, H., de Waal Malefyt, R. and de Vries, J.E. (1992b). Regulation of germ-line epsilon transcription and induction of epsilon switching in cloned EBV-transformed and malignant human B cell lines by cytokines and CD4$^+$ T cells. J. Immunol. 148, 2291–2299.

Gauchat, J.F., Henchoz, S., Mazzei, G., Aubry, J.P., Brunner, T., Blasey, H., Life, P., Talabot, D., Flores-Romo, L., Thompson, J. *et al.* (1993). Induction of human IgE synthesis in B cells by mast cells and basophils. Nature 365, 340–343.

Gazzinelli, R., Xu, Y., Hieny, S., Cheever, A. and Sher, A. (1992). Simultaneous depletion of CD4($^+$) and CD8($^+$) T lymphocytes is required to reactivate chronic infection with *Toxoplasma gondii*. J. Immunol. 149, 175–180.

Geha, R. L.-D. (1987). Regulation of the human allergic response. Int. Arch. Allergy Appl. Immunol. 82, 3–4.

Geha, R.S., Helm, B. and Gould, H. (1985). Inhibition of Prausnitz–Kustner reaction by an immunoglobulin epsilon-chain fragment synthesized in *E. coli*. Nature 315, 577–578.

Gould, H.J., Helm, B.A., Marsh, P.J. and Geha, R.S. (1987). Recombinant human IgE. Int. Arch. Allergy Appl. Immunol. 82, 3–4.

Greenbaum, L.A., Horowitz, J.B., Woods, A., Pasquali, T., Reich, E.-P. and Bottomly, K. (1988). Autocrine growth of

CD4+ T cells. Differential effects of IL-1 on helper and inflammatory T cells. J. Immunol. 140, 1555–1560.

Harries, M.G., Kemeny, D.M., Youlten, L. et al. (1984). Skin and radioallergosorbent tests in patients with sensitivity to bee and wasp venom. Clin. Allergy 14, 407–412.

Harriman, W., Volk, H., Defranoux, N. and Wabl, M. (1993). Immunoglobulin class switch recombination. Annu. Rev. Immunol. 11, 361–384.

Hattevig, G., Kjellman, B. and Bjorksten, B. (1987). Clinical symptoms and IgE responses to common food proteins and inhalants in the first 7 years of life. Clin. Allergy 17, 571–578.

Hellman, L. (1994). Profound reduction in allergen sensitivity following treatment with a novel allergy vaccine. Eur. J. Immunol. 24, 415–420.

Helm, B., Marsh, P., Vercelli, D., Padlan, E., Gould, H. and Geha, R. (1988). The mast cell binding site on human immunoglobulin E. Nature 331, 180–183.

Henchoz, S., Gauchat, J.F., Aubry, J.P., Graber, P., Pochon, S. and Bonnefoy, J. Y. (1994). Stimulation of human IgE production by a subset of anti-CD21 monoclonal antibodies: requirement of a co-signal to modulate epsilon transcripts. Immunology 81, 285–290.

Hidaka, T., Kitani, A., Hara, M., Harigai, M., Suzuki, K., Kawaguchi, Y., Ishizuka, T., Kawagoe, M. and Nakamura, H. (1992). IL-4 down-regulates the surface expression of CD5 on B cells and inhibits spontaneous immunoglobulin and IgM-rheumatoid factor production in patients with rheumatoid arthritis. Clin. Exp. Immunol. 89, 223–229.

Holmes, B.J., Diaz-Sanchez, D., Lawrence, R.A., Maizels, R.M. and Kemeny, D.M. (1994). The effect of CD8+ T cell depletion on the primary and established IgE response. FASEB J. 8, A763.

Hsieh, C.S., Macatonia, S.E., Tripp, C.S., Wolf, S.F., O'Garra, A. and Murphy, K.M. (1993). Development of T(H) 1 CD4(+) T cells through IL-12 produced by Listeria-induced macrophages. Science 260, 547–549.

Ichiki, T., Takahashi, W. and Watanabe, T. (1992). The effect of cytokines and mitogens on the induction of C epsilon germline transcripts in a human Burkitt lymphoma B cell line. Int. Immunol. 4, 747–754.

Ishioka, G.Y., Colon, S., Miles, G., Grey, H.M. and Chesnut, R.W. (1989). Induction of class I MHC-restricted, peptide-specific cytolytic T lymphocytes by peptide priming in vivo. J. Immunol. 143, 1094–1100.

Ishizaka, K. (1988). IgE-binding factors and regulation of the IgE antibody response. Annu. Rev. Immunol. 6, 513–534.

Ishizaka, K. and Ishizaka, T. (1967). Identification of γE-antibodies as a carrier of reaginic activity. J. Immunol. 99, 1187–1198.

Jabara, H.H., Fu, S.M., Geha, R.S. and Vercelli, D. (1990). CD40 and IgE: Synergism between anti-CD40 monoclonal antibody and interleukin 4 in the induction of IgE synthesis by highly purified human B cells. J. Exp. Med. 172, 1861–1864.

Jabara, H.H., Ahern, D.J., Vercelli, D. and Geha, R.S. (1991). Hydrocortisone and IL-4 induce IgE isotype switching in human B cells. J. Immunol. 147, 1557–1560.

Jenkins, M.K., Chen, C., Jung, G., Mueller, D.L. and Schwartz, R.H. (1990). Inhibition of antigen-specific proliferation of type 1 murine T cell clones after stimulation with immobilized anti-CD3 monoclonal antibody. J. Immunol. 144, 16–22.

Johansson, S.G. and Bennich, H. (1967). Immunological studies of an atypical (myeloma) immunoglobulin. Immunology 13, 381–394.

Joseph, M., Auriault, C., Capron, A. et al. (1983). A new function for platelets: IgE-dependent killing of schistosomes. Nature 303, 810–812.

Jung, S., Siebenkotten, G. and Radbruch, A. (1994). Frequency of immunoglobulin E class switching is autonomously determined and independent of prior switching to other classes. J. Exp. Med. 179, 2023–2026.

Kam, J.C., Szefler, S.J., Surs, W., Sher, E.R. and Leung, D.Y.M. (1993). Combination of IL-2 and IL-4 reduces glucocorticoid receptor-binding affinity and T cell response to glucocorticoids. J. Immunol. 151, 3640–3466.

Kang, S.M., Beverly, B., Tran, A.C., Brorson, K., Schwartz, R.H. and Lenardo, M.J. (1992). Transactivation by AP-1 is a molecular target of T cell clonal anergy. Science 257, 1134–1138.

Katz, D.H. (1980). New concepts concerning pathogenesis of the allergic phenotype and prospects for control of IgE antibody synthesis. Int. Arch. Allergy Appl. Immunol. 66, 25–30.

Kemeny, D.M., Urbanek, R., Ewan, P., McHugh, S., Richards, D., Patel, S. and Lessof, M.H. (1989). The subclass of IgG antibody in allergic disease: II. The IgG subclass of antibodies produced following natural exposure to dust mite and grass pollen in atopic and non-atopic individuals. Clin. Exp. Allergy 19, 545–549.

Kemeny, D.M., Diaz-Sanchez, D. and Holmes, B.J. (1992). CD8+ T cells in allergy. Eur. J. Allergy Clin. Immunol. 47, 12–21.

Kemeny, D.M., Diaz-Sanchez, D., Noble, A., Staynov, D. and Lee, T.H. (eds) (1993). CD8+ T cell control of T$_H$1 and T$_H$2 T cell development and IgE production. In "New Advances in Cytokines". Ares Serono Symposia, Raven Press, New York.

Kemeny, D.M., Noble, A., Holmes, B.J. and Diaz-Sanchez, D. (1994). Immune regulation: A new job for the CD8+ T cell. Immunol. Today 15, 107–110.

Kimata, H., Yoshida, A., Ishioka, C., Lindley, I. and Mikawa, H. (1992). Interleukin 8 (IL-8) selectively inhibits immunoglobulin E production induced by IL-4 in human B cells. J. Exp. Med. 176, 1227–1231.

Kohler, I., Alliger, P., Minty, A., Caput, D., Ferrara, P., Holl-Neugebauer, B., Rank, G. and Rieber, E.P. (1994). Human interleukin-13 activates the interleukin-4-dependent transcription factor NF-IL4 sharing a DNA binding motif with an interferon-gamma-induced nuclear binding factor. Febs. Lett. 345, 187–192.

Korthauer, U., Graf, D., Mages, H.W., Briere, F., Padayachee, M., Malcolm, S., Ugazio, A.G., Notarangelo, L.D., Levinsky, R.J. and Kroczek, R.A. (1993). Defective expression of T cell CD40 ligand causes X-linked immunodeficiency with hyper-IgM [see comments]. Nature 361, 539–541.

Kuhn, R., Rajewsky, K. and Muller, W. (1991). Generation and analysis of interleukin-4 deficient mice. Science 254, 707–710.

Ledermann, F., Heusser, C., Schlienger, C. and Le Gros, G. (1992). Interleukin-3-treated non-B, non-T cells switch activated B cells to IgG1/IgE synthesis. Eur. J. Immunol. 22, 2783–2787.

Le Gros, G., Ben-Sasson, S.Z., Conrad, D.H., Clark-Lewis, I., Finkelman, F.D., Plaut, M. and Paul, W.E. (1990). IL-3 promotes production of IL-4 by splenic non-B, non-T cells in response to Fc receptor cross-linkage. J. Immunol. 145, 2500–2506.

Li, X. F., Takiuchi, H., Zou, J.P., Katagiri, T., Yamamoto, N., Nagata, T., Ono, S., Fujiwara, H. and Hamaoka, T. (1993). Transforming growth factor-beta (TGF-beta)-mediated immunosuppression in the tumor-bearing state: enhanced production of TGF-beta and a progressive increase in TGF-beta susceptibility of anti-tumor CD4($^+$) T cell function. Jpn. J. Cancer Res. 84, 315–325.

Lichtman, A.H., Chin, J., Schmidt, J.A. and Abbas, A.K. (1988). Role of interleukin 1 in the activation of T lymphocytes. Proc. Natl. Acad. Sci. USA 85, 9699–9703.

Luo, H., Hofstetter, H., Banchereau, J. and Delespesse, G. (1991). Cross-linking of CD23 antigen by its natural ligand (IgE) or by anti-CD23 antibody prevents B lymphocyte proliferation and differentiation. J. Immunol. 146, 2122–2129.

Maggi, E., Parronchi, P., Manetti, R., Simonelli, C., Piccinni, M.P., Rugiu, F. S., De-Carli, M., Ricci, M. and Romagnani, S. (1992). Reciprocal regulatory effects of IFN- gamma and IL-4 on the in vitro development of human T_H1 and T_H2 clones. J. Immunol. 148, 2142–2147.

Mandler, R., Finkelman, F.D., Levine, A.D. and Snapper, C.M. (1993). IL-4 induction of IgE class switching by lipopolysaccharide-activated murine B cells occurs predominantly through sequential switching. J. Immunol. 150, 407–418.

Manetti, R., Parronchi, P., Giudizi, M.G., Piccinni, M.P., Maggi, E, Trinchieri, G. and Romagnani, S. (1993). Natural killer cell stimulatory factor (interleukin 12 [IL-12]) induces T helper type 1 (T_H1)-specific immune responses and inhibits the development of IL-4-producing Th cells. J. Exp. Med. 177, 1199–1204.

Marsh, D.G., Lichtenstein, L.M. and Norman, P.S. (1972). Induction of IgE-mediated immediate hypersensitivity to group I rye grass pollen allergen and allergoids in non-allergic man. Immunology 22, 1013–1028.

Mathieson, P.W., Stapleton, K.J., Oliveira, D.B. and Lockwood, C.M. (1991). Immunoregulation of mercuric chloride-induced autoimmunity in Brown Norway rats: a role for CD8$^+$ T cells revealed by in vivo depletion studies. Eur. J. Immunol. 21(9): 2105–2109.

McMenamin, C. and Holt, P.G. (1993). The natural immune response to inhaled soluble protein antigens involves major histocompatibility complex (MHC) class I-restricted CD8$^+$ T cell-mediated but MHC class II-restricted CD4$^+$ T cell-dependent immune deviation resulting in selective suppression of immunoglobulin E production. J. Exp. Med. 178, 889–899.

McMenamin, C. and Holt, P.G. (1994). Regulation of IgE responses to inhaled antigen in mice by antigen-specific $\gamma\delta$ T cells. Science 265, 1869–1871.

Metzger, H. (1992). The receptor with high affinity for IgE. Immunol. Rev. 125, 37–48.

Minty, A., Chalon, P., Derocq, J.M., Dumont, X., Guillemot, J.C., Kaghad, M., Labit, C., Leplatois, P., Liauzun, P., Miloux, B. et al. (1993). Interleukin-13 is a new human lymphokine regulating inflammatory and immune responses. Nature 362, 248–250.

Mosmann, T.R. (1992). T lymphocyte subsets, cytokines, and effector functions. Ann. New York Acad. Sci. 664, 89–92.

Mosmann, T.R. and Moore, K.W. (1991). The role of IL-10 in crossregulation of T(H)1 and T(H)2 responses. Parasitol. Today 7, A49–A53.

Mosmann, T.R., Cherwinski, H., Bond, M.W., Giedlin, M.A. and Coffman, R.L. (1986). Two types of murine helper T cell clones. 1. Definition according to profiles of lymphokine activities and secreted proteins. J. Immunol. 135, 2348–2357.

Mudde, G.C., Hansel, T.T., Van-Reijsen, F.C., Osterhoff, B.F. and Bruijnzeel-Koomen-CAFM (1990). IgE: an immunoglobulin specialized in antigen capture? Immunol. Today 11, 440–443.

Munoz, E., Beutner, U., Zubiaga, A. and Huber, B.T. (1990a). IL-1 activates two separate signal transduction pathways in T helper type II cells. J. Immunol. 144, 964–969.

Munoz, E., Zubiaga, A.M., Merrow, M., Sauter, N.P. and Huber, B.T. (1990b). Cholera toxin discriminates between T helper 1 and 2 cells in T cell receptor-mediated activation: role of cAMP in T cell proliferation. J. Exp. Med. 172, 95–103.

Munoz, E., Zubiaga, A.M. and Huber, B.T. (1992). Interleukin-1 induces c-fos and c-jun gene expression in T helper type II cells through different signal transmission pathways. Eur. J. Immunol. 22, 2101–2106.

Noble, A., Kemeny, D.M. and Staynov, D.Z. (1993a). Generation of T_H1- and T_H2-like cells in vitro is regulated by IL-4 and IFNγ. J. Immunol. 150, 274.

Noble, A., Staynov, D. and Kemeny, D.M. (1993b). Generation of rat T_H2-like cells in vitro is IL-4 dependent and is inhibited by IFNγ. Immunology 79, 562–567.

Noble, A., Diaz-Sanchez, D., Staynov, D., Lee, T. and Kemeny, D.M. (1993c). Elimination of IgE regulatory rat CD8$^+$ T cells in vivo increases the co-ordinate expression of the T_H2 cytokines IL-4, IL-5 and IL-10. Immunology 80, 326–329.

Noble, A., MacAry, P.A. and Kemeny, D.M. (1995). IL-4 and IFNγ regulate the growth and differentiation of CD8$^+$ T cells into subpopulations with distinct cytokine profiles. J. Immunol. 155, 2928–2937.

Otten, G.R. and Germain, R.N. (1991). Split anergy in a CD8($^+$) T cell: receptor-dependent cytolysis in the absence of interleukin-2 production. Science 251, 1228–1231.

Parronchi, P., Tiri, A., Macchia, D., De-Carli, M., Biswas, P., Simonelli, C., Maggi, E., Del-Prete, G., Ricci, M. and Romagnani, S. (1990). Noncognate contact-dependent B cell activation can promote IL-4-dependent in vitro human IgE synthesis. J. Immunol. 144, 2102–2108.

Parronchi, P., Macchia, D., Piccinni, M.P., Biswas, P., Simonelli, C., Maggi, E., Ricci, M., Ansari, A.A. and Romagnani, S. (1991). Allergen- and bacterial antigen-specific T cell clones established from atopic donors show a different profile of cytokine production. Proc. Natl. Acad. Sci. USA 88, 4538–4542.

Parronchi, P., De-Carli, M., Manetti, R., Simonelli, C., Sampognaro, S., Piccinni, M.P., Macchia, D., Maggi, E., Del-Prete, G. and Romagnani, S. (1992). IL-4 and IFN (alpha and gamma) exert opposite regulatory effects on the development of cytolytic potential by T_H1 or T_H2 human T cell clones. J. Immunol. 149, 2977–2983.

Pene, J., Rousset, F., Briere, F., Chretien, I., Wideman, J., Bonnefoy, J.Y. and De-Vries, J.E. (1988a). Interleukin 5 enhances interleukin 4-induced IgE production by normal human B cells. The role of soluble CD23 antigen. Eur. J. Immunol. 18, 929–935.

Pene, J., Rousset, F, Briere, F., Chretien, I., Bonnefoy, J.Y., Spits, H., Yokota, T., Arai, N., Arai, K., Banchereau, J. and De-Vries, J. (1988b). IgE production by normal human lymphocytes is induced by interleukin 4 and suppressed by interferons gamma and alpha and prostaglandin E-2. Proc. Natl. Acad. Sci. USA 85, 6880–6884.

Petit-Frere, C., Dugas, B., Braquet, P. and Mencia-Huerta, J.M. (1993). Interleukin-9 potentiates the interleukin-4 induced IgE and IgG1 release from murine B lymphocytes. Immunology 79, 146–151.

Platts-Mills, T.A.E. (1987). The biological role of allergy. In "Allergy: An International Textbook", pp. 1–48. John Wiley, Chichester.

Plaut, M., Pierce, J.H., Watson, C.J., Hanley-Hyde, J., Nordan, R. and Paul, W.E. (1989). Mast cell lines produce lymphokines in response to cross-linkage of Fc epsilon RI or to calcium ionophores. Nature 339, 64–67.

Podoba, J.E. and Stevenson, M.M. (1991). CD4($^+$) and CD8($^+$) T lymphocytes both contribute to acquired immunity to blood-stage Plasmodium chabaudi AS. Infect. Immun. 59, 51–58.

Punnonen, J. and de Vries, J.E. (1994). IL-13 induces proliferation, Ig isotype switching, and Ig synthesis by immature human fetal B cells. J. Immunol. 152, 1094–1102.

Punnonen, J., de Waal Malefyt, R., van Vlasselaer, P., Gauchat, J. and de Vries, J.E. (1993a). IL-10 and viral IL-10 prevent IL-4-induced IgE synthesis by inhibiting the accessory cell function of monocytes. J. Immunol. 151, 1280–1289.

Punnonen, J., Aversa, G., Cocks, B. G., McKenzie-ANJ, Menon, S., Zurawski, G., De-Waal-Malefyt, R. and De-Vries, J.E. (1993b). Interleukin 13 induces interleukin 4-independent IgG4 and IgE synthesis and CD23 expression by human B cells. Proc. Natl. Acad. Sci. USA 90, 3730–3734.

Purkerson, J.M. and Isakson, P.C. (1992). Interleukin 5 (IL-5) provides a signal that is required in addition to IL-4 for isotype switching to immunoglobulin (Ig) G1 and IgE. J. Exp. Med. 175, 973–982.

Rademecker, M., Bekhti, A., Poncelet., E. and Salmon., J. (1974). Serum IgE levels in protozoal and helminthic infections. Int. Arch. Allergy Appl. Immunol. 47, 285.

Ravetch, J.V. and Kinet, J.P. (1991). Fc receptors. Annu. Rev. Immunol. 9 457–492.

Rieger, A., Wang, B., Kilgus, O., Ochiai, K., Maurer, D., Fodinger, D., Kinet, J.P and Stingl, G. (1992). Fc epsilon RI mediates IgE binding to human epidermal Langerhans cells. J. Invest. Dermatol. 99.

Rocken, M., Muller, K.M., Saurat, J.H., Muller, I., Louis, J.A., Cerottini, J.C. and Hauser, C. (1992). Central role for TCR/CD3 ligation in the differentiation of CD4$^+$ T cells toward a Th1 or Th2 functional phenotype. J. Immunol. 148, 47–54.

Romagnani, S. (1992). Induction of TH1 and TH2 responses: a key role for the "natural" immune response? Immunol. Today 13, 379–381.

Rousset, F., Garcia, E. and Banchereau, J. (1991). Cytokine-induced proliferation and immunoglobulin production of human B lymphocytes triggered through their CD40 antigen. J. Exp. Med. 173, 705–710.

Salgame, P., Abrams, J.S., Clayberger, C., Goldstein, H., Convit, J., Modlin, R.L. and Bloom, B.R. (1991). Differing lymphokine profiles of functional subsets of human CD4 and CD8 T cell clones. Science 254, 279–282.

Schatz, D.G., Oettinger, M.A. and Baltimore, D. (1989). The V(D)J recombination activating gene, RAG-1. Cell 59, 1035–1048.

Schwartz, R.H. (1990). A cell culture model for T lymphocyte clonal anergy. Science 248, 1349–1356.

Seder, R.A., Boulay, J.L., Finkelman, F., Barbier, S., Ben-Sasson, S.Z., Le Gros, G. and E., P.W. (1992). CD8$^+$ T cells can be primed in vitro to produce IL-4. J. Immunol. 148, 1652–1656.

Sedgwick, J.D. and Holt, P.G. (1985). Induction of IgE-secreting cells and IgE isotype-specific suppressor T cells in the respiratory lymph nodes of rats in response to antigen inhalation. Cell. Immunol. 94, 182–194.

Sher, A., Gazzinelli, R.T., Oswald, I.P., Clerici, M., Kullberg, M., Pearce, E.J., Berzofsky, J. A., Mosmann, T. R., James, S. L., Morse-HC, I. and Shearer, G.M. (1992). Role of T cell derived cytokines in the downregulation of immune responses in parasitic retroviral infection. Immunol. Rev. 127, 183–204.

Siebenkotten, G., Esser, C., Wabl, M. and Radbruch, A. (1992). The murine IgG1/IgE class switch program. Eur. J. Immunol. 22, 1827–1834.

Smith-Gill, S.J., Hamel, P.A., Lovoie, T.B. and Dorrington, K.J. (1987). Contributions of immunoglobulin heavy and light chains to antibody specificity for lysozyme and two haptens. J. Immunol. 139, 4135–4144.

Snapper, C.M. and Paul, W.E. (1987). Interferon- gamma and B cell stimulatory factor-1 reciprocally regulate Ig isotype production. Science 236, 944–947.

Snapper, C.M., Finkelman, F.D. and Paul, W.E. (1988). Differential regulation of IgG1 and IgE synthesis by interleukin 4. J. Exp. Med. 167, 183–196.

Spriggs, M.K., Artnitage, R.J., Strockbine, L., Clifford, K.N., Macduff, B.M., Sato, T.A., Maliszewski, C.R. and Fanslow, W.C. (1992). Recombinant human CD40 ligand stimulates B cell proliferation and immunoglobulin E secretion. J. Exp. Med. 176, 1543–1550.

Subauste, C.S., Koniaris, A.H. and Remington, J.S. (1991). Murine CD8($^+$) cytotoxic T lymphocytes lyse Toxoplasma gondii-infected cells. J. Immunol. 147, 3955–3959.

Sutton, B.J. and Gould, H.J. (1993). The human IgE network. Nature 366, 421–428.

Swain, S.L. (1991). Regulation of the development of distinct subsets of CD4$^+$ T cells. Res. Immunol. 14, 14–18.

Swain, S.L., Weinberg, A.D., English, M. and Huston, G. (1990). IL-4 directs the development of T$_H$2-like helper effectors. J. Immunol. 145, 3796–3806.

Swain, S.L., Huston, G., Tonkonogy, S. and Weinberg, A. (1991). Transforming growth factor-beta and IL-4 cause helper T cell precursors to develop into distinct effector helper cells that differ in lymphokine secretion pattern and cell surface phenotype. J. Immunol. 147, 2991–3000.

Tada, T., Okumura, K., Platteau, B. and Bazin, H. (1975). Half-lives of two types of rat homocytotropic antibodies in circulation and in the skin. Int. Arch. Allergy Appl. Immunol. 48, 116–131.

Thorpe, S.C., Kemeny, D.M., Panzani, R. and Lessof, M.H. (1988). Allergy to castor bean. I. Its relationship to sensitization to common inhalant allergens (atopy). J. Allergy Clin. Immunol. 82, 62–66.

Thorpe, S.C., Murdoch, R.D. and Kemeny, D.M. (1989). The effect of the castor bean toxin, ricin, on rat IgE and IgG responses. Immunology 68, 307–311.

Thyphronitis, G., Tsokos, G.C., June, C.H., Levine, A.D. and Finkelman, F.D. (1989). IgE secretion by Epstein–Barr virus-infected purified human B lymphocytes is stimulated by interleukin 4 and suppressed by interferon gamma. Proc. Natl. Acad. Sci. USA 86, 5580–5584.

Thyphronitis, G., Katona, I.M., Gause, W.C. and Finkelman, F.D. (1993). Germline and productive Cε gene expression during in vivo IgE responses. J. Immunol. 151, 4128–4136.

Urban, J., Jr., Katona, I. M., Paul, W.E. and Finkelman, F.D. (1991). Interleukin 4 is important in protective immunity to a gastrointestinal nematode infection in mice. Proc. Natl. Acad. Sci. USA 88, 5513–5517.

Urbanek, R., Kemeny, D.M. and Richards, D. (1986). Sub-class of IgG anti-bee venom antibody produced during bee venom immunotherapy and its relationship to long-term protection from bee stings and following termination of venom immunotherapy. Clin. Allergy 16, 317–322.

van der Stoep, N., Korver, W. and Logtenberg, T. (1994). In vivo and in vitro IgE isotype switching in human B lymphocytes: evidence for a predominantly direct IgM to IgE class switch program. Eur. J. Immunol. 24, 1307–1311.

van Ommen, R., Vredendaal, A.E. and Savelkoul, H.F. (1994). Suppression of polyclonal and antigen-specific murine IgG1 but not IgE responses by neutralizing interleukin-6 in vivo. Eur. J. Immunol. 24, 1396–1403.

van Vlasselaer, P., Punnonen, J. and de Vries, J.E. (1992). Transforming growth factor-beta directs IgA switching in human B cells. J. Immunol. 148, 2062–2067.

Van Reijsen, F.C., Bruijnzeel-Koomen CAFM, Kalthoff, F.S., Maggi, E, Romagnani, S., Westland-JKT and Mudde, G.C. (1992). Skin-derived aeroallergen-specific T cell clones of T$_H$2 phenotype in patients with atopic dermatitis. J. Allergy Clin. Immunol. 90, 184–192.

Vercelli, D., Jabara, H.H., Arai, K., Yokota, T. and Geha, R.S. (1989a). Endogenous interleukin 6 plays an obligatory role in interleukin 4-dependent human IgE synthesis. Eur. J. Immunol. 19, 1419–1424.

Vercelli, D., Jabara, H.H., Arai, K.I. and Geha, R.S. (1989b). Induction of human IgE synthesis requires interleukin 4 and T/B cell interactions involving the T cell receptor/CD3 complex and MHC class II antigens. J. Exp. Med. 169, 1295–1307.

Waldmann, T.A. (1969). Disorders of immunoglobulin metabolism. N. Engl. J. Med. 281, 1170.

Wierenga, E.A., Snoek, M., Bos, J.D., Jansen, H.M. and Kapsenberg, M.L. (1990a). Comparison of diversity and function of house dust mite-specific T lymphocyte clones from atopic and non-atopic donors. Eur. J. Immunol. 20, 1519–1526.

Wierenga, E.A., Snoek, M., De-Groot, C., Chretien, I., Bos, J.D., Jansen, H.M. and Kapsenberg, M.L. (1990b). Evidence for compartmentalization of functional subsets of CD4($^+$) T lymphocytes in atopic patients. J. Immunol. 144, 4651–4656.

Wierenga, E.A., Snoek, M., Jansen, H.M., Bos, J.D., Van-Lier-RAW and Kapsenberg, M.L. (1991). Human atopen-specific types 1 and 2 T helper cell clones. J. Immunol. 147, 2942–2949.

Williams, I.R. and Unanue, E.R. (1990). Costimulatory requirements of murine T$_H$1 clones. The role of accessory cell-derived signals in responses to anti-CD3 antibody. J. Immunol. 145, 85–93.

Williams, M.E., Lichtman, A.H. and Abbas, A.K. (1990). Anti-CD3 antibody induces unresponsiveness to IL-2 in T$_H$1 clones but not in T$_H$2 clones. J. Immunol. 144, 1208–1214.

Williams, M.E., Kullberg, M.C., Barbieri, S., Caspar, , Berzofsky, J.A., Seder, R.A. and Sher, A. (1993). Fc epsilon receptor-positive cells are a major source of antigen-induced interleukin-4 in spleens of mice infected with Schistosoma mansoni. Eur. J. Immunol. 23, 1910–1916.

Wu, C.Y., Brinkmann, V., Cox, D., Heusser, C. and Delespesse, G. (1992). Modulation of human IgE synthesis by transforming growth factor-beta. Clin. Immunol. Immunopathol. 62, 277–284.

Yamamura, M., Uyemura, K., Deans, R.J., Weinberg, K., Rea, T.H., Bloom, B.R. and Modlin, R.L. (1991). Defining protective responses to pathogens: cytokine profiles in leprosy lesions. Science 254, 277–279.

Yunginger, J.W. and Gleich, G.J. (1973). Seasonal changes in IgE antibodies and their relationship to IgG antibodies during immunotherapy for ragweed hayfever. J. Clin. Invest. 52, 1268.

Zhang, K., Saxon, A. and Max, E.E. (1992). Two unusual forms of human immunoglobulin E encoded by alternative RNA splicing of epsilon heavy chain membrane exons. J. Exp. Med. 176, 233–243.

Zhang, K., Max, E.E., Cheah, H.K. and Saxon, A. (1994a). Complex alternative RNA splicing of epsilon-immunoglobulin transcripts produces mRNAs encoding four potential secreted protein isoforms. J. Biol. Chem. 269, 456–462.

Zhang, K., Mills, F.C. and Saxon, A. (1994b). Switch circles from IL-4-directed epsilon class switching from human B lymphocytes. Evidence for direct, sequential, and multiple step sequential switch from mu to epsilon Ig heavy chain gene. J. Immunol. 152, 3427–3435.

Zheng, Y., Shopes, B., Holowka, D. and Baird, B. (1992). Dynamic conformations compared for IgE and IgG1 in solution and bound to receptors. Biochemistry 31, 7446–7456.

Zubiaga, A.M., Munoz, E. and Huber, B.T. (1992). IL-4 and IL-2 selectively rescue Th cell subsets from glucocorticoid-induced apoptosis. J. Immunol. 149, 107–112.

Zurawski, G. and de Vries, J.E. (1994). Interleukin 13, an interleukin 4-like cytokine that acts on monocytes and B cells, but not on T cells. Immunol. Today 15, 19–26.

4. Mast Cells and Basophils: Their Role in Initiating and Maintaining Inflammatory Responses

Peter Bradding, Andrew F. Walls *and* Martin K. Church

Immunopharmacology of the Respiratory System
ISBN 0−12−352325−7

1. Introduction

The mast cell is well established as a major initiating cell of the early phase of allergic responses. On cross-linkage of its membrane-bound IgE receptors by specific allergen, the mast cell releases into the local environment the granule-associated mediator histamine (H) and the newly generated mediators prostaglandin D_2 (PGD$_2$) and leukotriene C_4 (LTC$_4$). These mediators, the biological properties of which are summarized in Table 4.1, are responsible for most of the early events which characterize allergic reactions of the lung, nose, eye, intestine and skin. It has recently been established that the mast cell is also able to contribute to the chronic inflammatory events of allergic disease by the secretion of cytokines. After a brief description of mast cell biology, it is the intention of this review to concentrate on the role of mast cells in initiating and maintaining allergic inflammation by secretion of cytokines, and their possible role in the repair process.

2. Mast Cell Development

The first description of the mast cell was made in 1863 by Von Recklinhausen, although it was not until 1878 that the cell was named by Paul Ehrlich (Ehrlich, 1878).

While still a medical student at Freiburg University, Ehrlich noticed that the granules of the cell appeared purple in colour when stained with blue aniline dyes. This change in colour, or metachromasia, we now know to represent the interaction of the dyes with the highly acidic heparin proteoglycan which is present in the granules. Ehrlich, however, supposed that the granules contained phagocytosed materials, or nutrients, and hence named the cells *mastzellen*, or well-fed cells, from where we have derived the English name mast cell. Ehrlich also described the association of mast cells with blood vessels, inflamed tissues, nerves and neoplastic foci, and provided the first description of mast cell degranulation. In addition, he also described the basophil separately as a metachromatically staining cell which circulated in the blood (Ehrlich, 1879). Although this was originally thought to be a circulating mast cell, there is now a plethora of evidence that the mast cell and the basophil are only distantly related, being derived from different stem cells.

Although H was discovered independently by chemical synthesis by Windaus and Vogt in 1907, and by extraction from tissues by von Kutscher in 1910, and shown to mimic many of the symptoms of the anaphylactic reaction by Dale and Laidlaw in 1910 (see also Dale and Laidlaw, 1911), we had to wait until 1953 before the studies of Riley and West positively identified the mast

Table 4.1 Human mast cell mediators and their biological effects

Mediator	Effects
Performed	
Histamine	Bronchoconstriction, tissue oedema, mucus secretion, fibroblast proliferation, collagen synthesis,[*] endothelial proliferation
Heparin	Anticoagulant, storage matrix for mast cell mediators, fibroblast activation, protects growth factors from degradation and potentiates their action, endothelial cell migration
Tryptase	Generates C3a and bradykinin, degrades neuropeptides, increases BHR,[*] indirectly activates collagenase, fibroblast proliferation, bone remodelling
Chymase	Mucus secretion,[*] extracellular matrix degradation
Newly generated	
PGD$_2$	Bronchoconstriction, tissue oedema, mucus secretion
LTC$_2$	Bronchoconstriction, tissue oedema, mucus secretion
PAF	Bronchoconstriction, tissue oedema, mucus secretion, neutrophil and eosinophil chemotaxis
Cytokines	
IL-4	IgE regulation; VCAM-1 expression; TH$_2$ cell development; fibroblast proliferation,[*] chemotaxis and protein secretion; capillary endothelial cell proliferation
IL-5	Eosinophil growth, adhesion, chemotaxis, activation, survival
IL-6	Acute phase protein response, immunoglobulin secretion, T cell activation
TNFα	Adhesion molecule expression, inflammatory cell activation and chemotaxis, fibroblast proliferation and collagenase production, endothelial cell proliferation, increases BHR[*]

[*] Animal data.

cell as the source of H released during allergic reactions. The final piece of this preliminary jigsaw had to wait until the late 1960s when Johansson and Bennich in Sweden and Ishizaka and colleagues in the United States independently isolated the allergic, or reaginic, antibody as immunoglobulin E (IgE; Bennich *et al.*, 1968). This antibody binds with high affinity to both mast cells and basophils, and when cross-linked by specific antibody, or allergen, initiates the generation of mast cell mediators.

Mast cells and basophils have many features in common, including the presence of acidic proteoglycan in the granule, their ability to store and release H and their ability to bind IgE with high affinity to a specific IgE RI receptor. However, there are also many differences including nuclear morphology, location within the body, mediator content and synthesis and responses to non-immunological stimulants and drugs. It is now considered that basophils are terminally differentiated leucocytes which mature in the bone marrow from precursors closely related to those of eosinophils, and are released into the bloodstream as fully mature cells (Galli, 1990). Furthermore, culture of basophils *in vitro* does not give rise to mast cells even under conditions favourable for mast cell growth (Seder *et al.*, 1991a). While basophils are not normally found in extravascular compartments, they may migrate there during late-phase allergic responses when they may be responsible for some of the symptoms at that time (Bochner and Lichtenstein, 1992).

In contrast, mast cells leave the bone marrow and circulate in the blood as progenitors and it is not until they enter the tissues that they undergo their terminal differentiation into mature mast cells (Fig. 4.1). A major factor necessary for the maturation of mast cells is the stromal cell-derived factor called stem cell factor (SCF), a ligand for the c-*kit* receptor. The most elegant studies showing the necessity of this factor have been performed by Kitamura and colleagues using mice. They have produced two strains of mouse, W/W^v and $S1/S1^d$, which both show macrocytic anaemia, sterility, lack of hair pigmentation and very few mast cells (Kitamura *et al.*, 1989). However, when mast cell precursors from $S1/S1^d$ are transplanted into W/W^v mice normal mast cells develop, but when the inverse experiment is performed no mast cells are found. These results suggest that $S1/S1^d$ mice have normal mast cell precursors, but are different in a factor in the microenvironment necessary for their maturation, while W/W^v mice have a normal microenvironment but their mast cell precursors are abnormal. Further genetic studies have shown W/W^v mast cell precursors to be deficient in the c-*kit* receptor while $S1/S1^d$ mice do not produce SCF (Kitamura *et al.*, 1978, 1989; Kitamura and Go, 1979; Sonoda *et al.*, 1984; Kanakura *et al.*, 1988).

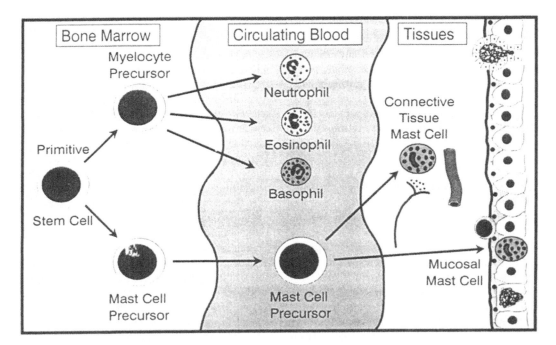

Figure 4.1 The development of mast cells from bone marrow precursors into mature cells in the tissues. Also illustrated are granulocytes, including basophils which mature in the bone marrow before being released into the blood as fully mature cells.

3. Mast Cell Heterogeneity

Studies on mast cells of all species have shown that they are not a homogeneous population, but differ markedly between species and within a particular species. The earliest suggestion of mast cell heterogeneity stems back to the histochemical studies of Maximow in 1906. However, it was not until 1966 that Enerback published a series of papers showing that the mast cells of the rat intestinal mucosa could be stained metachromatically after fixation in Carnoy's fixative, but not after fixation in formalin. In contrast, the mast cells of the connective tissues stained equally well after both types of fixation (Enerback, 1966a–d). This difference in staining has since been shown to be due to the different proteoglycan content of the two cells in rodents (Razin et al., 1982; Enerback et al., 1985). Furthermore, the studies of Enerback (1966a) showed that intestinal mast cells did not respond to compound 48/80, a property which was not typical of the peritoneal mast cells previously studied with this agent. Thus the mast cells of the mucosa became known as mucosal mast cells (MMCs), or atypical mast cells, whereas those of the peritoneal cavity were termed connective tissue mast cells (CTMCs). An indication that the development and physiological function of these mast cell subtypes may be different came from the observation that MMCs proliferated during parasitic infection (Miller et al., 1986) and are decreased in number with suppression of T cell function (King et al., 1985). In contrast, the numbers of CTMCs are not affected by either of these events. These observations strongly suggest that the MMC is associated functionally with the immune system, whereas the CTMC is not.

Extrapolation of Enerback's rodent mast cell staining criteria to those of humans by Stroebel and colleagues (Strobel et al., 1981) suggested a similar subdivision of mast cells in humans. However, two further studies, both of which confirmed the presence of formalin-sensitive and formalin-insensitive subpopulations in humans, failed to find an absolute relationship of histochemical subtype with anatomical site, both subtypes, for example, being found in the skin (Befus et al., 1985; Marshall et al., 1987). The relative unsatisfactoriness of differential histochemical methods to distinguish between human mast cell subtypes led to the use of antibodies raised against the two major proteases identified in human mast cells. These are tryptase (Glenner and Cohen, 1960; Schwartz et al., 1981b), which is present in all mast cells, and chymase (Schwartz et al., 1981b; Schechter et al., 1983), which is present in mast cells predominantly found in connective tissues (Irani et al., 1986). However, the realization that, as with staining, the use of anatomical location to define mast cell subtype was unreliable led to it being abandoned in 1989 in favour of immunocytochemical subtyping into MC_Ts (those mast cells which contain only tryptase) and MC_{TC}s (those mast cells which contain both tryptase and chymase). However,

it should be stated here that MC_Ts are preferentially located at mucosal surfaces (Irani et al., 1989b), increase in number in allergic disease (Irani et al., 1987a, 1988, 1989a) and are reduced in number in acquired and chronic immunodeficiency syndromes (Irani et al., 1987b), suggesting that, like rodent MMCs, they are acting as an arm of the immune system. In contrast, MC_{TC}s are found predominantly in submucosal and connective tissues, are not increased in numbers in areas of heavy lymphocytic infiltration (Irani et al., 1988, 1989a) and are not decreased in number in immunodeficiency syndromes (Irani et al., 1987b). The biological role of this cell is less clear but its associations with fibrotic disease (Walls et al., 1990c) and angiogenesis (Kessler et al., 1976; Rakusan and Campbell, 1991; Meininger and Zetter, 1992; Duncan et al., 1992; Sorbo et al., 1994) indicate that it is likely to play a role in tissue reconstruction.

The obvious question which arises at this point is the relationship between immunocytochemical heterogeneity and functional heterogeneity, viz. their responsiveness to secretagogues and their modulation by pharmacological agents. All mast cells bear in their membrane $Fc_\varepsilon RI$, receptors capable of binding with high affinity the Fc portion of IgE. Cross-linkage of two or more IgE molecules to bring their receptors into juxtaposition initiates a sequence of biochemical events which results in degranulation to release H, proteases and heparin, and synthesis of PGD_2 and LTC_4 from the membrane-associated phospholipid, arachidonic acid (AA). When we simulated this in the laboratory by cross-linking the IgE with anti-IgE, we found that mast cells dispersed from different tissues responded in a quantitatively different manner, both in the extent of H release and the time it took to release their H (Lowman et al., 1988c, Lau et al., 1995; Fig. 4.2). This bore no relationship to the predominating mast cell type, MC_T or MC_{TC}. It is likely to represent, however, the effect of the local environment on the maturation and priming of the mast cell. For example, it has been established that stromal cell-derived SCF can prime mast cells for H release (Bischoff and Dahinden, 1992; Columbo et al., 1992; Teixeira and Hellewell, 1993). To examine the heterogeneity of eicosanoid production, we compared dispersed lung, skin, colon mucosa and colon muscle mast cells, all stimulated with optional concentrations of anti-IgE (Campbell and Robinson, 1988; Rees et al., 1988; Robinson et al., 1989; Benyon et al., 1989; Church et al., 1989). The results showed that all the cell types produced equivalent amounts of PGD_2, $100–150$ pmol 10^{-6} mast cells, but lung mast cells produced considerably more leukotriene, around 55 pmol 10^{-6} mast cells, compared with skin mast cells, around 6 pmol 10^{-6} mast cells.

In addition to IgE-dependent stimulation, mast cells may be activated for mediator release by non-immunological secretagogues such as neuropeptides, anaphylatoxins and xenobiotics, including morphine,

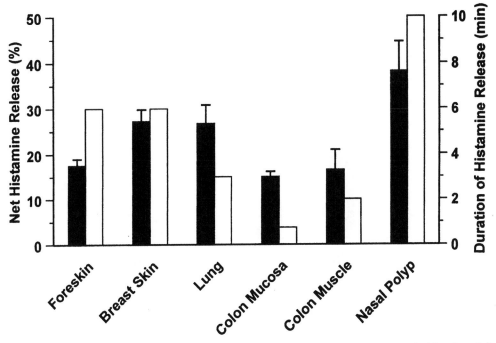

Figure 4.2 The release of H from human mast cells dispersed from different tissues (shaded bars) and the time taken for that H to be released (open bars). This figure was constructed using data from Benyon *et al.* (1987), Lowman *et al.* (198c), Rees *et al.* (1988) and Lau *et al.* (1995).

codeine, muscle relaxants and the H releaser, compound 48/80. Of these, perhaps the most studied is the neuropeptide substance P, which causes a wheal and flare response when injected intradermally into human skin (Hagermark *et al.*, 1978). We have confirmed, using dispersed skin mast cells, that substance P interacts directly with activation sites on the mast cell membrane to induce mediator release (Lowman *et al.*, 1988b). However, we also showed that the characteristics of substance P skin mast cell activation were quite different from those of IgE-dependent activation. Details of these differences are contained in Table 4.2. Of more relevance to this discussion, however, is the relationship between functional and immunocytochemical heterogeneity. Figure 4.3 shows that skin mast cells secrete H in response to substance P, while those of the lung and colon do not (Rees

Table 4.2 Similarities and differences between anti-IgE-induced and substance P-stimulated histamine release from human skin mast cells

	Anti-IgE	Substance P
Spectrum of mediators	Histamine, PGD$_2$, LTC$_4$	Histamine only
Kinetics of histamine release	Slow, around 5 min	Rapid, around 15 s
Effect of immunological densensitization or removal of cell surface IgE	Blockade of cell activation and mediator release	No effect
Effect of the neuropeptide antagonist SPA	No effect	Reduction of cell activation and mediator release
Calcium requirements	Extracellular and intracellular calcium	Intracellular calcium only
Dependency on glucose and oxidative phosphorylation	Complete dependency	Complete dependency
Nature of degranulation	Compound exocytosis	Compound exocytosis

This table was compiled using data from Benyon *et al.* (1986, 1987, 1989) and Church *et al.* (1989).

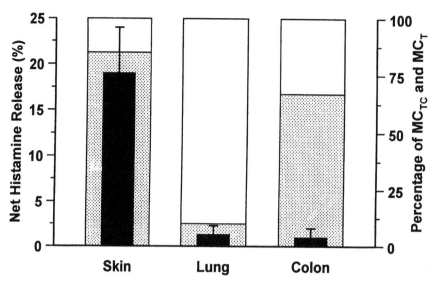

Figure 4.3   H release from human skin, lung and colon mast cells induced by substance P and the approximate proportions of MC$_{TC}$s and MC$_T$s in the mast cell preparations. This figure was constructed using data from Lowman et al. (1988c) and Rees et al. (1988).

et al., 1988). Examination of the protease content of the mast cells shows that the lung contains primarily MC$_T$s, whilst the skin and colon are composed primarily of MC$_{TC}$s. Thus we must conclude that the responsiveness to substance P does not follow immunocytochemical subtyping.

Another example of functional heterogeneity may be explored with the complement anaphylatoxin C5a. When injected into the skin, C5a causes an anti-histamine sensitive wheal and flare response (Wuepper *et al.*, 1972; Swerlick *et al.*, 1988) suggestive of a mast cell H releasing activity. We have recently confirmed this is dispersed human skin mast cells (El Lati *et al.*, 1994). The characteristics of H release are similar to those described for substance P, with the exception that the substance P antagonist, SPA, does not modulate release suggesting that C5a acts at a different cell surface activation site. In comparison with the skin, human lung mast cells do not respond to C5a (Schulman *et al.*, 1988). This work has recently been extended (P. Valent *et al.*, personal communication) by the observations that human skin mast cells express the C5a receptor (CD88) whereas mast cells of the lung and uterus do not. As the majority of the uterine mast cells are of the MC$_{TC}$ subtype, this provides another example of divergence between immunocytochemical and functional heterogeneity.

The final example of functional heterogeneity which we will illustrate relates to drug modulation of mediator release. Using mast cells dispersed from human lung, skin, adenoids, tonsils and colon, we have shown each to be similarly sensitive to β-stimulants and theophylline-like drugs, concentrations of 10^6 M and 5×10^4 M,

respectively, being effective inhibitors of mediator release (Church and Hiroi, 1987; Okayama and Church, 1992). In contrast, glucocorticoids do not inhibit mast cell degranulation or production of newly generated mediators (Schleimer *et al.*, 1983; Cohan *et al.*, 1989). With anti-allergic drugs, such as sodium cromoglycate and nedocromil sodium, heterogeneity is seen. In lung mast cells, high concentrations of both drugs, up to 100 μM, are required to cause modest inhibition. Furthermore, human lung mast cells rapidly become refractory to the modulatory effects of these drugs (Church and Hiroi, 1987; Okayama *et al.*, 1992; Okayama and Church, 1992). A different profile was seen with colonic mast cells (Rees *et al.*, 1988; Okayama *et al.*, 1992), tonsillar mast cells (Okayama *et al.*, 1992) and mast cells obtained from the airways by bronchoalveolar lavage (BAL; Flint *et al.*, 1985a), being both more sensitive to sodium cromoglycate and not exhibiting tachyphylaxis. In contrast, human skin mast cells are refractory to the modulatory actions of anti-allergic drugs (Clegg *et al.*, 1985; Lowman *et al.*, 1988a; Church *et al.*, 1989; Okayama *et al.*, 1992). Thus, again, there is no clear association between immunocytochemical and functional heterogeneity.

From our present knowledge base, it is logical to conclude that human mast cells mature into one or other of two basic phenotypes, the development of MC$_T$s being dependent on factors derived from the immune system while the MC$_{TC}$s appear to be independent of the immune system and probably have a role in tissue reconstruction. However, the expression of cell surface receptors on mast cells appears to be determined by the local environment and to be independent of phenotype.

4. Preformed Mast Cell Mediators

The secretory granule of the human mast cell is a complex matrix, including H, proteases and proteoglycan. Electron microscopy of MC_Ts shows them to have crystalline intragranular matrices which are seen as scrolls, crystals or whorls (Caulfield *et al.*, 1980). In contrast, MC_{TC}s, which contain chymase and carboxypeptidase in addition to tryptase, contain more electron dense granules in which the crystalline structure, although often just visible, is clouded by the sheer volume of protease (Caulfield *et al.*, 1990). Furthermore, the presence of a further crystalline form, the crystal form, has been reported by Dvorak and Rissel (1991) and Dvorak *et al.* (1991). When mast cell activation occurs, the granules swell, lose their crystalline nature as they become solubilized, and are expelled into the local environment by compound exocytosis (Benveniste *et al.*, 1972; Caulfield *et al.*, 1980, 1990) (Fig. 4.4). The granule matrix is then further solubilized by ion exchange, particularly with sodium (Uvnas, 1967), in the extracellular environment.

The mediator most readily associated with the mast cell is the diamine, H. This is synthesized in the mast cell granule by decarboxylation of histidine, mainly by histidine decarboxylase and, to a lesser extent, by the non-specific decarboxylase, dopa decarboxylase (Schayer, 1963; Beaven, 1978). H is stored in the acid pH of the granules at around 100 mM (equivalent to about 1–4 pg cell^{-1}) by ionic linkage with proteoglycans and proteases (Johnson *et al.*, 1980; Lagunoff and Richard, 1983). Once in the extracellular environment, H exerts its potent effects, which include increased vasopermeability between endothelial cells, vasodilatation, contraction of bronchial and gastrointestinal smooth muscle, and increased mucus production. However, H is a short-lived mediator, being rapidly metabolized by histamine-*N*-methyltransferase and diamine oxidase (histaminase), with the former playing the major role in humans.

The backbone of the crystalline mast cell granule is proteoglycan which, in human mast cells, is mainly heparin with chondroitin E (Stevens *et al.*, 1988a; Thompson *et al.*, 1988). Basophils, in contrast, do not contain heparin, the primary proteoglycan of their granule being chondroitin A (Metcalfe *et al.*, 1984). Proteoglycan comprises a single chain peptide core which, in humans, is 17.6 kDa containing a glycosaminoglycan attachment region of alternating Ser-Gly residues (Stevens *et al.*, 1988a). The specific glycosaminoglycans attached to these sites determine the nature of the proteoglycan. The relative homogeneity of proteoglycan throughout all

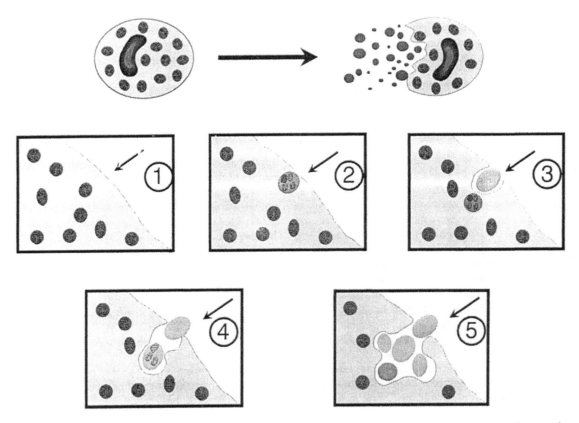

Figure 4.4 The release of preformed mast cell mediators by compound exocytosis. Note the progressive way in which successive granules merge with each other and with the cell membrane.

human mast cells, unlike the rat whose mast cell subtypes express markedly different proteoglycans, explains the inability to extrapolate the distinguishing histochemical techniques (Enerback, 1966c,d) to humans.

Within the granule, the acid nature of the sulphate groups of the glycosaminoglycans provides binding sites for H, neutral proteases and acid hydrolases, which it keeps in an inactive state. Thus the proteoglycan may be viewed as a granular storage matrix. Once released, heparin and, to a lesser effect, chondroitin E may affect the stability or function of other mast cell mediators, e.g. its stabilization of the active tetramer of tryptase (Schwartz and Bradford, 1986). Other actions of these proteoglycans include anti-coagulant, anti-complement and anti-kallikrein effects, the ability to sequester eosinophil major basic protein, enhancement of binding of collagen to fibronectin, and numerous growth factor-enhancing activities.

The major mast cell protease, present in all mast cells regardless of subtype, is tryptase. Tryptase is a tetrameric serine protease of around 130 kDa (Schwartz et al., 1981a; Smith et al., 1984; Walls et al., 1990a) which is stored in a fully active form in the granule (Alter et al., 1990). When released into the extracellular environment, the neutral pH allows tryptase to become enzymatically functional. While there appear to be no endogenous inhibitors of tryptase, it is likely to be an enzyme with a very local effect only. This is because in the absence of heparin, tryptase rapidly dissociates into inactive monomers with altered secondary and ternary structures (Schwartz and Bradford, 1986; Schwartz et al., 1990). The difficulty of extracting active tryptase from tissues, its relative instability, and the failure, as yet, to clone the enzyme in its fully active tetrameric form has meant that our knowledge of its function is limited. In early studies, it was shown to cleave the bronchodilator peptides, vasoactive intestinal polypeptide (VIP), peptide histidine methionine (PHM) and the vasodilator calcitonin gene-related peptide (CGRP; Tam and Caughey, 1990; Walls et al., 1992b). These effects, together with its failure to cleave the bronchoconstrictor, substance P, have led to suggestions that an imbalance of peptides may be a factor in bronchial hyperresponsiveness. Furthermore, the report by Sekizawa et al. (1989) that tryptase sensitizes bronchial smooth muscle to contractile agents, strengthens the view that tryptase may play a role in asthma. Further actions of tryptase include a kallikrein-like activity (Proud et al., 1988; Walls et al., 1992a), cleavage of matrix components, including 75 kDa gelatinase/type IV collagen, fibronectin (Lohi et al., 1992) and type VI collagen (Kielty et al., 1993), and activation of stromelysin which may, in turn, cleave other matrix components (Gruber et al., 1989). In addition, tryptase is a mitogen for fibroblasts (Ruoss et al., 1991; Hartmann et al., 1992) and in a human epithelial cell line may induce its proliferation, stimulate it to release the granulocyte chemoattractant IL-8 and up-regulate its

expression of intercellular adhesion molecule-1 (ICAM-1; Cairns et al., unpublished observations).

Chymase is a monomeric protease of 30 kDa (Fraki and Hopsu-Havu, 1972; Wintroub et al., 1986a; Urata et al., 1990) which is stored in the same secretory granules as tryptase (Craig et al., 1988), but is released in a macromolecule complex distinct from tryptase (Goldstein et al., 1992), suggesting that there is no physicochemical bonding between them. It is the presence of chymase in a subset only of mast cells that has allowed the differentiation into MC_Ts and MC_{TC}s. Within the lung, there are a negligible number of MC_{TC}s in the epithelial layer, but they are found in large numbers close to submucosal glands and, to a lesser extent, associated with smooth muscle (Matin et al., 1992). Like tryptase, chymase is stored within the granule in its fully active form (Huntley et al., 1985; Goldstein et al., 1992) so that is needs no further processing before release. The enzymatic activities of chymase include the degradation of neurotensin (Kinoshita et al., 1991), but not substance P or VIP (Urata et al., 1990) and the cleavage of angiotensin I (AI) to angiotensin II (AII; Reilley et al., 1982; Powers et al., 1995; Urata et al., 1990; Kinoshita et al., 1991). In fact, chymase is more active than angiotensin-converting enzyme (ACE) in this activity which has led to much interest in chymase from cardiac mast cells. Chymase may also contribute to the purported role of mast cells in tissue remodelling by cleaving type IV collagen (Sage et al., 1979) and splitting the dermal–epidermal junction (Briggman et al., 1984). Actions pertinent to mucosal inflammation include the activation of interleukin-1β (IL-1β) to IL-1 (Mizutani et al., 1991), the degradation of IL-4 (Tounon de Lara et al., 1994) and the stimulation of secretion from cultured submucosal gland cells (Sommerhoff et al., 1989).

Carboxypeptidase is a unique 34.5 kDa metalloproteinase which removes the carboxyl terminal residues from a range of peptides, including angiotensin, leu^5-enkephalin, kinetensin, neuromedin N and neurotensin (Goldstein et al., 1989, 1991; Bunnett et al., 1992). As carboxypeptidase is released together with chymase, cathepsin G and proteoglycan in a 400–500 kDa complex, it is likely to act in concert with the other enzymes to degrade proteins.

5. Newly Generated Mediators

Immunological activation of mast cells induces the liberation of AA within the membrane. This phospholipid is then rapidly oxidized down either of two pathways: the cyclooxygenase pathway to form PGD_2 or the lipoxygenase pathway to form LTC_4. These are the only two eicosanoids made by the human mast cell (Robinson et al., 1989).

PGD_2 is a potent bronchoconstrictor agent which is

rapidly degraded to another bronchoconstrictor agent, $9\alpha,11\beta$-PGF_2 (Hardy et al., 1985; Liston and Roberts, 1985; Beasley et al., 1987). Both of these substances are thought to exert the majority of their actions by the occupation of thromboxane (TX) receptors (Beasley et al., 1989b). In addition to its bronchoconstrictor effect, PGD_2 is chemokinetic for human neutrophils (Goetzl and Pickett, 1981), augments LTB_4-induced neutrophilia in the skin (Soter et al., 1983), and is a powerful inhibitor of platelet aggregation.

The physiological effects of the sulphidopeptide leukotrienes, including LTC_4, are potent contraction of bronchial smooth muscle, contraction of arterial, arteriolar and intestinal smooth muscle, enhanced permeability of post-capillary venules and enhanced bronchial mucus secretion. Because of their potent effects on the airways, leukotrienes have long been regarded as important molecules in the pathogenesis of asthma. Because of this, and the multicellular provenance of leukotrienes many comprehensive reviews have been written on these molecules to which the reader is referred (Ford et al., 1994; Chanarin and Johnston, 1994; Valone et al., 1994).

6. Mast Cell Cytokines

It is now established that murine mast cells are a source of several multifunctional cytokines. Initial studies demonstrated that many Abelson murine leukaemia virus (Ab-MuLV)-transformed mouse mast cell lines constitutively express granulocyte–macrophage colony-stimulating factor (GM-CSF) and IL-4 mRNA, and release GM-CSF and IL-4 bioactivity (Chung et al., 1986; Brown et al., 1987). A few lines also express mRNA for IL-3. In addition, Brown et al. (1987) demonstrated constitutive expression of IL-4 mRNA in non-transformed IL-3-dependent mast cell lines, although they could not detect IL-4 bioactivity in the cell supernatants. Further studies demonstrated that in response to $Fc_{\epsilon}RI$ activation, non-transformed murine mast cell lines or primary cultures of bone marrow-derived mast cells have the potential to synthesize and secrete many cytokines including IL-1, IL-2, IL-3, IL-4, IL-5, IL-6, GM-CSF, interferon γ (IFNγ), and three members of the intercrine family of cytokines, macrophage inflammatory protein (MIP)-1α, MIP-1β, and T cell activation antigen 3 (TLA-3), (Plaut et al., 1989; Wodnar-Filpowicz et al., 1989; Burd et al., 1989). Gorden et al. (1990) subsequently demonstrated that mast cells grown in vitro or freshly isolated mouse peritoneal mast cells constitutively contained tumour necrosis factor α (TNFα) bioactivity and could be induced by IgE-dependent stimulation to generate high levels of TNFα mRNA. The observations in the above experiments that not all mast cell lines and cultures produced the same pattern of cytokines, and that within

primary cultures of bone marrow-cultured mast cells (BMCMCs) not all cells produce the cytokine mRNA (Gurish et al., 1991) raises the question of whether mast cell heterogeneity exists with regard to cytokine production in addition to protease production.

Information on human mast cells is more limited. Klein et al. (1989) showed that degranulation of mast cells in human skin organ cultures induced expression of endothelial leucocyte adhesion molecule-1 (ELAM-1) on vascular endothelial cells, an effect which was abrogated by prior administration of either cromolyn sodium, an inhibitor of mast cell mediator release, or neutralizing antibodies to TNFα. The same group subsequently demonstrated that human dermal mast cells contain sizeable quantities of TNFα within granules, which can be released rapidly upon degranulation (Walsh et al., 1991b). Storage of TNFα, with IgE-dependent release and mRNA induction, has also been demonstrated in human lung mast cells (Ohkawara et al., 1992).

Studies from our own group have now demonstrated that human mast cells are also a source of IL-4, IL-5 and IL-6. Human mast cells isolated from human skin and lung contain immunoreactivity for IL-4 which is lost following IgE-dependent activation, indicative of release (Bradding et al., 1992). IL-4 mRNA is not expressed constitutively but is induced following 20 h of culture with SCF and anti-IgE (Church et al., 1994). The presence of IL-4 in mast cells in situ has been confirmed using immunohistochemistry on nasal and bronchial biopsies from normal subjects, patients with allergic perennial and seasonal rhinitis and allergic asthma (Bradding et al., 1992, 1993, 1994, 1995). Using two monoclonal antibodies (mAbs) to human IL-4 which identify different epitopes, an interesting pattern has emerged. Mast cells staining with one mAb to IL-4 (3H4) are clearly increased in number in these allergic mucosal diseases compared to normal subjects. Furthermore this mAb gives a pericellular staining pattern in disease suggesting it may detect secreted IL-4 (Bradding et al., 1993, 1994). Furthermore, strong correlations between 3H4[+] mast cell numbers and eosinophil numbers are present in both allergic asthma and allergic rhinitis. Expression of IL-4 immunoreactivity with this anti-IL-4 mAb is clearly suppressed by the potent topical corticosteroid fluticasone propionate following natural seasonal allergen exposure (Bradding et al., 1995). In contrast a second monoclonal antibody (4D9) gives a granular cytoplasmic staining pattern in both normals and patients, and stains the same number of cells in both normals and patients, suggesting that it detects stored IL-4 (Bradding et al., 1993, 1994). These studies have also identified IL-5, IL-6 and TNFα localized to mast cells although the percentage of mast cells staining for these cytokines is significantly less than for IL-4. We have recently found IL-5 mRNA in purified mast cells preincubated with SCF for 16 h and then challenged with anti-IgE (Okayama and Church, 1995), while Ying et al. (1994) have described expression of mRNA for IL-4 and IL-5 in about 20% and 10%, respectively, of nasal mucosal mast cells following allergen challenge. We have found no difference in the number of mast cells staining for IL-5 or IL-6 in allergic mucosal disease, but in asthma there was a significant increase in the number of TNFα-positive cells (Bradding et al., 1994). IL-4-positive mast cells are also increased in the nasal and bronchial epithelium of patients with rhinitis and asthma, respectively, but IL-5-, −6 and TNFα-positive cells are only present in the submucosa. Overall in both the nasal mucosa and bronchus, approximately 75%, 10%, 35% and 35% of mast cells contain IL-4, IL-5, IL-6 and TNFα, respectively. Some mast cells contain combinations of cytokines while others appear to produce only one cytokine. These observations, plus the predominance of IL-6 positive cells compared to others amongst the submucosal glands, suggest that heterogeneity exists among human mast cells with regards to cytokine production in addition to the well-established heterogeneity of protease production.

IL-8 is a further cytokine which has now been identified as a product of human mast cells. Möller et al. (1993) showed that mast cells in human skin cultures express IL-8 immunoreactivity after 16 h incubation with anti-IgE but do not express it constitutively. This lack of stored IL-8 may explain the apparent discrepancy with our own investigation of IL-8 immunoreactivity in the nasal mucosa of normal subjects and patients with allergic perennial rhinitis (McNamee et al., 1991), where IL-8 was localized predominantly to the nasal epithelium and was not found in mast cells.

6.1   REGULATION OF MAST CELL CYTOKINE PRODUCTION

The studies of mast cell cytokine production described above have shown that maximal induction of cytokine synthesis and release usually occurs in response to IgE-dependent activation. In common with many cell types, there is evidence that FcεRI on mast cells is coupled to the phospholipase C effector system that controls two distinct signal transduction pathways, one regulated by Ca^{2+} ions and the other by protein kinase C (PKC). Exocytotic degranulation is associated with an increased cytoplasmic level of Ca^{2+} ions, and activation of mast cells can be therefore achieved by the use of calcium ionophores which raise intracellular calcium concentrations through a receptor-independent mechanism. Alternative mast cell stimuli include phorbol-12-myristate-13-acetate (PMA) which activates PKC and induces mediator secretion from basophils and rodent mast cells but not from human mast cells, and concanavalin A (Con A), a lectin which can stimulate mast cells by cross-linking of cell-bound IgE and/or cell surface glycoproteins.

In addition to triggering with anti-IgE, most studies of murine mast cells have demonstrated that activation with calcium ionophores also results in significant cytokine

synthesis and release (Plaut et al., 1989; Wodnar-Filpowicz et al., 1989; Burd et al., 1990), the study by Burd et al. (1989) being the exception to this. PMA induced high levels of TNFα mRNA, and some GM-CSF and IL-6 mRNA, but failed to induce IL-1 or IL-3 mRNA (Wodnar-Filpowicz et al., 1989; Burd et al., 1989; Gordon et al., 1990). Subsequently Burd et al. (1990) demonstrated that in BMCMCs the calcium ionophore A-23187 or PMA could in fact induce IL-1, TCA3, MIP-1α and MIP-1β in addition to IL-6. Interestingly, mast cell activation with A23187 resembled that seen with IgE-dependent stimulation and was associated with H release, while PMA cytokine induction occurred independently of mast cell degranulation. This study also found that induction of mRNA for IL-2, IL-3 and IL-5 required the synergistic effects of both PMA and A23187. Whether the differences in these two studies by Burd et al. (1989, 1990) reflect the different mast cell types studied or methodological problems is unclear. Furthermore, Burd et al. (1989) showed that Con A induced IL-6 mRNA and the release of IL-6 bioactivity, but failed to have any effect on IL-1 or IL-4.

The kinetics of mRNA induction and cytokine release also suggests there may be some differences in the regulation of the various mast cell cytokines. Gordon et al. (1990) showed that TNFα mRNA is rapidly induced following IgE-dependent activation reaching maximal levels at 30–60 min, while levels of mRNA for IL-3 and IL-5 in the same cells were highest at about 2 h. In contrast, Wodnar-Filpowicz et al. (1989) found that IL-3 and GM-CSF mRNA were readily detectable by 30 min after activation and had decayed by 2 h, but also observed that secreted IL-3 peaked at 30 min while secreted GM-CSF peaked at 2 h. Wershil et al. (1990) demonstrated that in an IL-3-independent mast cell line, following activation with the calcium ionophore ionomycin, IL-3 and IL-4 mRNA were both detectable after 30 min, but that IL-3 mRNA continued to increase over a 24 h period while IL-4 mRNA peaked at 4 h and had significantly diminished by 24 h. Secreted bioactivity for these two cytokines paralleled the changes in mRNA. Finally Burd et al. (1990) reported that mRNA for IL-1 and IL-6 was detectable within 30 min following mast cell activation with anti-IgE or A23187, but was not detected until 60–90 min following induction with PMA. Our own studies with human mast cell cells suggest that there are differences in the kinetics of mRNA expression for IL-4 and IL-5. Challenge with anti-IgE leads to transient expression of IL-4 mRNA after 6 h, while IL-5 mRNA appears after only 2 h and remains detectable over a further 24 h (Okayama and Church, 1995).

Evidence also exists that transcriptional and post-transcriptional regulation of mast cell cytokine mRNA may differ for individual cytokines. Thus Wodnar-Filpowicz et al. (1989) demonstrated that in a murine mast cell line, induction of IL-3 mRNA with A23187 was controlled primarily at the post-transcriptional level, while induction of GM-CSF mRNA was subject to both transcriptional and post-transcriptional regulation.

How many of these apparent differences in the signal transduction and kinetics for the expression of individual cytokines are real or secondary to differing experimental conditions awaits further clarification. It is also likely that some of these differences are characteristics of the genes in general rather than specifically limited to their expression in mast cells. However it has recently been shown that murine Ab-MuLV-transformed and IL-3-dependent mast cell lines contain an active transcription enhancer in the second intron of the IL-4 gene which is inactive in a murine IL-4-producing T cell line (Henkel *et al.*, 1992). This suggests that the intronic enhancer is mast cell specific and that regulation of IL-4 transcription in mast cells and T cells may be distinct.

Overall, these results suggest that different regulatory mechanisms may exist in mast cells for the transcription of various cytokines. If these findings apply *in vivo*, then stimuli which favour either activation of PKC or mobilization of intracellular calcium may have distinct effects on the subsequent profile of cytokines released, and hence the type of inflammatory response. Furthermore if heterogeneity exists amongst mast cells in the cytokines they produce, this would provide further evidence that there are subsets of mast cells with specific effector roles. It must be stressed however that the full relevance of these findings to human mast cells is yet to be determined.

6.2   NON-B, NON-T CELL/BASOPHIL CYTOKINES

In mice, a population of splenic and bone marrow non-B, non-T cells expressing $Fc_\epsilon RI$ are capable of producing IL-4, and in the case of the former, also IL-3, following IgE-dependent activation (Ben-Sasson *et al.*, 1990; Conrad *et al.*, 1990). The production of IL-4 by these cells is potentiated by either preincubation with IL-3 or prior infection with *Nippostrongylus brasiliensis*. Subsequent characterization of these IL-4-producing cells has demonstrated that they are negative for the expression of c-*kit* and resemble basophils morphologically, suggesting that they are of basophil lineage (Seder *et al.*, 1991a,b). Human bone marrow non-B, non-T cells produce IL-4 and IL-5 mRNA, and release IL-4 but not IL-5 protein, following exposure to anti-IgE (Piccinni *et al.*, 1991). The induction of IL-4 mRNA occurred after 48 h and was potentiated by IL-3.

A recent study by Brunner *et al.* (1993) has confirmed that mature human basophils synthesize and release IL-4. For optimal secretion, priming with IL-3 for 18–48 h before activation with anti-IgE was required, although low amounts of IL-4 were occasionally detected following incubation with either IL-3 or anti-IgE alone. No spontaneous IL-4 production was detected, and other cytokines known to enhance basophil mediator release, namely IL-5, GM-CSF and nerve growth factor (NGF),

were inactive. IgE-dependent synthesis and release of IL-4 by IL-3-primed basophils occurred within 2–6 h, and no IL-4 was stored preformed. Stimulation of basophils with a combination of PMA and ionomycin, or ionomycin alone, resulted in IL-4 production of greater or comparable magnitude to that induced by sequential stimulation with IL-3 and anti-IgE. Arock *et al.* (1993) have subsequently confirmed that normal basophils produce IL-4 following activation, and in addition have shown that leukaemic basophils express IL-4 constitutively.

6.3   BIOLOGICAL PROPERTIES OF HUMAN MAST CELL CYTOKINES

Each of the cytokines so far identified as human mast cell products are likely to be involved in the pathogenesis of allergic mucosal inflammation. IL-4 activates B cells for immunoglobulin secretion through up-regulation of cell surface major histocompatibility complex (MHC) class II antigen (Rousset *et al.*, 1988), CD23 ($Fc_\epsilon RII$; Vercelli *et al.*, 1988) and CD40 (Clark *et al.*, 1989), and plays a pivotal role in the isotype switching of B cells to IgE synthesis (Del Prete *et al.*, 1988). IL-4 specifically increases expression of vascular cell adhesion molecule-1 (VCAM-1; Thornhill and Haskard, 1990) involved in the VLA-4-dependent recruitment of T cells and eosinophils (Thornhill *et al.*, 1991; Schleimer *et al.*, 1992), and induces the expression of the low affinity IgE receptor ($Fc_\epsilon RII$, CD23) on monocytes (te Velde *et al.*, 1988). In addition, IL-4 induces fibroblast chemotaxis (Postlethwaite and Seyer, 1991) and collagen secretion (Postlethwaite *et al.*, 1992). Possibly the most important effect of IL-4 is its ability to induce the development of the T_H2 phenotype of T cells (Le Gros *et al.*, 1990; Swain *et al.*, 1990), which itself produces IL-4, IL-5 and IL-6 (reviewed by Romagnani, 1991). The presence of IL-4 at the onset of an immunological response may therefore dictate whether a cell-mediated (T_H1) or humoral (T_H2) response develops.

The effects of IL-5 in humans are almost exclusively limited to eosinophils. It is a growth and differentiation factor (Clutterbuck *et al.*, 1989), and activator (Lopez *et al.*, 1988) for eosinophils, and in addition prevents their programmed cell death, prolonging survival (Yamaguchi *et al.*, 1991). IL-5 promotes eosinophil adhesion to vascular endothelium through a CD11/CD18-dependent mechanism (Walsh *et al.*, 1991a), and both primes eosinophils for chemotaxis in response to other mediators (Warringa *et al.*, 1992a) as well as being directly chemotactic itself (Wang *et al.*, 1989). As a consequence, IL-5 is considered as a pivotal cytokine in allergen- and parasite-mediated eosinophilic responses.

IL-6 has activities on a wide range of cellular processes including T cell activation (Tosato and Pike, 1988) and stimulation of immunoglobulin production by B cells, and thus it enhances IL-4-dependent IgE synthesis (Jabara *et al.*, 1988). IL-6 is also the most important cytokine responsible for the production of acute-phase proteins by hepatocytes during inflammatory responses. IL-8 belongs to the intercrine family of cytokines and is secreted by a wide variety of cells including T cells, macrophage/monocytes, endothelium and epithelial cells (Schroder and Christophers, 1989; Smyth et al., 1991; Standiford et al., 1991; Marini et al., 1992). It is a potent chemoattractant for neutrophils (Kunkel et al., 1991) and is also chemotactic for eosinophils after priming with IL-3, IL-5 or GM-CSF (Warringa et al., 1991, 1992).

TNFα (cachectin) is another cytokine implicated in the pathogenesis of asthma. When administered by inhalation or intravenously to animals TNFα increases bronchial responsiveness (Kips et al., 1992; Wheeler et al., 1990). It is a chemoattractant for neutrophils and monocytes (Ming et al., 1987), increases microvascular permeability, and enhances both mast cell mediator release (Van Overveld et al., 1992) and eosinophil cytotoxicity (Silberstein and David, 1986; Slungaard et al., 1990). In addition, it has the capacity to up-regulate the leucocyte endothelial cell adhesion molecules E-selectin, VCAM-1 and ICAM-1 (Bevilacqua et al., 1989; Osborn et al., 1989; Leung et al., 1991) involved in the recruitment of neutrophils, eosinophils, monocytes and T cells into inflammatory zones. TNFα also stimulates fibroblast proliferation and secretion of matrix proteins, collagenase and cytokines including IL-6 (Dayer et al., 1985; Kohase et al., 1986; Postlethwaite and Seyer, 1990; Mielke et al., 1990; Warringa et al., 1992).

7.   Mast Cells and Basophils in Disease

It is recognized that in addition to contributing to the pathophysiology of allergy, mast cells also play important roles in a number of other physiological, pathological and immunological processes including tissue remodelling, wound repair, pathological fibrosis, angiogenesis, clotting and host reactions to certain neoplasms. The ubiquitous distribution of mast cells throughout connective tissues, along epithelial surfaces and in close proximity to blood vessels, makes their products available to a large number of different cell types including fibroblasts, glandular epithelial cells, nerves, vascular endothelial cells, smooth muscle cells and other cells of the immune system. This tissue distribution, and the vast array of lipid mediators, proteases, proteoglycans and cytokines identified as potential products of murine and human mast cells, explains how this type of cell could mediate so many diverse effects.

7.1   ALLERGIC MUCOSAL INFLAMMATION

7.1.1   Asthma

Asthma is characterized by the presence of bronchial

obstruction which is potentially reversible either spontaneously or with pharmacological intervention, and is manifest clinically by wheeze, dyspnoea, chest tightness and cough. Exacerbations may be triggered by a number of different stimuli, one or more of which may predominate in any individual. The major pathological processes involved in the airflow obstruction of asthma are mucosal oedema due to increased vascular permeability, smooth muscle contraction, excessive mucus secretion and airway inflammation.

The presence of airway inflammation is thought to underlie much of the altered airway physiology of asthma. Severe inflammatory changes in relationship to asthma death have been recognized for many years (Huber and Koessler, 1922; Dunnill, 1960; Hogg, 1977), but the recent development of fibreoptic bronchoscopy as a research tool has allowed histological investigation of bronchial biopsies from patients with milder disease (Beasley et al., 1989a). Interestingly in mild asthmatics and even in those who are asymptomatic, significant inflammation is present with eosinophil infiltration and activation. Mast cell numbers are not increased but there is morphological evidence of continuous degranulation (Beasley et al., 1989a; Djukanovic et al., 1990). Basophils are rarely seen (Beasley et al., 1989a). There is also epithelial disruption, and type 3 and 5 collagen deposition in the sub-basement membrane (Roche et al., 1989).

Fibreoptic bronchoscopy has also allowed the study of BAL from asthmatic subjects, and again gives evidence that active inflammation is present (Beasley et al., 1989a). Several groups have shown increased numbers of mast cells, eosinophils and lymphocytes in BAL from unchallenged asthmatics compared to normal controls (Flint et al., 1985a; Casale et al., 1987; Kirby et al., 1987; Godard et al., 1987; Kelly et al., 1989) and increased levels of inflammatory mediators such as H (Casale et al., 1987; Kirby et al., 1987) and LTE_4 (Lam et al., 1988). Levels of the mast cell-specific protease tryptase are also elevated providing further evidence of ongoing mast cell degranulation (Wenzel et al., 1988).

7.1.1.1   Experimental Allergen-induced Asthma

Approximately 90% of subjects with asthma under the age of 30 are atopic (Smith, 1974), most frequently with reactions to the house dust mite *Dermatophagoides pteronyssinus* or *D. farinae*, whereas in subjects developing asthma for the first time over the age of 40, the prevalence of atopy is no greater than in the general population. The recognition of the early and late asthmatic reaction (EAR and LAR, respectively) in response to bronchial allergen challenge has been used as a model with which to gain further information on the pathological processes involved in clinical asthma. Similar responses may also be seen in the skin and nose following intracutaneous and nasal allergen challenge, respectively, in atopic subjects.

Following allergen challenge, up to 50% of asthmatic subjects exhibit a dual bronchoconstrictor response. There is a rapid tall in pulmonary function (e.g FEV_1) at 10–20 min which gradually recovers over the Following 2 h, defined as the EAR. Between 4 and 6 h there is a further fall in FEV_1, the LAR. This may last up to 12 h and in some individuals may be followed by recurring airway obstruction for several days or even weeks (Booij Noord et al., 1972).

7.1.1.2 The Early Asthmatic Response

During the EAR there is release of a wide range of vasoactive and spasmogenic mediators most of which originate from mast cells resident in the airway mucosa. When IgE bound to the high affinity IgE Fc receptor on mast cells is cross-linked by allergen, a series of membrane and cytoplasmic events utilizing Ca^{2+} and energy-dependent mechanisms culminate in the secretion of preformed granule-derived mediators, and the synthesis and release of newly formed lipid products (Table 4.1). The measurement of mediators in blood, urine and BAL has provided strong evidence that the EAR is predominantly due to the effects of released H, PGD_2 and the suplidopeptide leukotrienes LTC_4, LTD_4 and LTE_4 (slow reacting substances of anaphylaxis), the latter two being generated from LTC_4 extracellularly (Arm and Lee, 1990). Attenuation of the EAR following administration of specific receptor antagonists to these compounds provides further evidence of their role (Rafferty et al., 1987; Beasley et al., 1989a; Taylor et al., 1991).

In addition to these autacoids, other mast cell products are likely to contribute to the EAR. Tryptase and chymase are preformed proteases specific for mast cells (Schwartz et al., 1981a; Wintroub et al., 1986b) which are also released following IgE cross-linkage. Tryptase levels are raised at baseline in asthmatic subjects, and rise during the EAR following allergen provocation (Wenzel et al., 1988). Tryptase may generate C3 and bradykinin from their protein precursors, which may act as secondary mediators of smooth muscle contraction and vascular permeability, whilst chymase is a potent secretagogue for bovine airway submucosal glands (Sommerhoff et al., 1989).

The significance of the mast cell as a source of these mediators is supported by the observation that sodium cromoglycate, an inhibitor of mast cell degranulation, completely abolishes the early reaction (Pepys et al., 1968).

7.1.1.3 The Late Asthmatic Response

In contrast to the acute mediator-induced bronchoconstriction and mucosal oedema characteristic of the EAR, there is now considerable evidence that the LAR is associated with inflammatory cell accumulation and activation. This situation is generally considered to be analogous to that seen in chronic airway inflammation, but some caution is needed in extrapolating the results of applying a single large dose of allergen to the airways, in the presence of natural disease, to the natural disease itself. The characteristic cell infiltrating the bronchial mucosa during the LAR is the eosinophil. During the period between the EAR and LAR a transit eosinopenia occurs which is thought to be a reflection of eosinophil recruitment into the lung. BAL recovered during the LAR reveals an increased number of eosinophils and elevated levels of eosinophil basic proteins (De Monchy et al., 1985; Metzger et al., 1987; Diaz et al., 1989; Sedgwick et al., 1991), while bronchial biopsy shows eosinophil infiltration with increased numbers of both total and activated cells (Bentley et al., 1993).

Neutrophil numbers increase transiently following both allergen challenge (Diaz et al., 1989) and exposure in sensitized subjects to occupational agents such as toluene di-isocyanate (TDI; Fabbri et al., 1987) but are not increased in stable asthmatic subjects (Wardlaw et al., 1988; Bradley et al., 1991). Similarly, basophils, which are not usually prominent in the "steady-state" allergic bronchial mucosa (Beasley et al., 1989a) have been recovered from BAL fluid during the LAR, albeit in very low numbers when compared with the number of recovered eosinophils (Liu et al., 1991). Subsequent mediator release and tissue damage following inflammatory cell infiltration and activation are thought to account for the ensuing airway obstruction and associated increase in bronchial hyperresponsiveness which accompanies the LAR.

7.1.2 Mast Cell Mediators and Asthma Pathophysiology

The biological activities of a number of mast cell mediators may explain a number of the pathological features present in asthma (Fig. 4.5).

7.1.2.1 Epithelial Damage

As described above, the epithelium is fragile and denuded even in mild asthmatics. Often the basal cell layer is left intact suggesting the point of weakness is between this layer and the surface epithelium. Whether this abnormality is primary or secondary is unclear although evidence suggests the latter may be the case. Although it has been demonstrated that eosinophil products such as major basic protein (MBP), eosinophil cationic protein (ECP) and eosinophil peroxidase (EPO) are cytotoxic to the respiratory epithelium (Irani et al., 1989b; Ayars et al., 1989), mast cell products could also be important in this process, although there has been little study in this area. Superoxide produced following mast cell degranulation (Henderson and Kaliner, 1978) may generate highly reactive oxygen species such as hydrogen peroxide, the hydroxyl radical (OH^-) and oxygen free radicals which are able to damage cell membranes, and studies in dogs have shown that proteolytic enzymes such as chymase may also weaken intracellular bonds resulting in release of

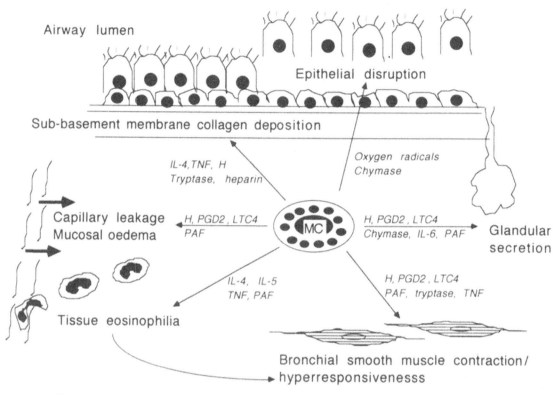

Figure 4.5   The potential contribution of mast cell mediators to the pathology of asthma.

epithelial cells from the basal layer (Briggman *et al.*, 1984).

7.1.2.2 Epithelial Sub-basement Membrane Thickening
A characteristic histological feature of asthma is thickening of the sub-basement membrane due to type III and V collagen deposition in the lamina reticularis (Roche *et al.*, 1989). The most likely origin for this collagen is proliferating myofibroblasts whose number correlates with the collagen thickness (Brewster *et al.*, 1990). The stimulus for this is not yet clear, but is probably under the control of fibrogenic cytokines which include IL-4 and TNFα, and possibly mast cell tryptase, heparin and H. The effects of mast cells on fibroblasts are described in more detail below.

7.1.2.3 Mucosal Oedema and Plasma Leakage
Mucosal oedema contributes to the airway narrowing present in asthma. Contraction of endothelial cells in post-capillary venules leads to the formation of gaps which allow the outflow of plasma. This occurs in response to mediators released from inflammatory cells such as H, PGs, LTs, and platelet activating factor (PAF) which probably act directly on endothelial cells and bradykinin which may act via neural reflexes. In addition to obstructing the airway lumen, exuded plasma may

markedly enhance mucus viscosity. Albumin may increase mucus viscosity through the formation of various protein–glycoprotein complexes (List *et al.*, 1978) which in combination with impaired ciliary motility will promote mucostasis.

7.1.2.4 Mucus Production
Extensive mucus plugging is a characteristic finding in patients dying from asthma, but is also likely to contribute to airways obstruction in stable disease (Cutz *et al.*, 1978). Mucus production in asthmatic airways comes from both hyperplastic goblet cells in the airway epithelium and hypertrophic submucosal glands. The latter are innervated by the parasympathetic nervous system. In addition to cholinergic stimulation, a number of mediators released from inflammatory cells also stimulate mucus secretion. In order of potency these include LTD$_4$, LTC$_4$, hydroxyeicosatetraenoic acid (HETE), prostanoids and H (Shelhamer *et al.*, 1980; Marom *et al.*, 1981, 1982). Canine mast cell chymase is also a potent mucus secretagogue when added to cultures of bovine airway glands (Sommerhoff *et al.*, 1989).

7.1.2.5 Bronchial Hyperresponsiveness
A characteristic pathophysiological feature of asthma is bronchial hyperresponsiveness (BHR), the exaggerated bronchoconstrictor response of the airways to a wide

range of specific and non-specific stimuli (Boushey et al., 1980). Stimuli may act directly, such as the pharmacological compounds H and methacholine, or indirectly such as adenosine, exercise, cold air, bradykinin and sulphur dioxide. BHR is usually expressed as the concentration of H or methacholine required to produce a 20% fall in FEV_1 (PC_{20}). It is on this background of BHR that natural insults such as exercise, cold air or irritant smoke may induce airway obstruction. The cause of airway narrowing is generally assumed to be bronchoconstriction, although oedema and mucus production may contribute. With respect to asthma, several studies have suggested a cause and effect relationship between inflammation and BHR. Following allergen challenge, BHR increases at the time of the late reaction, the time when there is an influx of inflammatory cells, particularly eosinophils, into the airway mucosa. This increase in BHR may last for days or even weeks (Cartier et al., 1982). BAL studies have shown a positive correlation between the number of activated eosinophils and BHR. Furthermore, levels of eosinophil products such as MBP, numbers of epithelial cells, numbers of mast cells, and levels of H in bronchial lavage also correlate with the degree of BHR (Wardlaw et al., 1988; Beasley et al., 1989a).

Several factors are probably related to the development of BHR in the presence of inflammation. The presence of bronchospastic and mucogenic mediators released by a variety of inflammatory cells, including mast cells, either spontaneously or following exposure to allergen probably contribute. TNFα induces BHR in mice and sheep, apparently independent of its effects on inflammatory cell recruitment (Wheeler et al., 1990; Kips et al., 1992). Similarly, tryptase has been shown to induce BHR to H in dogs (Sekizawa et al., 1989). Tryptase may also contribute to the development of BHR and bronchoconstriction through its degradation of bronchodilator neuropeptides (Caughey et al., 1988; Tam and Caughey, 1990).

7.1.3 Allergic Rhinitis

The diagnosis of allergic rhinitis relies on the presence of the classical symptoms of nasal and palatal itch, rhinorrhoea, sneezing and blockage. These symptoms follow exposure to airborne allergens in the environment, due to the presence of specific IgE antibodies demonstrable by skin prick testing. Symptoms may be either seasonal, commonly due to sensitivity to grass, tree or shrub pollens, or perennial, usually due to sensitivity to the house dust mite *Dermatophagoides pteronyssinus* or animal dander, or combinations of allergens. No firm definition of allergic rhinitis exists, and as with asthma, many patients present with symptoms when no identifiable allergy is present (non-allergic rhinitis).

7.1.3.1 Nasal Allergic Mucosal Inflammation
The mucosal inflammatory changes seen in allergic rhinitis are very similar to those described for asthma above.

In patients with seasonal allergic rhinitis (SAR) mast cell numbers increase in the nasal epithelium during the pollen season (Enerback et al., 1986; Howarth et al., 1991), but studies of mast cell numbers in the submucosa have given conflicting results (Viegas et al., 1987; Howarth et al., 1991; Bradding et al., 1995). There is ultrastructural evidence of mast cell degranulation (Howarth et al., 1991) and increased levels of LTC_4 but not H in nasal lavage fluid (Volovitz et al., 1988). In perennial allergic rhinitis (PAR) biopsy studies of the nasal mucosa have demonstrated that mast cell numbers in the submucosa do not differ from normal controls (Drake Lee et al., 1991; Bradding et al., 1993), but are increased in the nasal epithelium (Bradding et al., 1993). Submucosal and epithelial eosinophil numbers are increased in both PAR (Bradding et al., 1993), and SAR during the pollen season (Pipkorn et al., 1988; Lozewicz et al., 1990; Bradding et al., 1995).

7.1.3.2 Experimental Allergen-induced Rhinitis
Nasal challenge with a relevant allergen in patients with a history of allergic rhinitis results in an early-phase response (EPR) and then in about 40% of subjects a late phase response (LPR). During the EPR, typical rhinitic symptoms are associated with the release of a similar spectrum of inflammatory mediators due to mast degranulation, similar to those described above for asthma (Naclerio et al., 1983; Creticos et al., 1984; Castells and Schwartz, 1988). During the LPR which begins at about 6 h post-allergen challenge, symptoms recur in association with tissue infiltration by CD4$^+$ and CD25$^+$ T cells, eosinophils and neutrophils (Varney et al., 1992), and also low numbers of basophils (Bascom et al., 1988). This phase is associated with the release of a second wave of mediators (Naclerio et al., 1985). As mentioned with respect to asthma, the effects of a single dose of allergen delivered in high concentration to the nasal mucosa should probably be interpreted with caution when applying them to chronic disease. A better model for studying the mechanisms of allergic airway inflammation is probably that of Pipkorn et al. (1989) who performed controlled daily nasal allergen challenges for 7 days during winter months to a group of patients with seasonal allergic rhinitis. This study demonstrated eosinophil recruitment after 48 hours which then persisted at the same level for the remainder of the week, and migration of mast cells into the epithelium at Day 6.

As with asthma, the presence of inflammatory cell infiltration and mediator release may explain the associated symptoms. Nasal obstruction results from vascular engorgement induced by H, PGD_2, LTC_4 and LTD_4, and from increased vascular permeability producing mucosal oedema mediated by H, LTC_4 and LTD_4. H produces itching, sneezing and rhinorrhoea by direct mechanisms, whilst the generation of kinins by tryptase and kininogenase stimulates afferent sensory nerve endings to produce sneeze and itch (Howarth et al., 1991).

7.1.4 Do Mast Cells Initiate Late Phase Inflammatory Reactions?

The mast cell contains a virtual pharmacopoeia of biological substances, many of which are capable of inducing the typical pathophysiological changes of bronchial and nasal allergic mucosal disease, and, as described above, there is clear evidence of mast cell degranulation and mediator release in asthma and rhinitis. The recent identification of mast cells as a source of multifunctional cytokines which are released in response to IgE-dependent activation suggests that mast cells may also mediate several aspects of the allergic inflammatory response.

With respect to allergic inflammation, most studies have investigated the role of mast cell cytokines in the rodent model of the IgE-dependent cutaneous LPR. In mice and rats, this LPR is similar to that seen in humans although the infiltrating cells are predominantly neutrophils, which may in part be a reflection of the general eosinopenia present in normal "non-atopic" rodents. Studying mice that are normal, genetically mast cell deficient, and genetically mast cell deficient but subsequently replenished with mast cells from normal mice, has demonstrated that the early and late increase in vascular permeability and tissue swelling following IgE-dependent passive cutaneous anaphlylactic reactions is mast cell dependent (Wershil et al., 1987). Furthermore, nearly all the late leucocyte infiltration is also mast cell dependent (Wershil et al., 1991). Higher levels of TNFα were found in mast cell-reconstituted sites than mast cell-deficient sites (Gordon and Galli, 1991), and the local administration of neutralizing antibodies to TNFα reduced leucocyte infiltration by about 50% (Galli et al., 1990). The recruitment of the remaining leucocytes was likely to be due to other mast cell-derived cytokines. Other studies using this mouse model have also demonstrated that intradermal administration of PMA induces mast cell-dependent neutrophil infiltration (Wershil et al., 1988), while intradermal injection of substance P induces mast cell-dependent eosinophil infiltration (Matsuda et al., 1989). Similarly, in rats, intracutaneous injection of anti-IgE antibodies or compound 48/80, an agent which specifically induces mast cell degranulation, resulted in both EPR and LPR (Tannenbaum et al., 1980). This late-phase cellular infiltration was reproduced by intracutaneous injection of mast cell granules, and was shown to be due to an unidentified granule constituent (Tannenbaum et al., 1980). These studies provide strong evidence that mast cells and their products including TNFα are directly involved in late phase inflammatory cell recruitment in rodent systems.

The relevance of mast cell cytokines to the inflammatory response during parasite infection however is less clear. Nogami et al. (1990) demonstrated that in contrast to T cell deficiency, mast cell deficiency did not prevent marked pulmonary eosinophilia following transnasal infection with Ascaris suum.

Although the role of the mast cell in the early response to allergen is well defined and has been described already, studying mast cell dependency in human LPRs, and particularly in chronic allergic inflammation, is more difficult. It has been recognized for many years that human LPRs are IgE dependent (Dolovich et al., 1973; Solley et al., 1976), but this does not necessarily implicate mast cells solely in the pathogenesis as other inflammatory cells such as macrophages, B cells and possibly eosinophils, can be activated via IgE bound to its low affinity receptor (FcεRII, CD23). Several studies however have demonstrated that intradermal injection of compound 48/80 which produces degranulation of the MC$_{TC}$ subset of mast cells, initiates an LPR associated with inflammatory cell infiltration, therefore suggesting that mast cells are directly involved (Atkins et al., 1973; Solley et al., 1976; James et al., 1981). However, Dolovich et al. (1973) could only induce an immediate skin response with this compound.

The study by Klein et al. (1989), demonstrating that exposure of skin organ cultures to anti-IgE, morphine sulphate, compound 48/80 or A-23187 resulted in expression of E-selectin on endothelial cells within 2 h, and that this effect could be inhibited by prior administration of cromolyn sodium or anti-TNFα antibodies, provided indirect evidence that mast cell-derived TNFα is likely to be involved in cutaneous LPRs. Leung et al. (1991) also reached the conclusion that resident cells within the skin rather than newly recruited cells were responsible for TNFα-induced E-selectin expression following intradermal allergen challenge. Walsh et al. (1991b) subsequently showed that among normal dermal cells, mast cells are the predominant cell type that express both TNFα protein and TNFα mRNA, and that induction of E-selectin is a direct consequence of release of mast cell TNFα.

Taken together. these findings indicate that mast cell-derived TNFα, and probably other mast cell-derived cytokines such as IL-4 and IL-5, are likely to be involved in human LPRs, and therefore by inference, chronic human allergic inflammation. Many resident and recruited inflammatory cells have the potential to secrete further cytokines in response to mast cell cytokine release. This suggests that the response to release of mast cell cytokines may extend far beyond the changes mediated directly by mast cell cytokines themselves, and has led to the concept of the mast cell–leucocyte cytokine cascade (Galli, 1993).

7.1.5 Do Basophils Contribute to the Allergic Inflammatory Response?

The lack of a specific marker for basophils has hampered studying them in allergic diseases, and their relative importance is uncertain. They are only present in very low numbers in peripheral blood, and are not found in normal non-inflamed tissues, indicating that they are recruited to sites of inflammation by mediators from other cell types. Several studies have described the presence of basophils at sites of allergic inflammation using metachromatic staining and morphological criteria. In both nasal and cutaneous LPRs induced in otherwise asymptomatic patients, basophil recruitment has been reported although the number of infiltrating basophils compared to neutrophils and eosinophils is tiny (Bascom *et al.*, 1988; Charlesworth *et al.*, 1989a). In patients with symptomatic allergic rhinitis, small numbers of basophils were reported to be present in nasal secretions, while small numbers of mast cells were seen in nasal mucosal scrapings, but neither was present in normal subjects (Hastie *et al.*, 1979). In patients with symptomatic asthma, however, basophils were almost undetectable in BAL at baseline, but comprised 1% of cells recovered by BAL following allergen challenge, compared to 38% for eosinophils (Heaney *et al.*, 1994). One of the standard morphological criteria for identifying metachromatically staining cells as basophils is the presence of a multilobed nucleus. However, with the introduction of immuno-histochemistry to identify mast cells using mAbs to mast cell-specific tryptase, it has become apparent that some mast cells, particularly immature mast cells (Irani *et al.*, 1992; Dvorak *et al.*, 1993) and those cells residing in the epithelium (Walls *et al.*, 1990c), may also have a lobulated nucleus. This has also been described for up to 15% of mast cells identified by electron microcopy (Kawanami *et al.*, 1985). As the epithelial mast cell population is most likely to account for those cells recovered by lavage procedures, the identification of basophils in the above studies may be an overestimate.

Indirect evidence for the participation of basophils in allergic responses stems from the pattern of mediator release observed during allergen-induced LPRs. In both the nose and the skin, a secondary rise occurs in H and LTC$_4$, but not PGD$_2$, or tryptase (Miller, 1984; Naclerio *et al.*, 1985; Bascom *et al.*, 1988; Shalit *et al.*, 1988; Charlesworth *et al.*, 1989b). It has been suggested that this discriminates between mast cells and basophils as the source of these mediators in the LPR as basophils only produce H and LTC$_4$, while mast cells produce tryptase and PGD$_2$ in addition. Inhibition of this late H rise by corticosteroids also suggests that it may not be due to mast degranulation (Bascom *et al.*, 1988; Charlesworth *et al.*, 1991). It cannot be assumed, however, that mast cells always release their entire armamentarium of mediators on non-immunological stimulation. An example of this is the non-immunological activation of human skin mast cells which causes the preferential release of H, with negligible levels of newly generated prostanoids (Benyon *et al.*, 1989). Furthermore, eosinophils which accumulate in the LPR in large numbers also produce LTC$_4$.

Thus, current evidence suggests that the basophil is involved in allergic inflammatory reactions, but proof will have to await the development of a specific immuno-cytochemical marker for this cell.

7.1.6 IgE Production

As highlighted above, allergic mucosal and cutaneous responses are dependent on the presence of IgE. Until recently it was believed that immunoglobulin secretion by B cells was dependent on two types of signals, the first delivered by various cytokines, and the second delivered by contact with T cells. With respect to IgE synthesis, *in vitro* studies first demonstrated that IL-4 is an essential cofactor required for the production of IgE by B cells, an activity which is antagonized by IFNγ (Del Prete *et al.*, 1988; Pene *et al.*, 1988). IL-4 induces isotype switching to IgE mRNA within B cells (Lebman and Coffman, 1988; Gauchat *et al.*, 1990), but before they can produce IgE they also require a second signal from direct T–B cell contact. This second signal involves either a cognate interaction between B cell MHC class II antigen and the T cell receptor/CD3 complex (Vercelli *et al.*, 1989), or a non-cognate T–B cell interaction (Parronchi *et al.*, 1990) which probably involves B cell activation via CD40 (Spriggs *et al.*, 1992; Castle *et al.*, 1993). The requirement for T cells can be replaced by stimulation of B cells with anti-CD40 (Zhang *et al.*, 1991) or infection of B cells with the Epstein–Barr virus (EBV; Thyphronitis *et al.*, 1989). An alternative signal may also be provided by hydrocortisone (Wu *et al.*, 1991; Nusslein *et al.*, 1992).

The recognition that both mast cells and basophils produce IL-4 suggested that they have the potential to influence IgE production by B cells. Gauchat *et al.* (1993) tested this hypothesis recently using isolated human lung mast cells and peripheral blood basophils. They demonstrated convincingly that both mast cells and basophils express CD40 ligand (CD40L), and are able to provide the cell–cell signal required for IgE production through the interaction of this ligand with CD40 expressed on B cells. Furthermore, in their experimental fluid phase system, basophils induced IgE synthesis without the addition of exogenous IL-4 following stimulation with ionomycin and PMA for 3 h. In contrast, lung mast cells which were incubated with SCF (c-*kit* ligand) in the absence of anti-IgE for 72 h required the addition of exogenous IL-4 in order to induce IgE secretion. These observations, therefore, open a completely new field in mast cell biology.

7.2 MAST CELLS IN INFLAMMATION AND TISSUE REMODELLING/REPAIR

Tissue remodelling in normal healthy tissue is dependent upon continuous turnover of connective tissue elements. This requires dissolution of structural matrix proteins, and laying down of new structural components. Acute tissue injury by immunological, mechanical, physical or chemical stimuli is followed by the acute inflammatory response which classically consists of early vascular changes producing active hyperaemia and oedema, followed later by emigration of leucocytes, which are predominantly neutrophils in the early stages. The close proximity of mast cells to blood vessels suggests that their preformed and newly generated lipid mediators are likely to contribute to this response. It can be seen that this series of events is not dissimilar to those which follow allergen-induced mast cell degranulation. If the tissue insult is short-lived, resolution usually occurs with removal of cellular and tissue debris by phagocytes and secreted enzymes, and repair and remodelling of the extracellular matrix by connective tissue cells. This involves both the dissolution and synthesis of new matrix proteins by recruited fibroblasts, and in large wounds for instance, in growth of new blood vessels (neovascularization). However if the tissue insult persists, the features of chronic inflammation are likely to develop with infiltration by mononuclear cells and the development of tissue fibrosis. Mast cells are ubiquitous in human tissues and release many products that have the potential to influence these acute and chronic inflammatory processes.

Wound healing provides one of the clearest examples of the stages involved in the host response to injury and repair. Primary union of clean surgical wounds begins with the acute inflammatory (exudative) phase of response to tissue injury, with removal of clot and dead tissue by neutrophils and later macrophages. Within 3–5 days capillaries and fibroblasts (granulation tissue) grow into the wound beneath the epithelium, and collagen formed by the fibroblasts begins to bind the wound edges together, reaching a maximum in 2 or 3 weeks. Remodelling of the scar by fibroblasts with lysis and synthesis of collagen may then continue for up to a year. In open gaping wounds which heal by secondary intention, the vascular and fibroblastic proliferation which together make up the granulation tissue are much more abundant, and healing takes much longer.

In experimental skin wounds in rats, mast cells almost "disappear" near the wound edge in the first few days, possibly as a result of degranulation. Mast cell numbers then increase 2-fold over baseline numbers by Day 10, and then gradually return to normal (Wichman, 1955; Persinger et al., 1983). Murine studies have shown that accumulation of mast cells in surgically induced wounds in mast cell-deficient skin grafted onto a normal compatible host is dependent on the migration of stromal cells from the subcutaneous tissue of the normal host into the wound. The mast cells which appear in this situation are derived from circulating precursors, rather than migration of mature cells (Matsuda and Kitamura, 1981). However mast cell deficient mice are able to form scars by secondary intention suggesting that mast cells are not essential for this process (Matsuda and Kitamura, 1981).

As mast cells are normally found in skin, this recruitment of mast cells may simply reflect the restoration of normal tissue homeostasis, although they may play an intermediate role. Mast cells are also found in keloids and hypertrophic scars (Smith et al., 1987), both of which are considered to be abnormal types of wound healing, and it has been suggested that this provides further evidence of their involvement in fibrotic responses.

7.2.1 Mast Cells and Fibrosis

Fibrosis is often a normal response to tissue injury where the damaged cells of an organ are unable to regenerate. An example of this is seen in the heart following myocardial infarction, where the infarcted tissue is replaced by a strong fibrous scar. However there are many pathological conditions where the development of fibrosis is detrimental. The fibrotic tissue response of chronic inflammation is often associated with increased numbers of mast cells, and has been most extensively studied in pulmonary fibrosis and systemic sclerosis.

7.2.2 Pulmonary Fibrosis

Marked mast cell hyperplasia has been described in the fibrous alveolar septae of patients with pulmonary fibrosis resulting from several diverse aetiologies (Kawanami et al., 1985). These mast cells often showed reduced numbers of granules and disorganized granule content. suggesting there was partial ongoing degranulation. Basophils in contrast were rarely seen. In addition, elevated H levels, as well as elevated H releasing factor(s), have been described in BAL fluid from patients with pulmonary fibrosis (Broide et al., 1990). In animal models increased numbers of mast cells and elevated lung H content has been documented in pulmonary fibrosis induced by bleomycin (Goto et al., 1984), radiation (Watanabe et al., 1974) and asbestos (Wagner et al., 1984). In the rat bleomycin model mast cell numbers identified by metachromatic staining with toluidine blue decreased significantly by Day 7 suggesting degranulation, but by Day 14 significant mast cell hyperplasia was apparent and persisted for at least 50 days. A similar pattern was also observed with lung H content. The increased mast cell population was comprised predominantly of the connective tissue type, which as described above, appear to be dependent on fibroblasts for their growth. In the murine model of silica-induced pulmonary inflammation, mast cell hyperplasia is also seen (Suzuki et al., 1993), but mast cell-deficient mice develop significantly less severe lung lesions, and have a reduced BAL fluid neutrophilia and protein content, providing evidence that mast cells contribute to this response.

7.2.3 Systemic Sclerosis (Scleroderma)

Systemic sclerosis is a multi-system connective tissue disorder characterized by intimal proliferation and fibrosis in small arteries and arterioles, and degenerative changes with fibrosis in the skin and certain internal organs. The term scleroderma is best reserved for cutaneous involvement. The aetiology remains unclear although autoimmune mechanisms have been implicated as auto-antibodies are commonly present.

Studies by Hawkins *et al.* (1985) and Nishioka *et al.* (1987) showed increased numbers of mast cells were present in clinically involved skin in early cutaneous systemic sclerosis but not in later disease. Seibold *et al.* (1990) however found increased numbers of mast cells in both involved and uninvolved skin in both early and late disease, and also provided evidence of increased mast cell degranulation in early but not late disease, suggesting that increases in mast cell numbers and degranulation precede clinically apparent dermal fibrosis. A study by Falanga and Julien (1990) found elevated H levels in the plasma of patients with systemic sclerosis, providing evidence of mast cell activation.

Animal models have also provided useful information for the role of mast cells in cutaneous fibrosis. Chronic murine graft-versus-host disease is associated with skin changes very similar to those seen in human systemic sclerosis (Jaffee and Claman, 1983). Over the first 3–4 weeks stainable mast cells gradually "disappear" from the skin due to activation and loss of their granules, before gradually returning (Claman *et al.*, 1985, 1986; Choi *et al.*, 1987; Giorno *et al.*, 1987). The tight skin (TSK) mouse has a genetically transmitted connective tissue disease whose skin lesions resemble those of systemic sclerosis. This cutaneous fibrosis is associated with increased mast cell numbers and increased numbers of degranulated mast cells (Walker *et al.*, 1985). Of particular interest was the observation that treatment of TSK mice with oral sodium cromoglycate was associated with a decrease in degranulation of dermal mast cells, and a significant decrease in the subcutaneous fibrous layer (Walker *et al.*, 1985).

7.3 MAST CELLS, MEDIATORS AND CONNECTIVE TISSUE CELLS

In addition to finding mast cell infiltration and mediator release at sites of wound healing, tissue repair and fibrosis, studies of the effects of mast cell products on the cells involved in these processes, notably fibroblasts and endothelial cells, have supported the hypothesis that mast cells play an active role.

Initial studies with mast cell cultures or granules demonstrated that mast cells appear to have several important effects on connective tissue elements. Subba Rao *et al.* (1983) showed that cultured rat embryonic skin fibroblasts phagocytose rat mast cell granules added to the culture medium or released by co-cultured mast cells, and that this is followed by secretion of collagenase and p-hexosaminidase. Similarly Yoffe *et al.* (1984) demonstrated that granules derived from mast cells purified from dog mastocytomas induced a 10- to 50-fold stimulation of collagenase production by human

adherent rheumatoid synovial cells. Franzen and Norrby (1994) showed that mast cell degranulation in rat mesentery was followed by proliferation of both fibroblasts and mesothelial cells, supported by Roche (1985b) who also found that rat mast cell granules induced fibroblast proliferation.

Evidence also exists for a role for mast cells in angiogenesis. Studies of tumour angiogenesis on the chick chorioallantoic membrane showed an accumulation of mast cells prior to the appearance of new vessels, suggesting that the mast cell might play an intermediate role in tumour angiogenesis (Kessler *et al.*, 1976). Azizkhan *et al.* (1980) showed that mast cell contents stimulate migration of bovine capillary endothelial cells, and attributed this effect to heparin. Roche (1985a) and Marks *et al.* (1986) found that rat mast cell granules were mitogenic for human microvascular endothelial cells, although the first experiment suggested this was predominantly due to heparin, while the second found that this response was primarily due to H. Interestingly endothelial cells have also been seen to phagocytose mast cell granules (Greenburg and Burnstock, 1983).

The identification, isolation and study of specific mast cell mediators has subsequently provided some indication of which of them may be responsible for the above effects. As already mentioned, heparin has been implicated in endothelial cell migration and proliferation. Although it has been suggested that this may be a direct effect, the fact that heparin binds many growth factors, and probably other cytokines, with high affinity, means that some of these effects may actually be indirect. In addition, heparin protects acidic and basic fibroblast growth factor (FGF) from degradation (Gospodarowicz and Cheng, 1986), and releases them from basement membranes where they are stored (Folkman *et al.*, 1988), and may thus potentiate fibroblast activation and endothelial cell proliferation indirectly.

H has also been shown to promote fibroblast growth (Boucek and Nobel, 1973; Jordana *et al.*, 1988) in humans and collagen synthesis in guinea-pigs (Sandberg, 1962; Hatamochi *et al.*, 1985). TNFα induces both endothelial cell (Piguet *et al.*, 1990) and fibroblast proliferation (Vilcek *et al.*, 1986; Piguet *et al.*, 1990), and also stimulates adherent rheumatoid synovial cells and fibroblasts to secrete collagenase (Dayer *et al.*, 1985), a similar response to that seen with intact mast cell granules (Yoffe *et al.*, 1984). IL-4 is a potent mitogen for capillary endothelium (Toi *et al.*, 1991) and also stimulates fibroblast proliferation in mice (Monroe *et al.*, 1988), is a chemoattractant for human fibroblasts (Postlethwaite and Seyer, 1991) and also induces human fibroblasts to secrete collagen types I and III and fibronectin (Postlethwaite *et al.*, 1992).

Mast cell neutral proteases have also been implicated in tissue remodelling. Canine tryptase induces proliferation of Chinese hamster and rat fibroblasts and also potentiates their response to basic FGF (Ruoss *et al.*, 1991).

Through an indirect mechanism tryptase also activates collagenase (Gruber *et al.*, 1989), while mast cell chymase has been implicated in matrix degradation (Briggman *et al.*, 1984).

Taken together, the identification of mast cell hyperplasia and mediator release at sites of tissue fibrosis and wound healing, observations in animal models, and study of the actions of mast cell products, has provided much circumstantial evidence that mast cells are involved in tissue remodelling, healing and fibrosis. It is unlikely that mast cells are essential in these responses, but more likely that they augment them. Complex interactions between different connective tissue components, mast cells and other inflammatory cells are likely to operate, and are unlikely to be fully delineated in humans *in vivo*. It seems reasonable to hypothesize however that initial mast cell mediator release has the potential to activate fibroblasts, which may then promote the recruitment and proliferation of further mast cells, explaining the mast cell hyperplasia often witnessed at sites of chronic inflammation.

7.4 RHEUMATOID DISEASE

Rheumatoid disease is a multisystem connective tissue disease with the inflammatory process characteristically involving several joints. In the early stages of rheumatoid arthritis synovial inflammation is associated with infiltration by macrophages, neutrophils, T cells, B cells and plasma cells. This latter cell type accounts for the local production of rheumatoid factors. As the disease progresses, secondary irreversible changes occur in other joint structures. Subchondral erosions develop at the joint margin and a thin layer of granulation tissue (pannus) grows over the joint cartilage, which is gradually invaded by pannus and eroded by cartilage-degrading enzymes from inflammatory cells (Kobayashi and Ziff, 1975). Ultimately the granulation tissue becomes fibrosed.

Several lines of evidence suggest that mast cells may be involved in the pathogenesis of these particular changes in rheumatoid disease. Mast cell numbers are increased in both synovial fluid (Malone *et al.*, 1986) and rheumatoid synovial tissue including both extraosseus pannus and intracortical invasive tissue (Crisp *et al.*, 1984; Godfrey *et al.*, 1984; Bromley *et al.*, 1984; Gruber *et al.*, 1986). This is associated with an increase in total H content of synovial tissue and increased H concentrations in synovial fluid (Frewin *et al.*, 1986). Tryptase is also detectable in rheumatoid synovial fluid (Brodeur *et al.*, 1991). Following intra-articular administration of glucocorticoids mast cell hyperplasia diminishes in association with improvement in symptoms and clinical signs (Malone *et al.*, 1987). These findings, in addition to the observation that basophils are not present

(Godfrey *et al.*, 1984; Malone *et al.*, 1986), suggest that mast cells which are present are releasing their mediators, and these are likely to contribute to the inflammatory process. These mast cell changes however are not specific for rheumatoid disease, being found in a variety of other arthritides, and may therefore represent a non-specific response to synovial injury.

In a murine model of inflammatory arthritis mast cell-deficient mice developed similar acute and chronic inflammatory response to antigen as their normal littermates, but ultimately developed less cartilage destruction, suggesting mast cells may play a role later in the disease process (van den Broek *et al.*, 1988).

Although IgE rheumatoid factors have been identified in both the serum of patients with rheumatoid arthritis (Zuraw *et al.*, 1981) and rheumatoid synovial fluid (De Clerck *et al.*, 1991; Burastero *et al.*, 1993), it is quite likely that IgE-independent modes of mast cell activation operate in non-allergic inflammatory processes. These might include C3a and C5a generated by activation of the complement cascade (Moxley and Ruddy, 1985; Brodeur *et al.*, 1991) by immune complexes containing rheumatoid factors, substance P, and as yet unidentified mediators of mast cell secretion. From the above descriptions of the different mast cell mediators and cytokines it is easy to imagine how these may participate in the development of joint pathology.

8. Conclusions

From the evidence provided in this review, it can be seen that the mast cell is a multifunctional cell with role in both health and disease. It is also becoming apparent that mast cell precursors may, under the influence of factors in their local environment, mature into cells which are adapted for a particular role within that tissue. In health, it is becoming evident that mast cells may play a major role in angiogenesis and tissue reconstruction. In disease, H, PGD_2 and LTC_4 are obviously crucially involved in the mediation of the early phase response to allergen challenge in many tissues while cytokine generation gives mast cells a likely role in the initiation and maintenance of allergic inflammation (Fig. 4.6). What is more slowly becoming obvious is that mast cell mediators, particularly proteases, may also play a role in inflammatory diseases of connective tissues, for example rheumatoid arthritis and scleroderma.

9. Acknowledgments

P.B. was supported by an MRC Programme Grant.

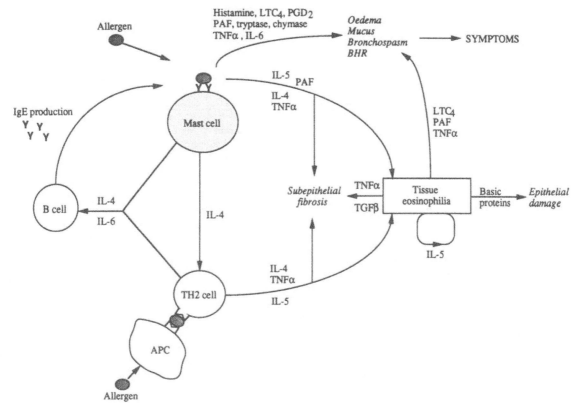

Figure 4.6 The possible position of the mast cell in the initiation and maintenance of allergic mucosal inflammatory responses.

10. References

Alter, S.C., Kramps, J.A., Janoff, A. and Schwartz, L.B. (1990). Interactions of human mast cell tryptase with biological protease inhibitors. Arch. Biochem. Biophys. 276, 26–31.

Arm, J.P. and Lee, T.H. (1990). Lipoxygenase mediators. In "Immunology and Allergy Clinics of North America: Allergic Inflammatory Mediators and Bronchial Hyperresponsiveness" (eds C.W. Bierman and T.H. Lee), pp. 373–381. W.B. Saunders Co., Philadelphia.

Arock, M., Merle-Béral, H., Dugas, B., Ouaaz, F., Le Goff, L., Vouldoukis, I., Mencia-Huerta, J.-M., Schmitt, C., Leblond-Missenard, V., Debre, P. and Mossalayi, M.D. (1993). IL-4 release by human leukemic and activated normal basophils. J. Immunol. 151, 1441–1447.

Atkins, P.C., Green, G.R. and Zweiman, B. (1973). Histologic studies of human skin test responses to ragweed, compound 48/80 and histamine. J. Allergy Clin. Immunol. 51, 263–273.

Ayars, G.H., Altman, L.C., McManus, M.M., Agosti, J.M., Baker, C., Luchtel, D.L., Loegering, D.A. and Gleich, G.J. (1989). Injurious effect of the eosinophil peroxide–hydrogen peroxide–halide system and major basic protein on human nasal epithelium in vitro. Am. Rev. Respir. Dis. 140, 125–131.

Azizkhan, R.G., Azizkhan, J.C., Zetter, B.R., and Folkman, J.

(1980). Mast cell heparin stimulates migration of capillary endothelial cells in vitro. J. Exp. Med. 152, 931–944.

Bascom, R., Wachs, M., Naclerio, R.M., Pipkorn. U., Galli. S.J. and Lichtenstein, L.M. (1988). Basophil influx occurs after nasal antigen challenge: effects of topical corticosteroid pretreatment. J. Allergy Clin. Immunol. 81, 580–589.

Beasley, C.R.W., Robinson, C., Featherstone, R.L., Varley, J.G., Hardy, C.C., Church, M.K. and Holgate, S.T. (1987). 9 alpha, 11 beta-prostaglandin F2, a novel metabolite of prostaglandin D2 is a potent contractile agonist of human and guinea pig airways. J. Clin. Invest. 79, 978–983.

Beasley, C.R.W., Roche, W.R., Roberts, J.A. and Holgate, S.T. (1989a). Cellular events in the bronchi in mild asthma and after bronchial provocation. Am. Rev. Respir. Dis. 139, 806–817.

Beasley, C.R.W., Featherstone, R.L., Church, M.K., Rafferty, P., Varley, J.G., Harris, A. and Holgate, S.T. (1989b). Effect of a thromboxane receptor antagonist on PGD2- and allergen-induced bronchoconstriction. J. Appl. Physiol. 66, 1685–1693.

Beasley, C.R.W., Featherstone, R.L., Church, M.K., Rafferty, P., Varley, J.G., Harris, A., Robinson, C. and Holgate, S.T. (1989c). Receptor antagonism of bronchoconstrictor prostanoids in vitro and in vivo by GR32191: implications for the contribution of these mediators to immediate allergen-induced bronchoconstriction in asthma. J. Appl. Physiol. 66, 1685–1693.

Beasley, C.R.W., Roche. W.R., Roberts, J.A. and Holgate, S.T. (1989d). Cellular events in the bronchi in mild asthma and after bronchial provocation. Am. Rev. Respir. Dis. 139, 806–817.

Beaven, M.A. (1978). Histamine: its role in physiological and pathological processes. Monogr. Allergy 13, 1–10.

Befus, A.D., Goodacre, R., Dyck, N. and Bienenstock, J. (1985). Mast cell heterogeneity in man. I. Histologic studies of the intestine. Int. Arch. Allergy Appl. Immunol. 76, 232–236.

Ben-Sasson, S.Z., Le Gros, G.S., Conrad, D.H., Finkelman, F.D. and Paul, W.E. (1990). Cross-linking Fc receptors stimulates splenic non-B non-T cells to secrete interleukin-4 and other cytokines. Proc. Natl. Acad. Sci. USA 87, 1421–1425.

Bennich, H., Ishizaka, K., Johansson, S.G.O., Rowe, D.S., Stanworth, D.R. and Terry, W.D. (1968). Immunoglobulin E, a new class of human immunoglobulin. Bull. WHO 38, 151–152.

Bentley, A.M., Meng, Q., Robinson, D.S., Hamid, Q., Kay, A.B. and Durham, S.R. (1993). Increases in activated T lymphocytes, eosinophils and cytokine mRNA expression for interleukin-5 and granulocyte/macrophage colony-stimulating factor after allergen inhalation challenge in atopic asthmatics. Am. J. Respir. Cell Mol. Biol. 8, 35–42.

Benveniste, J., Henson, P.M., and Cochrane, C.G. (1972). Leucocyte-dependent H release from rabbit platelets: the role of IgE, basophils and a platelet activating factor. J. Exp. Med. 136, 1356–1377.

Benyon, R.C., Church, M.K., Clegg, L.S. and Holgate, S.T. (1986). Dispersion and characterization of mast cells from human skin. Int. Arch. Allergy Appl. Immunol. 79, 332–334.

Benyon, R.C., Lowman, M.A. and Church, M.K. (1987). Human skin mast cells: their dispersion, purification and secretory characteristics. J. Immunol. 138, 861–867.

Benyon, R.C., Robinson, C. and Church, M.K. (1989). Differential release of H and eicosanoids from human skin mast cells activated by IgE-dependent and non-immunological stimuli. Br. J. Pharmacol. 97, 898–904.

Bevilacqua, M.P., Stengelin, S., Gimbrone, M.A., Jr. and Seed, B. (1989). Endothelial leukocyte adhesion molecule 1: an inducible receptor for neutrophils related to complement regulatory proteins and lectins. Science 243, 1160–1165.

Bischoff, S.C. and Dahinden, C.A. (1992). c-kit Ligand: a unique potentiator of mediator release by human lung mast cells. J. Exp. Med. 175, 237–244.

Bochner, B.S. and Lichtenstein, L.M. (1992). Mechanisms of basophil recruitment in allergic diseases. Clin. Exp. Allergy 22, 973–975.

Booij Noord, H., de Vries. K., Sluiter, H.J. and Orie. N.G.M. (1972). Late bronchial obstructive reaction to experimental inhalation of house dust mite extract. Clin. Allergy 2, 43–61.

Boucek, R.K. and Nobel, N.L. (1973). Histamine, norepinephrine and bradykinin stimulation of fibroblast growth and modification of serotonin response. Proc. Soc. Exp. Biol. Med. 144, 929–933.

Boushey, H.A., Holtzman, M.J., Sheller, R. and Nadel, J.A. (1980). State of the art. Bronchial hyperreactivity. Am. Rev. Respir. Dis. 121, 389–413.

Bradding, P., Feather, I.H., Howarth, P.H., Mueller, R.,

Roberts, J.A., Britten, K., Bews, J.P.A., Hunt, T.C., Okayama, Y., Heusser, C.H., Bullock, G.R., Church, M.K. and Holgate, S.T. (1992). Interleukin 4 is localized to and released by human mast cells. J. Exp. Med. 176, 1381–1386.

Bradding, P., Feather, I.H., Wilson, S., Bardin, P.G., Heusser, C.H., Holgate, S.T. and Howarth, P.H. (1993). Immunolocalization of cytokines in the nasal mucosa of normal and perennial rhinitic subjects: the mast cell as a source of IL-4, IL-5 and IL-6 in human allergic mucosal inflammation. J. Immunol. 151, 3853–3865.

Bradding, P., Roberts. J.A., Britten. K.M., Montefort. S., Djukanovic, R., Heusser, C.H., Howarth. P.H. and Holgate. S.T. (1994). Interleukins (IL)-4,-5,-6 and TNFα in normal and asthmatic airways: evidence for the human mast cell as an important source of these cytokines [Abstract]. Am. J. Respir. Cell Biol. 10, 471–480.

Bradding, P., Feather, I.H., Wilson, S., Holgate, S.T. and Howarth, P.H. (1995). Cytokine immunoreactivity in seasonal rhinitis: regulation by a topical corticosteroid. Am. J. Respir. Crit. Care Med. 151, 1900–1906.

Bradley, B.L., Azzawi, M., Jacobson, M., Assoufi, B., Collins, J.V., Irani, A.A., Schwartz, L.B., Durham, S.R., Jeffery, P.K. and Kay, A.B. (1991). Eosinophils, T-lymphocytes, mast cells, neutrophils and macrophages in bronchial biopsy specimens from atopic subjects with asthma – comparison with biopsy specimens from atopic subjects without asthma and normal control subjects and relationship to bronchial hyperresponsiveness. J. Allergy Clin. Immunol. 88, 661–674.

Brewster, C.E., Howarth, P.H., Djukanovic, R., Wilson, J., Holgate, S.T. and Roche, W.R. (1990). Myofibroblasts and subepithelial fibrosis in bronchial asthma. Am. J. Respir. Cell Mol. Biol. 3, 507–511.

Briggman, R.A., Schechter, N.M., Fraki, J.E. and Lazarus, G.S. (1984). Degradation of the epidermal-dermal junction by a proteolytic enzyme from human skin and human polymorphonuclear leukocytes. J. Exp. Med. 160, 1027–1042.

Brodeur, J.P., Ruddy, S., Schwartz, L.B. and Moxley, G. (1991). Synovial fluid levels of complement SC5b–9 and fragment Bb are elevated in patients with rheumatoid arthritis. Arthritis Rheum. 34, 1531–1537.

Broide, D.H., Smith. C.M. and Wasserman, S.I. (1990). Mast cells and pulmonary fibrosis. Identification of a histamine releasing factor in bronchoalveolar lavage. J. Immunol. 145, 1838–1844.

Bromley, M., Fisher, W.D. and Woolley, D.E. (1984). Mast cells at sites of cartilage erosion in the rheumatoid joint. Ann. Rheum. Dis. 43, 76–79.

Brown, M.A., Pierce, J.H., Watson, C.J., Falco, J., Ihle, J.N. and Paul, W.E. (1987). B cell stimulatory factor-1/interleukin-4 mRNA is expressed by normal and transformed mast cells. Cell 50, 809–818.

Brunner, T., Heusser, C.H. and Dahinden, C.A. (1993). Human peripheral blood basophils primed by interleukin-3 (IL-3) produce IL-4 in response to immunoglobulin E receptor stimulation. J. Exp. Med. 177, 605–611.

Bunnett, N.W., Goldstein, S.M. and Nakazato, P. (1992). Isolation of a neuropeptide-degrading carboxypeptidase from the human stomach. Gastroenterology 102, 76–87.

Burastero, S.E., Lo Pinto, G., Goletti, D., Cutolo, M., Burlando, L. and Falagiani, P. (1993). Rheumatoid arthritis with monoclonal IgE rheumatoid factor. J. Rheumatol. 20, 489–494.

Burd, P.R., Rogers. H.W., Gordon, J.R., Martin, C.A., Jayaraman, S., Wilson, S.D., Dvorak. A.M., Galli, S.J. and Dorf, M.E. (1989). Interleukin–3 dependent and independent cell lines stimulated with IgE and antigen express multiple cytokines. J. Exp. Med. 170, 245–258.

Burd, P.R., Costa. J.J., Metcalfe, D.D. and Siebenlist, U. (1990). Cytokine gene expression occurs via multiple pathways in primary murine bone marrow mast cells (BMMC). FASEB J. 4, A1705.

Campbell, A.M. and Robinson, C. (1988). Further studies on IgE-mediated eicosanoid release from human dispersed lung cells [Abstract]. Br. J. Pharmacol. 95(Suppl), 674P.

Cartier, A., Thomson, N.C., Frith, P.A., Roberts, R. and Hargreave, F.E. (1982). Allergen-induced increase in bronchial responsiveness to H: relationship to the late asthmatic response and change in airway calibre. J. Allergy Clin. Immunol. 70, 170–177.

Casale, T.B., Wood, D., Richerson, H.B., Trapp, S., Metzger, W.J., Zavala, D. and Hunninghake, G.W. (1987). Bronchoalveolar lavage fluid H levels in allergic asthmatics are associated with methacholine bronchial hyperresponsiveness. J. Clin. Invest. 79, 1197–1203.

Castells, M.C. and Schwartz, L.B. (1988). Tryptase levels in nasal-lavage fluid as an indicator of the immediate allergic response. J. Allergy Clin. Immunol. 82, 348–355.

Castle, B.E., Kishimoto, K., Stearns. C., Brown. M.L., and Kehry, M.R. (1993). Regulation of expression of the ligand for CD40 on T helper lymphocytes. J. Immunol. 151, 1777–1788.

Caughey, G.H., Leidig, F., Viro, N.F. and Nadel, J.A. (1988). Substance P and vasoactive intestinal peptide degradation by mast cell tryptase and chymase. J. Pharmacol. Exp. Ther. 244, 133–137.

Caulfield, J.P., Lewis, R.A., Hein, A. and Austen, K.F. (1980). Secretion of dissociated human pulmonary mast cells: evidence for solubilization of granule contents before discharge. J. Cell Biol. 85, 299–311.

Caulfield, J.P., El Lati, S.G., Thomas, G. and Church, M.K. (1990). Dissociated human skin mast cells degranulate in response to anti-IgE and substance P. Lab. Invest. 63, 502–510.

Chanarin, N. and Johnson, S.L. (1994). Leukotrienes as a target in asthma therapy. Drugs 47, 12–24.

Charlesworth, E.N., Hood, A.F., Soter, N.A., Kagey Sobotka, A., Norman, P.S. and Lichtenstein, L.M. (1989a). Cutaneous late-phase response to allergen. Mediator release and inflammatory cell infiltration. J. Clin. Invest. 83, 1519–1526.

Charlesworth, E.N., Kagey-Sobotka, A., Norman, P.S. and Lichtenstein, L.M. (1989b). Effects of cetirizine on mast cell mediator release and cellular traffic during the cutaneous late phase response. J. Allergy Clin. Immunol. 83, 905–912.

Charlesworth, E.N., Kagey Sobotka, A., Schleimer, R.P., Norman, P.S. and Lichtenstein, L.M. (1991). Prednisone inhibits the appearance of inflammatory mediators and the influx of eosinophils and basophils associated with the cutaneous late-phase response to allergen. J. Immunol. 146, 671–676.

Choi, K.L., Giorno. R. and Claman, H.N. (1987). Cutaneous mast cell depletion and recovery in murine graft-vs-host disease. J. Immunol. 138, 4093–4101.

Chung, S.W., Wong, P.M.C., Shen-Ong, G., Ruscetti, S.,

Ishizaka, T. and Eaves, C.J. (1986). Production of granulocyte-macrophage colony-stimulating factor by Abelson virus-induced tumorigenic mast cell lines. Blood 68, 1074–1081.

Church, M.K. and Hiroi, J. (1987). Inhibition of IgE-dependent H release from human dispersed lung mast cells by anti-allergic drugs and salbutamol. Br. J. Pharmacol. 90, 421–429.

Church, M.K., Benyon, R.C., Rees, P.H., Lowman, M.A., Campbell, A.M., Robinson, C. and Holgate, S.T. (1989). Functional heterogeneity of human mast cells. In "Mast Cell and Basophil Differentiation and Function in Health and Disease" (eds S.J. Galli and K.F. Austen), pp. 161–170. Raven Press, New York.

Church, M.K., Okayama, Y. and Bradding, P. (1994). The role of the mast cell in acute and chronic allergic inflammation. In: Cells and Cytokines in Lung Inflammation, 13–21. Edited by Chignard. M., Pretolani, M., Renesto, P. and Vargaftig, B.B. New York Acad. Sci., New York.

Claman, H.N., Jaffee, B.D., Huff, J.C. and Clark. R.A. (1985). Chronic graft-versus-host disease as a model for scleroderma. II. Mast cell depletion with deposition of immunoglobulins in the skin and fibrosis. Cell Immunol. 94, 73–84.

Claman, H.N., Choi, K.L., Suiansky, W. and Vatter, A.E. (1986). Mast cell "disappearance" in chronic murine graft-vs-host disease (GVHD)-ultrastructural demonstration of "phantom mast cells". J. Immunol. 137, 2009–2013.

Clark, E.A., Shu, G.L., Luscher, B., Draves, K.E., Banchereau, J., Ledbetter, J.A. and Valentine, M.A. (1989). Activation of human B cells. Comparison of the signal transduced by IL-4 to four different competence signals. J. Immunol. 143, 3873–3880.

Clegg, L.S., Church, M.K. and Holgate, S.T. (1985). Histamine secretion from human skin slices induced by anti-IgE and artificial secretagogues and the effects of sodium cromoglycate and salbutamol. Clin. Allergy 15, 321–328.

Clutterbuck, E.J., Hirst, E.M. and Sanderson, C.J. (1989). Human interleukin-5 (IL-5) regulates the production of eosinophils in human bone marrow cultures: comparison and interaction with IL-1, IL-3, IL-6 and GMCSF. Blood 73, 1504–1512.

Cohan, V.L., Undem, B.J., Fox, C.C., Adkinson, N.F., Lichtenstein, L.M. and Schleimer, R.P. (1989). Dexamethasone does not inhibit the release of mediators from human mast cells residing in airway, intestine or skin. Am. Rev. Respir. Dis. 140, 951–954.

Columbo, M., Horowitz, E.M., Botana, L.M., NfacGlashan, D.W.J., Bochner, B.S., Gillis, S., Zsebo, K.M., Galli, S.J. and Lichtenstein, L.M. (1992). The human recombinant c-kit receptor ligand, rhSCF, induces mediator release from human cutaneous mast cells and enhances IgE-dependent mediator release from both skin mast cells and peripheral blood basophils. J. Immunol. 149, 599–608.

Conrad, D.H., Ben-Sasson, S.Z., Le Gros, G.S., Finkelman, F.D. and Paul, W.E. (1990). Infection with Nippostrongylus brasiliensis or injection of anti-IgD antibodies markedly enhances Fc-receptor-mediated IL-4 production by splenic non-B non-T cells. J. Exp. Med. 171, 1497–1508.

Craig, S.S., Schechter, N.M. and Schwartz, L.B. (1988). Ultrastructural analysis of human T and TC mast cells identified by immunoelectron microscopy. Lab. Invest. 58, 682–691.

Creticos, P.S., Peters, S.P., Adkinson, N.F., Naclerio, R.M.,

Hayes, E.C., Norman, P.S. and Lichtenstein, L.M. (1984). Peptide leukotriene release after antigen challenge in patients sensitive to ragweed. N. Engl. J. Med. 310, 1626–1630.

Crisp, A.J., Chapman, C.M., Kirkham, S.E., Schiller, A.L. and Krane, S.M. (1984). Articular mastocytosis in rheumatoid arthritis. Arthritis Rheum. 27, 845–851.

Cutz, E., Levison, H. and Cooper, D.M. (1978). Ultrastructure of airways in children. Histopathology 2, 407–421.

Dale, H.H. and Laidlaw, P.P. (1910). The physiological action of beta-imidazolethylamine. J. Physiol. 41, 318–344.

Dale, H.H. and Laidlaw, P.P. (1911). Further observations on the action of beta-imidazolyethylamine. J. Physiol. 43, 182–195.

Dayer, J.M., Beutler, B. and Cerami, A. (1985). Cachectin/tumor necrosis factor stimulates collagenase and prostaglandin E2 production by human synovial cells and dermal fibroblasts. J. Exp. Med. 162, 2163–2168.

De Clerck, L.S., Struyf, N.J., Bridts, C.H. and Stevens, W.J. (1991). Activation of inflammatory cells by immune complexes containing IgE in serum and synovial fluid of patients with rheumatoid arthritis: a study using flow cytometric analysis. Ann. Rheum. Dis. 50, 379–382.

De Monchy, J.G.R., Kauffman, H.F., Venge, P., Koeter, G.H., Jansen, H.M., Sluiter, H.J. and de Vries, K. (1985). Broncho-alveolar eosinophilia during allergen-induced late asthmatic reactions. Am. Rev. Respir. Dis. 131, 373–376.

Del Prete, G., Maggi, E., Parronchi, P., Chretien, I., Tiri, A., Macchia, D., Ricci, M., Banchereau, J., De Vries, J.E. and Romagnani, S. (1988). IL-4 is an essential cofactor for the IgE synthesis induced in vitro by human T cell clones and their supernatants. J. Immunol. 140, 4193–4198.

Diaz, P., Gonzales, M.C., Galleguillos, F.R., Ancic, P., Cromwell, O., Shepherd. D., Durham. S.R., Gleich, G.J. and Kay, A.B. (1989). Leucoctes and mediators in bronchoalveolar lavage during allergen-induced late-phase asthmatic reactions. Am. Rev. Respir. Dis. 139,

Djukanovic, R., Wilson, J.W., Britten, K.M., Wilson, S.J., Walls, A.F., Roche, W.R., Howarth, P.H. and Holgate, S.T. (1990). Quantitation of mast cells and eosinophils in the bronchial mucosa of symptomatic atopic asthmatics and healthy control subjects using immunocytochemistry. Am. Rev. Respir. Dis. 142, 863–871.

Dolovich, J., Hargreave, F.E., Chalmers, R., Shier, K.J., Gauldie, J. and Bienenstock, J. (1973). Late cutaneous allergic responses in isolated IgE-dependent reactions. J. Allergy Clin. Immunol. 59, 38–46.

Drake Lee, A.B., Moriarty, B. and Smallman, L.A. (1991). Mast cell numbers in the mucosa of the inferior turbinate in patients with perennial allergic rhinitis: a light microscopic study. J. Laryngol. Otol. 105, 736–738.

Duncan, J.I., Brown, F.I., McKinnon, A., Long, W.F., Williamson, F.B. and Thompson, W.D. (1992). Patterns of angiogenic response to mast cell granule constituents. Int. J. Microcirc. Clin. Exp. 11, 21–33.

Dunnill, M.S. (1960). The pathology of asthma with special reference to changes in the bronchial mucosa. J. Clin. Pathol. 13, 27–33.

Dvorak, A.M. and Kissel, S. (1991a). Granule changes in human skin mast cells characteristic of piecemeal degranulation and associated with recovery during wound healing. J. Leuk. Biol. 49, 197–210.

Dvorak, A.M., Massey, W., Warner, J.A., Kissel, S., Kagey-Sobotka, A. and Lichtenstein, L.M. (1991). IgE-mediated anaphylactic degranulation of isolated human skin mast cells. Blood 77, 569–578.

Dvorak, A.M., Furitsu, T. and Ishizaka, T. (1993). Ultrastructural morphology of human mast cell progenitors in sequential cocultures of cord blood cells and fibroblasts. Int. Arch. Allergy Immunol. 100, 219–229.

Ehrlich, P. (1878). Beitrage zur Theorie und Praxis der Histologischen Farbung. Doctoral Thesis, University of Leipzig.

Ehrlich, P. (1879). Uber die Specifischen Granulationen des Blutes. Arch. Anat. Physiol. Phys. Abt. 571–577.

El Lati, S.G., Dahinden, C.A. and Church, M.K. (1994). Complement peptides C3a and C5a induce mediator release from dissociated human mast cells. (Abstract) J. Invest. Dermatol. 102, 803–806.

Enerback, L. (1966a). Mast cells in the gastrointestinal mucosa. III. Reactivity towards compound 48/80. Acta Path. Microbiol. Scand. 66, 313–322.

Enerback, L. (1966b). Mast cells in the gastrointestinal mucosa. IV. Monoamine storing capacity. Acta Path. Microbiol. Scand. 67, 365–379.

Enerback, L. (1966c). Mast cells in the gastrointestinal mucosa. II. Dye binding and metachromatic properties. Acta Path. Microbiol. Scand. 66, 303–312.

Enerback, L. (1966d). Mast cells in the gastrointestinal mucosa. I. Effects of fixation. Acta Path. Microbiol. Scand. 66, 289–302.

Enerback, L., Kolsef, S.O., Kusche, M., Hjerpe, A. and Lindahl, U. (1985). Glycosaminoglycans in rat mucosal mast cells. Biochem. J. 227, 661–668.

Enerback, L., Pipkorn, U. and Olofsson, A. (1986). Intraepithelial migration of mucosal mast cells in hay fever: ultrastructural observations. Int. Arch. Allergy Appl. Immunol. 81, 289–297.

Fabbri, L.M., Boschetto, P., Zocca, E., Milani, G.F., Licata, B., Pivirotto, F. and Mapp, C.E. (1987). Bronchoalveolar neutrophilia during late asthmatic reactions induced by toluene diisocyanate. Am. Rev. Respir. Dis. 136, 36–42.

Falanga, V. and Julien, J.M. (1990). Observations in the potential role of transforming growth factor-beta in cutaneous fibrosis. Systemic sclerosis. Ann. NY Acad. Sci. 593, 161–171.

Flint, K.C., Leung, K.B.P. Hudspith, B.N., Brostoff, J., Pearce, F.L. and Johnson, N.M.I. (1985a). Bronchoalveolar mast cells in extrinsic asthma: a mechanism for the initiation of antigen specific bronchoconstriction. Br. Med. J. 291, 923–926.

Flint, K.C., Leung, K.B.P., Pearce, F.L., Hudspith, B.N., Brostoff, J. and Johnson, N.M.I. (1985b). Human mast cells recovered from bronchoalveolar lavage: their morphology, H release and effects of sodium cromoglycate. Clin. Sci. 68, 427–432.

Folkman, J., Klagsbrun, M., Sasse, J., Wadzinski, M., Ingber, D. and Vlodavsky, I. (1988). A heparin-binding angiogenic protein – basic fibroblast growth factor – is stored within basement membrane. Am. J. Pathol. 130, 393–400.

Ford Hutchinson, A.W., Gresser, M. and Young, R.N. (1994). 5-lipoxygenase. Annu. Rev. Biochem. 63, 383–417.

Fraki, J.E. and Hopsu-Havu, V.K. (1972). Human skin proteases. Fractionation and characterisation. Arch. Derm. Forsch. 243, 52–66.

Franzen, L. and Norrby, K. (1994). Local mitogenic effect of tissue mast cell secretion. Cell Tissue Kinet. 13, 635–642.

Frewin, D.B., Cleland, L.G., Jonsson, J.R. and Robertson, P.W. (1986). Histamine levels in human synovial fluid. J. Rheumatol. 13, 13–14.

Galli, S.J. (1990). New insight into "the riddle of the mast cells": microenvironmental regulation of mast cell development and phenotypic heterogeneity. Lab. Invest. 62, 5–33.

Galli, S.J. (1993). New concepts about the mast cell. N. Engl. J. Med. 328, 257–265.

Gauchat, J.F., Lebman, D.A., Coffman, R.L., Gascan, H. and De Vries, J.E. (1990). Structure and expression of germline epsilon transcripts in human B cells induced by interleukin-4 to switch to IgE production. J. Exp. Med. 172, 463–473.

Gauchat, J.-F., Henchoz. S., Mlazzei, G., Aubry, J.-P., Brunner, T., Blasey, H., Life, P., Talabot, D., Flores-Romo, L., Thompson, J., Kishi, K., Butterfield, J., Dahinden, C.A. and Bonnefoy, J.-Y. (1993). Induction of human IgE synthesis in B cells by mast cells and basophils. Nature 365, 340–343.

Giorno, R., Choi, K.L. and Claman, H.N. (1987). Simultaneous in situ detection of IgE receptors and cytoplasmic granules in murine cutaneous mast cells. J. Immunol. Methods 99, 163–166.

Glenner, G.G. and Cohen, L.A. (1960). Histochemical demonstration of a species-specific trypsin-like enzyme in mast cells. Nature 185, 846–847.

Godard, P., Bousquet, J., Lebel, B. and Michel, F.B. (1987). Bronchoalveolar lavage in the asthmatic. Bull. Eur. Physiopathol. Respir. 23, 73–83.

Godfrey, H.P., llardi, C., Engber, W. and Graziano, F.M. (1984). Quantitation of human synovial mast cells in rheumatoid arthritis and other rheumatoid diseases. Arthritis Rheum. 27, 852–856.

Goetzl, E.J. and Pickett, W.C. (1981). Novel structural determinants of the human neutrophil chemotactic activity of leukotriene B. J. Exp. Med. 153, 482–487.

Goldstein, S.M., Kaempfer, C.E., Kealey, J.T. and Wintroub, B.U. (1989). Human mast cell carboxypeptidase. Purification and characterization. J. Clin. Invest. 83, 1630–1636.

Goldstein, S.M., Leong, J. and Bunnett, N.W. (1991). Human mast cell proteases hydrolyze neurotensin, kinetensin and Leu5-enkephalin. Peptides 12, 995–1000.

Goldstein, S.M., Leong, J., Schwartz, L.B. and Cooke, D. (1992). Protease composition of exocytosed human skin mast cell protease-proteoglycan complexes: tryptase resides in a complex distinct from chymase and carboxypeptidase. J. Immunol. 148, 2475–2482.

Gordon, J.R. and Galli, S.J. (1991). Release of both preformed and newly synthesized tumor necrosis factor alpha (TNF-alpha)/cachectin by mouse mast cells stimulated via the Fc epsilon RI. A mechanism for the sustained action of mast cell-derived TNF-alpha during IgE-dependent biological responses. J. Exp. Med. 174, 103–107.

Gordon, J.R., Burd, P.R. and Galli, S.J. (1990). Mast cells as a source of multifunctional cytokines. Immunol. Today 11, 458–464.

Gospodarowicz, D. and Cheng, J. (1986). Heparin protects basic and acidic FGF from inactivation. J. Cell Physiol. 128, 475–484.

Goto T., Befus, D. Low, R., and Bienenstock, J. (1984). Mast cell heterogeneity and hyperplasia in bleomycin-induced pulmonary fibrosis of rats. Am. Rev. Respir. Dis. 130, 797–802.

Greenburg, G. and Burnstock, G.A. (1983). A novel cell-to-cell interaction between mast cells and other cells. Exp. Cell Res. 147, 1–13.

Gruber, B.L., Poznansky, M., Boss, E., Partin, J., Gorevic, P. and Kaplan, A.P. (1986). Characterization and functional studies of rheumatoid synovial mast cells – activation by secretagogues, anti-IgE and a H-releasing lymphokine. Arthritis Rheum. 29, 944–955.

Gruber, B.L., Marchese, M.J., Suzuki, K., Schwartz, L.B., Okada, Y., Nagese, H. and Ramamurthy, N.S. (1989). Synovial procollagenase activation by human mast cell tryptase dependance upon matrix metalloproteinase 3 activation. J. Clin. Invest. 84, 1657–1662.

Gurish, M.F., Ghildyal, N., Arm, J.P., Austen, K.F., Avraham, S., Reynolds, D. and Stevens, R.L. (1991). Cytokine mRNA are preferentially increased relative to secretory granule protein mRNA in mouse bone marrow-derived mast cells that have undergone IgE-mediated activation and degranulation. J. Immunol. 146, 1527–1533.

Hagermark, O., Hokfelt, T. and Pernow, B. (1978). Flare and itch induced by substance P in human skin. J. Invest. Dermatol. 71, 233–235.

Hardy, C.C., Robinson, C., Lewis, R.A. Tattersfield, A.E. and Holgate, S.T. (1985). The airway and cardiovascular responses to inhaled prostaglandin 12 in normal and asthmatic man. Am. Rev. Respir. Dis. 131, 18–21.

Hartmann, T., Ruoss, S.J., Raymond. W.W., Seuwen, K. and Caughey, G.H. (1992). Human tryptase as a potent, cell-specific mitogen: role of signaling pathways in synergistic responses. Am. J. Physiol. Lung Cell. Mol. Physiol. 262, L528–L534.

Hastie, R., Heroy, J.H. and Levy, D.A. (1979). Basophil leukocytes and mast cells in human nasal secretions and scrapings studied by light microscopy. Lab. Invest. 40, 554–561.

Hatamochi, A., Fujiwara, K. and Ueki, H. (1985). Effects of H on collagen synthesis by cultured fibroblasts derived from guinea pig skin. Arch. Dermatol. Res. 277, 60–64.

Hawkins, R.A., Claman, H.N., Clark, R.A.F. and Steigerwald, J.C. (1985). Increased dermal mast cell populations in progressive systemic sclerosis: a link in chronic fibrosis? Ann. Intern. Med. 102, 182–186.

Heaney, L.G., Cross, L.J.M., Stanford, C.F. and Ennis, M. (1994). Differential reactivity of human bronchoalveolar lavage mast cells to substance P. Agents Actions 41, C19–C21.

Henderson, W.R. and Kaliner, M.A. (1978). Immunologic and non-immunologic generation of superoxide from mast cells and basophils. J. Clin. Invest. 61, 187–196.

Henkel, G., Weiss, D.L., McCoy, R., Delougherty, T., Tara, D. and Brown, M.A. (1992). A DNase I-hypersensitive site in the second intron of the murine IL-4 gene defines a mast cell-specific enhancer. J. Immunol. 149, 3239–3946.

Hogg, J.C. (1977). Pathologic abnormalities in asthma. In "Asthma: Physiology, and Treatment, Second International Symposium" (eds L.M. Lichtenstein and K.F. Austen), pp. 1–120, Academic Press, New York.

Howarth, P.H., Wilson, S., Lau, L. and Rajakulasingam, K. (1991). The nasal mast cell and rhinitis. Clin. Exp. Allergy 21 (Suppl 2), 3–8.

Huber, H.L. and Koessler, K.K. (1922). The pathology of bronchial asthma. Arch. Intern. Med. 30, 689–695.

Huntley, J.F., Newlands, G.F.J., Gibson, S., Ferguson, A. and Miller, H.R.P. (1985). Histochemical demonstration of chymotrypsin like serine esterases in mucosal mast cells in four species including man. J. Clin. Pathol. 38, 375–384.

Irani, A.A., Schechter, N.M., Craig, S.S., DeBlois, G. and Schwartz, L.B. (1986). Two types of human mast cells that have distinct neutral protease compositions. Proc. Natl. Acad. Sci. USA 83, 4464–4468.

Irani, A.A., Craig, S.S., DeBlois, G., Elson, C.O., Schechter, N.M. and Schwartz, L.B. (1987a). Deficiency of the tryptase-positive, chymase-negative mast cell type in gastrointestinal mucosa to patients with defective T lymphocyte function. J. Immunol. 138, 4381–4386.

Irani, A.A., Golzar, N., DeBlois, G., Gruber, B.L. and Schwartz, L.B. (1987b). Distribution of mast cell subtypes in rheumatoid arthritis and osteoarthritis synovia. Arthritis Rheum. 30, 66.

Irani, A.A., Butrus, S.I. and Schwartz, L.B. (1988). Distribution of T and TC mast cell subsets in vernal conjunctivitis (VC) and giant papillary conjunctivitis [Abstract]. NER Allergy Proc. 9, 451.

Irani, A.A., Sampson, H.A. and Schwartz, L.B. (1989a). Mast cells in atopic dermatitis. Allergy 44 (Suppl 9), 31–34.

Irani, A.M., Bradford, T.R., Kepley, C.L., Schechter, N.M. and Schwartz, L.B. (1989b). Detection of MCT and MCTC types of human mast cells by immunohistochemistry using new monoclonal anti-tryptase and anti-chymase antibodies. J. Histochem. Cytochem. 37, 1509–1515.

Irani, A.A., Nilsson, G., Miettinen, U., Craig, S.S., Ashman, L.K., Ishizaka, T., Zsebo, K.M. and Schwartz, L.B. (1992). Recombinant human stem cell factor stimulates differentiation of mast cells from dispersed human fetal liver cells. Blood 80, 3009–3021.

Jabara, H.H., Ackerman, S.J., Vercelli, D., Yokota, T., Arai, K., Abrams, J., Dvorak, A.M., Lavigne, M.C., Banchereau, J., De Vries, J. et al. (1988). Induction of interleukin-4-dependent IgE synthesis and interleukin-5-dependent eosinophil differentiation by supernatants of a human helper T cell clone. J. Clin. Immunol. 8, 437–446.

Jaffee, B.D. and Claman, H.N. (1983). Chronic graft versus host disease (GVHD) as a model for scleroderma. 1. Description of model systems. Cell Immunol. 77, 1–12.

James, M.P., Kennedy, A.R. and Eady, R.A.J. (1981). A microscopic study of inflammatory reactions in human skin induced by histamine and compound 48/80. J. Invest. Dermatol. 78, 406–413.

Johnson, R.G., Carty, S.E., Fingerhoff, B.J. and Scarpa, A. (1980). The internal pH of mast cell granules. FEBS Lett. 120, 75–79.

Jordana, M., Befus, A.D., Newhouse, M.T., Bienenstock, J. and Gauldie, J. (1988). Effect of histamine on proliferation of normal human adult lung fibroblasts. Thorax 43, 552–558.

Kanakura, Y., Thompson, H., Nakano, T., Yamamura, T., Asai, H., Kitamura, Y., Metcalfe, D.D. and Galli, S.J. (1988). Multiple bidirectional alterations of phenotype and changes in proliferative potential during the in vitro and in vivo passage of clonal mast cell populations derived from mouse peritoneal mast cells. Blood 72(3), 877–885.

Kawanami. O., Ferrans, V.J., Fulmer, J.D. and Crystal, R.G. (1985). Ultrastructure of pulmonary mast cells in patients with fibrotic lung disorders. Lab. Invest. 40, 717–734.

Kelly, C.A., Stenton, S.C., Ward, C., Bird, G., Hendrick, D.J. and Walters, E.H. (1989). Lymphocyte subsets in bronchoalveolar lavage fluid obtained from stable asthmatics and their correlations with bronchial responsiveness. Clin. Exp. Allergy 19, 169–175.

Kessler, D.A., Langer, R.S., Pless, N.A. and Folkman, J. (1976). Mast cells and tumor angiogenesis. Int. J. Cancer 18, 703–709.

Kielty, C.M., Lees, M., Shuttleworth, C.A. and Woolley, D.E. (1993). Catabolism of intact type VI collagen myofibrils: susceptibility to degradation by serine esterases. Biochem. Biophys. Res. Commun. 191, 1230–1236.

King, S.J., Miller, H.R., Newlands, G.F. and Woodbury, R.G. (1985). Depletion of mucosal mast cell protease by glucocorticosteroids: effect on intestinal anaphylaxis in the rat. Proc. Natl. Acad. Sci. USA 82, 1214–1218.

Kinoshita, A., Urata, H., Bumpus, F.M. and Husain, A. (1991). Multiple determinants for the high substrate specificity of an angiotensin-II-forming chymase from the human heart. J. Biol. Chem. 266, 19192–19197.

Kips, J.C., Tavernier, J. and Pauwels, R.A. (1992). Tumor necrosis factor causes bronchial hyperresponsiveness in rats. Am. Rev. Respir. Dis. 145, 332–336.

Kirby, J.G., Hargreave F.E., Gleich G.J. and O'Byrne, P.M. (1987). Bronchoalveolar cell profiles of asthmatic and non-asthmatic subjects. Am. Rev. Respir. Dis. 136, 379–383.

Kitamura, Y. and Go, S. (1979). Decreased production of mast cells in Sl/Sld anemic mice. Blood 53, 492–497.

Kitamura, Y., Go, S. and Hatanaka, K. (1978). Decrease of mast cells in W/Wv mice and their increase by bone marrow transplantation. Blood 52, 447–452.

Kitamura, Y., Nakayama, H. and Fujita, J. (1989). Mechanism of mast cell deficiency in mutant mice of W/Wv and Sl/Sld genotype. In "Mast Cell and Basophil Differentiation and Function in Health and Disease" (eds S.J. Galli and K.F. Austen), pp 15–25. Raven Press, New York.

Klein, L.M., Lavker, R.M., Matis, W.L. and Murphy, G.F. (1989). Degranulation of human mast cells induces an endothelial antigen central to leukocyte adhesion. Proc. Natl. Acad. Sci. USA 86, 8972–8976.

Kobayashi, L. and Ziff, M. (1975). Electron microscopic studies of the cartilage-pannus junction in rheumatoid arthritis. Arthritis Rheum. 18, 475–483.

Kohase, M., Henriksen DeStefano, D., May, L.T., Vilcek, J. and Sehgal, P.B. (1986). Induction to beta 2-interferon by tumor necrosis factor: a homeostatic mechanism in the control of cell proliferation. Cell 45, 659–666.

Kunkel, S.L., Standiford, T., Kasahara, K. and Strieter, R.M. (1991). Interleukin-8 (IL-8): the major neutrophil chemotactic factor in the lung. Exp. Lung Res. 17, 17–23.

Lagunoff, D. and Richard, A. (1983). Evidence for control of mast cell granule protease in situ by low pH. Exp. Cell Res. 144, 353

Lam, S., Chan, H., LeRiche, J.C., Chan-Yeung, M. and Salari, H. (1988). Release of leukotrienes in patients with bronchial asthma. J. Allergy Clin. Immunol. 81, 711–717.

Lau, L.C.K., Church, M.K., Walls, A.F. and Howarth, P.H. (1995). Dispersion and characterisation of mast cells from human nasal polyp and the effects of salbutamol and IBMX on histamine secretion induced by anti-IgE. Submitted.

Le Gros, G., Ben Sasson, S.Z., Seder, R., Finkelman, F.D. and Paul, W.E. (1990). Generation of interleukin 4 (IL-4)-

producing cells *in vivo* and *in vitro*: IL-2 and IL-4 are required for *in vitro* generation of IL-4-producing cells. J. Exp. Med. 172, 921–929.

Lebman, D.A. and Coffman, R.L. (1988). Interleukin 4 causes isotype switching to IgE in T cell-stimulated clonal B cell cultures. J. Exp. Med. 168, 853–862.

Leung, D.Y., Pober, J.S. and Cotran, R.S. (1991). Expression of endothelial-leukocyte adhesion molecule-1 in elicited late phase allergic reactions. J. Clin. Invest. 87, 1805–1809.

List, S.J., Findlay, B.P., Forstner, C.G. and Forstner, J.F. (1978). Enhancement of the viscosity of mucus by serum albumin. Biochem. J. 175, 565–571.

Liston, T.E. and Roberts, L.J. (1985). Transformation of prostaglandin D2 to 9alpha, 11 beta-(15S)-trihydroxy-prosta-(5Z,13E)-dien-1-oic acid (9alpha, 11 beta-prostaglandin F2): a unique biologically active prostaglandin produced enzymatically *in vivo* in humans. Proc. Natl. Acad. Sci. USA 82, 6030–6034.

Liu, M.C., Hubbard, W.C., Proud, D., Stealey, B.A., Galli, S.J., Kagey Sobotka, A., Bleecker, E.R. and Lichtenstein, L.M. (1991). Immediate and late inflammatory responses to ragweed antigen challenge of the peripheral airways in allergic asthmatics. Cellular, mediator and permeability changes. Am. Rev. Respir. Dis. 144, 51–58.

Lohi, J., Harvima, I. and Keski-Oja, J. (1992). Pericellular substrates of human mast cell tryptase: 72,000 Dalton gelatinase and fibronectin. J. Cell. Biochem. 50, 337–349.

Lopez, A.F., Sanderson, C.J., Gamble, J.R., Campbell, H.D., Young, I.G. and Vadas, M.A. (1988). Recombinant human interleukin-5 is a selective activator of human eosinophil function. J. Exp. Med. 167, 219–224.

Lowman M.A., Benyon, R.C. and Church, M.K. (1988a). Human skin mast cells: Effects of salbutamol and sodium cromoglycate on histamine release induced by anti-IgE and substance P. Skin Pharmacol. 1, 63–64.

Lowman, M.A., Benyon, R.C. and Church, M.K. (1988b). Characterization of neuropeptide-induced histamine released from human dispersed skin mast cells. Br. J. Pharmacol. 95, 121–130.

Lowman, M.A., Rees, P.H., Benyon, R.C. and Church, M.K. (1988c). Human mast cell heterogeneity: Histamine release from mast cells dispersed from skin, lung, adenoids, tonsils and intestinal mucosa in response to IgE-dependent and non-immunological stimuli. J. Allergy Clin. Immunol. 81, 590–597.

Lozewicz, S., Gomez, E., Clague, J., Gatland, D. and Davies, R.J. (1990). Allergen-induced changes in the nasal mucous membrane in seasonal allergic rhinitis: effect of nedocromil sodium. J. Allergy Clin. Immunol. 85, 125–131.

Malone, D.G., Irani, A.M., Schwartz, L.B., Barrett, K.E. and Metcalfe, D.D. (1986). Mast cell numbers and histamine levels in synovial fluids from patients with diverse arthritides. Arthritis Rheum. 29, 956–963.

Malone, D.G., Wilder, R.L., Saavedra Delgadoz, A.M. and Metcalfe, D.D. (1987). Mast cell numbers in rheumatoid synovial tissues. Correlations with quantitative measures of lymphocytic infiltration and modulation by antiinflammatory therapy. Arthritis Rheum. 30, 130–137.

Marini, M., Vittori, E., Hollemborg, J. and Mattoli, S. (1992). Expression of the potent inflammatory cytokines, granulocyte-macrophage-colony-stimulating factor and interleukin-6 and interleukin-8, in bronchial epithelial cells of patients with asthma. J. Allergy Clin. Immunol. 89, 1001–1009.

Marks, R.M., Roche, W.R., Czerniecki, M., Penny, R. and Nelson, D.S. (1986). Mast cell granules cause proliferation of human microvascular endothelial cells. Lab. Invest. 55, 289–294.

Marom, Z., Shelhamer, J.H. and Kaliner, M.A. (1981). Effects of arachidonic acid, monohydroxyeicosantetraenoic acid and prostaglandins on the release of mucus glycoproteins from human airways *in vitro*. J. Clin. Invest. 161, 657–658.

Marom, Z., Shelhamer, J.H., Bach, M.K., Morton, D.R. and Kaliner, M.A. (1982). Slow reacting substances, leukotrienes C4 and D4, increase the release of mucus from human airways *in vitro*. Am. Rev. Respir. Dis. 126, 449–451.

Marshall, J.S., Ford, G.P. and Bell, E.B. (1987). Formalin sensitivity and differential staining of mast cells in human dermis. Br. J. Dermatol. 117, 29–36.

Matin, R., Tam, E.K., Nadel, J.A. and Caughey, G.H. (1992). Distribution of chymase-containing mast cells in human bronchi. J. Histochem. Cytochem. 40. 781–786.

Matsuda, H. and Kitamura, Y. (1981). Migration of stromal cells supporting mast cell differentiation into open wounds produced in the skin of mice. Exp. Hematol. 9, 38–43.

Matsuda, H., Kawakita K., Kiso, Y., Nakano, T. and Kitamura, Y. (1989). Substance P induces granulocyte infiltration through degranulation of mast cells. J. Immunol. 142, 927–931.

Maximow, A. (1906). Uber die zellformen des lockeren bindegewebes. Arch. F. Mikr. Anat. (Bonn) 67, 680–757.

McNamee, L.A., Fattah, L.I., Baker, T.J., Bains, T.J. and Hissey, P.H. (1991). Production, characterization and use of monoclonal antibodies to human interleukin 5 in an enzyme-linked immunosorbent assay. J. Immunol. Methods 141, 81.

Meininger, C.J. and Zetter, B.R. (1992). Mast cells and angiogenesis. Semin. Cancer Biol. 3, 73–79.

Metcalfe, D.D., Bland, C.E. and Wasserman, S.I. (1984). Biochemical and functional characterisation of proteoglycans isolated from basophils of patients with chronic myelogenous leukemia. J. Immunol. 132, 1943.

Metzger, W.J., Zavala, D., Richerson, H.B., Moseley, P., Iwamota, P., Monick, M., Sjoedrdsma, K. and Hunninghake, G.W. (1987). Local allergen challenge and bronchoalveolar lavage of allergic asthmatic lungs: description of the model and local airway inflammation. Am. Rev. Respir. Dis. 135, 433–440.

Mielke, V., Bauman, J.G., Sticherling, M., Ibs, T., Zomershoe, A.G., Seligmann, K., Henneicke, H.H., Schroder, J.M., Sterry, W. and Christophers, E. (1990). Detection of neutrophil-activating peptide NAP/IL-8 and NAP/IL-8 mRNA in human recombinant IL-1 alpha- and human recombinant tumor necrosis factor-alpha-stimulated human dermal fibroblasts. An immunocytochemical and fluorescent *in situ* hybridization study. J. Immunol. 144, 153–161.

Miller, H.R.P. (1984). The protective mucosal response against gastrointestinal nematodes in ruminants and laboratory animals. Vet. Immunol. Immunopathol. 6, 167–259.

Miller, H.R.P., King, S.J., Gibson, S., Huntley, J.F., Newlands, G.F.J. and Woodbury, R.G. (1986). Intestinal mucosal mast cells in normal and parasitized rats. In "Mast Cell Differentiation and Heterogeneity" (eds A.D. Befus, J., Bienenstock and J.A. Denburg), pp. 239–255, Raven Press, New York.

Ming, W.J., Bersani, L. and Mantovani, A. (1987). Tumor necrosis factor is chemotactic for monocytes and polymorphonuclear leukocytes. J. Immunol. 138, 1469–1474.

Mizutani, H., Schechter, N.M., Lazarus, G.S., Black, R.A. and Kupper, T.S. (1991). Rapid and specific conversion of precursor interleukin 1 beta (IL-1 beta) to an active IL-1 species by human mast cell chymase. J. Exp. Med. 174, 821–825.

Möller, A., Lippert, U., Lessmann, D., Kolde, G., Hamann, K., Welker, P., Schadendorf, D., Rosenbach, T., Luger, T. and Czarnetzki, B.M. (1993). Human mast cells produce IL-8. J. Immunol. 151, 3261–3266.

Monroe, J.G., Haldar, S., Prystowsky, M.B. and Lammie, P. (1988). Lymphokine regulation of inflammatory processes: interleukin-4 stimulates fibroblast proliferation. Clin. Immunol. Immunopathol. 49, 292–298.

Moxley, G. and Ruddy, S. (1985). Elevated C3 anaphylatoxin levels in synovial fluids from patients with rheumatoid arthritis. Arthritis Rheum. 28, 1089–1095.

Naclerio, R.M., Meier, H.L., Kagey-Sobotka, A., Adkinson, N.F., Meyers, D.A., Norman, P.S. and Lichtenstein, L.M. (1983). Mediator release after nasal airway challenge with allergen. Am. Rev. Respir. Dis. 128, 597–602.

Naclerio, R.M., Proud, D., Togias, A.G., Adkinson, N.F., Meyers, D.A., Kagey-Sobotka, A., Plaut, M., Norman, P.S. and Lichtenstein, L.M. (1985). Inflammatory mediators in late antigen-induced rhinitis. N. Engl. J. Med. 313, 65–70.

Nishioka, K., Kobavashi, Y., Katayama, I. and Takijiri, C. (1987). Mast cell numbers in diffuse scleroderma. Arch. Dermatol. 123, 205–208.

Nogami, M., Suko, M., Okudaira, H., Miyamoto, T., Shiga, J., Ito, M. and Kasuya, S. (1990). Experimental pulmonary eosinophilia in mice by Ascaris suum extract. Am. Rev. Respir. Dis. 141, 1289–1295.

Nusslein, H.G., Trag, T., Winter, M., Dietz, A. and Kalden, J.R. (1992). The role of T cells and the effect of hydrocortisone on interleukin-4-induced IgE synthesis by non-T cells. Clin. Exp. Immunol. 90, 286–292.

Ohkawara, Y., Yamauchi, K., Tanno, Y., Tamura, G., Ohtani, H., Nagura, H., Ohkuda, K. and Takishima, T. (1992). Human lung mast cells and pulmonary macrophages produce tumor necrosis factor-α in sensitized lung tissue after IgE receptor triggering. Am. J. Respir. Cell Mol. Biol. 7, 385–392.

Okayama, Y. and Church, M.K. (1992). Comparison of the modulatory effect of ketotifen, sodium cromoglycate, procaterol and salbutamol in human skin, lung and tonsil mast cells. Int. Arch. Allergy Appl. Immunol. 97, 216–225.

Okayama, Y. and Church, M.K. (1995). Multiple cytokine mRNA expression in human mast cells stimulated via Fc$_\epsilon$RI. Int. Arch. Allergy Immunol, in press.

Okayama, Y., Benyon, R.C., Rees, P.H., Lowman, M.A., Hillier, K. and Church, M.K. (1992). Inhibition profiles of sodium cromoglycate and nedocromil sodium on mediator release from mast cells of human skin, lung, tonsil, adenoid and intestine. Clin. Exp. Allergy 22, 401–409.

Osborn, L., Hession. C., Tizard, R., Vassallo, C., Luhowskyj, S., Chi Rosso, G. and Lobb, R. (1989). Direct expression cloning of vascular cell adhesion molecule 1, a cytokine-induced endothelial protein that binds to lymphocytes. Cell 59, 1203–1211.

Parronchi, P., Tiri, A., Macchia, D., De Carli, M., Biswas, P., Simonelli, C., Maggi, E., Del Prete, G., Ricci, M. and Romagnani, S. (1990). Noncognate contact-dependent B cell activation can promote IL-4-dependent in vitro human IgE synthesis. J. Immunol. 144, 2102–2108.

Pene, J., Rousset, F., Briere, F., Chretien, I., Bonnefoy, J.Y., Spits, H., Yokota, T., Arai, N., Arai, K., Banchereau, J. et al. (1988). IgE production by normal human lymphocytes is induced by interleukin 4 and suppressed by interferons gamma and alpha and prostaglandin E2. Proc. Natl. Acad. Sci. USA 85, 6880–6884.

Pepys, J., Hargreave, F.E., Chan, M. and McCarthy, D.S. (1968). Inhibitory effects of disodium cromoglycate on allergen inhalation tests. Lancet ii, 134–137.

Persinger, M.A., Lepage, P. and Simard, J.P. (1983). Mast cell numbers in incisional wounds in rat skin as a function of distance, time and treatment. Br. J. Dermatol. 108, 179–184.

Piccinni, M.P., Macchia, D., Parronchi, P., Giudizi, M.G., Bani, D., Alterini, R., Grossi. A., Ricci, M., Maggi, E. and Romagnani, S. (1991). Human bone marrow non-B, non-T cells produce interleukin 4 in response to cross-linkage to Fc epsilon and Fc gamma receptors. Proc. Natl. Acad. Sci. USA 88, 8656–8660.

Piguet, P.F., Grau, G.E. and Vassalli, P. (1990). Subcutaneous perfusion of tumor necrosis factor induces local proliferation of fibroblasts, capillaries and epidermal cells, or massive tissue necrosis. Am. J. Pathol. 136, 103–110.

Pipkorn, U., Karlsson, G. and Enerback, L. (1988). The cellular response of the human allergic mucosa to natural allergen exposure. J. Allergy Clin. Immunol. 82, 1046–1054.

Pipkorn, U., Karlsson, G. and Enerback, L. (1989). Nasal mucosal response to repeated challenges with pollen allergen. Am. Rev. Respir. Dis. 140, 729–736.

Plaut, M., Pierce, J.H., Watson, C.J., Hanley-Hyde, J., Nordan, R.P. and Paul, W.E. (1989). Mast cell lines produce lymphokines in response to cross linkage of FcepsilonRI or to calcium ionophores. Nature 339, 64–67.

Postlethwaite, A.E. and Seyer, J.M. (1990). Stimulation of fibroblast chemotaxis by human recombinant tumor necrosis factor alpha (TNF-alpha) and a synthetic TNF-alpha 31–68 peptide. J. Exp. Med. 172, 1749–1756.

Postlethwaite, A.E. and Seyer, J.M. (1991). Fibroblast chemotaxis induction by human recombinant interleukin-4. Identification by synthetic peptide analysis of two chemotactic domains residing in amino acid sequences 70–88 and 89–122. J. Clin. Invest. 87, 2147–2152.

Postlethwaite, A.E., Holness, M.A., Katai, H. and Raghow, R. (1992). Human fibroblasts synthesize elevated levels of extracellular matrix proteins in response to interleukin 4. J. Clin. Invest. 90. 1479–1485.

Powers, J.C., Takumi, T., Harper, J.W., Minematsu, Y., Barker, L., Lincoln, D., Crumley, K.V., Fraki, J.E., Schechter, N.M., Lazarus, G.S., Nakajima, K., Nakashino, K., Neurath, H. and Woodbury, R.G. (1985). Mammalian chymotrypsin-like enzymes. Comparative reactivities of rat mast cell proteases, human and dog skin proteases and human cathepsin G with peptide-4-nitroanilide substrates and with peptide chloromethyl ketone and sulphonyl fluoride inhibitors. Biochemistry 24, 2048–2058.

Proud, D., Siekierski, E.S. and Bailey, G.S. (1988). Identification of human lung mast cell kininogenase as tryptase and relevance of tryptase kininogenase activity. Biochem. Pharmacol. 37, 1473–1480.

Rafferty, P., Beasley, C.R.W. and Holgate, S.T. (1987). The contribution of histamine to bronchoconstriction produced by inhaled allergen and adenosine 5'-monophosphate in asthma. Am. Rev. Respir. Dis. 136, 369–373.

Rakusan, K. and Campbell, S.E. (1991). Spatial relationship between cardiac mast cells and coronary capillaries in neonatal rats with cardiomegaly. Can. J. Physiol. Pharmacol. 69, 1750–1753.

Razin, E., Stevens, R.L., Akiyama, F., Schmid, K. and Austen, K.F. (1982). Culture from mouse bone marrow of a subclass of mast cells possessing a distinct chondroitin sulfate proteoglycan with glycosaminoglycans rich in N-acetylgalactosamine-4,6-disulphate. J. Biol. Chem. 257, 7229–7236.

Rees, P.H., Hillier, K. and Church, M.K. (1988). The secretory characteristics of mast cells isolated from the human large intestinal mucosa and muscle. Immunology 65, 437–442.

Reilley, C.F., Tewksbury, D.A., Schechter, N.M. and Travis, J. (1982). Rapid conversion of angiotensin I to angiotensin II by neutrophil and mast cell proteinases. J. Biol. Chem. 257, 8619–8622.

Riley, J.F. and West, G.B. (1953). The presence of histamine in tissue mast cells. J. Physiol. (Lond.) 120, 528–537.

Robinson, C., Benyon, R.C., Holgate, S.T. and Church, M.K. (1989) The IgE- and calcium-dependent release of eicosanoids and histamine from human cutaneous mast cells. J. Invest. Dermatol. 93, 397–404.

Roche, W.R. (1985a). Mast cells and tumour angiogenesis: the tumor-mediated release of an endothelial growth factor from mast cells. Int. J. Cancer 36, 721–728.

Roche, W.R. (1985b). Mast cells and tumors. The specific enhancement of tumor proliferation in vitro. Am. J. Pathol. 119, 57–64.

Roche, W.R., Beasley, R., Williams, J.H. and Holgate, S.T. (1989). Subepithelial fibrosis in the bronchi of asthmatics. Lancet 1, 520–524.

Romagnani, S. (1991). Type 1 T helper and type 2 T helper cells: functions, regulation and role in protection and disease. Int. J. Clin. Lab. Res. 21, 152–158.

Rousset, F., Malefijt, R.W., Slierendregt, B., Aubry, J.P., Bonnefoy, J.Y., Defrance, T., Banchereau, J. and De Vries, J.E. (1988). Regulation of Fc receptor for IgE (CD23) and class II MHC antigen expression on Burkitt's lymphoma cell lines by human IL-4 and IFN-gamma. J. Immunol. 140, 2625–2632.

Ruoss, S.J., Hartmann, T. and Caughey, G.H. (1991). Mast cell tryptase is a mitogen for cultured fibroblasts. J. Clin. Invest. 88, 493–499.

Sage, H., Woodbury, R.G. and Bornstein, P. (1979). Structural studies on human type IV collagen. J. Biol. Chem. 254, 9893–9900.

Sandberg, N. (1962). Accelerated collagen formation and histamine. Nature (Lond.) 194, 183–185.

Schayer, R.W. (1963). Histidine decarboxylase in man. Ann. NY Acad. Sci. 103, 164–178.

Schechter, N.M., Fraki, J.E., Geesin, J.C. and Lazarus, G.S. (1983). Human skin chymotryptic proteinase: isolation and relation to cathepsin G and rat mast cell protease. J. Biol. Chem. 258, 2973–2978.

Schleimer, R.P., Schulman, E.S., MacGlashan, D.W., Peters, S.P., Hayes, E.C., Adams, G.K., Lichtenstein, L.M. and Adkinson, N.F. (1983). Effects of dexamethasone on mediator release from human lung fragments and purified human lung mast cells. J. Clin. Invest. 71, 1830–1835.

Schleimer, R.P., Sterbinsky, S.A., Kaiser, J., Bickel, C.A., Klunk, D.A., Tomioka, K., Newman, W., Luscinskas, F.W., Gimbrone, M.A., Jr., McIntyre, B.W. and Bochner, B.S. (1992). IL-4 induces adherence of human eosinophils and basophils but not neutrophils to endothelium: association with expression of VCAM-1. J. Immunol. 148, 1086–1092.

Schroder, J.M. and Christophers, E. (1989). Secretion of novel and homologous neutrophil-activating peptides by LPS-stimulated human endothelial cells. J. Immunol. 142, 244–251.

Schulman, E.S., Post, T.J., Henson, P.M., and Gicias, P.C. (1988). Differential effects of complement peptides. C5a and C5a des arg on human basophil and lung mast cell H release. J. Clin. Invest. 81, 918–923.

Schwartz, L.B. and Bradford, T.R. (1986). Regulation of tryptase from human lung mast cells by heparin. Stabilization of the active tetramer. J. Biol. Chem. 261, 7372–7379.

Schwartz, L.B., Lewis, R.A. and Austen, K.F. (1981a). Tryptase from human pulmonary mast cells: purification and characterization. J. Biol. Chem. 256, 11939–11943.

Schwartz, L.B., Riedel, C., Caulfield, J.P., Wasserman, S.I. and Austen, K.F. (1981b). Cell association of complexes of chymase, heparin proteoglycan and protein after degranulation of rat mast cells. J. Immunol. 126, 2071–2078.

Schwartz, L.B., Bradford, T.R., Lee, D.C. and Chlebowski, J.F. (1990). Immunologic and physicochemical evidence for conformational changes occurring on conversion of human mast cell tryptase from active tetramer to inactive monomer. Production of monoclonal antibodies recognizing active tryptase. J. Immunol. 144, 2304–2311.

Seder, R.A., Paul, W.E., Ben Sasson. S.Z., LeGros, G.S., Kagey-Sobotka, A., Finkelman, F.D., Pierce, J.H. and Plaut, M. (1991a). Production of interleukin-4 and other cytokines following stimulation of mast cell lines and in vivo mast cells/basophils. Int. Arch. Allergy Appl. Immunol. 94, 137–140.

Seder, R.A., Paul, W.E., Dvorak, A.M., Sharkis, S.J., Kagey-Sobotka, A., Niv, Y., Finkelman, F.D., Barieri, S.A., Galli, S.J. and Plaut, M. (1991b). Mouse splenic and bone marrow preparation that express high-affinity epsilon receptor and produce interleukin 4 are highly enriched basophils. Proc. Natl. Acad. Sci. USA 88, 2836–2839.

Sedgwick, J.B., Calhoun, W.J., Gleich, G.J., Kita, H., Abrams, J.S., Schwartz, L.B., Volovitz, B., Ben-Yaakov, M. and Busse, W.W. (1991). Immediate and late airway response of allergic rhinitis patients to segmental antigen challenge: characterization of eosinophil and mast cell mediators. Am. Rev. Respir. Dis. 144, 1274–1281.

Seibold, J.R., Giorno, R.C. and Claman, H.N. (1990). Dermal mast cell degranulation in systemic sclerosis. Arthritis Rheum. 33, 1702–1709.

Sekizawa, K., Caughey, G.H., Lazarus, S.C., Gold, W.M. and Nadel, J.A. (1989). Mast cell tryptase causes airway smooth muscle hyperresponsiveness in dogs. J. Clin. Invest. 83, 175–179.

Shalit, M., Schwartz, L.B., Golzar, N., von Allman, C., Valenzano, M., Fleekop, P., Atkins, P.C. and Zweiman, B. (1988). Release of histamine and tryptase in vivo after prolonged cutaneous challenge with allergen in humans. J. Immunol. 141, 821–826.

Shelhamer, J.H., Marom, Z. and Kaliner, M.A. (1980). Immunologic and neuropharmacologic stimulation of mucous glycoprotein release from human airways *in vitro*. J. Clin. Invest. 66, 1400–1408.

Silberstein, D.S. and David, J.R. (1986). Tumor necrosis factor enhances eosinophil toxicity to Schistosoma mansoni larvae. Proc. Natl. Acad. Sci. USA 83, 1055–1059.

Slungaard, A., Vercellotti, G.M., Walker, G., Nelson, R.D. and Jacob, H.S. (1990). Tumor necrosis factor alpha/cachectin stimulates eosinophil oxidant production and toxicity towards human endothelium. J. Exp. Med. 171, 2025–2041.

Smith, C.J., Smith, J.C. and Finn, M.C. (1987). The possible role of mast cells (allergy) in the production of keloid and hypertrophic scarring. J. Burn Care Rehabil. 8, 126–131.

Smith, J.M. (1974). Incidence of atopic disease. Med. Clin. North Am. 3–24.

Smith, T.J., Houghland, M.W. and Johnson, D.A. (1984). Human lung tryptase: purification and characterization. J. Biol. Chem. 259, 11046–11051.

Smyth, M.J., Zachariae, C.O., Norihisa, Y., Ortaldo, J.R., Hishinuma, A. and Matsushima, K. (1991). IL-8 gene expression and production in human peripheral blood lymphocyte subsets. J. Immunol. 146, 3815–3823.

Solley, G.O., Gleich, G.J., Jordan, R.E. and Schroeter, A.L. (1976). The late phase of the immediate wheal and flare skin reaction: its dependence on IgE antibodies. J. Clin. Invest. 58, 408–420.

Sommerhoff, C.P., Caughey, G.H., Finkbeiner, W.E., Lazarus, S.C., Basbaum, C.B. and Nadel, J.A. (1989). Mast cell chymase. A potent secretagogue for airway gland serous cells. J. Immunol. 142, 2450–2456.

Sonoda, T., Kitamura, Y., Haku, Y., Hara, H. and Mori, K.J. (1984). Proliferation of peritoneal mast cells in the skin of W/Wv mice that genetically lack mast cells. J. Exp. Med. 160, 138–151.

Sorbo, J., Jakobsson, A. and Norrby, K. (1994). Mast-cell histamine is angiogenic through receptors for histamine(1) and histamine(2). Int. J. Exp. Pathol. 75, 43–50.

Soter, N.A., Lewis, R.A., Corey, E.J. and Austen, K.F. (1983). Local effects of synthetic leukotrienes (LTC4, LTD4, LTE4 and LTB4) in human skin. J. Invest. Dermatol. 80, 115–119.

Spriggs, M.K., Armitage, R.J., Strockbine, L., Clifford, K.N., Macduff, B.M., Sato, T.A., Maliszewski, C.R. and Fanslow, W.C. (1992). Recombinant human CD40 ligand stimulates B cell proliferation and immunoglobulin E secretion. J. Exp. Med. 176, 1543–1550.

Standiford, T.J., Kunkel, S.L., Kasahara, K., Milia, M.J., Rolfe, M.W. and Strieter, R.M. (1991). Interleukin-8 gene expression from human alveolar macrophages: the role of adherence. Am. J. Respir. Cell Mol. Biol. 5, 579–585.

Stevens, R.L., Avraham, S., Gartner, M.C., Bruns, G.A., Austen, K.F. and Weis, J.H. (1988a). Isolation and characterization of a cDNA which encodes the peptide core of the secretory granule proteoglycan of human promyelocytic leukemia HL-60 cells. J. Biol. Chem. 263, 7287–7291.

Stevens, R.L., Fox, C.C., Lichtenstein, L.M. and Austen, K.F. (1988b). Identification of chondroitin sulfate E proteoglycans and heparin proteoglycans in the secretory granules of human lung mast cells. Proc. Natl. Acad. Sci. USA 85, 2284–2287.

Strobel, S., Miller, H.R.P. and Ferguson, A. (1981). Human intestinal mucosal mast cells: evaluation of fixation and staining techniques. J. Clin. Pathol. 34, 851–858.

Subba Rao, P.V., Friedman, M.M., Atkins, F.M. and Metcalfe, D.D. (1983). Phagocytosis of mast cell granules by cultured fibroblasts. J. Immunol. 130, 341–349.

Suzuki, N., Horiuchi, T., Ohta, K., Yamaguchi, M., Ueda, T., Takizawa, H., Hirai, K., Shiga, J., Ito, K. and Miyamoto, T. (1993). Mast cells are essential for the full development of silica-induced pulmonary inflammation: a study with mast cell-deficient mice. Am. J. Respir. Cell Mol. Biol. 9, 475–483.

Swain, S.L., Weinberg, A.D., English, M. and Huston, G. (1990). IL-4 directs the development of TH2-like helper effectors. J. Immunol. 145, 3796–3806.

Swerlick, R.A., Yancey, K.B., Thomas. J. and Lawley, T.J. (1988). A direct *in vivo* comparison of the inflammatory properties of human C5a and C5a des Arg in human skin. J. Immunol. 140, 2376–2381.

Tam, E.K. and Caughey, G.H. (1990). Degradation of airway neuropeptides by human lung tryptase. Am. J. Respir. Cell Mol. Biol. 3, 27–32.

Tannenbaum, S., Oertel, H., Henderson, W. and Kaliner, M.A. (1980). The biological activity of mast cell granules. I. Elicitation of inflammatory responses in rat skin. J. Immunol. 125, 325–335.

Taylor, I.K., O'Shaughnessy, K.M., Fuller, R.W. and Dollery, C.T. (1991). Effect of cysteinyl-leukotriene receptor antagonist ICI 204.219 on allergen-induced bronchoconstriction and airway hyperreactivity in atopic subjects [see comments]. Lancet 337, 690–694.

Teixeira, M.M. and Hellewell, P.G. (1993). Suppression by intradermal administration of heparin of eosinophil accumulation but not oedema formation in inflammatory reactions in guinea-pig skin. Br. J. Pharmacol. 110, 1496–1500.

te Velde, A.A., Klomp, J.P., Yard, B.A., De Vries, J.E. and Figdor, C.G. (1988). Modulation of phenotypic and functional properties of human peripheral blood monocytes by IL-4. J. Immunol. 140, 1548–1554.

Thompson, H.L., Schulman, E.S. and Metcalfe, D.D. (1988). Identification of chondroitin sulfate E in human lung mast cells. J. Immunol. 140, 2708–2713.

Thornhill, M.H. and Haskard, D.O. (1990). IL-4 regulates endothelial cell activation by IL-1, tumor necrosis factor, or IFN-gamma. J. Immunol. 145, 865–872.

Thornhill, M.H., Wellicome, S.M., Mahiouz, D.L., Lanchbury, J.S., Kyan Aung, U. and Haskard, D.O. (1991). Tumor necrosis factor combines with IL-4 or IFN-gamma to selectively enhance endothelial cell adhesiveness for T cells. The contribution of vascular cell adhesion molecule-1-dependent and -independent binding mechanisms. J. Immunol. 146, 592–598.

Thyphronitis, G., Tsokos, G.C., June, C.H., Levine, A.D. and Finkelman, F.D. (1989). IgE secretion by Epstein–Barr virus-infected purified human B lymphocytes is stimulated by interleukin 4 and suppressed by interferon gamma. Proc. Natl. Acad. Sci. USA 86, 5580–5584.

Toi, M., Harris, A.L. and Bicknell, R. (1991). Interleukin-4 is a potent mitogen for capillary endothelium. Biochem. Biophys. Res. Commun. 174, 1287–1293.

Tosato, G. and Pike, S.E. (1988). Interferon-beta 2/interleukin

6 is a co-stimulant for human T lymphocytes. J. Immunol. 141, 1556–1562.

Tounon de Lara, J.M., Okayama, Y., McEuen, A.R., Heusser, C.H., Church, M.K., and Walls, A.F. (1994). Release and inactivation of interleukin-4 by mast cells. In "Cells and Cytokines in Lung Inflammation" (eds M. Chignard, M. Pretolani, P. Renesto and B.B. Vargaftig), pp. 50–58, New York Acad. Sci., New York.

Urata, H., Kinoshita, A., Misono, K.S., Bumpus, F.M. and Husain, A. (1990). Identification of a highly specific chymase as the major angiotensin II forming enzyme in the human heart. J. Biol. Chem. 265, 22348–22357.

Uvnas, B. (1967). Mode of binding and release of histamine in mast cell granules of the rat. Fed. Proc. 26, 919–221.

Valone, F.H., Boggs, J.M. and Goetzl, E.J. (1994). Lipid mediators of hypersensitivity and inflammation. In "Allergy: Principles and Practice" 4th edn. (eds E. Middleton, C.C. Reed, E.F. Ellis, N.F. Adkinson, J.W. Yangurgen and W.W. Busse), pp. 302–319. C.W. Mosby, St Louis, MI.

van den Broek, M.F., van den Berg, W.B. and van de Putte, L.B. (1988). The role of mast cells in antigen induced arthritis in mice. J. Rheumatol. 15, 544–551.

Van Overveld, F.J., Jorens, P.G., Rampart, M., De Backer, W. and Vermeire, P.A. (1992). Tumor necrosis factor—a novel stimulus for human skin mast cells to secrete histamine and tryptase. Agents Actions 36 (Suppl. C), C256-C259.

Varney, V.A., Jacobson, M.R., Sudderick, R.M., Robinson, D.S., Irani, A.M., Schwartz, L.B., Mackay, I.S., Kay, A.B. and Durham, S.R. (1992). Immunohistology of the nasal mucosa following allergen-induced rhinitis. Identification of activated T lymphocytes, eosinophils and neutrophils. Am. Rev. Respir. Dis. 146. 170–176.

Vercelli, D., Jabara, H.H., Lee, B.W., Woodland, N., Geha, R.S. and Leung, D.Y. (1988). Human recombinant interleukin 4 induces Fc epsilon R2/CD23 on normal human monocytes. J. Exp. Med. 167, 1406–1416.

Vercelli, D., Jabara, H.H., Arai, K. and Geha, R.S. (1989). Induction of human IgE synthesis requires interleukin 4 and T/B cell interactions involving the T cell receptor/CD3 complex and MHC class II antigens. J. Exp. Med. 169, 1295–1307.

Viegas, M., Gomez, E., Brooks, J., Gatland, D. and Davies, R.J. (1987). Effect of the pollen season on nasal mast cells. Br. Med. J. Clin. Res. Ed. 294, 414.

Vilcek, J., Palombella, V.J., Henriksen DeStefano, D., Swenson, C., Feinman, R., Hirai, M. and Tsujimoto, M. (1986). Fibroblast growth enhancing activity of tumor necrosis factor and its relationship to other polypeptide growth factors. J. Exp. Med. 163, 632–643.

Volovitz, B., Osur, S.L., Bernstein, J.M. and Ogra, P.L. (1988). Leukotriene C4 release in upper respiratory mucosa during natural exposure to ragweed in ragweed-sensitive children. J. Allergy Clin. Immunol. 82, 414–418.

von Kutscher, F. (1910). Die Physiologische Wirkung einer Secalbase und des Imidazolylathamins. Zentralb. Physiol. 24, 163–165.

Von Recklinhausen F. (1863). Uber Eiter und Bindewehskorperchen. Virchows Arch. Pathol. Anat. 28, 157–175.

Wagner, M.M.F., Edwards, R.E., Moncraft, C.B. and Wagner, J.C. (1984). Mast cells and inhalation of asbestos in rats. Thorax 39, 539–544.

Walker, M., Harley, R., Maize, J., DeLustro, F. and LeRoy, E.C. (1985). Mast cells and their degranulation in the Tsk mouse model of scleroderma. Proc. Soc. Exp. Biol. Med. 180, 323–328.

Walls, A.F., Bennett, A.R., McBride, H.M., Glennie, M.J., Holgate, S.T. and Church, M.K. (1990a). Production and characterization of monoclonal antibodies specific for human mast cell tryptase. Clin. Exp. Allergy 20, 581–589.

Walls, A.F., Jones, D.B., Williams, J.H., Church, M.K. and Holgate, S.T. (1990b). Immunohistochemical identification of mast cells in formaldehyde fixed tissue using monoclonal antibodies. J. Pathol. 162, 119–126.

Walls, A.F., Roberts, J.A., Godfrey, R.C., Church, M.K. and Holgate, S.T. (1990c). Histochemical heterogeneity of human mast cells: disease-related differences in mast cells recovered by bronchoalveolar lavage. Int. Arch. Allergy Appl. Immunol. 92, 233–241.

Walls, A.F., Bennett, A.R., Sueiras-Diaz, J. and Olsson, H. (1992a). The kininogenase activity of human mast cell tryptase. Biochem. Soc. Trans. 20, 260S

Walls, A.F., Brain, S.D., Desai, A., Jose, P.J., Hawkings, E., Church, M.K. and Williams, T.J. (1992b). Human mast cell tryptase attenuates the vasodilator activity of calcitonin gene-related peptide. Biochem. Pharmacol. 43, 1243–1248.

Walsh, G.M., Wardlaw, A.J., Hartnell, A., Sanderson, C.J. and Kay, A.B. (1991a). Interleukin-5 enhances the in vitro adhesion of human eosinophils, but not neutrophils, in a leucocyte integrin (CD11/18)-dependent manner. Int. Arch. Allergy Appl. Immunol. 94, 174–178.

Walsh, L.J., Trinchieri, G., Waldorf, H.A., Whitaker, D. and Murphy, G.F. (1991b). Human dermal mast cells contain and release tumor necrosis factor alpha, which induces endothelial leukocyte adhesion molecule 1. Proc. Natl. Acad. Sci. USA 88, 4220–4224.

Wang, J.M., Rambaldi, A., Biondi, A., Chen, Z.G., Sanderson, C.J. and Mantovani, A. (1989). Recombinant human interleukin 5 is a selective eosinophil chemoattractant. Eur. J. Immunol. 19, 701–705.

Wardlaw, A.J., Dunnette, S., Gleich, G.J., Collins, J.V. and Kay, A.B. (1988). Eosinophils and mast cells in bronchoalveolar lavage in subjects with mild asthma. Relationship to bronchial hyperreactivity. Am. Rev. Respir. Dis. 137, 62–69.

Warringa, R.A., Koenderman, L., Kok, P.T., Kreukniet, J. and Bruijnzeel, P.L. (1991). Modulation and induction of eosinophil chemotaxis by granulocyte–macrophage colony-stimulating factor and interleukin-3. Blood 77, 2694–2700.

Warringa, R.A., Schweizer, R.C., Maikoe, T., Kuijper, P.H., Bruijnzeel P.L. and Koendermann, L. (1992a). Modulation of eosinophil chemotaxis by interleukin-5. Am. J. Respir. Cell Mol. Biol. 7, 631–636.

Warringa, R.A.J., Mengelers, H.J.J., Kuijper, P.H.M., Raaijmakers, J.A.M., Bruijnzeel, P.L.B. and Koenderman, L. (1992b). In vivo priming of platelet-activating factor-induced eosinophil chemotaxis in allergic asthmatic individuals. Blood 79, 1836–1841.

Watanabe, S., Watanabe, K., Ohishi, T., Aiba, M. and Kageyama, K. (1974). Mast cells in the rat alveolar septa undergoing fibrosis after ionising irradiation. Lab. Invest. 31, 555–567.

Wenzel, S.E., Fowler, A.A. and Schwartz, L.B. (1988). Activation of pulmonary mast cells by bronchoalveolar allergen challenge. In vivo release of histamine and tryptase in atopic subjects with and without asthma. Am. Rev. Respir. Dis. 137, 1002–1008.

Wershil, B.K., Mekori, Y.A., Murakami, T. and Galli, S.J. (1987). 125I-fibrin deposition in IgE-dependent immediate hypersensitivity reactions in mouse skin. Demonstration of the role of mast cells using genetically mast cell-deficient mice locally reconstituted with cultured mast cells. J. Immunol. 139, 2605–2614.

Wershil, B.K., Murakami, T. and Galli, S.J. (1988). Mast cell-dependent amplification of an immunologically nonspecific inflammatory response. Mast cells are required for the full expression of cutaneous acute inflammation induced by phorbol 12-myristate 13-acetate. J. Immunol. 140, 2356–2360.

Wershil, B.K., Levine, A.D., Nauert, I.P., Gordon, J.R. and Galli, S.J. (1990). Ionomycin stimulation of a growth factor-independent mouse mast cell induces differential IL-3 and IL-4 gene expression and release of product. FASEB J. 4, A1943.

Wershil, B.K., Wang, Z.S., Gordon, J.R. and Galli, S.J. (1991). Recruitment of neutrophils during IgE-dependent cutaneous late phase reactions in the mouse is mast cell-dependent. Partial inhibition of the reaction with antiserum against tumor necrosis factor-alpha. J. Clin. Invest. 87, 446–453.

Wheeler, A.P., Jesmok, G. and Brigham, K.L. (1990). Tumor necrosis factor's effects on lung mechanics, gas exchange and airway reactivity in sheep. J. Appl. Physiol. 68, 2542–2549.

Wichman, B.E. (1955). The mast cell count during the process of wound healing. Acta Path. Microbiol. Scand. 108 Suppl. 1–35.

Windaus, A. and Vogt, W. (1907). Synthese des imidazolylathylamins. Ber. Dtsch. Chem. Ges. 3, 3691–3695.

Wintroub, B.U., Kaempfer, C.E., Schechter, N.M. and Proud, D. (1986a). Human lung mast cell chymotrypsin-like enzyme: identification and partial characterization. J. Clin. Invest. 77, 196–201.

Wintroub, B.U., Kaempter, C.E., Schechter, N.M. and Proud. D. (1986b). Human lung mast cell chymotrypsin-like enzyme: identification and partial characterization. J. Clin. Invest. 77, 196–201.

Wodnar-Filpowicz. A., Heusser, C.H. and Moroni, C. (1989). Production of the haemopoietic growth factors GM-CSF and interleukin-3 by mast cells in response to IgE receptor-mediated activation. Nature 339, 150–152.

Wu, C.Y., Sarfati, M., Heusser, C., Fournier, S., Rubio Trujillo, M., Peleman, R. and Delespesse, G. (1991). Glucocorticoids increase the synthesis of immunoglobulin E by interleukin 4-stimulated human lymphocytes. J. Clin. Invest. 87, 870–877.

Wuepper, K.D., Bokisch, V., Muller-Eberhard, H.J. and Stoughton, R.B. (1972). Cutaneous responses to human C3 anaphylatoxin in man. Clin. Exp. Immunol. 11, 13–20.

Yamaguchi, Y., Suda, T., Ohta, S., Tominaga, K., Miura, Y. and Kasahara, T. (1991). Analysis of the survival of mature human eosinophils: interleukin–5 prevents apoptosis in mature human eosinophils. Blood 78, 2542–2547.

Ying, S., Durham, S.R., Jacobson, M.R., Rak, S., Masuyama, . K., Lowhagen, O., Kay, A.B. and Hamid, Q.A. (1994). T lymphocytes and mast cells express messenger RNA for interleukin-4 in the nasal mucosa in allergen-induced rhinitis. Immunology 82, 200–206.

Yoffe, J., Taylor, D.J. and Woolley, D.E. (1980 Mast cell products stimulate collagenase and prostaglandin E production by cultures of adherent rheumatoid synovial cells. Biochem. Biophys. Res. Commun. 122, 270–276.

Zhang, K., Clark, E.A. and Saxon, A. (1991). CD40 stimulation provides an IFN-gamma-independent and IL-4-dependent differentiation signal directly to human B cells for IgE production. J. Immunol. 146, 1836–1842.

Zuraw, B.L., O'Hair, C.H., Vaughan, J.H., Mathison, D.A., Curd, J.G. and Katz, D.H. (1981). Immunoglobulin E-rheumatoid factor in the serum of patients with rheumatoid arthritis, asthma and other diseases. J. Clin. Invest. 68, 1610–1613.

5. Eosinophils: Effector Leukocytes of Allergic Inflammatory Responses

Kaiser G. Lim *and* Peter F. Weller

1. Eosinophils and the Pathophysiology of Asthma

In 1866, Paul Erlich identified the eosinophil as a distinct granulocyte based on its ability to take up eosin. The presence of eosinophils in asthmatic sputum was reported in 1889, and a positive association between circulating eosinophil levels and asthma was made in 1922 (Huber and Koessler, 1922). Circumstantial evidence linking eosinophils and asthma began to accrue when eosinophilic infiltration of airways was found to be a regular feature of fatal asthma and of mucosal biopsies obtained from asthmatic patients (Laitinen *et al.*, 1985; Djukanovic *et al.*, 1990a). Eosinophils are rarely seen in bronchial lavages and biopsies of normal subjects (Djukanovic *et al.*, 1990b; Merchant *et al.*, 1992; Ollerenshaw and Woolcock, 1992; Laitinen *et al.*, 1993b).

The histologic findings in asthma consist of massive epithelial shedding, hyperplasia of smooth muscle fibers, mucus glands and goblet cells, basement membrane thickening (Brewster *et al.*, 1990), and eosinophil infiltra-tion of bronchial cartilages, interstitium, and epithelium (Laitinen *et al.*, 1985; Beasley *et al.*, 1989; Djukanovic *et al.*, 1990a). By immunohistochemistry, secreted eosinophil cationic protein (EG2) and anti-major basic protein (MBP) localized to sites of bronchial epithelial damage providing evidence of *in vivo* eosinophil degranulation (Wardlaw *et al.*, 1988). There were ultrastructural signs of cellular activation with electron lucency of specific eosinophil granular matrix and loss of the crystalline core. Corkscrew-shaped mucus, epithelial cell clumps, eosinophilia and needle-like Charcot–Leyden crystals were present in asthmatic sputa. The latter is now known to be composed of eosinophil-derived lysophospholipase (Weller *et al.*, 1980). In a study of serial biopsies from the bronchial mucosa of an asthmatic patient at various stages of activity, the appearance and disappearance of eosinophilic infiltration coincided with the clinical activity and resolution, respectively (Laitinen *et al.*, 1991). These and other findings hint at a causal role for the eosinophil in asthma.

Asthma is well known to be associated with an elevated eosinophil count (Horn *et al.*, 1975; Ädelroth *et al.*,

1986). Elevated peripheral eosinophil counts in asthmatics have correlated with decreases in specific airway conductance, forced expiratory volume (FEV$_1$), maximum mid-expiratory flow rate (Horn *et al.*, 1975), bronchial hyperreactivity (BHR) to histamine (H; Taylor and Luksza, 1987) and clinical severity scores (Aas; Bousquet *et al.*, 1990). When circulating eosinophils were isolated from asymptomatic asthmatics, a large proportion of cells (35%) was recovered with centrifugal density less than 1.082 g ml^{-1} compared to normal subjects (10%; Frick *et al.*, 1989). Alteration of centrifugal density of eosinophils is one of the phenotypic responses to cellular activation (Hansel *et al.*, 1990; Fukuda and Makino, 1992). Not only is the number of cells increased in asthma, circulating eosinophils may be activated intravascularly as well.

Nocturnal exacerbations in asthma have been associated with significant nocturnal elevation in peripheral eosinophil counts well above the normal nocturnal elevations (Calhoun *et al.*, 1992). This nocturnal elevation has been related to an increasing susceptibility of developing a late asthmatic response (LAR) after allergen inhalational challenge in asthmatic subjects (Mohiuddin and Martin, 1990). When allergen was administered in the evening, nine out of ten subjects developed an LAR while only four of the same ten subjects developed LAR when the allergen was given in the morning.

Although eosinophilic infiltration of the airway is virtually diagnostic of asthma in the right clinical context, there are reports of fatal asthma without airway eosinophilia (Gleich *et al.*, 1987; Strunk, 1993; Sur *et al.*, 1993). Initially thought to be a constant feature of asthmatic exacerbations, peripheral blood eosinophilia was not uniformly present in all asthmatic patients examined (Bruijnzeel *et al.*, 1987). Still others have noted peripheral eosinophilia (>5%) to be predictive of respiratory disorders only in young subjects with positive allergy skin reactions (Burrows *et al.*, 1980). Eosinophilia alone, without a positive skin test, did not appear to be related to ventilatory impairment (Burrows *et al.*, 1980); neither is sputum eosinophilia specific for asthma. Fifty-seven per cent of asthmatics and 58% of wheezing chronic bronchitics had more than 80% eosinophils in the sputum (Vieira and Prolla, 1979). Sputum eosinophilia (>40%) has been found in emphysema, chronic bronchitis without wheeze, tuberculosis and idiopathic pulmonary fibrosis (Vieira and Prolla, 1979). Thus while eosinophilia in pulmonary secretions is almost uniform in asthma, other disease processes, not associated with the physiologic alterations of asthma, may also be accompanied by eosinophilic infiltration.

In an allergen inhalation model of asthma, subjects who develop the LAR have an initial drop in circulating eosinophil count followed by a rise at 48 h post-challenge (Cookson *et al.*, 1989). The initial drop may reflect the recruitment of circulating eosinophils to the lung where they participate in the development of a late phase response. The rise at 48 h may represent a bone marrow response of eosinophilpoiesis as the airway inflammation persists and mediators spill over into the systemic circulation. This is supported by the finding that circulating eosinophil/basophil progenitor cells increase only in atopic asthmatics who develop the LAR post-challenge (Gibson *et al.*, 1990). The intensity of the local inflammation is an important factor since only subjects who exhibit both early asthmatic responses (EARs) and LARs have a higher percentage of hypodense circulating eosinophils post-challenge (Frick *et al.*, 1989). During steroid treatment, blood eosinophil level decreased significantly within the first day of clinical exacerbations in chronic asthmatics (Charles *et al.*, 1979; Baigelman *et al.*, 1983). The sputum eosinophilia decreased significantly only by the third day of therapy, lagging behind that of the peripheral count (Brown 1958; Baigelman *et al.*, 1983). The lag between blood and sputum eosinophilia may indicate differences in the initiation, intensity and resolution of the local airway inflammation in asthma.

The development of LAR is positively related to early recruitment of eosinophils and eosinophil degranulation in the airway after allergen inhalation challenge (De Monchy *et al.*, 1985; Rossi *et al.*, 1991). The numbers of degranulated eosinophils per millimeter of basement membrane, the absolute numbers of eosinophils, and epithelial cells and the amounts of MBP on lavage have been shown to have significant inverse correlations with BHR (Wardlaw *et al.*, 1988; Azzawi *et al.*, 1990; Bentley *et al.*, 1992). The variable with the best correlation was the level of MBP (Wardlaw *et al.*, 1988). In studies using allergic cynomolgus monkeys, the influx of eosinophils with release of MBP and eosinophil peroxidase (EPO) was accompanied by the development of BHR (Gundel *et al.*, 1989, 1992a,b). Eosinophil infiltration showed positive correlation with the incidence of epithelial tight junction opening the degree of epithilial intercellular space widening, and epithelial cell shedding (Wardlaw *et al.*, 1988; Ohashi *et al.*, 1992). All three were negatively related to BHR. These studies suggest that in addition to the eosinophil influx, the degree of eosinophil activation and degranulation may be important in the development of BHR. The current hypothesis is that eosinophil-mediated epithelial damage plays a significant role in the development of BHR in asthma.

Using local allergen instillation into subsegmental bronchi of atopic nonasthmatics, MBP, eosinophil catonic protein (ECP), eosinophil-derived neurotoxin (EDN) and EPO were measured at 105 ng ml^{-1} 52.5 ng ml^{-1}, 43.1 ng ml^{-1} and 164.2 ng ml^{-1}, respectively, post-challenge (Sedgwick *et al.*, 1991). Bronchoalveolar lavage (BAL) leukotriene (LT) C$_4$ (Wenzel *et al.*, 1990; Sedgwick *et al.*, 1991) was elevated post-challenge and correlated positively with the eosinophil count (Sedgwick *et al.*, 1991). The eosinophil is a likely cellular source of the BAL LTC$_4$ since a corresponding increase in mast cell tryptase was not detected (Sedgwick *et al.*, 1991).

Bronchovascular permeability, as measured by BAL albumin, was positively related with the eosinophil count post challenge (Collins *et al.*, 1993). Thus eosinophils and their degranulation products have a role in the pathogenesis of bronchospasm, BHR and mucosal edema in asthma (Sedgwick *et al.*, 1991; Collins *et al.*, 1993).

2. Mechanism of Selective Eosinophil Recruitment into the Airways

The mechanisms by which eosinophils selectively accumulate in areas of allergic inflammation involve: (1) interactions via adhesion molecules; (2) eosinophil chemoattractants; and (3) prolonged tissue survival by delaying apoptosis. There is evidence to indicate that circulating eosinophils in asthmatics are primed intravascularly with enhanced adherence and transmigration capacity (Walsh *et al.*, 1991b; Moser *et al.*, 1992a,b). Freshly isolated eosinophils from allergic asthmatic donors adhere and migrate spontaneously across interleukin-1 (IL-1)- and tumour necrosis factor α (TNFα)-stimulated human umbilical vein endothelial cells (HUVEC), while those from non-atopic donors adhere but do not transmigrate (Moser *et al.*, 1992b). Eosinophils preincubated with IL-3, IL-5 and granulocyte–macrophage colony-stimulating factor (GM-CSF) have enhanced *in vitro* adhesion to endothelial cells and can be made to transmigrate (Walsh *et al.*, 1991b; Moser *et al.*, 1992b).

Increased percentages of circulating light density eosinophils were observed in asymptomatic asthmatics and in asthmatics who developed a late phase response after allergen challenge (Fukuda *et al.*, 1985; Frick *et al.*, 1989). Hypodensity in eosinophils can be generated immediately and non-selectively with platelet-activating factor (PAF), F-met-leu-phe (FMLP), A23187 and serum-opsonized zymosan (SOZ; Kloprogge *et al.*, 1989). Hypodensity can also be generated selectively over days with exposure to the hematopoietic cytokines IL-5 (Owen *et al.*, 1987), IL-3 (Rothenberg *et al.*, 1988) and GM-CSF (Rothenberg *et al.*, 1989). IL-5 and GM-CSF have been identified in the peripheral blood (Walker *et al.*, 1991a; Corrigan *et al.*, 1993) and BAL fluid of asthmatics (Broide *et al.*, 1992a; Walker *et al.*, 1992). IL-5 has the distinct feature of being specific for eosinophils (Clutterbuck *et al.*, 1989). When eosinophils are cultured with these cytokines, they have increased survival, enhanced cytotoxicity to schistosomula, and enhanced LTC$_4$ generation (Silberstein *et al.*, 1986; Rothenberg *et al.*, 1988, 1989; Weller 1992). IL-3, IL-5 and GM-CSF modulate and induce eosinophil chemotaxis to PAF, IL-8 and FMLP (Wang *et al.*, 1989; Warringa *et al.*, 1991). Pre-exposure of eosinophils to IL-5 dramatically potentiates the transendothelial migration response to RANTES *in vitro* (Ebisawa *et al.*, 1994). In addition, IL-5-cultured eosinophils degranulated more readily to

secretory IgA and IgG (Kita *et al.*, 1992). IL-5 is chemotactic only for eosinophils (Wang *et al.*, 1989; Sehmi *et al.*, 1992). It primes the chemotactic response of eosinophils from non-atopics to PAF, FMLP and LTB$_4$ (Sehmi *et al.*, 1992). This effect is not observed in eosinophils obtained from atopic subjects, suggesting *in vivo* desensitization to IL-5 (Sehmi *et al.*, 1992). Thus eosinophils primed intravascularly by specific cytokines have increased adherence and transmigration capacity, and augmented response to non-specific chemoattractants.

2.1 EOSINOPHIL–ENDOTHELIAL INTERACTIONS

The predominance of eosinophils in asthmatic airways is the result of a concerted recruitment process initiated in the airways and communicated through the endothelium to the peripheral circulation. Eosinophils, lymphocytes and monocytes, but not neutrophils, express very late activation antigen (VLA)-4 $\alpha_4\beta_1$ (Bochner *et al.*, 1991; Dobrina *et al.*, 1991; Kyan-Aung *et al.*, 1991; Walsh *et al.*, 1991a; Weller *et al.*, 1991) and $\alpha_4\beta_7$ (Erle *et al.*, 1994). By interacting with vascular cell adhesion molecule-1 (VCAM-1) on endothelial cells, a selective pathway is provided for the recruitment of lymphocytes, eosinophils and monocytes (Berman and Weller, 1992; Erle *et al.*, 1994). Cytokines detectable in the airways of symptomatic asthmatics include IL-5, GM-CSF, IL-4, TNFα and IL-1 (Walker *et al.*, 1991a; Broide *et al.*, 1992a). IL-4, TNFα and IL-1 are capable of inducing VCAM-1 in HUVEC (Thornhill *et al.*, 1991; Briscoe *et al.*, 1992). Interferon γ (IFNγ) and TNFα in combination with IL-3, GM-CSF or IL-5 can induce expression of intercellular adhesion molecule-1 (ICAM-1) on eosinophils (Czech *et al.*, 1993).

In an allergic sheep model, monoclonal antibody (mAb) to α_{4j}-integrin (HP1/2) administered intravenously and via aerosol attenuated BHR without altering the cellular infiltration (Abraham *et al.*, 1994). Using an mAb to ICAM-1 (R6.5), daily intravenous treatment attenuated eosinophil infiltration and airway hyperresponsiveness in a cynomolgus monkey model of asthma using three alternate day antigen inhalations (Wegner *et al.*, 1990). This is in contrast to the ineffectiveness of intravenous R6.5 in reducing eosinophilic airway inflammation and BHR in a cynomolgus monkey model of established asthma (Gundel *et al.*, 1992b). Treatment with steroid followed by treatment with intravenous R6.5 prevented the recurrence of airway inflammation and BHR. The different response to R6.5 in these two primate models of asthma is instructive since anti-ICAM-1 antibody had no effect on established airway inflammation. With steroid therapy and resolution of inflammation, R6.5 would then be able to block further recruitment of inflammatory cells into the airways. The ability of aerosolized mAbs to adhesion molecules to reduce BHR without decreasing cellular infiltrate suggests effects other than blocking leukocyte–endothelial adhesion (Anwar *et al.*, 1993; Abraham *et al.*, 1994). The expression of adhesion molecules on basal epithelial cells may provide additional signals to the infiltrating cells and influence their function (Dri *et al.*, 1991). The use of mAbs against ICAM-1 and α_4-integrins in asthma may be of therapeutic importance in the future if cellular infiltration and BHR can be abrogated.

There is data to suggest that *in vivo* the endothelium may participate to prevent rapid washout of particular chemoattractants (Tanaka *et al.*, 1993b). Recently molecular mechanisms have been described that permit cytokines to act as adhesion triggers (Tanaka *et al.*, 1993a). Cytokines [transforming growth factor β (TGFβ) and GM-CSF] and chemokines (IL-8 and MIP-1β) that bind to glycosaminoglycan side chains of proteoglycans are immobilized on endothelial surfaces (Rot, 1992; Tanaka *et al.*, 1993b). When they encounter a receptive leukocyte, the cytokine can trigger integrin-mediated adhesion. It has been shown that MIP-1β can trigger adhesion of T cell subsets to VCAM-1 via β_1-integrins (Tanaka *et al.*, 1993a). The regional differences in endothelium allow differential binding of chemokines. These chemokines in turn display selective chemotactic recruitment of leukocyte subsets. This is another way by which the type of leukocyte entering a tissue can be regulated.

Immunostaining with antibodies against ICAM-1, E-selectin and VCAM-1 in asthmatics and normal bronchial tissues after allergen inhalational challenge revealed constitutive expression of ICAM-1, E-selectin and VCAM (Montefort *et al.*, 1992; Bentley *et al.*, 1993). An increased basal epithelial expression of ICAM-1 in asthmatics as well as increased endothelial expression of ICAM-1 and E-selectin was observed in intrinsic asthmatics. Allergen challenge did not significantly increase the overall endothelial expression of ICAM-1, E-selectin and VCAM-1 (Bentley *et al.*, 1993). The finding of constitutive expression of ICAM-1, VCAM and E-selectin in the endothelium of asthmatics and normal subjects suggests that *in vivo* expression of adhesion molecules in the lung is different from *in vitro* cultures of HUVEC (Montefort *et al.*, 1992; Bentley et al., 1993).

2.2   EOSINOPHILS AND CHEMOATTRACTANTS

BAL studies have shown that there is an increase in CD4$^+$ lymphocytes, monocytes and eosinophils in asthmatic airways (Walker *et al.*, 1991b). Among the multitude of agents capable of inducing eosinophil migration (Resnick and Weller, 1993), lymphocyte chemotactic factor (LCF; Resnick and Weller, 1993), RANTES (Kameyoshi *et al.*, 1992), MIP-1α (Rot *et al.*, 1992), MCP-3 (Dahinden *et al.*, 1994), C5a and PAF (Wardlaw

et al., 1986) are especially potent eosinophil chemoattractants (Wardlaw *et al.*, 1986; Morita *et al.*, 1989). Many cytokines, including IL-5, IL-2 and GM-CSF, are present in the airways (Walker *et al.*, 1991a; Broide *et al.*, 1992a). They may function both as proinflammatory growth factors and as chemoattractants. Preliminary data suggest that LCF, MIP-1α and RANTES may be present as well in asthmatic airways (Alam *et al.*, 1994; Cruikshank *et al.*, 1994a). LCF has the ability to induce migration of CD4$^+$ cells like lymphocytes, eosinophils and monocytes (Rand *et al.*, 1991a; Cruikshank *et al.*, 1994b). RANTES is preferentially chemotactic for human CD4$^+$ memory T lymphocytes, monocytes and eosinophils (Schall *et al.*, 1990), but has not demonstrated chemoattractant activity *in vitro* for CD4$^+$ naive T lymphocytes, CD8$^+$ cytotoxic lymphocytes, B lymphocytes or neutrophils (Schall *et al.*, 1990). RANTES induced eosinophil transendothelial migration *in vitro* without affecting neutrophil transmigration (Ebisawa *et al.*, 1994). The intradermal injection of human RANTES in experimental animals has resulted in the recruitment of lymphocytes (Murphy *et al.*, 1994), monocytes and eosinophils (Meurer *et al.*, 1993). MCP-3 combines the properties of RANTES, a chemoattractant, and MCP-1, a highly effective stimulus of mediator release (Dahinden *et al.*, 1994). RANTES and MIP-1α significantly induced the production of reactive oxygen species by human eosinophils (Kapp *et al.*, 1994). Eosinophils, upon activation, express IL-2 receptor β-subunit (CD25) (Rand *et al.*, 1991b) and are chemokinetic toward IL-2. The low molar concentration needed to elicit the migratory response suggests that the high affinity IL-2 receptor is involved. Selective recruitment of leukocyte subsets may in part be regulated through the elaboration and release of cytokines and mediators that are specific for leukocyte subsets.

The process of adherence and transmigration across a stimulated endothelium may serve to activate the eosinophil. Sputum eosinophils from asthmatics were found to express significantly higher levels of CD11b and CD11c than blood eosinophils (Hansel *et al.*, 1991). ICAM-1 and human histocompatibility antigen class II (HLA-DR) were detected in sputum but not in blood eosinophils (Hansel *et al.*, 1991). BAL fluid eosinophils from asthmatics show a different surface phenotype from blood eosinophils. There was up-regulation of ICAM-1, LFA-3 (CD58), HLA-DR, CD11b (Kroegel *et al.*, 1994), CD11c, CD67 and CD63 with down-regulation of L-selectin (Sedgwick *et al.*, 1992; Mengelers *et al.*, 1994). *In vitro*, the up-regulation of CD11b and shedding of L-selectin has been noted in eosinophils that have undergone transmigration (Ebisawa *et al.*, 1992). This up-regulation of CD11b expression was associated with an increased capacity to generate superoxide after stimulation with opsonized zymosan (Walker *et al.*, 1993). ICAM-1 and HLA-DR were not induced by transmigration. The difference in the function and immunophenotype of eosinophils recovered from the airways and blood indicates that eosinophils undergo further activation after they leave the vascular compartment.

3. Eosinophil Effector Mechanism

3.1 THE ROLE OF EOSINOPHIL GRANULAR PROTEINS

The effector role of the eosinophil in the pathogenesis of asthma is mediated in part by the release of eosinophil-specific granular proteins. The four principal basic proteins in the secondary granules of the eosinophil are MBP, EPO, ECP and EDN.

Extracellular MBP can be detected in mucus plugs, along damaged bronchial epithelial surfaces, and in necrotic areas beneath the basement membrane in the lung tissue of fatal asthmatics (Filley *et al.*, 1982). Respiratory epithelial damage is most likely the result of eosinophil infiltration, degranulation and MBP release (Gleich *et al.*, 1988). MBP is detected in BAL fluid in asthmatics and is increased after antigen challenge (Wardlaw *et al.*, 1988).

When MBP was applied to guinea-pig tracheal rings, it produced ciliostasis at $10 \, \mu g \, ml^{-1}$, extensive epithelial damage at $50{-}100 \, \mu g \, ml^{-1}$, and mucosal sloughing at greater than $250 \, \mu g \, ml^{-1}$ (Frigas *et al.*, 1980). MBP can disrupt the plasma membrane. It changed the ordered state of acidic lipids on the liposomes via electrosatic and hydrophobic interactions (Abu-Ghazaleh *et al.*, 1992). This resulted in the fusion and lysis of liposomes. MBP interaction with the glycocalyx of cells is another mechanism for the cytotoxic effect of MBP (Gleich *et al.*, 1994). MBP can disrupt ciliary activity by inhibiting the ATPase activity of axonemes (Hastie *et al.*, 1987). MBP is highly cationic and induced bronchoconstriction and a dose-related increase in BHR when instilled into the airways of cynomolgus monkeys (Gundel *et al.*, 1991). This effect appears to be independent of its cytotoxic property and has been attributed to the cationic charge of MBP (Uchida *et al.*, 1993). This ability to induce BHR can be mimicked by other polycationic proteins and abrogated with polyanionic polymers to neutralize the charge (Uchida *et al.*, 1993). Other properties of MBP include inducing PGE_2 production and chloride and water transport in canine tracheal epithelium (Jacoby *et al.*, 1988), activating neutrophil chemiluminescence (Moy *et al.*, 1990), macrophage superoxide generation (Rankin *et al.*, 1992), and lysosomal enzyme release (Moy *et al.*, 1990). MBP, ECP and EPO are all capable of stimulating mast cell H release (Zheutlin *et al.*, 1984). Only native MBP can induce H release from human basophils in a non-cytolytic, IgE-independent fashion (Zheutlin *et al.*, 1984).

Eosinophil cationic protein can be detected by immunohistochemistry in fatal cases of asthma (Venge *et al.*, 1988). The small amount of ECP normally detected in serum is related to the peripheral eosinophil count (Spry, 1988). In mild asthmatics, the serum ECP is elevated (Ädelroth *et al.*, 1990). BAL ECP level increased significantly after antigen provocation (Sedgwick *et al.*, 1991). Intratracheal instillation of ECP produced patchy epithelial injury in the bronchial tree of rabbits (Dahl *et al.*, 1987). It can inhibit proteoglycan degradation in fibroblast (Hernnäs *et al.*, 1992), stimulate mucus secretion by airway epithelial cells (Lundgren *et al.*, 1991), induce basophil H release and inhibit T cell proliferation *in vitro* (Peterson *et al.*, 1986). A proposed mechanism of the cytotoxic action of ECP is as a pore forming protein (Young *et al.*, 1986), although this has not been fully established.

EPO can cause a dose-dependent lysis of epithelial cells (Ayars *et al.*, 1989) and endothelial cells (Slungaard and Mahoney, 1991; Yoshikawa *et al.*, 1993). Paradoxically, EPO can oxidize LTC_4 and LTD_4 to neutralize their smooth muscle contractile effect (Henderson *et al.*, 1982; Weller *et al.*, 1991).

3.2 EOSINOPHILS AND LIPID MEDIATORS IN ASTHMA

3.2.1 Eosinophils and Leukotrienes

Peptidoleukotrienes are present in the sputum (Lam *et al.*, 1988), urine (Taylor *et al.*, 1989), BAL fluid (Wardlaw *et al.*, 1989), plasma (Okubo *et al.*, 1987), and nasal secretions (Ferreri *et al.*, 1988) of asthmatic patients. Lung specimens (Dahlen *et al.*, 1983) and peripheral blood leukocytes (Mita *et al.*, 1986) from allergic asthmatics release leukotrienes after specific antigen challenge. In allergen-induced LAR a significant rise in LTC_4 level was detected in BAL fluid at 6 h post-challenge only in dual responders and not in single responders (Diaź *et al.*, 1989). Urinary LTE_4, a metabolite of LTC_4, is elevated many fold in aspirin-sensitive asthmatics given aspirin (Lee, 1992). Local segmental bronchoprovocation with antigen in a group of allergic rhinitics revealed BAL fluid LTC_4 levels of 134 ± 45 and $1308 \pm 664 \, ng \, ml^{-1}$ at 10 min and at 48 h, respectively. BAL fluid LTC_4 at 48 h correlated significantly with BAL eosinophil count (Sedgwick *et al.*, 1991). LTC_4 was the predominant sulfidopeptide leukotriene found in BAL fluid from atopic asthmatics (Wenzel *et al.*, 1990). Baseline LTC_4 levels of $64 \pm 18 \, pg \, ml^{-1}$ increased to $616 \pm 193 \, pg \, ml^{-1}$ 5 min after allergen challenge in atopic asthmatics (Wenzel *et al.*, 1990).

The cells that are known sources of leukotrienes are mast cells (Scott and Kaliner, 1993), basophils, monocytes (LTB_4 and LTC_4; Czop and Austen, 1985) and eosinophils (Weller *et al.*, 1983). Leukotrienes cause bronchoconstriction (Bisgaard *et al.*, 1983; Smith *et al.*,

1985), mucosal edema, mucus secretion (Marom *et al.*, 1982), and BHR (O'Hickey *et al.*, 1991). Inhaled LTC$_4$ and LTD$_4$ are approximately 6000 times more potent than H and produce a more sustained response in asthmatics (Weiss *et al.*, 1983). Airways of asthmatics were approximately 14-, 25-, 6-, 9- and 219-fold more responsive to H, methacholine, LTC$_4$, LTD$_4$ and LTE$_4$, respectively, than normal subjects (Arm *et al.*, 1990). LTE$_4$ inhalation can induce eosinophilic infiltration into asthmatic airways (Laitinen *et al.*, 1993a).

Eosinophils preferentially elaborate LTC$_4$ and contain a specific glutathione-S-transferase, LTC$_4$ synthetase, to form LTC$_4$ from LTA$_4$. Eosinophils elaborate LTC$_4$ *in vitro* when stimulated by calcium ionophore (Weller *et al.*, 1983; Shaw *et al.*, 1984), immunoglobulins (Shaw *et al.*, 1985; Moqbel *et al.*, 1990), PAF and FMLP (Tamura *et al.*, 1988; Takafugi *et al.*, 1991). Eosinophil LTC$_4$ generation by C5a and PAF can be augmented by preincubation with IL-3 and IL-5 for 90 min (Takafugi *et al.*, 1991). Hypodense peripheral blood eosinophils release more LTC$_4$ than normal density eosinophils (Kajita *et al.*, 1985; Hodges *et al.*, 1988).

3.2.2 Eosinophils and Platelet-activating Factor

PAF has a wide range of biological effects that has made it one of the most studied mediators in asthma. PAF exhibits the following properties: (1) stimulates chemotaxis of eosinophils and neutrophils (Wardlaw *et al.*, 1986); (2) increases airway vascular permeability and promotes mucosal edema (Evans *et al.*, 1987); (3) induces bronchoconstriction (Rubin *et al.*, 1987); and (4) increases airway hyperresponsiveness (Smith, 1991; Page, 1992). PAF is detectable in plasma and nasal lavages during allergen-induced bronchoprovocation (Miadonna, 1989; Chan-Yeung, 1991). There is considerable tachyphylaxis with PAF. Clinical trials with PAF antagonist in asthma have been disappointing. The contribution of PAF in asthma may not be very dominant since PAF antagonists fail to significantly decrease the early and late phase asthmatic responses of subjects post-allergen (Freitag *et al.*, 1993; Kuitert *et al.*, 1993).

On a per cell basis, the eosinophil is considered one of the more potent PAF producers (Spry *et al.*, 1992). Immunoglobulin-mediated stimulation may be an important physiological stimulus of PAF generation in eosinophils. Eosinophils produced more PAF than neutrophils when stimulated through Fcγ R with IgG-sepharose beads (Cromwell *et al.*, 1990). Eosinophils synthesize PAF in response to FMLP, C5a, A23187, and unopsonized zymosan. After stimulation with A23187, 1–2 pg of PAF per 10^6 eosinophils can be detected extracellularly (Lee *et al.*, 1984) and 35 ng of PAF intracellularly per 10^6 eosinophils (Cromwell *et al.*, 1990). The majority of PAF generated

remained cell associated suggesting an autocrine role to activate eosinophils (Burke *et al.*, 1990).

4. Regulation of Eosinophil Function

4.1 EOSINOPHILS AND THE CYTOKINE NETWORK

There is increasing recognition that cytokines play an important role in the pathogenesis of asthma (Barnes, 1994). They constitute part of the intercellular messages that control the traffic of cells and molecules. Cytokines regulate cell–cell interactions and the cellular responses to allergic inflammation. Different cytokines alone and in combinations activate distinct, and at times overlapping sets of functions that contribute to the overall pathophysiology of asthma (Table 5.1). Cytokines are produced by immune and inflammatory cells that are activated in asthma. This results in the recruitment, prolonged survival, and activation of these cells. The ability of the eosinophil to elaborate and release granule-derived proteins and lipid mediators is well recognized (Table 5.2). The significance of eosinophil-derived cytokines is not clearly appreciated (Moqbel, 1994; Weller, 1994). While many cytokines may be derived from mast cells (Bradding *et al.*, 1994), lymphocytes and macrophages, the eosinophil's ability to synthesize, store and release cytokines increases the versatility of the eosinophil in allergic inflammatory reactions. Eosinophil-derived cytokines may function primarily in an autocrine fashion and also act as paracrine mediators to influence the activities of surrounding cells. Given the extensive and early eosinophil infiltration in asthma, eosinophil-derived cytokines may potentially contribute to the inflammatory amplification loop, the recruitment of cells, the presentation of antigen to T lymphocytes, and the modulation of tissue repair.

Table 5.1 Cytokines present in asthmatic airway BAL fluid

Cytokine	Amount (pg ml^{-1})	Reference
TNFα	578 ± 917	Broide *et al.* (1992a)
IFNγ	85 ± 1.6	Walker *et al.* (1992)
IL-1β	266 ± 270	Broide *et al.* (1992a)
	57 ± 5.9	Borish *et al.* (1992)
IL-2	1.4 ± 2.8	Broide *et al.* (1992a)
	36.5 ± 14.5	Walker *et al.* (1992)
IL-4	28.9 ± 7.2	Walker *et al.* (1992)
IL-5	2.12 ± 0.56	Walker *et al.* (1992)
	654 ± 416*	Sedgwick *et al.* (1991)
IL-6	225 ± 327	Broide *et al.* (1992a)
GM-CSF	24 ± 41	Broide *et al.* (1992a)

* Rhinitic non-asthmatics after allergen challenge.

Table 5.2 Cytokines elaborated by eosinophils

Cytokine	Source/condition	Method of Detection	Reference
TGFα	Oral tumors	In situ hybridization	Wong et al. (1990)
	HES	Immunocytochemistry	
TGFβ	Eosinophilic donors	Northern blot	Wong et al. (1991)
		In situ hybridization	
		Immunocytochemistry	
	Nasal polyp	In situ hybridization	Ohno et al. (1992)
		Immunohistochemistry	
MIP-1α	HES	In situ hybridization	Costa et al. (1993)
	Eosinophilic and normal donor	Northern blot	
	Nasal polyp		
GM-CSF	IFN-γ and ionomycin	In situ hybridization	Moqbel et al. (1991)
	stimulation	Immunocytochemistry	
	Ionomycin stimulation	Survival blocked by anti-GM-CSF	Kita et al. (1991)
	BAL after endobronchial	In situ hybridization	Broide et al. (1992b)
	antigen challenge		
TNFα	HES,	In situ hybridization	Costa et al. (1993)
	Atopic donor	Northern blot	
	Normal donors	Immunocytochemistry	
	Nasal polyp	ELISA	
	HES	EM immunogold staining	Beil et al. (1993)
IL-1α	PMA stimulation	Northern blot	Weller et al. (1993)
	HES	In situ hybridization	
		Immunocytochemistry	
IL-3	Ionomycin stimulation	Survival blocked by anti-IL-3	Kita et al. (1991)
		ELISA	
IL-5	BAL cells of asthmatics	In situ hybridization	Broide et al. (1992b)
		EM immunogold staining	
	HES	In situ hybridization	Dubucquoi et al. (1994)
	Eosinophilic cystitis	EM immunogold-staining	
		Immunohistochemistry	
	HES heart disease	In situ hybridization	Desreumaux et al. (1993)
		Immunohistochemistry	
	Coeliac disease	In situ hybridization	Desreumaux et al. (1992)
IL-6	Constitutively expressed	In situ hybridization	Hamid et al. (1992)
	Increased by IFN-γ	Northern blot analysis	
		Immunocytochemistry	Moqbel et al. (1994)
	HES, asthmatic and normal	In situ hybridization	Melani et al. (1993)
	donors	RT-PCR	
IL-8	A23187 stimulation	RT-PCR	Braun et al. (1993)
		ELISA	
		Immunocytochemistry	

HES, hypereosinophilic syndrome.

4.2 EOSINOPHILS AND IMMUNOGLOBULIN RECEPTORS

Using BAL fluid/serum quotient of IgG, asthmatics often have quotients over 1.0, suggesting local production of IgG (Out et al., 1991). No specific deficiency or excess for any of the IgG subclasses has been detected. The concentration of secretory IgA in the BAL fluid of asthmatics is elevated compared to controls (median 0.48 vs 1.29 mg^{-1}; $p < 0.01$; Van De Graaf et al., 1991). No difference in IgG, IgA, IgM or IgE levels was reported after diluent and antigen challenge at 6 h (Van De Graaf et al., 1991).

Eosinophils express receptors for IgA (Abu-Ghazaleh et al., 1989), IgG (CD32, CD16; Hartnell et al., 1990, 1992; Valerius et al., 1990) and IgE (FcεRI and FcεRII-CD23; Capron et al., 1992; Gounni et al., 1994). Peripheral eosinophils from allergic rhinitic and asthmatic subjects have higher IgA receptor expression (Monteiro *et al.*, 1993). Differential release of EPO and ECP has been reported after incubating eosinophils with different subtypes of immunoglobulin (Tomassini *et al.*, 1991). EPO alone, EPO and ECP, and ECP alone were detected on stimulation with mAb against IgE, IgA and IgG, respectively. The selective release of various granular proteins depending on the Fc receptor engaged may be an important mechanism to tailor eosinophil response. After stimulation with IgA, IgE or IgG immune complexes, blood eosinophils release IL-5 as detected by ELISA and immunocytochemistry (Dubucquoi *et al.*, 1994). This potentially allows the eosinophil to survive longer in local areas of inflammation and to mount an effector response when antigens are focused by IgE, IgA or IgG.

5. Eosinophil–Lymphocyte Interactions

The presence of CD4 and HLA-DR on eosinophils enables them to serve as antigen-presenting cells in cognate T lymphocyte stimulation (Del Pozo *et al.*, 1992; Hansel *et al.*, 1992; Weller *et al.*, 1993; Mawhorter *et al.*, 1994). Eosinophils recovered from sputum and BAL were positive for HLA-DR and ICAM-1 (Hansel *et al.*, 1991; Mengelers *et al.*, 1994). Whether eosinophils serve as antigen-presenting cells in the amplification loop to sustain an allergic inflammatory response to chronic antigenic stimulation is currently unknown (Weller *et al.*, 1993).

Human eosinophils were found to elaborate chemoattractants for lymphocytes (Lim *et al.*, 1994). Eosinophils contained mRNA transcripts for LCF and RANTES detectable by RT-PCR amplification and released bioactive LCF neutralized both by antibodies to LCF and CD4, the LCF receptor on migrating lymphocytes (Lim *et al.*, 1994). In addition, eosinophils released RANTES protein detectable by ELISA and by functional lymphocyte migration assay. Thus, eosinophils are a source of cytokines capable of specifically affecting the function of CD4$^+$ lymphocytes. The recognition that eosinophils are a source of lymphocyte chemoattractant activity identifies an additional mechanism whereby eosinophils may contribute to lymphocyte responses. Since eosinophils are recruited early in certain allergic inflammatory immune responses, the release of LCF and RANTES by human eosinophils may enable eosinophils to recruit and activate not only other eosinophils but also CD4$^+$ lymphocytes (Lim *et al.*, 1994).

6. Conclusion

Eosinophils are important effector granulocytes in asthma. Eosinophils possess a unique array of biological functions. As effector granulocytes, they contribute to the inflammation in asthma by releasing cytotoxic granule proteins, lipid mediators, oxygen free radicals and cytokines. By these actions, eosinophils are involved in the initiation and perpetuation of airway inflammation in asthma. While many of their functions are well recognized, there are still gaps in our understanding of how eosinophils may interact with lymphocytes, macrophages, fibroblasts and endothelium in asthma.

7. References

Abraham, W.M., Sielczak, M.W., Ahmed, A., Cortes, A., Lauredo, I. T., Kim, J., Pepinsky, B., Benjamin, C.D., Leone, D.R., Lobb, R.R. and Weller, P.F. (1994). Anti-α4 intergrin mediates antigen-induced late bronchial responses and prolonged airway hyperresponsiveness in sheep. J. Clin. Invest. 93, 776–787.

Abu-Ghazaleh, R.I., Fujisawa, T., Mestecky, J., Kyle, R.A. and Gleich, G.J. (1989). IgA-induced eosinophil degranulation. J. Immunol. 142, 2393–2400.

Abu-Ghazaleh, R.I., Gleich, G.J. and Prendergast, F.G. (1992). Interaction of eosinophil granule major basic protein with synthetic lipid bilayers: a mechanism for toxicity. J. Membr. Biol. 128, 153–164.

Ädelroth, E., Morris, M.M., Hargreave, F.E. and O'Byrne, P.M. (1986). Airway responsiveness to leukotrienes C4 and D4 and to methacholine in patients with asthma and normal controls. N. Engl. J. Med. 315, 480–484.

Ädelroth, E., Rosenhall, L., Johansson, S., Linden, M. and Venge, P. (1990). Inflammatory cells and eosinophilic activity in asthmatics investigated by bronchoalveolar lavage. Am. Rev. Respir. Dis. 142, 91–99.

Alam, R., York, J., Boyars, M., Grant, J., Stafford, S., Forsythe, P. and Weido, A. (1994). The detection of the mRNA for MCP-1, MCP-3, RANTES, MIP-1a and IL-8 in bronchoalveolar lavage cells and the measurement of RANTES and MIP-1a in the lavage. Am. Rev. Respir. Dis. 149, A951.

Anwar, A.R.F., Moqbel, R., Walsh, G.M., Kay, A.B. and Wardlaw, A.J. (1993). Adhesion to fibronectin prolongs eosinophil survival. J. Exp. Med. 177, 839–843.

Arm, J.P., O'Hickey, S.P., Hawksworth, R.J., Fong, C.Y., Crea, A.E.G., Spur, B.W. and Lee, T.H. (1990). Asthmatic airways have a disproportionate hyperresponsiveness to LTE4, as compared with normal airways, but not to LTC4, LTD4, methacholine, and histamine. Am. Rev. Respir. Dis. 142, 1112–1118.

Ayars, G.H., Altman, L.C., McManus, M.M., Agosti, J.M., Baker, C., Luchtel, D.L., Loegering, D.A. and Gleich, G.J. (1989). Injurious effect of the eosinophil peroxide–hydrogen peroxide–halide system and major basic protein on human nasal epithelium *in vitro*. Am. Rev. Respir. Dis. 140, 125–131.

Azzawi, M., Bradley, B., Jeffrey, P.K., Frew, A.J., Wardlaw, A.J., Knowles, G., Assoufi, B., Collins, J.V., Durham, S. and Kay, A.B. (1990). Identification of activated T lymphocytes and eosinophils in bronchial biopsies in stable atopic asthma. Am. Rev. Resp. Dis. 142, 1407–1413.

Baigelman, W., Chodosh, S., Pizzuto, S. and Cupples, L. A. (1983). Sputum and blood eosinophils during corticosteroid treatment of acute exacerbations of asthma. Am. J. Med. 75, 929–936.

Barnes, P. (1994). Cytokines as mediators of chronic asthma. Am. J. Respir. Crit. Care Med. 150, S42–49.

Beasley, R., Roche, W.R., Roberts, J.A. and Holgate, S.T. (1989). Cellular events in the bronchi in mild asthma and after bronchial provocation. Am. Rev. Respir. Dis. 139, 806–817.

Beil, W.J., Weller, P.F., Tzizik, D.M., Galli, S.J. and Dvorak, A.M. (1993). Ultrastructural immunogold localization of tumor necrosis factor to the matrix compartment of human eosinophil secondary granules. J. Histochem. Cytochem. 41, 1611–1615.

Bentley, A.M., Menz, G., Storz, C., Robinson, D.S., Bradley, B., Jeffery, P.K., Durham, S.R. and Kay, A.B. (1992). Identification of T lymphocytes, macrophages and activated eosinophils in the bronchial mucosa in intrinsic asthma. Relationship to symptoms and bronchial responsiveness. Am. Rev. Respir. Dis. 146, 500–506.

Bentley, A.M., Durham, S.R., Robinson, D.S., Menz, G., Storz, C., Cromwell, O., Kay, A. B. and Wardlaw, A.J. (1993). Expression of endothelial and leukocyte adhesion molecules intercellular adhesion molecule-1, E-selectin and vascular cell adhesion molecule-1 in the bronchial mucosa in steady-state and allergen-induced asthma. J. Allergy Clin. Immunol. 92, 857–868.

Berman, J.S. and Weller, P.F. (1992). Airway eosinophils and lymphocytes in asthma. Birds of a feather? [editorial]. Am. Rev. Respir. Dis. 145, 1246–1248.

Bisgaard, H., Groth, S. and Dirksen, M. (1983). Leukotriene induces bronchoconstriction in man. Allergy 38, 441–443.

Bochner, B.S., Luscinskas, F.W., Gimbrone, M.A.J., Newman, W., Sterbinsky, S.A., Derse-Anthony, C.P., Klunk, D. and Schleimer, R.P. (1991). Adhesion of human basophils, eosinophils and neutrophils to interleukin 1-activated human vascular endothelial cells: contributions of endothelial cell adhesion molecules. J. Exp. Med. 173, 1553–1557.

Borish, L., Mascali, J.J., Dishuck, J., Beam, W.R., Martin, R.J. and Rosenwasser, L.J. (1992). Detection of alveolar macrophage-derived IL-1β in asthma. J. Immunol. 149, 3078–3082.

Bousquet, J., Chanez, P., Lacoste, J.Y., Barneon, G., Ghavanian, N., Enander, I., Venge, P., Ahlstedt, S., Simony-Lafontaine, J., Godard, P. et al. (1990). Eosinophilic inflammation in asthma. N. Engl. J. Med. 323, 1033–1039.

Bradding, P., Roberts, J.A., Britten, K.M., Montefort, S., Djukanovic, R., Mueller, R., Heuser, C.H., Howarth, P. H. and Holgate, S.T. (1994). Interleukin-4, -5, -6 and tumor necrosis factor-α in normal and asthmatic airways: evidence for the human mast cell as a source of these cytokines. J. Respir. Cell Mol. Biol. 10, 471–480.

Braun, R.K., Franchini, M., Erard, F., Rihs, S., De Vries, I.J.M., Blaser, K., Hansel, T. T. and Walker, C. (1993). Human peripheral blood eosinophils produce and release interleukin-8 on stimulation with calcium ionophore. Eur. J. Immunol. 23, 956–960.

Brewster, C.E.P., Howarth, P.H., Djukanovic, R., Wilson, J., Holgate, S.T. and Roche, W.R. (1990). Myofibroblasts and subepithelial fibrosis in bronchial asthma. Am. J. Resp. Cell. Mol. Biol. 3, 507–511.

Briscoe, D.M., Cotran, R.S. and Pober, J.S. (1992). Effects of tumor necrosis factor, lipopolysaccharide and IL-4 on the expression of vascular cell adhesion molecule-1 in vivo. Correlation with CD-3 T cell infiltration. J. Immunol. 149, 2954–2960.

Broide, D.H., Lotz, M., Cuomo, A.J., Coburn, D.A., Federman, E.C. and Wasserman, S.I. (1992a). Cytokines in symptomatic asthma airways. J. Allergy Clin. Immunol. 89, 958–967.

Broide, D.H., Paine, M.M. and Firestein, G.S. (1992b). Eosinophils express interleukin 5 and granulocyte macrophage-colony-stimulating factor mRNA at sites of allergic inflammation in asthmatics. J. Clin. Invest. 90, 1414–1424.

Brown, H.M. (1958). Treatment of chronic ashma with prednisolone – significance of eosinophils in the sputum. Lancet 11, 1245–1247.

Bruijnzeel, P.L., Hamalink, M.L., Prins, K., Remmert, G. and Meyling, F.H. (1987). Blood lymphocyte subpopulations in extrinsic and intrinsic asthmatics. Ann. Allergy 58, 179–182.

Burke, L.A., Crea, A.E., Wilkinson, J.R., Arm, J.P., Spur, B.W. and Lee, T.H. (1990). Comparison of the generation of platelet-activating factor and leukotriene C$_4$ in human eosinophils stimulated by unopsonized zymosan and by the calcium ionophore A23187: the effects of nedocromil sodium. J. Allergy Clin. Immunol. 85, 26–35.

Burrows, B., Hasan, F.M., Barbee, R.A., Halonen, M. and Lebowitz, M.D. (1980). Epidemiologic observations on eosinophilia and its relation to respiratory disorders. Am. Rev. Respir. Dis. 122, 709–719.

Calhoun, W.J., Bates, M.E., Schrader, L., Sedgwick, J.B. and Busse, W.W. (1992). Characteristics of peripheral blood eosinophils in patients with nocturnal asthma. Am. Rev. Respir. Dis. 145, 577–581.

Capron, M., Truong, M. J., Aldebert, D., Gruart, V., Suemura, M., Delespesse, G., Tourvieille, B. and Capron, A. (1992). Eosinophil IgE receptor and CD23. Immunol. Res. 11, 252–259.

Charles, T.J., Williams, S.J., Seaton, A., Bruce, C. and Taylor, W.H. (1979). Histamines, basophils and eosinophils in severe asthma. Clin. Sci. 57, 39–45.

Clutterbuck, E.J., Hirst, E.M. and Sanderson, C.J. (1989). Human interleukin-5 (IL-5) regulates the production of eosinophils in human bone marrow cultures: comparison and interaction with IL-1, IL-3, IL-6, and GM-CSF. Blood 73, 1504–1512.

Collins, D.S., Dupuis, R., Gleich, G.J., Bartemes, K.R., Koh, Y.Y., Pollice, M., Albertine, K.H., Fish, J.E. and Peters, S.P. (1993). Immunoglobulin E-mediated increase in vascular permeability correlates with eosinophilic inflammation. Am. Rev. Respir. Dis. 147, 677–683.

Cookson, W.O., Craddock, C.F., Benson, M.K. and Durham, S.R. (1989). Falls in peripheral eosinophil counts parallel the late asthmatic response. Am. Rev. Respir. Dis. 139, 458–462.

Corrigan, C.J., Haezku, A., Gemou-Engesaeth, V., Doi, S., Kikuchi, Y., Takatsu, K., Durham, R. and Kay, A. B. (1993). CD4 T-lymphocyte activation in asthma is accompanied by increased serum concentration of IL-5. Am. Rev. Respir. Dis. 147, 540–547.

Costa, J.J., Matossian, K., Beil, W.J., Wong, D.T.W., Gordon, J.R., Dvorak, A.M., Weller, P.F. and Galli, S.J. (1993). Human eosinophils can express the cytokines TNF-α and MIP-1α. J. Clin. Invest. 91, 2673–2684.

Cromwell, O., Wardlaw, A.J., Champion, A., Moqbel, R., Osei, D. and Kay, A.B. (1990). IgG-dependent generation of platelet-activating factor by normal and low density human eosinophils. J. Immunol. 145, 3862–3868.

Cruikshank, W., Melissa, F., Teran, L. and Center, D. (1994a). Early detection of a CD4$^+$ lymphocyte and eosinophil chemoattractant in bronchoalveolar lavage fluid from asthmatics following antigen challenge. Am. Thor. Soc. 149, A954.

Cruikshank, W.W., Center, D.M., Nisar, N., Wu, M., Theodore, A.C. and Kornfeld, H. (1994b). Molecular and functional analysis of a human chemoattractant lymphokine. Proc. Natl. Acad. Sci. USA 91, 5109–5113.

Czech, W., Krutmann, J., Budnik, A., Schopf, E. and Kapp, A. (1993). Induction of intercellular adhesion molecule 1 (ICAM-1) expression in normal human eosinophils by inflammatory cytokines. J. Invest. Dermatol. 100, 417–423.

Czop, J.K. and Austen, K.F. (1985). Generation of leukotrienes by human monocytes upon stimulation of their beta-glucan receptor during phagocytosis. Proc. Natl. Acad. Sci. USA 82, 2751–2755.

Dahinden, C.A., Geiser, T., Brunner, T., Von Tscharner, V., Caput, D., Ferrara, P., Minty, A. and Baggiolini, M. (1994). Monocyte chemotactic protein is a most effective basopil- and eosinophil-activating chemokine. J. Exp. Med. 179, 751–756.

Dahl, R., Venge, P. and Fredens, K. (1987). The eosinophil. In "Asthma: Basic Mechanisms and Clinical Management" (eds P.J. Barnes, I. Rodger and N. Thomson), pp. 115–130. Academic Press, London.

Dahlen, S.E., Hansson, G., Hedqvist, P., Bjork, T., Granstrom, E. and Dahlen, B. (1983). Allergen challenge of lung tissue from asthmatics elicits bronchial contraction that correlates with release of leukotrienes C4, D4 and E4. Proc. Natl. Acad. Sci. USA 80, 1712–1716.

Del Pozo, V., De Andrés, B., Martín, E., Cárdaba, B., Fernández, J.C., Gallardo, S., Tramón, P., Leyva-Cobian, F., Palomino, P. and Lahoz, C. (1992). Eosinophil as antigen-presenting cell: activation of T cell clones and T cell hybridoma by eosinophils after antigen processing. Eur. J. Immunol. 22, 1919–1925.

De Monchy, J.G.R., Kauffman, H.F., Venge, P., Koëter, G.H., Jansen, H.M., Sluiter, H.J. and De Vries, K. (1985). Bronchoalveolar eosinophilia during allergen-induced late asthmatic reactions. Am. Rev. Respir. Dis. 131, 373–376.

Desreumaux, P., Janin, A., Colombel, J.F., Prin, L., Plumas, J., Emiliee, D., Torpier, G., Capron, A. and Capron, M. (1992). Interleukin-5 messenger RNA expression by eosinophils in the intestinal mucosa of patients with coeliac disease. J. Exp. Med. 175, 293–296.

Desreumaux, P., Janin, A., Dubucquoi, S., Copin, M.C., Torpier, G., Capron, A., Capron, M. and Prin, L. (1993). Synthesis of interleukin-5 by activated eosinophils in patients with eosinophilic heart diseases. Blood 82, 1553–1560.

Diaz, P., Gonzalez, M.C., Galleguillos, F.R., Ancic, P., Cromwell, O., Shepherd, D., Durham, S.R., Gleich, G.J. and Kay, A.B. (1989). Leukocytes and mediators in bronchoalveolar lavage during allergen-induced late-phase asthmatic reactions. Am. Rev. Respir. Dis. 139, 1383–1389.

Djukanovic, R., Roche, W.R., Wilson, J.W., Beasley, C.R.W., Twentyman, O.P., Howarth, P.H. and Holgate, S.T. (1990a). Mucosal inflammation in asthma. Am. Rev. Respir. Dis. 142, 434–457.

Djukanovic, R., Wilson, J.W., Britten, K.M., Wilson, S.J., Walls, A.F., Roche, W.R., Howarth, P.H. and Holgate, S.T. (1990b). Quantitation of mast cells and eosinophils in the bronchial mucosa of symptomatic atopic asthmatics and healthy control subjects using immunohistochemistry. Am. Rev. Respir. Dis. 142, 863–871.

Dobrina, A., Menegazzi, R., Carlos, T.M., Nardon, E., Cramer, R., Zacchi, T., Harlan, J.M. and Patriarca, P. (1991). Mechanisms of eosinophil adherence to cultured vascular endothelial cells. Eosinophils bind to the cytokine-induced ligand vascular cell adhesion molecule-1 via the very late activation antigen-4 integrin receptor. J. Clin. Invest. 88, 20–26.

Dri, P., Cramer, R., Spessotto, P., Romano, M. and Patriarca, P. (1991). Eosinophil activation on biologic surfaces. Production of O_2^- in response to physiologic soluble stimuli is differentially modulated by extracellular matrix components and endothelial cells. J. Immunol. 147, 613–620.

Dubucquoi, S.P., Desreumaux, A., Janin, O., Klein, M., Goldman, J., Tavernier, A., Capron, A. and Capron, M. (1994). Interleukin 5 synthesis by eosinophils: association with granules and immunoglobulin-dependent secretion. J. Exp. Med. 179, 703–708.

Ebisawa, M., Bochner, B.S., Georgas, S.N. and Schleimer, R.P. (1992). Eosinophil transendothelial migration induced by cytokines. I. Role of endothelial and eosinophil adhesion molecules in IL-1β-induced transendothelial migration. J. Immunol. 149, 4021–4028.

Ebisawa, M., Yamada, T., Bickel, C., Klunk, D. and Schleimer, R. (1994). Eosinophil transendothelial migration induced by cytokines. III. Effect of the chemokine RANTES. J. Immunol. 153, 2153–2160.

Erle, D., Briskin, M., Butcher, E., Garcia-Pardo, A., Lazorovits, A. and Tidswell, M. (1994). Expression and function of the MadCAM-1 receptor, integrin $\alpha 4\beta 7$, on human leukocytes. J. Immunol. 153, 517–528.

Evans, T.W., Chung, K., Rogers, D.F. and Barnes, P.J. (1987). Effects of platelet-activating factor on airway vascular permeability: possible mechanisms. J. Appl. Physiol. 63, 479–484.

Ferreri, N.R., Howland, W.C., Stevenson, D.D. and Spiegelberg, H.L. (1988). Release of leukotrienes, prostaglandins, and histamine into nasal secretion of aspirin-sensitive asthmatics during reaction to aspirin. Am. Rev. Respir. Dis. 137, 847–854.

Filley, W.V., Holley, K.E., Kephart, G.M. and Gleich, G.J. (1982). Identification by immunofluorescence of eosinophil granule major basic protein in lung tissues of patients with bronchial asthma. Lancet 2, 11–15.

Freitag, A., Watson, R.M., Matsos, G., Eastwood, C. and O'Byrne, P.M. (1993). Effect of a platelet activating factor antagonist, WEB 2086, on allergen induced asthmatic responses. Thorax 48, 594–598.

Frick, W.E., Sedgwick, J.B. and Busse, W.W. (1989). The appearance of hypodense eosinophils in antigen-dependent late phase asthma. Am. Rev. Respir. Dis. 139, 1401–1406.

Frigas, E., Loegering, D.A. and Gleich, G.J. (1980). Cytotoxic effects of guinea pig eosinophil major basic protein on tracheal epithelium. Lab. Invest. 42, 35–43.

Fukuda, T. and Makino, S. (1992). Heterogeneity and activation. In "Eosinophils: Biological and Clinical Aspects" (eds

S. Makino and T. Fukuda), pp. 155–170. CRC Press, Boca Raton.

Fukuda, T., Dunnette, S.L., Reed, C.E., Ackerman, S.J., Peters, M.S. and Gleich, G.J. (1985). Increased numbers of hypodense eosinophils in the blood of patients with bronchial asthma. Am. Rev. Respir. Dis. 132, 981–985.

Gibson, P.G., Dolovich, J., Girgis-Gabardo, A., Morris, M.M., Anderson, M., Hargreave, F.E. and Denburg, J.A. (1990). The inflammaory response in asthma exacerbation: changes in circulating eosinophils, basophils and their progenitors. Clin. Exp. Allergy 20, 661–668.

Gleich, G.J., Motojima, S., Frigas, E., Kephart, G.M., Fujisawa, T. and Kravis, L.P. (1987). The eosinophilic leukocyte and the pathology of fatal bronchial asthma: evidence of pathologic heterogeneity. J. Allergy Clin. Immunol. 80, 412–415.

Gleich, G.J., Flavahan, N.A., Fujisawa, T. and Vanhoutte, P.M. (1988). The eosinophil as a mediator of damage to respiratory epithelium: A model for bronchial hyperreactivity. J. Allergy Clin. Immunol. 81, 776–781.

Gleich, G.J., Abu-Ghazaleh, R.I. and Glitz, D.G. (1994). Eosinophil granule proteins: structure and function. In "Eosinophils in Allergy and Inflammation" (eds G.J. Gleich and A.B. Kay), pp. 1–20. Marcel Dekker, New York.

Gounni, A.S., Lamkhioued, B., Tanaka, Y., Delaporte, E., Capron, A., Kinet, J.P. and Capron, M. (1994). High-affinity IgE receptor on eosinophils is involved in defence against parasites. Nature 367, 183–186.

Gundel, R.H., Gerritsen, M.E. and Wegner, C.D. (1989). Antigen-coated sepharose beads induce airway eosinophilia and airway hyperresponsiveness in cynomolgus monkeys. Am. Rev. Respir. Dis. 140, 629–633.

Gundel, R.H., Letts, L.G. and Gleich, G.J. (1991). Human eosinophil major basic protein induces airway constriction and airway hyperresponsiveness in primates. J. Clin. Invest. 87, 1470–1473.

Gundel, R., Wegner, C. and Letts, L. (1992a). Antigen-induced acute and late-phase responses in primates. Am. Rev. Respir. Dis. 146, 369–373.

Gundel, R.H., Wegner, C.D., Torcellini, C.A. and Letts, L.G. (1992b). The role of intercellular adhesion molecule-1 in chronic airway inflammation. Clin. Exp. Allergy 22, 567–569.

Hamid, Q., Barkans, J., Meng, Q., Ying, S., Abrams, J.S., Kay, A.B. and Moqbel, R. (1992). Human eosinophils synthesize and secrete interleukin-6, in vitro. Blood 80, 1496–1501.

Hansel, T., Pound, J. and Thompson, R. (1990). Isolation of eosinophils from human blood. J. Immunol. Meth. 127, 153–164.

Hansel, T.T., Braunstein, J.B., Walker, C., Blaser, K., Bruijnzeel, P.L.B., Virchow, J.C., Jr. and Virchow, C. (1991). Sputum eosinophils from asthmatics express ICAM-1 and HLA-DR. Clin. Exp. Immunol. 86, 271–277.

Hansel, T.T., De Vries, I.J.M., Carballido, J.M., Braun, R.K., Carballido-Perrig, N., Rihs, S., Blaser, K. and Walker, C. (1992). Induction and function of eosinophil intercellular adhesion molecule-1 and HLA-DR. J. Immunol. 149, 2130–2136.

Hartnell, A., Moqbel, R., Walsh, G.M., Bradley, B. and Kay, A.B. (1990). Fc gamma and CD11/CD18 receptor expression on normal density and low density eosinophils. Immunology 69, 264–270.

Hartnell, A., Kay, A.B. and Wardlaw, A.J. (1992). IFN-gamma induces expression of Fc gamma RIII (CD16) on human eosinophils. J. Immunol. 148, 1471–1478.

Hastie, A.T., Loegering, D.A., Gleich, G.J. and Kueppers, F. (1987). The effect of purified human eosinophil major basic protein on mammalian ciliary activity. Am. Rev. Respir. Dis. 135, 848–853.

Henderson, W.R., Jörg, A. and Klebanoff, S.J. (1982). Eosinophil peroxidase-mediated inactivation of leukotrienes B4, C4, and D4. J. Immunol. 128, 2609–2613.

Hernnäs, J., Sarnstrand, B., Lindroth, P., Peterson, C.G.P., Venge, P. and Malmström, A. (1992). Eosinophil cationic protein alters proteoglycan metabolism in human lung fibroblasts. Eur. J. Cell Biol. 59, 352–363.

Hodges, M.K., Weller, P.F., Gerard, N.P., Ackerman, S.J. and Drazen, J.M. (1988). Heterogeneity of leukotriene C4 production by eosinophils from asthmatic and normal subjects. Am. Rev. Respir. Dis. 138, 799–804.

Horn, B.R., Robin, E.D., Theodore, J. and Van Kessel, A. (1975). Total eosinophil counts in the management of bronchial asthma. N. Engl. J. Med. 292, 1152–1155.

Huber, H.L. and Koessler, K.K. (1922). The pathology of bronchial asthma. Arch. Intern. Med. 30, 689–760.

Jacoby, D.B., Ueki, I.F., Widdicombe, J.H., Loegering, D.A., Gleich, G.J. and Nadel, J.A. (1988). Effect of human eosinophil major basic protein on ion transport in dog tracheal epithelium. Am. Rev. Respir. Dis. 137, 13–16.

Kajita, T., Yui, Y., Mita, H., Taniguchi, N., Saito, H., Mishima, T. and Shida, T. (1985). Release of leukotriene C4 from human eosinophils and its relation to the cell density. Int. Arch. Allergy Appl. Immunol. 78, 406–410.

Kameyoshi, Y., Dorschner, A., Mallet, A.I., Christophers, E. and Schroder, J.M. (1992). Cytokine RANTES released by thrombin-stimulated platelets is a potent attractant for human eosinophils. J. Exp. Med. 176, 587–592.

Kapp, A., Zeck-Kapp, G., Czech, W. and Schopf, E. (1994). The chemokine RANTES is more than a chemoattractant: characterization of its effect on human eosinophil oxidative metabolism and morphology in comparison with Il-5 and GM-CSF. J. Invest. Dermatol. 102, 906–914.

Kita, H., Ohnishi, T., Okubo, Y., Weiler, D., Abrams, J. S. and Gleich, G.J. (1991). Granulocyte/macrophage colony-stimulating factor and interleukin 3 release from human peripheral blood eosinophils and neutrophils. J. Exp. Med. 174, 745–748.

Kita, H., Weiler, D.A., Abu-Ghazaleh, R., Sanderson, C.J. and Gleich, G.J. (1992). Release of granule proteins from eosinophils cultured with Il-5. J. Immunol. 149, 629–635.

Kloprogge, E., de Leeuw, A.J., de Monchy, J.G. and Kauffman, H.F. (1989). Hypodense eosinophilic granulocytes in normal individuals and patients with asthma: generation of hypodense cell populations in vitro. J. Allergy Clin. Immunol. 83, 393–400.

Kroegel, C., Liu, M.C., Hubbard, W.C., Lichtenstein, L.M. and Bochner, B.S. (1994). Blood and bronchoalveolar eosinophils in allergic subjects after segmental antigen challenge: surface phenotype, density heterogeneity, and prostanoid production. J. Allergy Clin. Immunol. 93, 725–734.

Kuitert, L.M., Hui, K.P., Uthayarkumar, S., Burke, W., Newland, A.C., Uden, S. and Barnes, N.C. (1993). Effect of the platelet-activating factor antagonist UK-74,505 on the early and late response to allergen. Am. Rev. Respir. Dis. 147, 82–86.

Kyan-Aung, U., Haskard, D.O. and Lee, T.H. (1991). Vascular cell adhesion molecule-1 in eosinophil adhesion to cultured human umbilical vein endothelial cells in vitro. Am. J. Respir. Cell Mol. Biol. 5, 445–450.

Laitinen, L.A., Heino, M., Laitinen, A., Kava, T. and Haahtela, T. (1985). Damage of the airway epithelium and bronchial reactivity in patients with asthma. Am. Rev. Respir. Dis. 131, 599–606.

Laitinen, L.A., Laitinen, A., Heino, M. and Haahtela, T. (1991). Eosinophilic airway inflammation during exacerbation of asthma and its treatment with inhaled corticosteroid. Am. Rev. Respir. Dis. 143, 423–427.

Laitinen, L.A., Haahtela, T., Spur, B.W., Laitinen, A., Vilkka, V. and Lee, T. H. (1993a). Leukotriene E4 and granulocytic infiltration into asthmatic airways. Lancet 341, 989–990.

Laitinen, L.A., Laitinen, A. and Haahtela, T. (1993b). Airway mucosal inflammation even in patients with newly diagnosed asthma. Am. Rev. Respir. Dis. 147, 697–704.

Lam, S., Chan, H., LeRiche, J.C., Chan-Yeung, M. and Salari, H. (1988). Release of leukotrienes in patients with bronchial asthma. J. Allergy Clin. Immunol. 81, 711–717.

Lee, T.-C., Lenihan, D.J., Malone, B., Roddy, L.L. and Wasserman, S.I. (1984). Increased biosynthesis of platelet-activating factor in activated human eosinophils. J. Biol. Chem. 259, 5526–5530.

Lee, T.H. (1992). Mechanism of aspirin sensitivity. Am. Rev. Respir. Dis. 145, S34–S36.

Lim, K., Wan, H., Resnick, M., Wong, D., Cruikshank, W., Center, D. and Weller, P. (1995). Human eosinophils elaborate lymphocyte chemoattractant cytokines: RANTES and lymphocyte chemoattractant factor. Int. Arch. Allergy Appl Immunol; in press.

Lundgren, J.D., Davey, R.T.J., Lundgren, B., Mullol, J., Marom, Z., Logun, C., Baraniuk, J., Kaliner, M.A. and Shelhamer, J.H. (1991). Eosinophil cationic protein stimulates and major basic protein inhibits airway mucus secretion. J. Allergy Clin Immunol. 87, 689–698.

Marom, Z., Shelhamer, J.H., Bach, M.K., Morton, D.R. and Kaliner, M. (1982). Slow-reacting substances leukotrienes C4 and D4, increase the release of mucus from human airway. Am. Rev. Respir. Dis. 126, 449–451.

Mawhorter, S.D., Kazura, J.W. and Boom, W.H. (1994). Human eosinophils as antigen-presenting cells: relative efficiency for superantigen and antigen-induced CD4+ T cell proliferation. Immunology 81, 584–591.

Melani, C., Mattia, G.F., Silvani, A., Care, A., Rivoltini, L., Parmiani, G. and Colombo, M.P. (1993). Interleukin-6 expression in human neutrophil and eosinophil peripheral blood granulocytes. Blood 81, 2744–2749.

Mengelers, H.J., Maikoe, T., Brinkman, L., Hooibrink, B., Lammers, J.J. and Koenderman, L. (1994). Immunophenotyping of eosinophils recovered from blood and BAL of allergic asthmatics. Am. J. Respir. Crit. Care Med. 149, 345–351.

Merchant, R.K., Schwartz, D.A., Helmers, R.A., Dayton, C.S. and Hunninghake, G. W. (1992). Bronchoalveolar lavage cellularity: the distribution in normal volunteers. Am. Rev. Respir. Dis. 146, 448–453.

Meurer, R., Van Riper, G., Feeney, W., Cunningham, P., Hora, D., Springer, M., MacIntyre, D. and Rosen, H.

(1993). Formation of eosinophilic and monocytic intradermal inflammatory sites in the dog by injection of human RANTES but not human monocyte chemoattractant protein 1, human macrophage inflammatory protein alpha, or human interleukin 8. J. Exp. Med. 178, 1913–1921.

Mita, H., Yiu, y., Yasueda, H., Kajita, T., Saito, T. and Shida, T. (1986). Allergen induced histamine released and immunoreactive-leukotriene C4 generation from leukocytes in mite sensitive asthmatic patients. Prostaglandins 31, 869–886.

Mohiuddin, A.A. and Martin, R.J. (1990). Circadian basis of the late asthmatic response. Am. Rev. Respir. Dis. 142, 1153–1157.

Montefort, S., Roche, W.R., Howarth, P.H., Djukanovic, R., Gratziou, C., Carroll, M., Smith, L., Britten, K.M., Haskard, D., Lee, T.H. and Holgate, S.T. (1992). Intercellular adhesion molecule-1 (ICAM-1) and endothelial leukocyte adhesion molecule-1 (ELAM-1) expression in the bronchial mucosa of normal and asthmatic subjects. Eur. Respir. J. 5, 815–823.

Monteiro, R.C., Hostoffer, R.W., Cooper, M.D., Bonner, J.R., Gartland, G.L. and Kubagawa, H. (1993). Definition of immunoglobulin A receptors on eosinophils and their enhanced expression in allergic individuals. J. Clin. Invest. 92, 1681–1685.

Moqbel, R. (1994). Eosinophils, cytokines, and allergic inflammation. Ann. NY Acad. Sci. 725, 223–233.

Moqbel, R., Macdonald, A.J., Cromwell, O. and Kay, A.B. (1990). Release of leukotriene C4 (LTC4) from human eosinophils following adherence to IgE- and IgG-coated schistosomula of Schistosoma mansoni. Immunology 69, 435–442.

Moqbel, R., Hamid, Q., Ying, S., Barkans, J., Hartnell, A., Tsicopoulos, A., Wardlaw, A.J. and Kay, A. B. (1991). Expression of mRNA and immunoreactivity for the granulocyte/macrophage colony-stimulating factor in activated human eosinophils. J. Exp. Med. 174, 749–752.

Moqbel, R., Lacy, P., Levi-Schaffer, F., Manna, M., North, J., Gomperts, B. and Kay, A.B. (1995). Interleukin-6 as a granule-associated pre-formed mediator in peripheral blood eosinophils from asthmatic subjects. Am. J. Resp. Crit. Care Med., in press.

Morita, E., Schroder, J.M. and Christophers, E. (1989). Differential sensitivities of purified human eosinophils and neutrophils to defined chemotaxins. Scand. J. Immunol. 29, 709–716.

Moser, R., Fehr, J. and Bruijnzeel, P.L. (1992a). IL-4 controls the selective endothelium-driven transmigration of eosinophils from allergic individuals. J. Immunol. 149, 1432–1438.

Moser, R., Fehr, J., Olgiati, L. and Bruijnzeel, P.L. (1992b). Migration of primed human eosinophils across cytokine-activated endothelial cell monolayers. Blood 79, 2937–2945.

Moy, J.N., Gleich, G. and Thomas, L.L. (1990). Noncytotoxic activation of neutrophils by eosinophil granule major basic protein: effect on superoxide anion generation and lysosomal enzyme release. J. Immunol. 145, 2626.

Murphy, W., Taub, D., Anver, M., Conlon, K., Oppenheim, J., Kelvin, D. and Longo, D. (1994). Human RANTES induces the migration of human T lymphocytes into the peripheral tissues of mice with severe combined immune deficiency. Eur. J. Immunol. 24, 1823–1827.

Ohashi, Y., Motojima, S., Fukuda, T. and Makino, S. (1992). Airway hyperresponsiveness, increased intracellular spaces of bronchial epithelium, and increased infiltration of eosinophils and lymphocytes in bronchial mucosa in asthma. Am. Rev. Respir. Dis. 145, 1469–1476.

O'Hickey, S.P., Hawksworth, R.J., Fong, C.Y., Arm, J.P., Spur, B.W. and Lee, T.H. (1991). Leukotrienes C4, D4, and E4 enhance histamine responsiveness in asthmatic airways. Am. Rev. Respir. Dis. 144, 1053–1057.

Ohno, I., Lea, R.G., Flanders, K.C., Clark, D.A., Banwatt, D., Dolovich, J., Denburg, J., Harley, C.B., Gauldie, J. and Jordana, M. (1992). Eosinophils in chronically inflamed human upper airway tissues express transforming growth factor beta 1 gene (TGF beta 1). J. Clin. Invest. 89, 1662–1668.

Okubo, T., Takahashi, H., Sumitano, M., Shimdoh, K. and Suzuki, S. (1987). Plasma levels of leukotrienes C4 and D4 during wheezing attack in asthmatic patients. Int. Arch. Allergy Appl. Immunol. 84, 149–155.

Ollerenshaw, S.L. and Woolcock, A.J. (1992). Characteristics of the inflammation in biopsies from large airways of subjects with asthma and subjects with chronic airflow limitation. Am. Rev. Respir. Dis. 145, 922–927.

Out, T.A., Van De Graaf, E.A., Van Den Berg, N.J. and Jansen, H.M. (1991). IgG subclasses in bronchoalveolar lavage fluid from patients with asthma. Scand. J. Immunol. 33, 719–727.

Owen, W.F., Jr., Rothenberg, M.F., Silberstein, D.S., Gasson, J.C., Stevens, R.L., Austen, K.F. and Soberman, R.J. (1987). Regulation of human eosinophil viability, density, and function by granulocyte/macrophage colony-stimulating factor in the presence of 3T3 fibroblasts. J. Exp. Med. 166, 129–141.

Page, C.P. (1992). Mechanisms of hyperresponsiveness: platelet-activating factor. Am. Rev. Respir. Dis. 145, S31–S33.

Peterson, C.G.B., Skoog, V. and Venge, P. (1986). Human eosinophil cationic proteins (ECP and EPX) and their suppressive effects on lymphocyte proliferation. Immunobiology 171, 1–13.

Rand, T.H., Cruikshank, W.W., Center, D.M. and Weller, P.F. (1991a). CD4-mediated stimulation of human eosinophils: lymphocyte chemoattractant factor and other CD4-binding ligands elicit eosinophil migration. J. Exp. Med. 173, 1521–1528.

Rand, T.H., Silberstein, D.S., Kornfeld, H. and Weller, P.F. (1991b). Human eosinophils express functional interleukin 2 receptors. J. Clin. Invest. 88, 825–832.

Rankin, J.A., Harris, P. and Ackerman, S.J. (1992). The effects of eosinophil-granule major basic protein on lung-macrophage superoxide anion generation. J. Allergy Clin. Immunol. 89, 746–751.

Resnick, M.B. and Weller, P.F. (1993). Mechanisms of eosinophil recruitment. Am. J. Respir. Cell Mol. Biol. 8, 349–355.

Rossi, G.A., Crimi, E., Lantero, S., Gianiorio, P., Oddera, S., Crimi, P. and Brusasco, V. (1991). Late-phase asthmatic reaction to inhaled allergen is associated with early recruitment of eosinophils in the airways. Am. Rev. Respir. Dis. 144, 379–383.

Rot, A. (1992). Endothelial cell binding of NAP/IL-8: role in neutrophil emigration. Immunol. Today 13, 291–294.

Rot, A., Krieger, M., Brunner, T., Bischoff, S.C., Schall, T.J. and Dahinden, C.A. (1992). RANTES and macrophage

inflammatory protein 1 α induce the migration and activation of normal human eosinophil granulocytes. J. Exp. Med. 176, 1489–1495.

Rothenberg, M.E., Owen, W.F., Jr., Silberstein, D.S., Woods, J., Soberman, R.J., Austen, K.F. and Stevens, R.L. (1988). Human eosinophils have prolonged survival, enhanced functional properties, and become hypodense when exposed to human interleukin 3. J. Clin. Invest. 81, 1986–1992.

Rothenberg, M.E., Petersen, J., Stevens, R.L., Silberstein, D.S., McKenzie, D.T., Austen, K.F. and Owen, W.F., Jr. (1989). IL-5-dependent conversion of normodense human eosinophils to the hypodense phenotype uses 3T3 fibroblasts for enhanced viability, accelerated hypodensity, and sustained antibody-dependent cytotoxicity. J. Immunol. 143, 2311–2316.

Rubin, A.E., Smith, L.J. and Patterson, R. (1987). The bronchoconstrictor properties of platelet-activating factor in humans. Am. Rev. Respir. Dis. 136, 1145–1151.

Schall, T.J., Bacon, K., Toy, K.J. and Goeddel, D.V. (1990). Selective attraction of monocytes and T lymphocytes of the memory phenotype by cytokine RANTES. Nature 347, 669–671.

Scott, T. and Kaliner, M. (1993). Mast cells in asthma. In "The Mast Cell in Health and Disease" (eds M.A. Kaliner and D.D. Metcalfe), pp. 575–608. Marcel Dekker, New York.

Sedgwick, J.B., Calhoun, W.J., Gleich, G.J., Kita, H., Abrams, J.S., Schwartz, L.B., Volovitz, B., Ben-Yaakov, M. and Busse, W.W. (1991). Immediate and late airway response of allergic rhinitis patients to segmental antigen challenge. Characterization of eosinophil and mast cell mediators. Am. Rev. Respir. Dis. 144, 1274–1281.

Sedgwick, J.B., Calhoun, W.J., Vrtis, R.F., Bates, M.E., McAllister, P.K. and Busse, W.W. (1992). Comparison of airway and blood eosinophil function after in vivo antigen challenge. J. Immunol. 149, 3710–3718.

Sehmi, R., Wardlaw, A.J., Cromwell, O., Kurihara, K., Waltmann, P. and Kay, A.B. (1992). Interleukin-5 selectively enhances the chemotactic response of eosinophils obtained from normal but not eosinophilic subjects. Blood 79, 2952–2959.

Shaw, R.J., Cromwell, O. and Kay, A.B. (1984). Preferential generation of leukotriene C_4 by human eosinophils. Clin. Exp. Immunol. 56, 716–722.

Shaw, R.J., Walsh, G.M., Cromwell, O., Moqbel, R., Spry, C.J. and Kay, A.B. (1985). Activated human eosinophils generate SRS-A leukotrienes following IgG-dependent stimulation. Nature 316, 150–152.

Silberstein, D.S., Owen, W.F., Gasson, J.C., DiPersio, J.F., Golde, D.W., Bina, J.C., Soberman, R., Austen, K.F. and David, J.R. (1986). Enhancement of human eosinophil cytotoxicity and leukotriene synthesis by biosynthetic (recombinant) granulocyte-macrophage colony-stimulating factor. J. Immunol. 137, 3290–3294.

Slungaard, A. and Mahoney, J.R.J. (1991). Bromide-dependent toxicity of eosinophil peroxidase for endothelium and isolated working rat hearts: a model for eosinophilic endocarditis. J. Exp. Med. 173, 117–126.

Smith, L.J. (1991). The role of platelet-activating factor in asthma. Am. Rev. Respir. Dis. 143, S100–S102.

Smith, L.J., Greenberger, P.A., Patterson, R., Krell, R.D. and Bernstein, P.R. (1985). The effect of inhaled leukotriene D4 in humans. Am. Rev. Respir. Dis. 131, 368–372.

Spry, C.J.F. (1988). "Eosinophils. A Comprehensive Review and Guide to the Scientific and Medical Literature". Oxford Medical Publications, Oxford.

Spry, C.J., Kay, A.B. and Gleich, G.J. (1992). Eosinophils 1992. Immunol. Today 13, 384–387.

Strunk, R.C. (1993). Death due to asthma. Am. Rev. Respir. Dis. 148, 550–552.

Sur, S., Crotty, T.B., Kephart, G.M., Hyma, B.A., Colby, T.V., Reed, C.E., Hunt, L.W. and Gleich, G.J. (1993). Sudden-onset fatal asthma. Am. Rev. Respir. Dis. 148, 713–719.

Takafugi, S., Bishchoff, S.C., Deweck, A.L. and Dahinden, C.A. (1991). IL-3 and IL-5 prime normal human eosinophils to produce leukotriene C_4 in response to soluble agonists. J. Immunol. 147, 3855–3861.

Tamura, N., Agrawal, D.K. and Townley, R.G. (1988). Leukotriene C_4 production from human eosinophils in vitro. Role of eosinophil chemotactic factors on eosinophil activation. J. Immunol. 141, 4291–4297.

Tanaka, Y., Adams, D., Hubsner, S. and Hirano, H. (1993a). T cell adhesion induced by proteoglycan-immobilized cytokine MIP-1β. Nature 361, 79–82.

Tanaka, Y., Adams, D. and Shaw, S. (1993b). Proteoglycans on endothelial cells present adhesion-inducing cytokines to leukocytes. Immunol. Today 14, 111–114.

Taylor, G.W., Taylor, I. and Black, P. (1989). Urinary leukotriene E4 after antigen challenge in acute asthma and in allergic rhinitis. Lancet 1, 584–588.

Taylor, K.J. and Luksza, A.R. (1987). Peripheral blood eosinophil counts and bronchial responsiveness. Thorax 42, 452–456.

Thornhill, M.H., Wellicome, S.M., Mahiouz, D.L., Lanchbury, J.S.S., Kyan-Aung, U. and Haskard, D.O. (1991). Tumor necrosis factor combines with IL-4 or IFNγ to selectively enhance endothelial cell adhesiveness for T cells. The contribution of vascular cell adhesion molecule-1-dependent and -independent binding mechanisms. J. Immunol. 146, 592–598.

Tomassini, M., Tsicopoulos, A., Tai, P.C., Gruart, V., Tonnel, A.-B., Prin, L., Capron, A. and Capron, M. (1991). Release of granule proteins by eosinophils from allergic and nonallergic patients with eosinophilia on immunoglobulin-dependent activation. J. Allergy Clin. Immunol. 88, 365–375.

Uchida, D.A., Coyle, A.J., Larsen, G.L., Ackerman, S.J., Weller, P.F., Freed, J. and Irvin, C.G. (1993). The effect of human eosinophil granule major basic protein on airway responsiveness in the rat in vivo. A comparison with polycations. Am. Rev. Respir. Dis. 147, 982–988.

Valerius, T., Repp, R., Kalden, J.R. and Platzer, E. (1990). Effects of IFN on human eosinophils in comparison with other cytokines. A novel class of eosinophil activators with delayed onset of action. J. Immunol. 145, 2950–2958.

Van De Graaf, E.A., Out, T.A., Kobesen, A. and Jansen, H.M. (1991). Lactoferrin and secretory IgA in the bronchoalveolar lavage fluid from patients with a stable asthma. Lung 169, 275–283.

Venge, P., Dahl, R., Fredens, K. and Peterson, C.G.B. (1988). Epithelial injury by human eosinophils. Am. Rev. Respir. Dis. 138, S54–S57.

Vieira, V.G. and Prolla, J.C. (1979). Clinical evaluation of eosinophils in the sputum. J. Clin. Pathol. 32, 1054–1057.

Walker, C., Virchow, J.C.J., Bruijnzeel, P.L. and Blaser, K. (1991a). T cell subsets and their soluble products regulate eosinophilia in allergic and nonallergic asthma. J. Immunol. 146, 1829–1835.

Walker, C., Virchow, J.C.J., Iff, T., Bruijnzeel, P.L. and Blaser, K. (1991b). T cells and asthma. 1. Lymphocyte subpopulations and activation in allergic and nonallergic asthma. Int. Arch. Allergy Appl. Immunol. 94, 241–243.

Walker, C., Bode, E., Boer, L., Hansel, T.T., Blaser, K. and J-C. Virchow, J. (1992). Allergic and nonallergic asthmatics have distinct patterns of T cell activation and cytokine production in peripheral blood and bronchoalveolar lavage. Am. Rev. Respir. Dis. 146, 109–115.

Walker, C., Rihs, S., Braun, R.K., Betz, S. and Bruijnzeel, P.L. (1993). Increased expression of CD11b and functional changes in eosinophils after migration across endothelial cell monolayers. J. Immunol. 150, 4061–4071.

Walsh, G.M., Mermod, J.J., Hartnell, A., Kay, A.B. and Wardlaw, A.J. (1991a). Human eosinophil, but not neutrophil, adherence to IL-1-stimulated human umbilical vascular endothelial cells is alpha 4 beta 1 (very late antigen-4) dependent. J. Immunol. 146, 3419–3423.

Walsh, G.M., Wardlaw, A.J., Hartnell, A., Sanderson, C.J. and Kay, A.B. (1991b). Interleukin-5 enhances the in vitro adhesion of human eosinophils, but not neutrophils, in a leukocyte integrin (CD11/18)-dependent manner. Int. Arch. Allergy Appl. Immunol. 94, 174–178.

Wang, J.M., Rambaldi, A., Biondi, A., Chen, Z.G., Sanderson, C.J. and Mantovani, A. (1989). Recombinant human interleukin 5 is a selective eosinophil chemoattractant. Eur. J. Immunol. 19, 701–705.

Wardlaw, A.J., Moqbel, R., Cromwell, O. and Kay, A.B. (1986). Platelet activating factor. A potent chemotactic and chemokinetic factor for human eosinophils. J. Clin. Invest. 78, 1701–1706.

Wardlaw, A.J., Dunnette, S., Gleich, G.J., Collins, J.V. and Kay, A.B. (1988). Eosinophils and mast cells in bronchoalveolar lavage in subjects with mild asthma. Relationship to bronchial hyperreactivity. Am. Rev. Respir. Dis. 137, 62–69.

Wardlaw, A.J., Hay, H., Cromwell, O., Collins, J.W. and Kay, A.B. (1989). Leukotrienes, LTC4 and LTB4 in bronchoalveolar lavage in bronchial asthma and other respiratory diseases. J. Allergy Clin. Immunol. 84, 19–26.

Warringa, R.A., Koenderman, L., Kok, P.T., Kreukniet, J. and Bruijnzeel, P.L. (1991). Modulation and induction of eosinophil chemotaxis by granulocyte-macrophage colony-stimulating factor and interleukin-3. Blood 77, 2694–2700.

Wegner, C.D., Gundel, R.H., Reilly, P., Haynes, N., Letts, G. and Rothlein, G. (1990). Intercellular adhesion molecule-1 (ICAM-1) in the pathogenesis of asthma. Science 247, 456–459.

Weiss, J.W., Drazen, J.M., McFadden, E.R., Jr., Weller, P., Corey, E.J., Lewis, R.A. and Austen, K.F. (1983). Airway constriction in normal humans produced by inhalation of leukotriene D: potency, time course and effect of acetyl salicylic acid. J. Am. Med. Assoc. 249, 2814–2817.

Weller, P.F. (1992). Eicosanoids, cytokines and other mediators elaborated by eosinophils. In "Eosinophils: Biological and Clinical Aspects" (eds. S. Makino and T. Fukuda), pp. 125–154. CRC Press, Boca Raton.

Weller, P.F. (1994). Eosinophils: structure and functions. Curr. Opin. Immunol. 6, 85–90.

Weller, P.F., Goetzl, E.J. and Austen, K.F. (1980). Identification of human eosinophil lysophospholipase as the constituent of Charcot–Leyden crystals. Proc. Natl. Acad. Sci. USA 77, 7440–7443.

Weller, P.F., Lee, C.W., Foster, D.W., Corey, E.J., Austen, K.F. and Lewis, R.A. (1983). Generation and metabolism of 5-lipoxygenase pathway leukotrienes by human eosinophils: predominant production of leukotriene C$_4$. Proc. Natl. Acad. Sci. USA 80, 7626–7630.

Weller, P.F., Rand, T.H., Goelz, S.E., Chi-Rosso,. G. and Lobb, R.J. (1991). Human eosinophil adherence to vascular endothelium mediated by binding to VCAM-1 and ELAM-1. Proc. Natl. Acad. Sci. USA 88, 7430–7433.

Weller, P.F., Rand, T.H., Barrett, T., Elovic, A., Wong, D.T. and Finberg, R.W. (1993). Accessory cell function of human eosinophils: HLA-DR dependent, MHC-restricted antigen-presentation and interleukin-1α formation. J. Immunol. 150, 2554–2562.

Wenzel, S.E., Larsen, G.L., Johnston, K., Voelkel, N.F. and Westcott, J.K. (1990). Elevated levels of leukotriene C4 in bronchoalveolar lavage fluid from atopic asthmatics after endobronchial challenge. Am. Rev. Respis. Dis. 142, 112–119.

Wong, D.T.W., Weller, P.F., Galli, S.J., Rand, T.H., Elovic, A., Chiang, T., Chou, M.Y., Gallagher, G.T., Matossian, K., McBride, J. and Todd, R. (1990). Human eosinophils express transforming growth factor α. J. Exp. Med. 172, 673–681.

Wong, D.T.W., Elovic, A., Matossian, K., Nagura, N., McBride, J., Gordon, J.R., Rand, T.H., Galli, S.J. and Weller, P.F. (1991). Eosinophils from patients with blood eosinophilia express transforming growth factor β_1. Blood 78, 2702–2707.

Yoshikawa, S., Kayes, S.G. and Parker, J.C. (1993). Eosinophils increase lung microvascular permeability via the peroxidase-hydrogen peroxide-halide system. Bronchoconstriction and vasoconstriction unaffected by eosinophil peroxidase inhibition. Am. Rev. Respir. Dis. 147, 914–920.

Young, J.D.-E., Peterson, C.G.B., Venge, P. and Cohn, Z.A. (1986). Mechanism of membrane damage mediated by human eosinophil cationic protein. Nature 321, 613–616.

Zheutlin, L.M., Ackerman, S.J., Gleich, G.J. and Thomas, L.L. (1984). Stimulation of basophil and rat mast cell H release by eosinophil granule-derived cationic proteins. J. Immunol. 133, 2180–2185.

6. Cytokine Regulation of Chronic Inflammation in Asthma

P.J. Barnes, K.F. Chung *and* I. Adcock

1. Introduction

The bronchial histopathology of patients who have died of asthma shows an intense infiltration of the bronchial mucosa with inflammatory cells, particularly eosinophils, macrophages, lymphocytes and to a lesser extent neutrophils. Deposition of eosinophil products in the bronchial epithelium and subepithelium is a particularly prominent feature (Filley *et al.*, 1982). Epithelial denudation, dilatation of blood vessels, mucosal oedema and hypertrophy of both submucosal glands and bronchial smooth muscle are other features. Many of these features of asthma deaths are also observed in milder and well-controlled asthmatics. Elevated numbers of eosinophils, moncytes/macrophages and activated lymphocytes are persistent features observed in bronchial biopsies obtained by fibreoptic bronchoscopy (Djukanovic *et al.*, 1990). An apparent thickening of the epithelial basement membrane in asthmatic bronchial mucosa has been described, but this is in fact the result of subepithelial deposition of reticular collagen, suggesting fibroblast activation (Roche *et al.*, 1989). How the chronic inflammatory changes observed in asthmatic airways are induced and sustained remains unclear.

There is increasing recognition that large peptide mediators, such as cytokines and growth factors, orchestrate and perpetuate the chronic inflammation of asthma (Kelley, 1990; Arai *et al.*, 1990; Cluzel and Lee, 1992). Cytokines are extracellular signalling proteins, usually less than 80 kD in weight and many are glycosylated. Cytokines usually have an effect on closely adjacent cells, and therefore function in a predominantly paracrine manner, although they may also act at a distance (endocrine) and may have effects on the cell of origin (autocrine). Cytokines may be regarded as a mechanism for cell–cell communication and are involved in cell growth and differentiation, inflammation, immunity and repair, which are all aspects of importance in our understanding of the cellular events in chronic asthma. Cytokines have been divided into various classes. Lymphocyte-derived cytokines were named lymphokines, and monocyte-derived products monokines, but it is now clear that there is a considerable overlap between the lymphokines and monokines in terms of their cell source and biological activities. The term interleukin (IL) has been used to denote molecules that act as a molecular messenger between leucocytes and denotes a range of cytokines, currently ranging from IL-1 to IL-15. However, these molecules also have effects on cells other than blood cells and are also produced by non-haemopoietic cells. Molecules such as interferons, growth factors or colony-stimulating factors, which regulate cellular activation and stimulate growth, are also included as cytokines. More recently, a new family of low molecular weight cytokines (8–10 kDa), the chemokines, has been described with the characteristic activity of possessing potent chemotactic activities for a wide range of cells including monocytes, eosinophils, neutrophils and lymphocytes.

Over 50 distinct cytokines have now been identified and each cytokine may have complex cellular effects. Cytokines are chiefly involved in modulating events within the local environment in which they are released, modulating the activities of cells within the vicinity of the cell source. An individual cytokine may stimulate the release of different cytokines from adjacent cells and may also influence the expression of its own receptors or receptors for different cytokines. Furthermore there may be interactions between different cytokines, so that one cytokine in the presence of another may have a different cellular effect. Cytokines may have overlapping effects with other cytokines and may have differing effects, depending on the maturity or previous history of the target cell. These complexities have made it difficult to unravel the effects of particular cytokines, particularly in the absence of specific antagonists for most cytokines. Cytokines may act as proinflammatory mediators, but also have the capacity to down-regulate inflammation. Each cytokine therefore has multiple effects, and these may depend on the presence of other cytokines. This has made it difficult to ascribe particular functions to individual cytokines and each cytokine should be regarded as part of a complex network.

Many cytokines have been implicated in the pathophysiology of asthma (Arai *et al.*, 1990; Cluzel and Lee, 1992). Cytokines are important in the production of immunoglobulin E (IgE), in the regulation of cellular infiltration of the airways, in the activation of inflammatory and resident cells, and in producing the structural changes that are a consequence of chronic inflammation.

The potential contribution of cytokines to the chronic inflammatory process of asthma has been made possible mainly by looking for the presence of cytokines using immunohistochemical techniques on airway mucosal tissues obtained from the proximal airways of patients with asthma. In addition, localization of cytokine mRNA by *in situ* hybridization or its detection by polymerase chain reaction (PCR) has also been used, although expression of mRNA may not necessarily mean that the protein is produced. The exact contribution of individual cytokines can best be surmised from studies of their effect in cells *in vitro* or in animals, particularly with the use of blocking antibodies, although extrapolation to the situation in disease must be made with some caution.

1.1 CYTOKINE INTERACTIONS

The effect of an individual cytokine may be difficult to predict because it may influenced by the presence of other cytokines released from the same cell or from target cells after activation by the cytokine (Fig. 6.1). The effects of cytokines are mediated by binding to cell surface high affinity receptors usually present in relatively low numbers. However, the number of cytokine receptors can be up-regulated with cell activation. Cytokines themselves may induce the expression of receptors which may change the responsiveness of both source cell and target cells. For example, interferon γ (IFNγ) decreases the expression of tumour necrosis factor α (TNFα) receptors on murine macrophages (Draper and Wietzerbin, 1991), whereas IL-1β increases the expression of the same receptors in rat macrophages (Shepherd, 1991). Some cytokines may stimulate their own production in an autocrine manner, whereas others stimulate the synthesis of different cytokines that have a feedback stimulatory effect on the first cytokine, resulting in an augmentation of its effects.

2. *Cellular Origins*

Every cell is capable of releasing cytokines under certain conditions, although each cell produces a characteristic

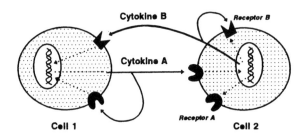

Figure 6.1 Interaction between cytokines. One cell may release a cytokine which then releases other cytokines from adjacent cells. Each cytokine may then affect the expression of cytokine receptors.

spectrum of cytokines. Cytokines are produced by immune and inflammatory cells which are activated in asthma, resulting in recruitment, survival, priming and activation of these cells.

2.1 MAST CELLS

Murine mast cells are capable of synthesizing IL-3, IL-4, IL-5, granulocyte–macrophage colony-stimulating factor (GM-CSF) and TNFα (Plaut *et al.*, 1989; Wodnar-Filipowicz *et al.*, 1989; Gordon *et al.*, 1990; Gordon and Galli, 1991) and human cells of the mast cell/basophil lineage are also capable of cytokine synthesis (Piccini *et al.*, 1991). Allergen-induced triggering of mast cells via high affinity IgE receptors may therefore lead to the local production of cytokines in the airways, resulting in activation of lymphocytes and eosinophils in the airways. Using a double immunostaining technique, IL-4, IL-5, IL-6 and TNFα have been shown to be localized to mast cells in bronchial mucosal biopsies from normal volunteers and from patients with asthma. A 7-fold increase in the number of mast cells staining for TNFα has been observed in biopsies obtained from asthmatics (Bradding *et al.*, 1994). These observations suggest that TNFα, IL-4, IL-5 and IL-6 are in stored form in the mast cell, ready to be released in response to allergen challenge. In resected human lung tissue, TNFα was also found to be localized to mast cells and alveolar macrophages following IgE receptor triggering (Ohkawara *et al.*, 1992). Greatly elevated levels of TNFα have been reported in bronchoalveolar lavage (BAL) fluid obtained from active versus quiescent asthmatics (Broide *et al.*, 1992a).

2.2 EOSINOPHILS

Eosinophils, which are a prominent feature of the asthmatic airway, have recently been shown to have the potential to synthesize a variety of cytokines in addition to secreting several distinct preformed cationic proteins and eicosanoids. Transforming growth factor α (TGFα) was the first cytokine identified in human blood eosinophils or in eosinophils at sites of tissue pathology and was also detected by *in situ* hybridization in blood eosinophils of patients with hypereosinophilia (Wong *et al.*, 1990). Subsequently, it has been shown that human tissue or blood eosinophils represent potential sources of TGFβ1 (Wong *et al.*, 1991; Okno *et al.*, 1992), GM-CSF (Moqbel *et al.*, 1991; Kita *et al.*, 1991), IL-3 (Kita *et al.*, 1991), IL-5 (Desreumaux *et al.*, 1992), IL-1α (Weller *et al.*, 1993), IL-6 (Hamid *et al.*, 1992), TNFα and MIP-1α (Costa *et al.*, 1993). Both IFNγ and IL-5 increase the gene expression of GM-CSF (Moqbel *et al.*, 1991). Production of IL-3, IL-5 and GM-CSF by eosinophils presumably would act as an autocrine mechanism to enhance tissue survival of eosinophils.

Whether eosinophils actually contribute to significant release of these cytokines in chronic inflammation of asthma is not known. Eosinophils express both IL-5 and GM-CSF mRNA after endobronchial challenge of asthmatic subjects with allergen (Broide *et al.*, 1992b). These observations suggest that eosinophils may influence inflammatory reactions through a broader spectrum of mechanisms than has been previously proposed. Thus eosinophils may be involved in activation of other cells and in the process of healing and repair.

2.3 T LYMPHOCYTES

T lymphocytes produce a large number of cytokines (lymphokines). Two patterns of cytokine expression have been distinguished according to studies of CD4+ T cell clones in mice: T_H1 cells synthesize IL-2, IFNγ and GM-CSF, whereas T_H2 cells synthesise IL-3, IL-4 and IL-5 and IL-10 (Mossman *et al.*, 1986, 1990; Chang *et al.*, 1990; Fig. 6.2). The events leading to differentiation of T-helper precursors into T_H1 and T_H2 cells are not known but may be related to the nature of the antigenic trigger. Tuberculin favours a T_H1-type response while allergens favour a T_H2 differentiation. T_H2 cells induce a strong differentiation and proliferation of B cells through the production of IL-4, IL-5 and IL-10. IL-4 switches the production of IgE whereas IFNγ from T_H1 cells blocks the production of IgE. In addition, T_H2 lymphocytes may modulate the activity of T_H1 cells, and vice versa. IL-10 appears to be inhibitory for the expression of T_H1 cells, whereas IFNγ is inhibitory for T_H2 cells, resulting in mutual inhibition (Romagnani, 1990).

In asthma, there is evidence for the preponderance of T_H2-like cells with the expression of mRNA for IL-3,

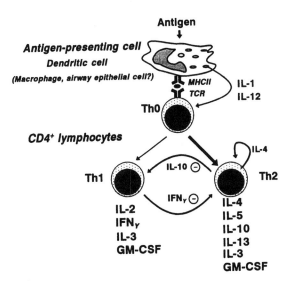

Figure 6.2 Subsets of T helper (CD4+) lymphocytes are now recognized which synthesize different patterns of cytokines. Antigen presentation to lymphocytes by antigen-presenting cells (predominantly dendritic cells in the airways) results in clonal expansion of T_H2 lymphocytes.

IL-4 and IL-5 in CD4$^+$ T lymphocytes in bronchial mucosal biopsies (Hamid *et al.*, 1991). *In situ* hybridization of cells from BAL fluid from atopic asthmatics demonstrated increased proportions of cells with signals for IL-2, IL-3, IL-4, IL-5 and GM-CSF mRNA when compared with non-smoking non-atopic control subjects (Robinson *et al.*, 1992). These cytokines were shown to be localized to T lymphocytes by immunomagnetic separation of the BAL cells (Robinson *et al.*, 1992). A significant enhancement of steady-state IL-5 transcript as measured by PCR in BAL cells from allergen-challenged as compared with saline-challenged control asthmatic subjects has also been reported. The cellular source for IL-5 mRNA increase was primarily from infiltrating mononuclear cells (Krishnaswamy *et al.*, 1993). Concentrated BAL fluid from intrinsic asthmatics showed detectable levels of IL-2 and IL-5 but, in contrast to atopic subjects, IL-4 was not detected (Walker *et al.*, 1992). Serum and peripheral blood T cell culture supernatants from subjects with asthma have been shown to support eosinophil survival *in vitro*. The T cell supernatant activity appeared to be mainly GM-CSF, derived from CD4$^+$ T cells (Walker *et al.*, 1991). Following allergen challenge, increased levels of GM-CSF were detected in BAL fluid, and increased expression of GM-CSF mRNA was found in lymphoctes recovered by BAL (Broide and Firestein, 1991). Similarly, in another study, there were increased numbers of lymphocytes expressing IL-4, IL-5 and GM-CSF mRNA in BAL following allergen challenge (Robinson *et al.*, 1992). These studies therefore support the notion that the lymphocyte, particularly the CD4$^+$ T cell, is an important source of cytokines of the T$_H$2 profile.

2.4 Mononuclear Phagocytes

Alveolar macrophages have the capacity of secreting large numbers of well-defined molecules which include cytokines, growth factors, eicosanoids, enzymes and enzyme inhibitors, clotting factors, reactive oxygen intermediates and nitric oxide. They can express a number of cytokines when studied *in vitro*, including IL-1, TNFα, IL-6, IL-8, GM-CSF, IFNγ, MIP-1α and platelet-derived growth factor (PDGF) (Kelley, 1990). There is evidence that alveolar macrophages lavaged from asthmatic airways have increased expression of IL-6 and TNFα, particularly after allergen challenge (Gosset *et al.*, 1991, 1992). Enhanced release of GM-CSF, TNFα, IL-1β and IL-8 from alveolar macrophages of patients with asthma has been reported (Spiteri *et al.*, 1992; Hallsworth *et al.*, 1994). One source of TNFα from IgE-triggered human lung is the alveolar macrophage (Hallsworth *et al.*, 1994). These studies suggest that alveolar macrophages are primed to release more cytokines. Macrophages may also be important as secretors of IL-1 receptor antagonist, which may be a down-regulator of inflammation (Arend *et al.*, 1990).

Macrophages may also be important for angiogenesis and wound repair and in this respect secrete fibroblast and epithelial cell growth factors such as TGFβ, fibroblast growth factor (FGF), epidermal growth factor (EGF), PDGF and insulin-like growth factor (IGF). The role of these growth factors in asthma is unclear.

2.5 Airway Structural Cells

Cytokines are also produced by structural cells within the airway, including epithelial cells, endothelial cells and fibroblasts. Local activation of these structural cells may lead to recruitment of inflammatory cells into the airway and these structural cells may play an important role in perpetuating inflammation in the airways (Fig. 6.3). Airway epithelial cells have recently been shown to produce a large number of cytokines and growth factors, including IL-6, IL-8, GM-CSF, TNFα, PDGF and IGF-1. These may be produced in response to cytokines released within the airway (Marini *et al.*, 1991; Cromwell *et al.*, 1992; Churchill *et al.*, 1992; Kwon *et al.*, 1993) or in the airway lumen, or may be triggered by the effect of luminal agents, including oxidants, inflammatory mediators or irritants. The air pollutant nitrogen dioxide (NO$_2$), which has been associated with increased airway responsiveness, induces the synthesis of GM-CSF, IL-8 and TNFα in cultured human airway epithelial cells (Devalia *et al.*, 1993), providing a link between air pollution and asthmatic inflammation. Virus infections may also increase cytokine expression in airway epithelial cells; influenza A virus has been found to increase the expression of IL-8 (Choi and Jacoby, 1992). There is evidence for increased expression of IL-6, IL-8, macrophage inflammatory protein-1 (MCP-1) and GM-CSF in airway epithelium of asthmatic patients (Marini *et al.*, 1992; Sousa *et al.*, 1994). The pattern of cytokines produced by epithelial cells may determine the nature of the ensuing inflammatory response. Since epithelial cells are likely to be a major cellular source of cytokines in the airway, these cells may play an important amplifying role in asthma and other inflammatory airways diseases. Inhaled steroids may act, at least in part, by inhibiting the synthesis of epithelial cytokines (Barnes and Pedersen, 1993).

2.6 Endothelial Cells

Endothelial cells are strategically located at the interface between circulating blood cells and tissues, and are an important source and target of cytokine action (Pober and Cotran, 1990). Cytokines such as IL-1 and TNFα enhance the adherence of white blood cells or modify endothelial permeability. Endothelial cells also express intercellular adhesion molecules-1 and -2 (ICAM-1, and -2) which are members of the immunoglobulin supergene family. Expression of ICAM-1 is increased by exposure of endothelial cells to the cytokines IL-1 and TNFα (Mantovani and Dejana, 1989). These cytokines

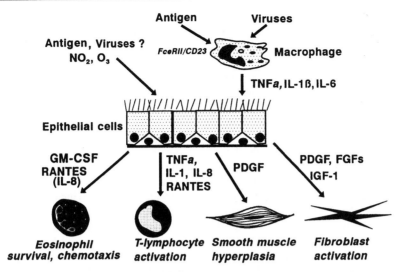

Figure 6.3 Airway epithelial cells may be an important source of cytokines and may amplify asthmatic inflammation through the release of cytokines and growth factors.

also augment the expression of other adhesion molecules such as E-selectin and vascular cell adhesion molecule-1 (VCAM-1). VCAM-1 is the ligand of the very late antigen (VLA-4), which is an integrin expressed on T lymphocytes (Elices *et al.*, 1990). Thus, cytokines IL-1 and TNFα may determine the migration of various cells from the circulation into inflammatory tissues by their action on endothelial cell production of various adhesion molecules. When exposed to inflammatory stimuli such as bacterial lipopolysaccharide, endothelial cells produce IL-1 and IL-6 in addition to monocyte chemotactic and activating factor (Sica *et al.*, 1990). Endothelial cells can also produce colony-stimulating factors (CSF) such as GM-CSF, which are involved in cell differentiation. Thus cytokines derived from endothelial cells may regulate the recruitment of leucocytes, their proliferation and differentiation, and also endothelial cell function in an autocrine fashion. Local endobronchial challenge instillation leads to an up-regulation of specific adhesion molecules such as an increase in endothelial intercellular adhesion molecules type 1 and E-selectin (Montefort *et al.*, 1994).

3. Cytokine Receptors

The effects of cytokines are mediated by specific surface receptors, several of which have now been cloned (Shepherd, 1991). Many of the cloned cytokine receptors have a primary structure which differs from the seven transmembrane spanning segments associated with G-protein-coupled receptors. Thus, the receptor for TNFα is a 55 kD protein which has recently been cloned and appears to have a single transmembrane spanning helical segment, an extracellular domain which binds TNFα and an intracellular domain (Sprang, 1990). The intracellular

domain is responsible for activation of transcription factors such as nuclear factor-xB (NF-xB) and activator protein-1 (AP-1). The structure of the receptor is analogous to the nerve growth factor receptor. A second soluble receptor for TNF has also been cloned (MW 75 kD), which differs markedly in sequence but has structural similarity to the 55 kD receptor. The two receptors may be linked to different intracellular pathways and may be differentially regulated (Sprang, 1990). Similarly two distinct receptors for IL-1 have been cloned (80 and 60 kD) and these appear to be differentially regulated and may also be coupled to different intracellular pathways (Dinarello, 1989).

The general first step in the signalling processes of cytokines is the dimerization of receptors following ligand binding. The cytoplasmic domains of these receptors may interact to produce a downstream signal cascade system. Molecular cloning has now revealed that although cytokines may be structurally diverse, their receptors may be grouped into various families which share structural homology. In the case of the receptors sharing gp 130 (e.g. IL-6, IL-13 and PDGF), signalling is triggered by formation of homo- or heterodimers of gp 130. Thus, the IL-6/IL-6R complex functions as a ligand to induce active dimers of the receptor signalling components (Taga *et al.*, 1989). This group of receptors belongs to the immunoglobulin superfamily, which includes T cell antigen receptors and also some cell surface adhesion molecules (Williams and Barclay, 1988). Another cytokine receptor superfamily, the haematopoietin receptor superfamily, includes receptors for IL-2, IL-3, IL-4, IL-5, IL-7, interferons (IFNs) and GM-CSF (Cosman *et al.*, 1990; Bazan, 1990). Prolactin and erythropoietin are also included in this family. The receptor proteins are orientated with an extracellular N-terminal domain and a single hydrophobic transmembrane spanning segment. There is striking homology in the extracellular ligand binding domain with four conserved cysteine residues. There is very close homology between the receptors for IL-3, IL-5 and GM-CSF, all of which stimulate growth of eosinophils, sharing the KH97 protein as their signal transducing protein. The receptor–ligand complex dimerizes with KH97 to produce an interaction of the cytoplasmic domains which is important for cellular signalling (Sakamaki *et al.*, 1992). This model, in which cytoplasmic signalling is initiated by the dimerization of the cytoplasmic domains of two receptor components, is also applicable to the IL-2R. The high affinity IL-2 binding site is composed of the IL-2R α, β and γ chains. The β and γ chains belong to the haemopoietic cytokine receptor family, whereas the α chain is unique to the IL-2R (Takeshita *et al.*, 1992). Heterodimerization of the β and γ chains is important for cell signalling and the α chain is involved in ligand binding. The receptor complexes for IL-4 and IL-13 might also be heterodimers that share the IL-2Rγ chain (Zurawski *et al.*, 1993). This may explain why these cytokines have overlapping biological activities.

For the chemokine family, six different chemokine receptors have been identified and cloned (Neote *et al.*, 1993) and are members of the superfamily of hepta-helical, rhodopsin-like, G-protein coupled receptors (Holmes *et al.*, 1991). Activation leads to stimulation of phosphoinositide hydrolysis resulting in an increase in intracellular calcium ion concentration and activation of protein kinase C. These receptors are mainly expressed on immune cells but two of them are expressed on viruses also. This family can be divided into: (1) leucocyte chemokine receptors specific for C-C or C-X-C chemokines, but not both; (2) herpes virus homologues of the leucocyte C-C and C-X-C chemokine receptors; and (3) the Duffy coat erythrocyte antigen, which may bind promiscuously to several C-C and C-X-C chemokines. Studies of [^{125}I]IL-8 binding have revealed more than one binding site with a single class of high affinity binding sites (Moser *et al.*, 1991). Two IL-8 receptors have been cloned, the IL-8A receptor expressed on a wide range of cells including T cells, monocytes and neutrophils and the IL-8B receptor showing a more restricted expression, confined primarily to myeloid cells. Receptors for MIP-1α and MIP-1β have been identified on human monocytes and peripheral blood T cells. RANTES and MCP-1 can also bind to the MIP-1α/MIP-1β shared receptor (Holmes *et al.*, 1991).

3.1 INTRACELLULAR PATHWAYS

The intracellular mechanisms involved in cytokine signalling are complex, although increasing evidence suggests that stimulation of various protein kinases leads to activation of transcription factors, which are nuclear proteins that regulate the expression of various target genes (Muegge and Durum, 1990; Fig. 6.4). Cell stimulation by various cytokines invariably induces the activation of

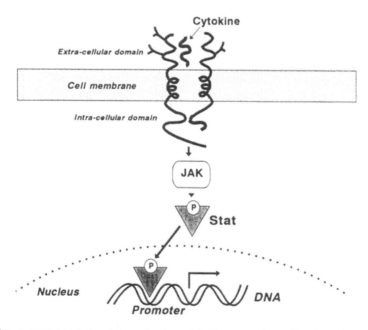

Figure 6.4 Cytokine receptor and signal transduction. Cytokine receptors often have two chains which cooperate in signal transduction. Activation of receptors by the binding of a cytokine results in the activation of various kinases, including JAK (just another kinase) which phosphorylates transcription factors such as members of the Stat family (signal transducer and activator of transcription), which bind to sequences in the promoter region of target genes, resulting in changes in transcription rate.

tyrosine kinases and tyrosine phosphorylation of cellular proteins (Venkitaraman and Cowling, 1992). Although some cytokine receptor components possess a tyrosine kinase domain, others do not and these may activate tyrosine kinases associated with the activated receptor complex. Some cytokine receptors (e.g. EGF receptors, PDGF receptors) have intrinsic protein tyrosine kinase (PTK) activity, leading to phosphorylation of cytosolic substrates which results in altered gene transcription. Activation of other cytokine receptors results in activation of protein kinase C (PKC), although this may be indirect. Many cytokines activate the transcription factors AP-1 and NF-xB, which regulate the gene transcription of many target genes, including enzymes, receptors and cytokines themselves. In human lung TNFα stimulates the activation of both AP-1 and NF-xB and this effect is blocked by corticosteroids (Adcock et al., 1992).

IL-3, TNFα and GM-CSF stimulation induces the tyrosine phosphorylation of JAK1 kinase, a non-receptor-type tyrosine kinase, and activates its kinase activity (Witthuhn et al., 1993). JAK1 directly associates with the IL-3R and the GM-CSFR prior to stimulation. This suggests that IL-3, TNFα and GM-CSF-induced dimerization of their receptors causes the associated JAK1 molecules to phosphorylate one another. There are at least two distinct signalling pathways elicited by the IL-2 receptor (Shibuyo et al., 1992). The tyrosine kinase-linked pathway is associated with the induction of the c-fos gene, the product of which forms a monomer of the AP-1 transcription factor, whereas the kinase-independent pathway is linked to cell proliferation, although this segregation varies with cell type (Asoa et al., 1993). This is also the case with the KH97 signal transducer associated with IL-3, IL-5 and GM-CSF. One region of KH97 causes the induction of c-myc and another is required for the activation of Ras, Raf and mitogen-activated protein (MAP) kinases as well as the induction of c-fos and c-jun, the gene products of which constitute AP-1 (Sato et al., 1993).

Studies on the signal transduction following IL-6R stimulation suggest the presence of two distinct pathways. The ultimate target of the first pathway is nuclear factor-IL-6 (NF-IL6), which is phosphorylated and activated via a Ras-MAP kinase cascade. NF-IL6 was initially identified as an IL-1 responsive element in the promotor region of the IL-6 gene (Akira et al., 1990) but is now known to bind to the promotor region of important immune and inflammatory response genes. The second pathway utilizes a transcription factor (STAT1) that is tyrosine phosphorylated by a gp130-activated tyrosine kinase and translocated to the nucleus. Similar signal transduction schemes may apply for other cytokines, e.g. the activation of IFN-stimulated gene factor 3 (ISGF3) by tyrosine phosphorylation and nuclear translocation, following IFN stimulation.

It has become increasingly evident that a combined effect of transcription factors is very important in gene regulation. NF-xB is a well-characterized transcription factor that is important in the inflammatory responses and is thought to be involved in the expression of many cellular genes that encode cytokines and immunoregulatory receptors (Lenardo and Baltimore, 1989). The combination of NF-IL6 and NF-xB elements is essential for IL-8 gene induction by IL-1, TNFα or phorbol esters (Mokaida et al., 1990). Many genes involved in the regulation of acute and chronic inflammation contain both NF-IL6 and NF-xB sites, and it may be possible that cooperative interactions between these two transcription factors play an important role in the expression of these genes. Some pathogenic effects of viral infection may be due to the constitutive activation of transcription factors such as NF-xB and NF-IL6 resulting from an interplay between these transcription factors and virally encoded gene products (Mahé et al., 1991).

Some of the characteristic features of cytokine actions, namely their multiple actions and redundancy, may now be explained on the basis of the molecular structure of their receptors and the subsequent activation of specific signal transduction pathways. Most cytokine receptors are composed of several chains, both ligand specific and general signal transducing common to several receptors. Upon ligand binding, activation homo- or hetero-dimerization occurs to produce association and activation of intracellular tyrosine kinases. This is followed either by the activation of the MAP kinase pathway and subsequent activation of various transcription factors or by the direct phosphorylation and activation of other transcription factors. Thus many divergent pathways induced by cytokines binding to their receptors all converge on a few specific nuclear factors such as AP-1, NF-xB, NF-IL6 and ISGF. The relative amounts of each factor activated or the inactivation of particular signal pathways may determine the exact transcriptional effect produced in a particular cell to a barrage of cytokine stimuli.

4. Cytokine Effects

The actions of cytokines are very diverse and there is considerable overlap between their effects. Although the cytokines should be considered as working in a network, an understanding of the potential effect of cytokines is best obtained by considering individual cytokines, particularly those likely to be involved in the asthmatic inflammation.

4.1 INTERLEUKIN-1

IL-1 was one of the first cytokines to be identified. It is secreted predominantly from monocytes and macrophages, but may be produced from several other cell types (Mattoli et al., 1991). IL-1 activity comprises two distinct gene products, IL-1α and IL-1β, which only have 26% homology, yet appear to have identical actions as they bind to the same receptor. It is likely that IL-1β is the secreted form of IL-1, whereas IL-1α remains associated with the cell membrane. Two types of IL-1 receptor have been identified, but it is not clear whether they are linked to different intracellular mechanisms or whether they may be differentially regulated. IL-1α is produced by macrophages from a precursor peptide and its formation involves a distinct processing enzyme. IL-1 has effects on macrophages/monocytes and T lymphocytes: it acts on monocyte/macrophages to induce its own synthesis as well as production of TNF and IL-6 (Lovett *et al.*, 1986; Navarro *et al.*, 1989), and it activates T lymphocytes to produce IL-2 and express IL-2 receptors (Kaye and Janeway, 1984) and induces the production of GM-CSF and IL-4 from activated T cells (Herrmann *et al.*, 1988). It also induces B cell proliferation and maturation and increased immunoglobulin synthesis (Romain and Lipsky, 1983; Abbas, 1987). IL-1β is also able to activate airway epithelial cells to produce IL-8 and GM-CSF (Cromwell *et al.*, 1992; Kwon *et al.*, 1993). There is evidence for increased production of IL-1 in asthmatic airways (Mattoli *et al.*, 1991). In rats, inhalation of IL-1β results in infiltration of neutrophils into the airways and increased airway responsiveness to inhaled bradykinin (Tsukagoshi *et al.*, 1993).

IL-1 receptor antagonist (ILRa) is produced by alveolar macrophages and specifically inhibits the effects of IL-1 (Gosset *et al.*, 1988; Arend *et al.*, 1990). Thus, ILRa has been demonstrated to block IL-1 activity both *in vivo* and *in vitro*. For example, in rabbits pretreatment with ILRa prevents death resulting from septic shock produced by injection of endotoxin. Normally IL-1Ra is secreted in excess, but it is possible that in inflammatory conditions the synthesis of IL-1 may exceed IL-1Ra, resulting in inflammation. In the guinea-pig, an IL-1 receptor antagonist inhibited airway hyperreactivity, pulmonary eosinophil accumulation and tumour necrosis factor generation induced by allergen challenge (Watson *et al.*, 1993).

4.2 Interleukin-2

IL-2 is produced by activated T_H1 lymphocytes and acts as an autocrine growth factor. On stimulation with IL-1, IL-2 synthesis is induced along with induction of IL-2 receptors. The IL-2 receptor consists of at least three components, referred to as IL-2Rα, β and γ, which have a high affinity for IL-2 when dimerized. IL-2 also results in the formation of activated natural killer (NK) cells. Infusion of IL-2 in Brown–Norway rats results in airway hyperresponsiveness (Renzi *et al.*, 1991). In acute asthma, it has been reported that there is an increase in the number of circulating T lymphocytes which express IL-2 receptors (CD45$^+$; Corrigan and Kay, 1990), but this has not been confirmed by other investigators (Brown *et al.*, 1991). An increase in the number of CD4$^+$ lymphocytes in the biopsies of both allergic and non-allergic asthmatic patients has been reported (Brown *et al.*, 1991; Bentley *et al.*, 1992).

4.3 Interleukin-3

IL-3 is produced by T lymphocytes, but also by mast cells. In rodents it is important for the development of mast cells and basophils and promotes eosinophil survival (Wasserman, 1990). Antibodies to IL-3 result in a fall in tissue numbers of mast cells in mice. IL-3 has been reported to protect mast cells from apoptosis, an event likely mediated by the maintenance of intracellular calcium levels (Mekori *et al.*, 1993). Whether IL-3 plays a critical role in mast cell expression in human tissues is not clear and other cytokines including IL-10 and stem cell factor (c-*kit* ligand) may also be important.

4.4 Interleukin-4

IL-4 is derived from T_H2-like T lymphocytes, but there is also evidence for its expression in mast cells of asthmatic patients (Bradding *et al.*, 1992). IL-4 plays a critical role in the switching of B lymphocytes to produce IgE, and may therefore be of critical importance in the development of atopy (Romagnani, 1990). It induces the expression of the low affinity IgE receptor (FcϵRII, CD23) on macrophages (Vercelli *et al.*, 1988). Although IL-4 is essential for isotype switching, other factors, such as IL-6 and TNFα, have synergistic effects. IL-4 has many other actions and may increase expression of the adhesion molecule VCAM-1 in endothelial cells, which may be involved in eosinophil adhesion in the bronchial circulation (Schleimer *et al.*, 1992). In addition, IL-4 induces fibroblast chemotaxis and activation (Postlethwaite *et al.*, 1991, 1992), and in concert with IL-3, IL-4 promotes the growth of human basophils and eosinophils (Farre *et al.*, 1990). Studies in mice have shown that IL-4 may contribute to eosinophil accumulation in the lungs induced by parasite antigen (Lukacs *et al.*, 1994). While some of the effects of IL-4 promote allergic inflammation, it may also have anti-inflammatory effects. For example, IL-4 inhibits the expression of IL-1, TNFα, IL-8 and MIP-1α from activated monocytes (Essner *et al.*, 1989; Standiford *et al.*, 1990, 1993).

4.5 Interleukin-5

IL-5 is of particular interest in the pathophysiology of asthma as it is associated with eosinophilic inflammation (Sanderson, 1992). IL-5 is produced by T_H2-type lymphocytes and there is evidence for increased expression in T lymphocytes in asthmatic airways (Hamid *et al.*, 1991). Endobronchial allergen challenge results in IL-5 mRNA expression in eosinophils in BAL (Broide *et al.*, 1992b) and an increase in IL-5 concentration (Sedgewick *et al.*, 1991; Ohnishi *et al.*, 1993a). Elevated IL-5 concentrations have also been reported in BAL fluid from symptomatic, but not asymptomatic asthmatics (Ohnishi et al., 1993b). IL-5 mRNA has been detected in the sputum and bronchial biopsies of patients with asthma but not in non-asthmatic controls using PCR amplification (Gelder et al., 1993a,b). Increased circulating levels of immunoreactive IL-5 have been measured in the serum of patients with exacerbations of asthma, which fell with corticosteroid treatment (Corrigan et al., 1993). IL-5 is important in the terminal differentiation of eosinophils, in promoting eosinophil survival and in priming and activation of eosinophils (Sanderson, 1992). In vitro, IL-5 is chemotactic for eosinophils and can increase the survival of mature eosinophils as well as stimulate human eosoinophil function. A monoclonal antibody (mAb) to IL-5 inhibits eosinophil infiltration into the airways of animals sensitized to allergen (Gulbenkian et al., 1992; Chand et al., 1992), and also reduces airway hyperresponsiveness (Mauser et al., 1992; van Oosterhoot et al., 1993). It is of interest that the gene coding for IL-5 is located at position $q31$ on chromosome 5, as are genes for IL-3 and GM-CSF.

4.6 INTERLEUKIN-6

IL-6 is produced by many different cells on activation and appears to be responsible for many of the effects which occur during the acute phase response (van Snick, 1990). There is evidence for increased release of IL-6 from alveolar macrophages from asthmatic patients after allergen challenge (Gosset et al., 1991) and increased basal release compared to non-asthmatic subjects (Broide et al., 1992a). IgE-dependent triggering stimulates the secretion of IL-6 in both blood monocytes and alveolar macrophages in vitro (Gosset et al., 1992). IL-6 acts as a co-stimulatory factor with other cytokines on a variety of inflammatory and immune cells.

4.7 INTERLEUKIN-10

IL-10, previously known as cytokine synthesis inhibitor factor (CSIF), was originally identified as a product of murine T_H2 clones that suppressed the production of cytokines by T_H1 clones responding to stimulation of antigen (Fiorentino et al., 1989). In humans, T_H0, T_H1 and T_H2-like CD4$^+$ T cell clones, activated monocytes and peripheral blood T cells including CD4$^+$ and CD8$^+$ T cells have the capacity to produce IL-10 (Spizt and de Waal Malefyt, 1992; Enk and Katz, 1992). IL-10 is a pleotropic cytokine that can exert either immunosuppressive or immunostimulatory effects on a variety of cell types. It is a potent inhibitor of monocyte/macrophage function, suppressing the production of a number of proinflammatory cytokines, including TNFα, IL-1, IL-6 and IL-8 (Fiorentino et al., 1991; de Waal Malefyt et al., 1991). By contrast, IL-10 up-regulates the monocyte expression of IL-1 receptor antagonist, an anti-inflammatory agent (de Waal Malefyt et al., 1992). IL-10

suppresses the synthesis of superoxide anions and reactive nitrogen intermediates by activated monocytes/macrophages. These results suggest that IL-10 deactivates macrophages and exhibits potent anti-inflammatory properties. Indeed, an IL-10 antibody enhances the release of cytokines from activated monocytes, suggesting that this cytokine may play an inhibitory role when the cell is stimulated (de Waal Malefyt et al., 1991). On the other hand, IL-10 acts on B cells to enhance their viability, cell proliferation, immunoglobulin secretion and class II MHC expression. IL-10 is also a growth co-stimulator for thymocytes and mast cells (Thompson-Snipes et al., 1991), as well as an enhancer of cytotoxic T cell development (Chen and Zbtnik, 1991). There is little information as to whether IL-10 is expressed in asthmatic airways. However, IL-10 has been shown to inhibit the late response and the influx of eosinophils and lymphocytes after allergen challenge in the Brown–Norway rat (Broide et al., 1992a). The potential role for IL-10 in asthmatic inflammation remains open.

4.8 INTERLEUKIN-12

IL-12, previously known as natural killer cell stimulatory factor (Kobayashi et al., 1989) and cytotoxic lymphocyte maturation factor (Stern et al., 1990), is a cytokine produced by macrophages and B cells that acts mainly as a promoter of cell-mediated immunity. It enhances the growth of activated T cells and Nk cells (Gately et al., 1991; Robertson et al., 1992; Perussia et al., 1992; Bertagnolli et al., 1992) and enhances cytotoxic T cell and NK activity (Kobayashi et al., 1989; Robertson et al., 1992; Gately et al., 1992). IL-12 stimulates NK cells and T cells to produce IFNγ (Kobayashi et al., 1989; Chan et al., 1991; Wolf et al., 1991; Schoenhaut et al., 1992), promotes the in vitro differentiation of mouse and human T cells that secrete IFNγ and TNFα (Chan et al., 1991; Perussia et al., 1992; Manetti et al., 1993; Hsieh et al., 1993), and inhibits the differentiation of T cells into IL-4-secreting cells (Manetti et al., 1993; Sypek et al., 1993). It indirectly inhibits IL-4-induced human IgE responses by IFNγ-dependent and -independent mechanisms in vitro (Kiniwa et al., 1992).

4.9 INTERLEUKIN-13

IL-13 is a recently described cytokine secreted by different human T cell subsets and is a potent modulator of human monocyte and B cell function (Minty et al., 1993). Both CD4$^+$ and CD8$^+$ T cell clones synthesize IL-13 in response to antigen-specific or polyclonal stimuli (Zurawski and De Vries, 1994). IL-13 has profound effects on human monocyte morphology, surface antigen expression, antibody-dependent cellular toxicity and cytokine synthesis (Minty et al., 1993; McKenzie et al., 1993). In human monocytes stimulated by LPS, the production of proinflammatory cytokines, chemokines and CSFs is inhibited by IL-13, while IL-1Ra secretion is increased (Zurawsli *et al.*, 1993). Production of IL-1α, IL-1β, IL-6, IL-8, MIP-1α, TNFα, IL-10 and GM-CSF is inhibited. This action of IL-13 is similar to that of IL-4 and IL-10. The suppressive effects of IL-13 and of IL-4 are not related to endogenous production of IL-10. Similar to IL-4, IL-13 decreases the transcription of IFNα and the α and β chains of IL-12. It is possible that IL-13 acts like IL-4 and suppresses the development of T$_H$1 cells by down-regulating IL-12 production by monocytes, thereby favouring the development of T$_H$2 cells (Swain *et al.*, 1990; Le Gros *et al.*, 1990; Hsieh *et al.*, 1994).

IL-13 induces the expression of CD23 on purified human B cells and acts as a switch factor directing IgE synthesis, similar to IL-4 (Punnone *et al.*, 1993; Cocks *et al.*, 1993). However, unlike IL-4, IL-13 does not activate human T cells. A potent receptor antagonist of the biological activity of IL-4, a mutant protein of human IL-4, antagonizes IL-13 actions such as blocking the proliferation of B cells and IgE synthesis (Aversa *et al.*, 1993). This mutant protein of human IL-4 may therefore have therapeutic potential for the treatment of allergies. There is currently little information on the IL-13 receptor and on the biological effects of IL-13 *in vivo*.

4.10   CHEMOKINES

Chemokines (alternatively known as intercrines, PF-4 superfamily of cytokines or SIS cytokines) are members of a superfamily of small (8–10 kDa), inducible, secreted, pro-inflammatory cytokines. The sequences of members of the chemokine family have a conserved four-cysteine motif, and are divided into two branches with the first two cysteines separated by another amino acid residue (C-X-C), or the first two cysteines are adjacent (C-C) (Oppenheim *et al.*, 1991). The C-X-C chemokines include IL-8, β-thromboglobulin (β-TG) and platelet factor 4 (PF$_4$), and the C-C chemokines include MIP-1α, MIP-1β, RANTES, MCP-1/MCAF, MCP-2 and MCP-3.

The chemokines are involved in a variety of immunoregulatory functions, acting primarily as chemoattractants and activators of specific leucocytes. The C-X-C chemokines are chemoattractants and activators for neutrophils, while C-C chemokines are chemoattractants and activators for monocytes and T lymphocytes. Some C-C chemokines, e.g. RANTES, are also chemoattractants for eosinophils or activators of H release from basophils (Rot *et al.*, 1992; Kameyoshi *et al.*, 1992). Chemokines therefore may recruit and activate leucocytes at sites of inflammation.

IL-8 (previously known as neutrophil activation protein-1) is a potent neutrophil chemoattractant cytokine, which may be released by a variety of airway cells (Kunkel *et al.*, 1991), including T cells, endothelial cells, macrophages, eosinophils and epithelial cells

(Nakamura *et al.*, 1991; Cromwell *et al.*; Kwon *et al.*, 1993) and fibroblasts (Rolff *et al.*, 1991) in response to a wide variety of pro-inflammatory stimuli such as exposure to IL-1, TNF and endotoxin. IL-8 is a chemoattractant for neutrophils, and causes degranulation of neutrophil-specific granules. It enhances the adherence of neutrophils to endothelial cells and the subendothelial matrix (Rot, 1992). Under certain experimental conditions, IL-8 has been reported to be chemotactic for T cells and eosinophils (Taub and Oppenheim, 1993). IL-8 expression has been shown to be increased in lung tissue of patients with pulmonary fibrosis (Carré *et al.*, 1991).

Of the C-C chemokines, RANTES and MIP-1α induce the migration and activation of human eosinophils (Rot *et al.*, 1992; Kameyoshi *et al.*, 1992). RANTES selectively attracts blood monocytes and T lymphocytes expressing the cell surface antigens CD4 and UCHL1, the CD45RO antigen expressed on memory T cells (Schall *et al.*, 1990). MIP-1α and β selectively induce the attraction of T lymphocytes activated by CD3, CD4 and CD8 without attracting unstimulated lymphocytes (Taub *et al.*, 1993). In addition, these chemokines increased the adhesion of activated lymphocytes to activated human umbilical cord vein endothelium (Taub *et al.*, 1993). MCP-3 shows chemotactic activity for monocytes and eosinophils but not for neutrophils (Van Damme *et al.*, 1992; Dahinden *et al.*, 1994). These chemokines can activate eosinophils, monocytes, basophils and mast cells. Thus, RANTES and MIP-1α induce eosinophil cationic protein release with superoxide anions (Rot *et al.*, 1992). In basophils, RANTES, MIP-1α and MCP-3 induce cytosolic free calcium concentration changes and H release, and leukotriene C$_4$ (LTC$_4$) formation (Dahinden *et al.*, 1994). MIP-1α also activates mast cells to release H (Alam *et al.*, 1992).

4.11   GRANULOCYTE–MACROPHAGE COLONY-STIMULATING FACTOR

GM-CSF is a pleotropic cytokine that can stimulate the proliferation, maturation and function of haematopoietic cells. It is produced by several airway cells, including macrophages, eosinophils, T lymphocytes and epithelial cells. There is evidence for increased expression of GM-CSF in the epithelial cells of asthmatic patients (Sousa *et al.*, 1993) and in T lymphocytes and eosinophils after endobronchial challenge with allergen. Using PCR it has been possible to detect mRNA for GM-CSF in the sputum of asthmatic patients (Gelder *et al.*, 1993a). Increased circulating concentrations have been detected in the circulation of patients with acute severe asthma (Brown *et al.*, 1991) and peripheral blood monocytes from asthmatic patients secrete increased amounts (Nakamura *et al.*, 1993). It has effects on several cell types and may be involved in priming inflammatory cells, such as neutrophils and eosinophils. It can prolong the survival of eosinophils in culture (Hallsworth *et al.*, 1992). GM-CSF can enhance the release of superoxide anions and sulphidopeptide leukotrienes from eosinophils (Silberstein *et al.*, 1986). It can also induce the synthesis and release of a number of cytokines, including IL-1 and TNFα from monocytes. It can also induce non-haematopoietic cells such as endothelial cells to migrate and proliferate (Bussolino *et al.*, 1989).

4.12 TUMOUR NECROSIS FACTOR

TNFα (previously known as cachectin) is produced by many cells, including macrophages, lymphocytes, mast cells and epithelial cells. There is evidence for increased expression in asthmatic airways (Ying *et al.*, 1991), and IgE triggering in sensitized lungs leads to increased expression in epithelial cells in both rat and human lung (Ohno *et al.*, 1990; Ohkawara *et al.*, 1992). TNFα is also present in the BAL fluid of asthmatic patients (Broide *et al.*, 1992a). It is also released from alveolar macrophages of asthmatic patients after allergen challenge (Gosset *et al.*, 1991). Furthermore, both blood monocytes and alveolar macrophages show increased gene expression of TNFα after IgE triggering *in vitro* and this effect is enhanced by IFNγ (Gosset *et al.*, 1992). TNFα potently stimulates airway epithelial cells to produce cytokines, including IL-8 and GM-CSF (Cromwell *et al.*, 1992; Kwan *et al.*, 1993), and also increases the expression of the adhesion molecule ICAM-1 (Tosi *et al.*, 1992). TNFα also has a synergistic effect with IL-4 and IFNγ to increase VCAM-1 expression on endothelial cells (Thornhill *et al.*, 1991). This has the effect of increasing the adhesion of inflammatory leucocytes, such as neutrophils and eosinophils, at the airway surface. Infusion of TNFα causes increased airway responsiveness in Brown–Norway rats (Kips *et al.*, 1992), and inhalation of TNFα in normal human subjects results in increased airway responsiveness at 24 h and an increase in sputum neutrophils (Yates *et al.*, 1993). TNFα may be an important mediator in initiating chronic inflammation by activating the secretion of cytokines from a variety of cells in the airways. TNFα may also have chronic inflammatory effects, is a potent inducer of angiogenesis, and induces the formation of new blood vessels at sites of chronic inflammation.

4.13 INTERFERONS

IFNγ has extensive and diverse immunoregulatory effects on various cells. It is produced by T_H1 lymphocytes and exerts an inhibitory effect on T_H2 cells (Romagnani, 1990). This suggests that IFNs may have therapeutic potential in allergic diseases. IFNγ has been shown to inhibit antigen-induced eosinophil recruitment in the mouse (Nakajima *et al.*, 1993). However, IFNγ may also have pro-inflammatory effects and may activate airway epithelial cells to release cytokines and express adhesion molecules (Look *et al.*, 1992). IFNγ has an amplifying effect on the release of TNFα from alveolar macrophages induced by IgE triggering or by endotoxin (Gifford and Lohmann-Matthess, 1987; Gosset *et al.*, 1992) and increases the expression of class I and class II MHC molecules on macrophages and epithelial cells. It increases the production of IL-1, PAF and hydrogen peroxide from monocytes, in addition to downregulating IL-8 mRNA expression, which is up-regulated by IL-2 (Billiau and Dijkmans, 1990; Sen and Lenggel, 1992; Gusella *et al.*, 1993).

4.14 GROWTH FACTORS

The role of peptide growth factors in asthma has not been extensively investigated, but there is growing recognition that chronic structural changes may occur in the airways in response to the chronic inflammation, such as the proliferation of myofibroblasts and the hyperplasia of airway smooth muscle. Imaging of the airways by high resolution computed tomography has revealed the presence of dilated and thickened intrapulmonary airways (Paganin *et al.*, 1992; Carr *et al.*, 1994), which may represent the chronic structural abnormalities of the chronic inflammation of asthma. The mechanisms by which these structural changes occur are unclear but several growth factors may be released from inflammatory cells in the airways, such as macrophages and eosinophils, but also by structural cells such as airway epithelium, endothelial cells and fibroblasts.

PDGF is released from many different cells in the airways and consists of two peptide chains, so that A-A, B-B or A-B dimers may be secreted by different cells. These dimers act on α (AA) or β (AB or BB) receptors (Rose *et al.*, 1986). Macrophages release PDGF on activation (Haynes and Shaw, 1992). PDGF may activate fibroblasts to proliferate and secrete collagen (Rose *et al.*, 1986) and may also stimulate proliferation of airway smooth muscle (Hirst *et al.*, 1992), which is mediated via a β receptor (Hirst *et al.*, 1993). Fibroblast growth factors (FGFs) may be important regulators of fibrogenesis in the airway and may be produced by various inflammatory cells including macrophages. FGFs have the capacity to stimulate angiogenesis and this is a feature of chronic asthma. Epithelial cells may be an important source of growth factors in the airway and epithelial cells release PDGF, TGFβ and IGF-1 (Kelley, 1990; Cambrey *et al.*, 1993). Supernatants from cultured human airway epithelial cells stimulate the proliferation of human lung fibroblasts and this activity appears to reside predominantly in IGF-1 (Cambrey *et al.*, 1993). TGFβ comprises a family of growth-modulating cytokines that have an important influence on the turnover of matrix proteins (Moses *et al.*, 1990). They may either inhibit or stimulate proliferation of fibroblasts depending on the presence of other cytokines. TGFβ induces the transcription of fibronectin which can function as a chemotactic agent and growth factor for human fibroblasts (Ignotz et al., 1986; Infeld et al., 1992). Lung fibroblasts themselves may be a source of TGFβ (Kelley et al., 1991), but it is also secreted by inflammatory cells, including eosinophils, airway smooth muscle cells (Kelley, 1990) and structural cells such as epithelial cells (Sacco et al., 1992). TGFβ is present in the epithelial lining fluid of the normal lower respiratory tract (Yamauchi et al., 1988) TGFβ mRNA and protein have been found to be abundantly expressed in human lung, with TGFβ$_1$ precursor being immunolocalized throughout the airway wall including the epithelium and in alveolar macrophages, and the mature protein localized mainly within the connective tissue of the airway wall (Aubert et al., 1994). However, the patterns of expression for both forms of TGFβ$_1$ were similar in lungs from normals and asthmatic subjects (Aubert et al., 1994). The fibrogenic cytokines may also be involved in the repair process of airway epithelial damage characteristic of asthma. TGFβ is a potent inducer of differentiation for normal epithelial cells (Masui et al., 1986), and PDGF can promote wound healing (Clark et al., 1989).

5. The Cytokine Network in Chronic Asthma

Cytokines play an integral role in the coordination and persistence of inflammation in asthma, although the precise role of each cytokine remains to be determined. Cytokines play a fundamental role in inducing and increasing the production of specific IgE by B lymphocytes. IL-4 plays a critical role in the switching of B lymphocytes from IgG to IgE production, but other cytokines, including TNFα and IL-6, may also be important (Romagnani, 1990). IL-4 also increases the expression of an inducible form of the low affinity receptor for IgE (FcεRII or CD23) on B cells and macrophages. This may account for the increased expression of CD23 on alveolar macrophages from asthmatic patients.

Cytokines may also play an important role in antigen presentation and may enhance or suppress the ability of macrophages to act as antigen-presenting cells. Normally airway macrophages are poor at antigen presentation and suppress T cell proliferative responses (possibly via release of cytokines such as IL-1Ra), but in asthma there is evidence for reduced suppression after exposure to allergen (Aubus et al., 1984; Spiteri et al., 1991). Both GM-CSF and IFNγ increase the ability of macrophages to present allergen and express HLA-DR (Fisher et al., 1988). IL-1 is important in activating T lymphocytes and is an important co-stimulator of the expansion of T$_H$2 cells after antigen presentation (Chang et al., 1990).

Airway macrophages may be an important source of "first wave" cytokines, such as IL-1, TNFα and IL-6, which may be released on exposure to inhaled allergens via FcεRII receptors. These cytokines may then act on epithelial cells to release a second wave of cytokines, including GM-CSF, IL-8 and RANTES, which then amplifies the inflammatory response and leads to influx of secondary cells, such as eosinophils, which themselves may release multiple cytokines (Fig. 6.3). Cytokines such as TNFα and IL-1 may also increase the expression of adhesion molecules, such as ICAM-1, on epithelial cells, leading to adherence of leucocytes at the airway surface (Tosi et al., 1992). Cytokines may also switch on the gene transcription of the inducible form of nitric oxide synthase (iNOS) (Nathan, 1992), which is expressed in asthmatic but not normal epithelium (Springall et al., 1993), resulting in increased nitric oxide formation which may contribute to epithelial damage in asthma.

Mast cells are activated directly by allergen via FcεRI receptors and have the capacity to produce several cytokines which may perpetuate the inflammatory response (Gordon et al., 1990). Antigen presentation via dendritic cells leads to proliferation of T$_H$2 cells, which release IL-3, IL-4 and IL-5, thus perpetuating mast cell activation and eosinophilic inflammation. The inflammatory cells activated in asthma release multiple cytokines which have the effect of perpetuating the inflammatory state, by acting in an autocrine manner and by recruiting and activating adjacent structural cells or recruiting inflammatory cells from the circulation.

Cytokines may exert an important regulatory effect on the expression of adhesion molecules, both on endothelial cells of the bronchial circulation and on airway epithelial cells. Thus IL-4 increases the expression of VCAM-1 on endothelial cells and this may be important in eosinophil and lymphocyte trafficking (Schleimer et al., 1992), and IL-1 and TNFα increase the expression of ICAM-1 in both vascular endothelium and airway epithelium (Tosi et al., 1992).

Several cells which are activated in asthmatic airways also have the capacity to induce the secretion of growth factors, such as PDGF, EGF and IGF-1, which may stimulate fibrogenesis by recruiting and activating fibroblasts or transforming myofibroblasts. There is particular interest in the possibility that epithelial cells may release growth factors, since collagen deposition tends to occur underneath the basement membrane of the airway epithelium (Brewster et al., 1990). Growth factors may also stimulate the proliferation and growth of airway smooth muscle cells. PDGF$_{AB}$, PDGF$_{BB}$ and EGF are potent stimulants of animal and human airway smooth muscle proliferation (Hirst et al., 1992) and these effects are mediated via activation of tyrosine kinase and PKC. Growth factors may also be important in the proliferation of mucosal blood vessels and in the goblet cell hyperplasia that are characteristic of the chronically inflamed asthmatic airway. Cytokines such as TNFα and FGFs may also play an important role in angiogenesis which is reported in chronic asthma.

6. *Therapeutic Implications*

The recognition that cytokines play a critical role in the pathophysiology of asthma has prompted a search for new therapies based on cytokines. There are several approaches, but as multiple cytokines are involved it is unlikely that a single specific cytokine inhibitor will be as effective as less specific therapies, such as inhaled steroids. There appears to be considerable redundancy in the cytokine network, so that other cytokines may assume the role of the blocked cytokine. The similarities in the general features of T cell regulation between mice and humans have made murine models very useful for sorting out the activities of cytokines and the effects of individual cytokines *in vivo*. Many approaches are available, including recombinant cytokines, neutralizing anti-cytokine antibodies, and transgenic and homologous recombinant ("gene knockout") mice, with increased or suppressed production of specific cytokines. For example, studies in transgenic "knockout" mice show that deletion of a specific cytokine may have therapeutic potential. Thus IL-4 "knockout" mice have normal B and T lymphocyte development, but fail to produce IgE (Kuhn *et al.*, 1991).

6.1 CYTOKINE ANTIBODIES

Experimentally specific antibodies to individual cytokines have been used to explore their role in inflammatory diseases. Thus a murine mAb to IL-5 (TRK-5) blocks the eosinophilia after allergen exposure in sensitized guinea-pigs and also blocks airway hyperresponsiveness (although only at higher doses) (Gulbenkian *et al.*, 1992; Chand *et al.*, 1992; van Oosterhoot *et al.*, 1993). It is unlikely that IL-5 mAbs will have therapeutic potential because of antigenicity and the problem of repeated dosing, but they may be useful in demonstrating the potential for the IL-5 blocking strategies. It is possible that with chronic blockade of IL-5 effects, other cytokines take over its role. However, mAbs to TNF are now under trial in severe rheumatoid arthritis and appear to have long-lasting effects (over weeks), which implies that treatment may only be required infrequently.

6.2 SYNTHESIS INHIBITORS

Corticosteroids may be effective in asthma by suppressing cytokine synthesis in inflammatory cells (Guyre *et al.*, 1988; Barnes and Pedersen, 1993), whereas cyclosporin, FK506 and rapamycin inhibit cytokine synthesis predominantly in T lymphocytes. There is evidence that steroids inhibit the synthesis of IL-5 in asthmatic airways at a transcriptional level (Robinson *et al.*, 1993). Airway epithelial cells may be an important target of inhaled steroid therapy and steroids are able to inhibit the gene expression of cytokine synthesis, including GM-CSF and IL-8, in human airway epithelial cells (Kwon *et al.*, 1993) and of GM-CSF, IL-6 and IL-8 in human fibroblasts

(Tobler *et al.*, 1992). Inhaled steroids also reduce the increased expression of GM-CSF in airway epithelial cells of asthmatic patients (Sousa *et al.*, 1993). Corticosteroids also inhibit the expression of MIP-1α mRNA and the release of MIP-1α protein from activated monocytes and alveolar macrophages *in vitro* (Berkman *et al.*, 1994).

Other drugs may also have an inhibitory effect on cytokine synthesis. Elevation of cyclic AMP has an inhibitory effect on IL-2 release from lymphocytes *in vitro* (Didier *et al.*, 1987) and this may be achieved by theophylline at high plasma concentrations. Selective phosphodiesterase inhibitors may also be effective and recently phosphodiesterase IV inhibitors have been demonstrated to inhibit the release of IL-4 and IL-5 from T lymphocytes *in vitro* (Essayan *et al.*, 1993).

6.3 RECEPTOR ANTAGONISTS

Another approach is to develop drugs that block cytokine receptors. Blocking antibodies have been developed to some cytokine receptors. The receptors for IL-3, IL-5 and GM-CSF share a common α subunit, which is a particularly attractive target. However, there are difficulties in developing specific receptor antagonists for cytokines, since they are large peptides and have a very high affinity for their receptors. The human IL-5 receptor has now been cloned (Murata *et al.*, 1992), and it might be possible to develop antagonists in the future by molecular modelling. Recently a peptide analogue of IL-4 has been reported to have antagonist activity with only minimal agonist activity (Kruse *et al.*, 1992). It may be possible to discover a cytokine receptor antagonist by random screening using a cloned cytokine receptor expressed in a suitable vector.

IL-1Ra has been cloned (Arend 1991), and in experimental allergic inflammation of the airways recombinant IL-1Ra has some inhibitory effect on allergic inflammation in animals (Selig and Tocker, 1992; Watson *et al.*, 1993). Studies of human recombinant IL-1Ra in asthma are currently in progress. Similar antagonists of other cytokines may be discovered which could lead to the development of future antagonists.

Soluble receptors have been identified for a number of cytokines, including TNFα and IL-5. It is possible that these might have therapeutic value by mopping up released cytokines, but as these are large protein molecules it is unlikely that they would be of value in chronic asthma.

6.4 ANTISENSE DRUGS

Strategies such as antisense nucleotides, which would inactivate the specific mRNA encoded by cytokine genes or cytokine receptor genes, may be a more fruitful approach in the future when the problems of stability, cell penetration and delivery have been resolved (Colman, 1990).

6.5 RECOMBINANT CYTOKINES AS THERAPY

Cytokines may have anti-inflammatory as well as pro-inflammatory effects. IFNγ inhibits the production of IgE by B lymphocytes in human lymphocyte clones *in vitro* (Romagnani, 1990) and inhibits the development of T$_H$2 clones by antigen stimulation (Maggi *et al.*, 1992), and this suggests that IFNγ may have therapeutic potential in asthma. IFNα, which shares many activities with IFNγ, has been used to reduce IgE and eosinophil levels in patients with the hyper-IgE and hyper-eosinophilic syndromes (Souillet *et al.*, 1989; Zielinski and Lawrence, 1990). However, recombinant IFNγ appears to have no therapeutic benefit in steroid-dependent asthmatic patients (Boguniewicz *et al.*, 1992). On the other hand, IFNγ also has pro-inflammatory actions and may enhance the release of pro-inflammatory cytokines from T lymphocytes and macrophages. Indeed administration of human recombinant IFNγ by inhalation appears to activate alveolar macrophages (Jaffe *et al.*, 1991). Therefore, the pleotropic nature of the action of certain cytokines makes it difficult to predict what the overall therapeutic effect would be in asthma.

7. Conclusions

Cytokines play a critical role in the induction, coordination and perpetuation of the immunological and inflammatory responses observed in asthma. Multiple cytokines have been implicated in asthma and they are derived from a variety of cells. IL-1 and TNFα may be important in initiating the inflammatory response, IL-4 in switching of B lymphocytes to IgE production and IL-5 in eosinophilic inflammation. The role of the more recently described cytokines, such as IL-12 and -13, in allergic inflammation and asthma is unclear. Many inflammatory cells are capable of expressing a range of cytokines. Inhibition of cytokines is an important approach to asthma treatment and it is likely that inhaled corticosteroids are effective, at least in part, by inhibition of cytokine synthesis in airway cells, such as epithelial cells and lymphocytes. In the future it may be possible to develop more specific cytokine inhibitors and as more is learned about specific classes of cytokines, inhibition of the actions of certain classes of cytokines may provide more lasting benefit.

8. References

Abbas, A.K. (1987). Cellular interactions in the immune response. The roles of B lymphocytes and interleukin-4. Am. J. Pathol. 129, 25–33.

Adcock, I.M., Gelder, C.M., Shirasaki, H., Yacoub, M. and Barnes, P.J. (1992). Effects of steroids on transcription factors in human lung. Am. Rev. Respir. Dis. 145, A834.

Akira, S., Isshiki, H., Sugita, T., Tanabe, O., Kinoshita, S., Nishio, Y., Nakajima, T., Hirano, T. and Kishimoto, T. (1990). A nuclear factor for IL-6 expression (NFIL-6) is a member of a C/EBP family. EMBO J. 9, 1897–1906.

Alam, R., Forsythe, P.A., Stafford, S., Lett-Brown, M.A. and Grant, J.A. (1992). Macrophage inflammatory protein-1a activates basophils and mast cells. J. Exp. Med. 176, 781–786.

Arai, K., Lee, F., Miyajima, A., Miyatake, S., Arai, N. and Yokota T. (1990). Cytokines: coordinators of immune and inflammatory responses. Ann. Rev. Biochem. 59, 783–836.

Arend, W.P. (1991). Interleukin 1 receptor antagonist. A new member of the interleukin 1 family. J. Clin. Invest. 88, 1445–1451.

Arend, W.P., Welgus, H.G., Thompson, R.C. and Eisenberg, S.P. (1990). Biological properties of recombinant human monocyte-derived interleukin 1 receptor antagonist. J. Clin. Invest. 85, 1654–1702.

Asoa, H., Takeshita, T., Ishii, N., Kumaki, S., Nakamura, M. and Sugamura, K. (1993). Reconstitution of functional interleukin-2 receptor complexes on fibroblast cells; involvement of the cytoplasmic domain of the t chain in two distinct signalling pathways. Proc. Natl. Acad. Sci. 90, 4127–4131.

Aubert, J.-D., Dalal, B.I., Bai, T.R., Roberts, C.R., Hayashi, S. and Hogg, J.C. (1994). Transforming growth factor β gene expression in human airways. Thorax 45, 225–232.

Aubus, P., Cosso, B., Godard, P., Miche, F.B. and Clot J. (1984). Decreased suppressor cell activity of alveolar macrophages in bronchial asthma. Am. Rev. Respir. Dis. 130, 875–878.

Aversa, G., Punnonen, J., Cocks, B.G., de Waal Malefyt, R., Vega, F., Zurawski, S.M., Zurawski, G. and De Vries, J.E. (1993). An interleukin 4 (IL-4) mutant protein inhibits both IL-4 or IL-13-induced human immunoglobulin G4 (IgG4) and IgE synthesis and B cell proliferation: support for a common component shared by IL-4 and IL-13 receptors. J Exp. Med. 178, 2213–2218.

Barnes, P.J. and Pedersen, S. (1993). Efficacy and safety of inhaled steroids in asthma. Am. Rev. Respir. Dis. 148, S1–S26.

Bazan, J.F. (1990). Structural design and molecular evolution of a cytokine receptor superfamily. Proc. Natl. Acad. Sci. USA 87, 6934–6938.

Bentley, A.M., Menz, G., Storz, C., Robinson, D.S., Bradley, B., Jeffery, P.K., Durham, S.R. and Kay A.B. (1992). Identification of T lymphocytes, macrophages and activated eosinophils in the bronchial mucosa of intrinsic asthma: relationship to symptoms and bronchial hyperresponsiveness. Am. Rev. Respir. Dis. 146, 500–506.

Berkman, N., Jose, P., Williams, T., Barnes, P.J. and Chung, K.F. (1995). Corticosteroid inhibition of macrophage inflammatory protein-1α expression in human monocytes and alveolar macrophages. Am. J. Physiol. in press.

Bertagnolli, M.M., Lin, B.-Y., Young, D. and Herrmann, S.H. (1992). IL-12 augments antigen-dependent proliferation of activated T lymphocytes. J. Immunol. 149, 3778–3782.

Billiau, A. and Dijkmans, R. (1990). Interferon-gamma: mechanism of action and therapeutic potential. Biochem. Pharmacol. 40, 1433–1439.

Boguniewicz, M., Schneider, L.C., Milgrom, H.N, Jaffe, H.S., Izu A.C., Buxalo, L.R. and Leung D.Y.M. (1992). Treatment of steroid-dependent asthma with recombinant interferon-γ. J. Allergy. Clin. Immunol. 89, 288.

Bradding, P., Feather, I.H., Howarth, P.H., Mueller, R., Roberts, J.A., Britten, K., Bews, J.P., Hunt, T.C., Okayama, Y. and Heusser, C.H. (1992). Interleukin 4 is localized to and released by human mast cells. J. Exp. Med. 176, 1381–1386.

Bradding, P., Roberts, J.A., Britten, K.M., Montefort, S., Djukanovic, R., Mueller, R., Heusser, C.H., Howarth, P.H. and Holgate, S.T. (1994). Interleukin-4, -5 and -6 and tumor necrosis factor -α in normal and asthmatic airways: evidence for the human mast cell as a source of these cytokines. Am. J. Respir. Cell Mol. Biol. 10, 471–480.

Brewster, C.E.P., Howarth, P.H., Djukanovic, R., Wilson, J., Holgate, S.T. and Roche, W.R. (1990). Myofibroblasts and subepithelial fibrosis in bronchial asthma. Am. J. Respir. Cell Mol. Biol. 3, 507–511.

Broide, D.H. and Firestein, G.S. (1991). Endobronchial allergen challenge in asthma. Demonstration of cellular source of granulocyte–macrophage colony stimulating factor by in situ hybridization. J. Clin. Invest. 88, 1048–1053.

Broide, D.H, Lotz, M. and Cuomo, A.J. (1992a). Cytokines in symptomatic asthma. J. Allergy Clin. Immunol. 89, 958–967.

Broide, D., Paine, M.M. and Firestein, G.S. (1992b). Eosinophils express interleukin 5 and granulocyte-macrophage colony-stimulating factor mRNA at sites of allergic inflammation in asthmatics. J. Clin. Invest. 90, 1414–1424.

Brown, P.A., Crompton, G.K. and Greening, A.P. (1991). Proinflammatory cytokines in acute asthma. Lancet 338, 590–593.

Bussolino, F., Wang, J.M., Defilippi, P., Turrini, F., Sanavio, F., Edgell, C.J., Aglietta, M., Arese, P. and Mantovani A. (1989). Granulocyte- and granulocyte-macrophage-colony stimulating factors induce human endothelial cells to migrate and proliferate. Nature 337, 471–473.

Cambrey, A.D., Kwon, O.J., McAnulty, R.J., Harrison, N.K., Barnes, P.J., Laurent, G.J. and Chung, K.F. (1993). Release of fibroblast proliferative activity from cultured human airway epithelial cells: a role for insulin-like growth factor 1 (IGF-1). Am. Rev. Respir. Dis. 147, A272.

Carr, D., Hibon, S. and Chung, K.F. (1994). High resolution computed tomography (HRCT) scanning to evaluate structural changes in lungs of patients with severe chronic asthma. Thorax 49, 430.

Carré, P.C., Mortenson, R.L., King, R.E., Noble, P.W., Sable, C.L. and Riches D.W.H. (1991). Increased expression of the interleukin-8 gene by alveolar macrophages in idiopathic pulmonary fibrosis. J. Clin. Invest. 88, 1802–1810.

Chan, S.H., Perussia, B., Gupta, J.W., Kobayashi, M., Pospisil, M., Young, H.A., Wolf, S.F., Young, D., Clark, S.C. and Trinchieri, G. (1991). Induction of interferon γ production by natural killer cell stimulatory factor: characterization of the responder cells and synergy with other inducers. J. Exp. Med. 173, 369–904.

Chand, N., Harrison, J.E., Rooney, S., Pillar, J., Jakubicki, R., Nolan, K., Diamantis, W. and Sofia, R.D. (1992). Anti IL-5 monoclonal antibody inhibits allergic late phase bronchial eosinophilia in guinea pigs: a therapeutic approach. Eur. J. Pharmacol. 211, 121–123.

Chang, T.L., Shea, C.H., Urioste, S., Thompson, R.C., Boom, W.H. and Abbas, A.K. (1990). Heterogeneity of helper/inducer T lymphocytes: lymphokine production and lymphokine responsiveness J. Immunol. 145, 2803–2808.

Chen, W.F. and Zlotnik, A. IL-10: A novel cytotoxic T cell differentiation factor. J. Immunol. 147, 520–534.

Choi, A.M.K. and Jacoby, D.B. (1992). Influenza virus A infection induces interleukin-8 gene expression in human airway epithelial cells. FEBS Lett. 309, 327–329.

Churchill, I., Friedman, B., Schleimer, R.P. and Proud, D. (1992). Production of granulocyte–macrophage colony-stimulating factor by cultured human tracheal epithelial cells. Immunology 75, 189–195.

Clark, J.G., Dedon, T.F., Wayner, E.A. and Carter, W.G. (1989). Effects of interferon-gamma on expression of cell surface receptors for collagen and deposition of newly synthesized collagen by cultured human lung fibroblasts. J. Clin. Invest. 83, 1505–1511.

Cluzel, M., and Lee, T.H. (1992). Cytokines. In "Asthma: Basic Mechanisms and Clinical Management" 2nd edn (eds P.J. Barnes, I.W. Rodger, N.C. Thomson, Vol. 2, pp. 315–331, Academic Press, London.

Cocks, B.G., de Waal Malefyt., R, Galizzi, J.P., De Vries, J.E. and Aversa, G. (1993). IL-13 induces proliferation and differentiation of human B cells activated by the CD40 ligand. Int. Immunol. 6, 657–663.

Colman, A. (1990). Antisense strategies in cell and developmental biology. J. Cell Sci. 97, 399–409.

Corrigan, C.J. and Kay, A.B.(1990). CD4+ T-lymphocyte activation in acute severe asthma. Relationship to disease severity. Am. Rev. Respir. Dis. 140, 970–977.

Corrigan, C.J., Haczku, A., Gemon-Engesaeth, V., Doi, S., Kikuchi, Y., Takatsu, K. et al (1993). CD4 T-lymphocyte activation in asthma is accompanied by increased serum concentration of interleukin-5: effect of glucocorticoid therapy. Am. Rev. Respir. Dis. 147, 540–547.

Cosman, D., Lyman, S.D., Idzerda, R.L., Beckman, M.P., Park, L.S., Goodwin, R.G. and March, C.J. (1990). A new cytokine receptor superfamily. Trends Biochem. Sci. 15, 265–270.

Costa, J.J., Matossian, K., Resnick, M.B., Beil, W.J., Wong, D.T.W., Gordon, J.R., Dvorak, A.M., Weller, P.F. and Galli, S.J. (1993). Human eosinophils can express the cytokines tumor necrosis factor-α and macrophage inflammatory protein-1α J. Clin. Invest. 91, 2673–2684.

Cromwell, O., Hamid, Q., Corrigan, C.J., Barkans, J., Meng, Q., Collins, P.D. and Kay, A.B. (1992). Expression and generation of interleukin-8, IL-6 and granulocyte colony-stimulating factor by bronchial epithelial cells and enhancement by IL-1β and tumor necrosis factor-α. Immunology 77, 330–337.

Dahinden, C.A., Geiser, T., Brunner, T., von Tscharner, V., Caput, D., Ferrara, P., Minty, A. and Baggiolini, M. (1994). Monocyte chemotactic protein 3 is a most effective bosophil- and eosinophil-activating chemokine. J. Exp. Med. 179, 751–756.

Desreumaux, P., Janin, A., Colombel, J.F., Prin, L., Plumas, J., Emilie, D., Torpier, G. and Capron, M. (1992). Interleukin 5 messenger RNA expression by eosinophils in the intestinal mucosa of patients with coeliac disease. J. Exp. Med. 175, 293–296.

Devalia, J.L., Rusznak, C., Sapsford, R.J., Calderon M. and Davies, R.J. (1992). Nitrogen dioxide (NO_2)-induced permeability and synthesis of inflammatory cytokines by human bronchial epithelial cell monolayers *in vitro*. J. Allergy Clin. Immunol. 91, 328.

de Waal Malefyt, R., Abrams, J., Bennett, B., Figdor, C.G. and De Vries, J.E. (1991). Interleukin 10 (IL-10) inhibits cytokine synthesis by human monocytes: an auto regulatory role of IL-10 produced by monocytes. J. Exp. Med. 179, 1209–1220.

de Waal Malefyt, R., Yssel, H., Roncarolo, M.G., Spits, H. and De Vries, J.E. (1992). Interleukin-10. Curr. Opin. Immunol. 4, 314–320.

Didier, M., Aussel, C., Ferrua, B. and Fehlman, M. (1987). Regulation of interleukin 2 synthesis by cAMP in human T cells. J. Immunol. 139, 1179–1184.

Dinarello, C.A. (1989). Interleukin-l and biologically related cytokines. Adv. Immunol. 44, 153–205.

Djukanovic, R., Roche, W.R., Wilson, J.W., Beasley, C.R.W., Twentyman, O.P., Howarth, P.H. *et al* (1990). Mucosal inflammation in asthma. Am. Rev. Respir. Dis. 142, 434–457.

Draper, J.C. and Wietzerbin, J. (1991). IFN-gamma decreases specific binding of tumor necrosis factor on murine macrophages. J. Immunol. 146, 1198–1203.

Elices, M.L., Osborn, L., Takada, Y., Crouse, C., Luhowskyj, S., Hemler, M.E. *et al*. VCAM-1 on activated endothelium interacts with the leukocyte integrin VLA-4 at a site distinct from the VLA-4/fibronectin binding site. Cell 60, 577–584.

Enk, A.H. and Katz, S.I. (1992). Identification and induction of keratinocyte-derived IL10. J. Immunol. 149, 92–95.

Essayan, D.M., Kage, Y., Sobotka, A., Lichtenstein, L.M. and Huang S.R. (1993). Modulation of allergen-induced cytokine gene expression and proliferation by phosphodiesterase inhibitors *in vitro*. J. Allergy Clin. Immunol. 91, 254.

Essner, R., Rhoades, K., McBride, W.H., Morton, D.L. and Economou, J.S. (1989). IL-4 down-regulates IL-1 and TNF gene expression in human monocytes. J. Immunol. 142, 3857–3861.

Favre, C., Saeland, S., Caux, C., Duvert, V. and De Vries, J.E. (1990). Interleukin 4 has basophilic and eosinophilic cells growth-promoting activity on cord blood cells. Blood 75, 67–73.

Filley, W.V., Holley, K.E., Kephart, G.M. and Gleich, G.J. (1982). Identification by immunofluorescence of eosinophil granule major basic protein in lung tissues of patients with bronchial asthma. Lancet 2, 11–16.

Fiorentino, D.F., Bond, M.W. and Mosmann, T.R. (1989). Two types of mouse helper T cells. IV. Th 2 clones secrete a factor that inhibits cytokine production by Th 1 clones. J. Exp. Med. 170, 2081.

Fiorentino, D.F., Zlotnik, A., Mossmann, T.R., Howard, M. and O'Garra, A. (1991). IL-10 inhibits cytokine production by activated macrophages. J. Immunol. 147, 3815–3822.

Fischer, H.G., Frosch, S., Reske, K. and Reske-Kunz, A.B. (1988). Granulocyte-macrophage colony-stimulating factor activates macrophages derived from bone marrow cultures to synthesis of MHC class II molecules and to augmented antigen presentation function. J. Immunol. 141, 3882–3888.

Gately, M.K., Desai, B., Wolitzky, A.G., Quinn, P.M., Dwyer, C.M., Podlaski, F.J., Familletti, P.C., Sinigaglia, F., Chizzonite, R., Gubler, U. and Stern, A.S. (1991). Regulation of human lymphocyte proliferation by a heterodimeric cytokine, IL-12 (cytotoxic lymphocyte maturation factor). J. Immunol. 147, 874–879.

Gately, M.K., Wolitzky, A.G., Quinn, P.M. and Chizzonite, R. (1992). Regulation of human cytolytic lymphocyte responses by interleukin-12. Cell Immunol. 143, 127–130.

Gelder, C.M., Morrison, J.F.S., Yates, D.H., Thomas, P.S. and Barnes, P.J. (1993a). Polymerase chain reaction cytokine profiles of induced sputum from asthmatic subjects. Am. Rev. Respir. Dis. 147, A786.

Gelder, C.M., Morrison, J.F.J., Southcott, A.M., Adcock, I.M., Kidney, J., Peters, M., O'Connor, B., Chung, K.F. and Barnes, P.J. (1993b). Cytokine mRNA profiles in asthmatic endobronchial biopsies. Am. Rev. Respir. Dis. 147, A786.

Gifford, G.E. and Lohmann-Matthess, M.L. (1987). Gamma interferon priming of mouse and human macrophages for induction of tumor necrosis factor production by bacterial lipopolysaccharide. J. Natl. Cancer Inst. 78, 121–124.

Gordon, J.R. and Galli, S.J. (1991). Release of both preformed and newly synthesized tumor necrosis factorα (TNFα)/cachectin by mouse mast cells stimulated by the FcRI. A mechanism for the sustained action of mast cell-derived TNFα during IgE-dependent biological responses. J. Exp. Med. 174, 103–107.

Gordon, J.R., Burd, P.R. and Galli, S.J. (1990). Mast cells as a source of multifunctional cytokines. Immunol. Toduy 11, 458–467.

Gosset, P., Lassalle, P., Tonnel, A.B., Dessaint, J.P., Wallaert, B., Prin, L., Pestel, J. and Capron, A. (1988). Production of an interleukin-l inhibitory factor by human alveolar macrophages from mammals and allergic asthmatic patients. Am. Rev. Respir. Dis. 138, 40–46.

Gosset, P., Tsicopoulos, A., Wallaert, B., Vannimenus, C., Joseph, M., Tonnel, A.B. and Capron, A. (1991). Increased secretion by tumor necrosis factor α and interleukin 6 by alveolar macrophages consecutive to the development of the late asthmatic reaction. J. Allergy Clin. Immunol. 88, 561–571.

Gosset, P., Tsicopoulos, A., Wallaert, B., Vannimenus, C., Joseph, M., Tonnel, A.B., and Capron A. (1992). Tumor necrosis factor α and interleukin–6 production by human mononuclear phagocytes from allergic asthmatics after IgE-dependent stimulation. Am. Rev. Respir. Dis. 146, 768–774.

Gulbenkian, A.R., Egan, R.W., Fernandez, X., Jones, H, Kreugner, W., Kung, T., Payvandi, F,, Sullivan, L., Zurcher, J.A. amd Watnick, A.S. (1992). Interleukin-5 modulates eosinophil accumulation in allergic guinea pig lung. Am. Rev. Respir. Dis. 146, 263–266.

Gusella, G.L., Musso, T., Bovco, M.C., Espinoza-Delgado, I., Matsushima, K. and Varesio, L. (1993). IL-2 up-regulates but IFNγ suppresses IL-8 expression in human monocytes. J. Immunol. 151, 2725–2732.

Guyre, P.M., Girard, M.T., Morganelli, P.M. and Manginiello, P.D. (1988). Glucocorticoid effects on the production and action of immune cytokines. J. Steroid Biochem. 30, 89–93.

Hallsworth, M.P., Litchfield, T.M. and Lee, T.H. (1992). Glucocorticosteroids   inhibit   granulocyte–macrophage colony-stimulating factor and interleukin-5 enhanced *in vitro* survival of human eosinophils. Immunology 75, 382–385.

Hallsworth, M.P., Soh, C.P.C., Lane, S.J., Arm, J.P. and Lee, T.H. (1994). Selective enhancement of GM-CSF, TNFα, IL-lb and IL-8 production by monocytes and macrophages of asthmatic subjects. Eur. Respir. J. 7, 1096–1102.

Hamid, Q., Azzawi, M., Ying, S., Moqbel, R., Wardlaw, A.J., Corrigan, C.J., Bradley, B., Durham, S.R., Collins, J.V., Jeffery, P.K., Quint, D. and Kay, A.B. (1991). Expression of mRNA for interleukins in mucosal bronchial biopsies from asthma. J. Clin. Invest. 87, 1541–1546.

Hamid, Q., Barkans, J., Meng, Q., Ying, S., Abrams, J.S., Kay, A.B. and Moqbel R. (1992). Human eosinophils synthesize and secrete interleukin-6 *in vitro*. Blood 80, 1496–1501.

Haynes, A.R. and Shaw, R.J. (1992). Dexamethasone-induced increase in platelet-derived growth factor (B) mRNA in human alveolar macrophages and myelomonocytic HL60 macrophage-like cells. Am. J. Respir. Cell Mol. Biol. 7, 198–206.

Herrmann, F., Oster, W., Meuer, S.C., Lindemann, A. and Mertelsmann, R.H. (1988). Interleukin-1 stimulates T-lymphocytes to produce granulocyte colony stimulating factor. J. Clin. Invest. 81, 1415–1418.

Hirst, S.J., Barnes, P.J. and Twort, C.H.L. (1992). Quantifying proliferation of cultured human and rabbit airway smooth muscle cells in response to serum and platelet-derived growth factor. Am. J. Respir. Cell Mol. Biol. 7, 574–581.

Hirst, S.J., Barnes, P.J. and Twort, C.H.C. (1995). Differential stimulation of the proliferative response of rabbit cultured tracheal smooth muscle cells by platelet-derived growth factor isoforms. Br. J. Pharmacol. 1995; in press.

Holmes, W.E., Lee, J., Kuang, W.J., Rice, G.C. and Wood, W.I. (1991). Structure and functional expression of a human interleukin-8 receptor. Science 253, 1278–1280.

Hsieh, C.-S., Macatonia, S.E., Tripp, C.S., Wolf, S.F., O'Garra, A. and Murphy, K.M. (1993). Development of TH1 CD4+ T cells through IL-12 produced by *Listeria*-induced macrophages, Science 260, 547–549.

Ignotz, J.B.C., Ignotz, R.A. and Massagne, J. (1986). Transforming growth factor-beta stimulates expression of fibronectin and collagen and their incorporation into the extracellular matrix. J. Biol. Chem. 261, 4337–4345.

Infeld, M.D., Brennan, J.A. and Davis, P.B. (1992). Human tracheobronchial epithelial cells direct migration of lung fibroblasts in three-dimensional collagen gels. Am. J. Physiol. 262, L535–L541.

Jaffe, H.A., Buhl, R., Mastrangeli, A., Holroyd, K.J., Saltini, C., Czerski, D., Jaffe, H.S., Kramer, S., Sherwin, S. and Crystal, R.G. (1991). Organ-specific cytokine therapy – local activation of mononuclear phagocytes by delivery of an aerosol of recombinant interferon-gamma to the human lung. J. Clin. Invest. 88, 297–302.

Kameyoshi, Y., Dorschner, A., Mallet, A.I., Christophers, E. and Schroder, J.M. (1992). Cytokine, RANTES released from thrombin-stimulated platelets is a potent attractant for human eosinophils. J. Exp. Med. 173, 587–592.

Kaye, J. and Janeway, C.A.Jr. (1984). Induction of receptors for interleukin 2 requires T cell Ag:Ia receptor crosslinking and interleukin 1. Lymph Res. 3, 175–182.

Kelley, J. (1990). Cytokines of the lung. Am. Rev. Respir. Dis. 141, 765–788.

Kelley, J., Fabisiak, J.P., Hawes, K. and Abscher, M. (1991). Cytokine signalling in lung: transforming growth factorβ secretion by lung fibroblasts. Am. J. Physiol. 260, L123–L128.

Kiniwa, M., Gately, M., Gubler, U., Chizzonite, R., Fargeas, C. and Delespesse, G. (1992). Recombinant interleukin-12 suppresses the synthesis of IgE by interleukin-4 stimulated human lymnphocytes. J. Clin. Invest. 90, 262–266.

Kips, J.C., Tavernier, J. and Pauwels, R.A. (1992). Tumor necrosis factor causes bronchial hyperresponsiveness in rats. Am. Rev. Respir. Dis. 145, 332–336.

Kita, H., Ohnishi, T., Okubo, Y., Welier, D., Abrams, J.S. and Gleich G.J. (1991). Granulocyte/macrophage colony-stimulating factor and interleukin-3 release from human peripheral blood eosinophils and neutrophils. J. Exp. Med. 174, 745–748.

Kobayashi, M., Fitz, L., Ryan, M., Hewick, R.M., Clark, S.C., Chan, S., Loudon, R., Sherman, F., Perussia, B. and Trinchieri G. (1989). Identification and purification of natural killer cell simulatory factor (NKSF), a cytokine with multiple biological effects on human lymphocytes. J. Exp. Med. 170, 827–832.

Krishnaswamy, G., Liu, M.C., Su, S.-N., Kumai, M., Xiao, H.-Q., Marsh, D.G. and Huang, S.-K. (1993). Analysis of cytokine transcripts in the bronchoalveolar lavage cells of patients with asthma. Am. J. Respir. Cell Mol. Biol. 9, 279–286.

Kruse, N., Tony, H.P. and Sebald, W. (1992). Conversion of human interleukin-4 into a high affinity antagonist by a single amino acid replacement. EMBO J. 11, 3237–3244.

Kuhn, R., Rajewsky, R. and Muller, W. (1991). Generation and analysis of interleukin-4 deficient mice. Science 254, 707–710.

Kunkel, S.L. Standiford, T., Kasaka, K. and Strieter, R.M. (1991). Interleukin-8 (IL-1): the major neutrophil chemotactic factor in lung. Exp. Lung Res. 17, 17–23.

Kwon, O.J., Collins, P.D., Au, B., Adcock, I.M., Yacoub, M., Chung, K.F. and Barnes, P.J. (1993). Glucocorticoid inhibition of TNFα-induced IL-8 gene expression in human primary cultured epithelial cells. Am. Rev. Respir. Dis. 147, A752.

Le Gros, G., Ben-Sasson, S.Z., Seder, R.A., Finkelman, F.D. and Paul, W.E. (1990). Generation of interleukin 4 (IL-4)-producing cells *in vivo* and *in vitro*: IL-2 and IL-4 are required for *in vitro* generation of IL-4-producing cells. J. Exp. Med. 172, 921–929.

Lenardo, M.J. amd Baltimore, D. (1989). NF-κB: a pleiotropic mediator of inducible and tissue specific gene control. Cell 58, 227–229.

Look, D.C., Rapp, S.R., Keller, B.T. and Holzman, M.J. (1992). Selective induction of intracellular adhesion molecule-1 by interferon-γ in human airway epithelial cells. Am. J. Physiol 263, L79–L87.

Lovett, D., Kozan, B., Hadam, M., Resch, K. and Gemsa, D. (1986). Macrophage cytotoxicity: interleukin 1 as a mediator of tumor cytostasis. J. Immunol. 136, 340–347.

Lukacs, N.W., Strieter, R.M., Chensue, S.W. and Kunkel, S.L. Interleukin-4-dependent pulmonary eosinophil infiltration in a murine model of asthma. Am. J. Respir. Cell Mol. Biol. 10, 526–530.

Maggi, E., Parronchi, P., Manetti, R., Simonelli, C., Piccinni, M.P., Rugiu, F.S., de Carli, M., Ricci, M. and Romagnani, S. (1992). Reciprocal regulatory effects of IFN-gamma and

IL-4 on the *in vitro* development of human T$_H$1 and T$_H$2 clones. J. Immunol. 148, 2142–2147.

Mahé, Y., Mukaida, N., Kuno, K., Akiyama, M., Ikeda, N., Matsushima, K. and Murakami, S. (1991). Hepatitis B virus X protein transactivates human interleukin–8 gene through acting on nuclear factor KB and CAAT/enhancer-binding protein-like *cis* elements. J. Biol. Chem. 266, 13759–13763.

Manetti, R., Parronchi, P., Giudizi, M.G., Piccinni, M.-P., Maggi, E., Trinchieri, G. and Romagnani, S. (1993). Natural killer cell stimulatory factor (interleukin 12 (IL-12)) induces helper type 1 (Th1)-specific immune responses and inhibits the development of IL-4-producing Th cells. J. Exp. Med. 177, 1199–1204.

Mantovani, A. and Dejana, E. (1989). Cytokines as communication signals between leukocytes and endothelial cells. Immunol. Today 10, 370–375.

Marini, M., Soloperto, M., Mezzetti, M., Fasoli, A. and Mattoli, S. (1991). Interleukin 1 binds to specific receptors on human bronchial epithelial cells and upregulates granulocyte–macrophage colony-stimulating factor synthesis and release. Am. J. Respir. Cell Mol. Biol. 4, 519–524.

Marini, M., Vittori, E., Hollemburg, J. and Mattoli, S. (1992). Expression of the potent inflammatory cytokines granulocyte–macrophage colony stimulating factor, interleukin-6 and interleukin-8 in bronchial epithelial cells of patients with asthma. J. Allergy Clin. Immunol. 82, 1001–1009.

Masui, T., Wakefield, L.M., Lechner, J.F., LaVeck, M.A., Sporn, M.B. and Harris, C.C. (1986). Type β transforming growth factor is the primary differentiation-inducing serum factor for normal human bronchial epithelial cells. Proc. Natl. Acad. Sci. USA 83, 2438–2442.

Mattoli, S., Mattoso, V.L., Soloperto, M., Allegra, L. and Fasoli, A. (1991). Cellular and biochemical characteristics of bronchoalveolar lavage fluid in symptomatic non allergic asthma. J. Allergy Clin. Immunol. 84, 794–803.

Mauser, P.J., Pitman, A., Fernandez, X., Zurcher, J., Watnick, A., Egan, R.W., Kreutner, W. and Adams, G.K. (1992). The effects of anti-IL-5 on antigen-induced airway hyperreactivity and pulmonary eosinophilia in guinea pigs. Am. Rev. Respir. Dis. 145, A859.

McKenzie, A.N.J., Culpepper, J.A., de Waal Malefyt, R., Brière F., Punnonen J., Aversa G., Sato, A., Dang, W., Cocks, B.G., Menon, S., De Vries, J.E., Banchereau, J. and Zurawski, G. (1993). Interleukin 13, a T-cell-derived cytokine that regulates human monocyte and B-cell function. Proc. Natl. Acad. Sci. USA 90, 3735–3739.

Mekori, Y.A., Oh, C.K. and Metcalfe, D.D. (1993). IL-3-dependent murine mast cells undergo apoptosis on removal of IL-3. Prevention of apoptosis by c-*kit* ligand. J. Immunol. 151, 3775–3784.

Minty, A., Chalon, P., Derocq, J.-M., Dumont, X., Guillemot, J.-C., Kaghad, M., Labit, C., Leplatois, P., Liauzun, P., Miloux, B., Minty, C., Casellas, P., Loison, G., Lupker, J., Shire, D., Ferrara, P. and Caput, D. (1993). Interleukin-13 is a new human lymphokine regulating inflammatory and immune responses. Nature 362, 248–250.

Montefort, S., Gratziou, C., Goulding, D., Polosa, R., Haskard, D.O., Howarth, P.H., Holgate, S.T. and Carroll, M.P. (1994). Bronchial biopsy evidence for leukocyte infiltration and upregulation of leukocyte-endothelial cell adhesion molecules 6 hours after local allergen challenge of sensitised asthmatic airways. J. Clin. Invest. 93, 1411–1421.

Moqbel, R., Hamid, Q., Ying, S., Barkans, J., Hartnell, A., Tsicopoulos, A., Wardlaw, A.J. and Kay, A.B. (1991). Expression of mRNA and immunoreactivity for the granulocyte/macrophage colony-stimulating factor in activated human eosinophils. J. Exp. Med. 174, 749–752.

Moser, B., Schumacher, C., von Tscharner, V., Clark-Lewis, I. and Baggiolini, M. (1991). Neutrophil-activating peptide 2 and gro/Melanoma growth-stimulating activity interact with neutrophil-activating peptide 1/interleukin 8 receptors on neutrophils. J. Biol. Chem. 266, 10666–10671.

Moses, H.L., Yang, E.L. and Pietenpol, J.A. (1990). TGFβ stimulation and inhibition of cell proliferation: new mechanistic insights. Cell 63, 245–247.

Mosmann, T.R., Cherwinski, H., Bond, M.W., Giedlin, M.A. and Coffman, R.L. (1986). Two types of murine helper T cell clone. A. Definition according to profiles of lymphokine activities and secreted proteins. J. Immunol. 136, 2348–2357.

Mossman, T.R. and Coffman, R.L. (1989). T$_H$1 and T$_H$2 cells: different patterns of lymphokine secretion lead to different functional properties. Annu. Rev. Immunol 7, 145–173.

Muegge, K. and Durum, S.K. (1990). Cytokines and transcription factors. Cytokine 2, 1–8.

Mukaida, N., Mahé, Y. and Matsushima, K. (1990). Cooperative interaction of nuclear factor kB and *cis*-regulatory enhancer binding protein-like factor bonding elements in activating the interleukin-8 gene by pro-inflammatory cytokines. J. Biol. Chem. 265, 21128–21133.

Murata, Y., Takaki, S., Migita, M., Kikuchi, Y., Tominga, A. and Takatsu, K. (1992). Molecular cloning and expression of the human interleukin 5 receptor. J. Exp. Med. 175, 341–351.

Nakajima, H., Iwamoto, I. and Yoshida, S. (1993). Aerosolized recombinant interferon-gamma prevents antigen-induced eosinophil recruitment in mouse trachea. Am. Rev. Respir. Dis. 148, 1102–1104.

Nakamura, H., Yoshimura, K., Jaffe, H.A. and Crystal, R.G. (1991). Interleukin-8 gene expression in human bronchial epithelial cells. J. Biol. Chem. 266, 19611–19617.

Nakamura, Y., Ozaki, T., Kamgi, T., Kawaji, K., Banno, K., Miki, S., Fujisawa, K., Yasuora, S. and Ogura, T. (1993). Increased granulocyte–macrophage colony-stimulating factor production by mononuclear cells from peripheral blood of patients with bronchial asthma. Am. Rev. Respir. Dis. 147, 87–91.

Nathan, C. (1992). Nitric oxide as a secretory product of mammalian cells. FASEB J. 6, 3051–3064.

Navarro, S., Debili, N., Bernaudin, J.F., Vainchenker, W. and Doly, J. (1989). Regulation of the expression of IL-6 in human monocytes. J. Immunol. 142, 4339–4345.

Neote, K., Digregorio, D., Mak, J.Y., Horak, R. and Schall, T.J. (1993). Molecular cloning, functional expression and signaling characteristics of a C-C chemokine receptor. Cell 72, 415–425.

Ohkawara, Y., Yamauchi, K., Tanno, Y., Tamura, H., Outani, H., Nagura, H., Ohkuda, K. and Takishima, T. (1992). Human lung mast cells and pulmonary macrophages produce tumor necrosis factor-α in sensitised lung tissue after IgE receptor triggering. Am. J. Respir. Cell Mol. Biol. 7, 385–392.

Ohnishi T., Kita, H., Weiler, D., Sur, S., Sedgwick, J.B., Calhoun, W.J., Busse, W.W., Abrams, J.S. and Gleich, G.J.

(1993a). IL-5 is the predominant eosinophil-active cytokine in the antigen-induced pulmonary late-phase reaction. Am. Rev. Respir. Dis. 147, 901–907.

Ohnishi, T., Sur, S., Collins, D.S., Fisa, J.E, Gleich, G.J. and Peters, S.P. (1993b). Eosinophil survival activity identified as interleukin-5 is associated with eosinophil recruitment and degranulation and lung injury twenty-four hours after segmental antigen lung challenge. J. Allergy Clin. Immunol. 92, 607–615.

Ohno, I., Ohkawara, Y., Yamauchi, K., Tanno, Y. and Takishima, T. (1990). Production of tumor necrosis factor with IgE receptor triggering from sensitized lung tissue. Am. J. Respir. Cell. Mol. Biol. 3, 285–289.

Ohno, I., Lea, R.G., Flanders, K.C., Clark, D.A., Banwatt, D., Dolovich, J., Denburg, J., Harley Gauldie, C.B. and Jordana, M. (1992). Eosinophils in chronically inflamed human upper airway tissues express transforming growth factor β1 gene. J. Clin. Invest 89, 1662–1668.

Oppenheim, J.J., Zacharide, C.O.C., Mukaida, N. and Matsushima, K. (1991). Properties of the novel proinflammatory supergene "intercrine" cytokine family. Annu. Rev. Immunol. 9, 617–648.

Paganin, F., Trussard, V., Seneterre, E. et al. (1992). Chest radiography and high-resolution computed tomography of the lungs in asthma. Am. Rev. Respir. Dis 146, 1084–1087.

Perussia, B., Chan, S.H., D'Andrea, A., Tsuju, K., Santoli, M., Pospisil, D., Young, S.F., Wolf, S.F. and Trinchieri G. (1992). Natural killer (NK) cell stimulatory factor or IL-12 has differential effects on the proliferation of TCR-αβ+, TCR-γδ+ T lymphocytes, and NK cells. J. Immunol. 149, 3495–3499.

Piccini, M.P., Macchia, D., Parronchi, P., Giudizi, M.G., Bani, D., Alerini, R., Grossi, A., Ricci, M., Maggi, E. and Romagnani, S. (1991). Human bone marrow non b, non T cells produce interleukin 4 in response to cross-linkage of Fcε and Fcγ receptors. Proc. Natl. Acad. Sci. USA 88, 8656–8660.

Plaut, M., Pierce, J.H., Watson, C.J., Hanley-Hyde, J., Nordan, R.P. and Paul, W.E. (1989). Mast cell lines produce lymphokines in response to cross-linkage of FcRI or to calcium ionophores. Nature 339, 64–67.

Pober, J.S. and Cotran, R.S. (1990). Cytokines and endothelial cell biology. Physiol. Rev. 70, 427–451.

Postlethwaite, A.E. and Seyer, J.M. (1991). Fibroblast chemotaxis induction by human recombinant interleukin-4: identification by synthetic peptide analysis of two chemotactic domains residing in aminoacid sequences 70–78 and 89–122. J. Clin. Invest. 87, 2147–2152.

Postlethwaite, A.E., Holness, M.A., Katai, H. and Raghow, R. (1992). Human fibroblasts synthesizes elevated levels of extracellular matrix proteins in response to interleukin 4. J. Clin. Invest. 90, 1479–1485.

Punnone, J., Aversa, G., Cocks, B.G. et al. (1993). Interleukin 13 induces interleukin 4-independent IgG4 and IgE synthesis and CD23 expression by human B cells. Proc. Natl. Acad. Sci. USA 90, 3730–3734.

Renzi, P.M., Sapienza, D.U.T, Wang, N.S. and Martin, J.G. (1991). Lymphokine-induced airway responsiveness in the rat. Am. Rev. Respir. Dis. 143, 375–379.

Robertson, M.J., Soiffer, R.J., Wolf, S.F., Manley, T.J., Donohue, C., Young, D., Herrmann, S.H. and Ritz, J. (1992). Responses of human natural killer (NK) cells to NK cell stimulatory factor (NKSF): cytolytic activity and proliferation of NK cells are differentially regulated by NKSF. J. Exp. Med. 175, 779–785.

Robinson, D.S., Hamid, Q., Ying, S., Tsicopoulos, A., Barkans, J., Bentley, A.M., Corrigan, C., Durham, S.R. and Kay, A.B. (1992). Predominant TH2-like bronchoalveolar T lymphocyte population in atopic asthma. N. Engl. J. Med. 326, 298–304.

Robinson, D.S., Hamid, Q., Ying, S., Bentley, A.M., Assoufi, B., North, J., Qui, M., Durham, S.R. and Kay, A.B. (1993). Prednisolone treatment in asthma is associated with modulation of bronchoalveolar lavage cell IL-4, IL-5 and IFNγ cytokine gene expression. Am. Rev. Respir. Dis. 148, 401–406.

Roche, W.R., Beasley, R., Williams, J.H. and Holgate, S.T. (1989). Subepithelial fibrosis in the bronchi of asthmatics. Lancet 1, 520–524.

Rolff, M.W., Kunkel, S.L., Standiford, T.J., Chensue, S.W., Allen, R.M., Evanoff, H.L., Phan, S.H. and Strieter, R.M. (1991). Pulmonary fibroblast expression of interleukin-8: a model for alveolar macrophage-derived cytokine networking. Am. J. Respir. Cell. Mol. Biol. 5, 493–501.

Romagnani, S. (1990). Regulation and dysregulation of human IgE synthesis. Immunol Today 11, 316–321.

Romain, P.L. and Lipsky, P.E. (1983). Immunoglobulin secretion during the autologous mixed lymphocyte reaction in man. J. Immunol. 130, 1146–51.

Rose, R., Raines, E.W. and Bowen-Pope, D.F. (1986). The biology of platelet-derived growth factor. Cell 46, 155–169.

Rot, A. (1992). Endothelial cell binding of NAP-1/IL-8: role in neutrophil emigration. Immunol. Today 13, 291–294.

Rot, A., Krieger, M., Brunner, T., Bischoff, S.C., Schall, T.J. and Dahinden, C.A. (1992). RANTES and macrophage inhibitory protein 1a induce the migration and activation of normal human eosinophil granulocytes. J. Exp. Med. 176, 1489–1495.

Sacco, O., Romberger, D., Rizzino, A., Beckman, J.D., Rennard, S.I. and Spurzem, J.R. (1992). Spontaneous production of transforming growth factor-β2 by primary cultures of bronchial epithelial cells. J. Clin. Invest. 90, 1379–1385.

Sakamaki, K., Miyajima, I., Kitamura, T. and Miyajima, A. (1992). Critical cytoplasmic domains of the common β subunit of the human GM-CSF, IL-3 and IL-5 receptors for growth signal transduction and tyrosine phosphorylation. EMBO J. 11, 3541–3549.

Sanderson, C.J. (1992). Interleukin-5, eosinophils and disease. Blood 1992, 3101–3109.

Sato, N., Sakamaki, K., Terada, N., Arai, K. and Miyajima, A. (1993). Signal transduction by the high-affinity GM-CSF receptor: two distinct cytoplasmic regions of the common beta subunit responsible for different signaling. EMBO J. 12, 4181–4189.

Schall, T.J., Bacon, K., Toy, K.J. and Goeddel, D.V. (1990). Selective attraction of monocytes and T lymphocytes of the memory phenotype of cytokine RANTES. Nature 347, 669–671.

Schleimer, R.P., Sterbinsky, C.A., Kaiser, C.A., Bickel, D.A., Klunk, K., Tomioka, K., Newman, W., Luscinskas, M.A., Gimbrone, M.A., McIntire, B.W. and Buchner, B.S. (1992). Interleukin-4 induces adherence of human eosinophils and basophils but not neutrophils to endothelium: association with expression of VCAM-1. J. Immunol. 148, 1086–1092.

Schoenhaut, D.S., Chua, A.O., Wolitzky, A.G., Quinn, P.M., Dwyer, C.M., McComas, W., Familletti, P.C., Gately, M.K. and Gubler, U. (1992). Cloning and expression of murine IL-12. J. Immunol. 148, 3433–3437.

Sedgewick, J.B., Calhoun, W.J., Gleich, G.J., Kita, H., Abrams J.S., Schwartz L.B., Volovitz, B., Ben-Yaacov, M. and Busse, B. (1991). Immediate and late airway response of allergic rhinitis patients to segmental antigen challenge. Characterization of eosinophil and mast cell mediators. Am. Rev. Respir. Dis. 144, 1274–1281.

Selig, W. and Tocker, J. (1992). Effect of interleukin-l receptor antagonist on antigen-induced pulmonary responses in guinea-pigs. Eur. J. Pharmacol. 213, 331–336.

Sen, G.C. and Lenggel, P. (1992). The interferon system: a bird's eye view of its biochemistry. J. Biol. Chem. 267, 5017–5019.

Shepherd, V.L. (1991). Cytokine receptors in lung. Am. J. Respir. Cell. Mol. Biol. 5, 403–410.

Shibuya, H., Yoneyama, M., Ninomiya-Tsui, J., Matsumoto, K. and Taniguchi, T. (1992).IL-2 and EGF receptors stimulate the hematopoietic cell cycle via different signalling pathways: demonstration of a novel role for c-myc. Cell 70, 57–67.

Sica, A., Wang, J.M., Coletta, F., Dejana, E., Mantovani, A., Oppenheim, J.J. et al. (1990). Monocyte chemotactic and activating factor gene expression induced in endothelial cells by IL-1 and TNF. J. Immunol. 144, 3034–3038.

Silberstein, D.S., Owen, W.F., Gasson, J.C., Di Pierso, J.F., Golde, D.W., Bina, J.C., Soberman, R.J., Austen, K.F. and David, J.R. (1986). Enhancement of human eosinophil cytotoxicity and leukotriene synthesis by biosynthetic (recombinant) granulocyte–macrophage colony stimulating factor. J. Immunol. 137, 3290–3294.

Souillet, G., Rousset, F. and De Vries, J.E. (1989). γ-interferon treatment of patient with hyper IgE syndrome. Lancet 1, 1384.

Sousa, A.R., Poston, R.N., Lane, S.J., Narhosteen, J.A. and Lee, T.H. (1993). Detection of GM-CSF in asthmatic bronchial epithelium and decrease by inhaled corticosteroids. Am. Rev. Respir. Dis. 147, 1557–1561.

Sousa, A.R., Lane, S.J., Nakhosteen, J.A., Yoshimura, T., Lee, T.H. and Poston, R.N. (1994). Increased expression of the monocyte chemoattractant protein-1 in bronchial tissue from asthmatic subjects. Am. J. Respir. Cell. Mol. Biol. 10, 142–147.

Spiteri, M.A., Knight, R.A., Wordsell, M., Barnes, P.J. and Chung K.F. (1991). Alveolar macrophage-induced suppression of T cell hyperresponsiveness in asthma is reversed following allergen exposure in vitro. Am. Rev. Respir. Dis. 143, A801.

Spiteri, M.A., Prior, C., Herold, M., Knight, R.A., Clarke, S.W. and Chung, K.F. (1992). Spontaneous release of IL-1, IL-6, TNFα and GMCSF by alveolar macrophages (AM) in bronchial asthma. Am. Rev. Respir. Dis. 145, A239.

Spits, H. and de Waal Malefyt, R. Functional characterization of human IL-10. Int. Arch. Allergy Appl. Immunol. 99, 9–15.

Sprang, S.R. (1990). The divergent receptors for TNF. Trends Biochem. Sci. 15, 366–368.

Springall, D.R., Hamid, Q.A., Buttery, L.K.D., Chanez, P., Howarth, P., Bousquet, J., Holgate, S.T. and Polak, J.M. (1993). Nitric oxide synthase induction in asthmatic human lung. Am. Rev. Respir. Dis. 147, A515.

Standiford, T.J., Strieter, R.M., Chensue, S.W., Westwick, J., Kasahara, K. and Kunkel, S.L. (1990). IL-4 inhibits expression of IL-8 from stimulated human monocytes. J. Immunol. 145, 1435–1439.

Standiford, T.J., Kunkel, S.L., Liebler, J.M., Burdick, M.D., Gilbert, A.R. and Strieter, R.M. (1993). Gene expression of macrophage inflammatory protein-lα from human blood monocytes and alveolar macrophages is inhibited by interleukin-4. Am. J. Cell. Respir. Biol. 9, 192–198.

Stern, A.S., Podlaski, F.J., Hulmes, J.D., Pan, Y.E., Quinn, P.M., Wolitzky, A.G., Familletti, P.C., Stremlo, D.L., Truitt, T., Chizzonite, R. and Gately, M.K. (1990). Purification to homogeneity and partial characterization of cytotoxic lymphocyte maturation factor from human b-lymphoblastoid cells. Proc. Natl. Acad. Sci. USA 87, 6808–6813.

Swain, S.L., Weinberg, A.D., English, M. and Huston, G. (1990). IL-4 directs the development of T$_H$2-like helper effectors. J. Immunol. 145, 3796–3806.

Sypek, J.P., Chung, C.L., Mayor, S.E.H., Subramanyam, J.M., Goldman, S.J., Sieburth, D.S., Wolf, S.F. and Schaub, R.G. (1993). Resolution of cutaneous leishmaniasis: interleukin 12 initiates a protective T helper Type 1 immune response. J. Exp. Med. 177, 1797–1802.

Taga, T., Hibi, M., Hirata, Y., Yamasaki, K., Yasukawa, K., Matsuda, T., Hirano, T. and Kishimoto, T. (1989). Interleukin-6 triggers the association of its receptor with a possible signal transducer, gp130. Cell 573, 581.

Takeshita, T., Asoa, H., Ohtani, K., Ishii, N., Kumaki, S., Tanaka, S., Munakata, H., Nakamura, M. and Sugamura, K. (1992). Cloning of the t chain of the human IL-2 receptor. Science 257, 379–382.

Taub, D.D. and Oppenheim, J.J. (1993). Review of the chemokine meeting the Third International Symposium of Chemotactic Cytokines. Cytokine 5, 175–179.

Taub, D.D., Conlon, K., Lloyd, A.R., Oppenheim, J.J. and Kelvin, D.J. (1993). Preferential migration of activated CD4+ and CD8+ T cells in response to MIP-1β and MIP-1b. Science 260, 355–357.

Thompson-Snipes, L.A., Dhar, V., Bond, M.W., Mosmann, T.R., Moore, K.W. and Rennick, D.M. (1991). Interleukin 10: a novel stimulatory factor for mast cells and their progenitors. J. Exp. Med. 173, 507–510.

Thornhill, M.H., Wellicome, S.M., Mahiouz, D.L., Lanchbury, J.S.S., Kyan-Aung, V. and Haskard, D.O. (1991). Tumor necrosis factor combines with IL-4 or IFNγ to selectively enhance endothelial cell adhesiveness for T cells. The contribution of vascular adhesion molecule-1-dependent and -independent binding mechanisms. J. Immunol. 146, 592–598.

Tobler, A., Meier, R., Seitz, M., Dewald, B., Baggiolini, M. and Fey, M.F. (1992). Glucocorticoids downregulate gene expression of GM-CSF, NAP-IIL-8 and IL-6, but not of M-CSF in human fibroblasts. Blood 79, 45–51.

Tosi, M.F., Stark, S.M., Smith, C.W., Hamedani, A., Gruenert, D.C. and Infeld, W. (1992). Induction of ICAM-1 expression on human airway epithelial cells by inflammatory cytokines: effects on neutrophil–epithelial cell adhesion. Am. J. Respir. Cell. Mol. Biol. 7, 214–221.

Tsukagoshi, H., Sakamoto, T., Xu, W., Barnes, P.J. and Chung, K.F. (1994). Effect of interleukin-1β on airway hyperresponsiveness and inflammation in sensitized and non-sensitized Brown Norway rats. J. Allergy Clin. Immunol. 93, 464–469.

Van Damme, J., Proost, P., Lenaerts, J.-P. and Opdenakker, G. (1992). Structural and functional identification of two human, tumor-derived monocyte chemotactic proteins (MCP-2 and MCP-3) belonging to the chemokine family. J. Exp. Med. 176, 59–64.

van Oosterhoot, A.J.M., Rudolf, A., Ladenius, C., Savelkoul, H.F.J., van Ark, I., Delsman, K.C. and Nijkamp, F.P. (1993). Effect of anti-IL-5 and IL-5 in airway hyperreactivity and eosinophils in guinea pigs. Am. Rev. Respir. Dis. 147, 548–552.

van Snick, J. (1990). Interleukin-6: an overview. Annu. Rev. Immunol. 8, 253–278.

Venkitaraman, A.R. and Cowling, R.J. (1992). Interleukin 7 receptor functions by recruiting the tyrosine kinase p59fyn through a segment of its cytoplasmic tail. Proc. Natl. Acad. Sci. USA 89, 12083–12087.

Vercelli, D., Jabara, H.H., Lee, B.W., Woodland, N., Geha, R.S. and Leung, D.Y. (1988). Human recombinant interleukin-4 induces FC4RII/CD23 on normal human monocytes. J. Exp. Med. 167, 1406–1416.

Walker, C., Virchow, J.C., Bruijnzeel, P.L.B. and Blaser, K. (1991). T cell subsets and their soluble products regulate eosinophilia in allergic and nonallergic asthma. J. Immunol. 146, 1829–1835.

Walker, C., Bode, E., Boer, L., Hansel, T.T., Blaser, K. and Virchow, J.-C.Jr. (1992). Allergic and non-allergic asthmatics have distinct patterns of T cell activation and cytokine production in peripheral blood and bronchoalveolar lavage. Am. Rev. Respir. Dis. 146, 109–115.

Wasserman, S.I. (1990). Mast cell biology. J. Allergy Clin. Immunol. 86, 590–599.

Watson, M.L., Smith, D., Bourne, A.D., Thompson, R.C. and Westwick, J. (1993). Cytokines contribute to airway dysfunction in antigen-challenged guinea pigs: inhibition of airway hyperreactivity, pulmonary eosinophil accumulation and tumor necrosis factor generation by pretreatment with an interleukin-l receptor antagonist. Am. J. Respir. Cell. Mol. Biol. 8, 365–369.

Weller, P.F., Rand, T.H., Barrett, T., Elovoic, A., Wong, D.T.W. and Finberg, R.W. (1993). Accessory cell function of human eosinophils: HLA-DR-dependent, MHC-restricted antigen presentation and interleukin-1β expression. J. Immunol. 150, 2554–2562.

Williams, A.F. and Barclay, A.N. (1988). The immunoglobulin superfamily – domains for cell surface recognition. Annu. Rev. Immunol. 6, 381–405.

Witthuhn, B.A., Quelle, F.W., Silvennoinen, O., Yi, T., Tang, B., Miura, O. and Ihle, J.N. (1993). JAK2 associates with the erythropoietin receptor and is tyrosine phosphorylated and activated following stimulation with erythropoietin. Cell 74, 227–236.

Wodnar-Filipowicz, A., Heusser, C.H. and Moroni, C. (1989). Production of the haemopoietic growth factors GM-CSF and interleukin-3 by mast cells in response to IgE receptor mediated activation. Nature 339, 150–152.

Wolf, S.F., Temple, P.A., Koboyashi, M., Young, D., Dicig, M., Lowe, L., Dzialo, R., Fitz, D., Ferenz, D., Hewick, R.M., Kelleher, K., Herrmann, S.L., Clark, S.C., Azzoni, L., Chan, S.J., Trinchieri, G. and Perussia, B. (1991). Cloning of cDNA for natural killer cell stimulatory factor, a heterodimeric cytokine with multiple biologic effects on T and natural killer cells. J. Immunol. 146, 3074–3079.

Wong, D.T.W., Weller, P.F., Galli, S.J., Elovic, A., Rand, T.H., Gallagher, G.T., Chiang, T., Chou, M.Y., Matossian, K., McBride, J. and Todd, R. (1990). Human eosinophils express transforming growth factor-alpha. J. Exp. Med. 172, 673–681.

Wong, D.T.W., Elovic, A., Matossian, K., Nagura, N., McBride, J., Chou, M.Y., Gordon, J.R., Rand, T.H., Galli, S.J. and Weller, P.F. (1991). Eosinophils from patients with blood eosinophilia express transforming growth factor bl. Blood 78, 2702–2707.

Yamauchi, K., Martinet, Y., Basset, P., Fells, G.A. and Crystal, R.G. (1988). High levels of transforming growth factor-β are present in the epithelial cell lining fluid of the normal human lower respiratory tract. Am. Rev. Respir. Dis. 137, 1360–1363.

Yates, D.H., Barnes, P.J. and Thomas, P.S. (1993). Tumor necrosis factor α alters human bronchial reactivity and induces inflammatory cell influx. Am. Rev. Respir. Dis. 147, A1011.

Ying, S., Robinson, D.S., Varney, V., Meng, Q., Tsicopoulos, A., Moqbel, R., Durham, S.R., Kay, A.B. and Hamid, Q. (1991). TNF alpha mRNA expression in allergic inflammation. Clin. Exp. Allergy 21, 745–750.

Zielinski, R.M. and Lawrence, W.D. (1990). Interferon-γ for the hypereosinophilic syndrome. Ann. Intern. Med. 113, 716–718.

Zurawsli, S.M., Vega, F.Jr, Huyghe, B. and Zurawski, G. (1993). Receptors for interleukin-13 and interleukin-4 are complex and share a novel component that functions in signal transduction. EMBO J. 12, 2663–2670.

Zurawsli, G. and De Vries J.E. (1994). Interleukin 13, an interleukin 4-like cytokine that acts on monocytes and B cells, but not on T cells. Immunol. Today 15, 19–26.

Zurawsli, S.M., Vega, F.Jr, Huyghe, B. and Zurawski, G. (1993). Receptors for interleukin-13 and interleukin-4 are complex and share a novel component that functions in signal transduction. EMBO J. 12, 2663–2670.

7. Neural Networks in the Lung

Stephanie A. Shore, Craig M. Lilly, Benjamin Gaston *and* Jeffrey M. Drazen

1. Introduction

The lung has a rich autonomic innervation that enables it to modulate the airway lumen available for airflow, the vascular resistance to blood flow, or the matching of these two flows in response to stimuli that impinge on the airway or vascular surface. The impulses travelling in these nerves fulfill a wide variety of sensory and effector functions that can be either physiologic or phlogistic in nature; a comprehensive review of all aspects of pulmonary neurofunction could easily fill many volumes. In keeping with the goal of this volume – i.e. to review

Immunopharmacology of the Respiratory System
ISBN 0–12–352325–7

specific aspects of pulmonary inflammation – we have chosen to focus on the neural functions most closely related to pulmonary inflammation. In this broad category the nerves that are most commonly considered "inflammatory" in nature are those which respond to stimuli by the release of specific peptide neurotransmitters. Although this class of nerves is not anatomically distinct, the nerve fibers involved are considered peptidergic and the neurotransmitters are termed neuropeptides. Within this class are various pulmonary neuropeptides that have both homeostatic and inflammatory roles (Barnes et al., 1991).

In this chapter, we will review our current understanding of the pulmonary effects of a specific class of prophlogistic neurotransmitters known as the tachykinins and of two homeostatic neurotransmitters, vasoactive intestinal peptide (VIP) and nitric oxide (NO·). We will pay special attention to the roles of these transmitters in inflammatory obstructive airway disorders. Each class of neurotransmitter will be considered separately.

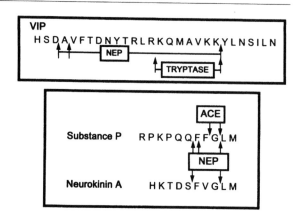

Figure 7.1 Amino acid sequences of VIP (upper box) and substance P and NkA (lower box). Note that the carboxy terminus of these peptides is amidated and amidation is required for receptor recognition. Cleavage sites for NEP, ACE and tryptase are shown.

2. Tachykinins

We will first consider the synthesis, release and localization of tachykinins in the airways. Next we will examine the impact of tachykinin degradation and tachykinin receptor expression on the physiological actions of these neurotransmitters in the airways. Finally, we will review the evidence for modulation of the effects of tachykinins in the inflammatory microenvironment (e.g. in asthma) and we will describe the interactions of tachykinins with mast cells and lymphocytes.

2.1 SYNTHESIS, LOCALIZATION AND RELEASE OF TACHYKININS IN THE AIRWAYS

2.1.1 Tachykinins and PPT Gene Products

The tachykinins are a family of small peptides (fewer than 50 residues), each of which has the same amino acid sequence, Phe-X-Gly-Leu-Met-NH₂, at the amidated carboxy terminus (Fig. 7.1). Although three primary mammalian tachykinins – substance P, neurokinin A (NkA) and neurokinin B (NkB) – have been recognized, only substance P and NkA have been identified in the lungs and airways. Tachykinins are synthesized in nerve cell bodies, appropriately processed, and then transported by axoplasmic flow to the terminal ramifications of axon dendrites, where they serve their neurotransmitter functions. Both substance P and NkA are derived from transcription and translation of the preprotachykinin I (PPT-I) gene (Nawa et al., 1984; Krause et al., 1987). NkB is derived from a separate gene,

PPT-II (Kotani et al., 1986). After transcription, PPT-I mRNA is alternatively spliced to yield at least three distinct forms of PPT-I mRNA, α, β and γ. Substance P is formed by translation of all three forms of mRNA, while NkA is produced by translation of only the β and γ forms. Differential post-translational processing of the β and γ forms of the PPT-I gene product can also result in the formation of three other NkA-related peptides: NkA$_{3–10}$, NPγ and NPK. NPγ and NPK are both N-terminally extended forms of NkA and result from alternative post-translational processing of the γ and β forms of PPT-I mRNA, respectively (MacDonald et al., 1989). Both NkA$_{3–10}$ and NPK have been demonstrated in extracts of guinea-pig and human lung tissue (Hua et al., 1985; Saria et al., 1988). NPγ has not yet been demonstrated in lung extracts but is present in other organs (Takeda et al., 1990).

2.1.2 Anatomic Localization of PPT Gene Products

Immunohistochemical analysis of lung tissues from animals and humans has revealed the presence of substance P in C-fiber afferent neurons in the airways. Within the lung, NkA is co-localized to the same nerve terminals as substance P and has a similar distribution (Sundler et al., 1985; Uchida et al., 1987). The cell bodies for the C-fiber neurons in the lungs lie in the nodose, jugular and thoracic dorsal root ganglia. The C-fibers supplying the trachea have a vagal origin, whereas those supplying the rest of the lung have both vagal and spinal origins (Lundberg et al., 1984a; Kummer et al., 1992). In the cat, substance P has been observed in cell bodies of neurons in airway ganglia, where it is usually co-localized with VIP (Dey et al., 1988), but in other species only peripheral neuronal processes have solely substance P immunoreactivity (Lundberg et al., 1984a). Substance P-immunoreactive neurons are found in the airways from the nose (Baraniuk *et al.*, 1991) down to the peripheral bronchioles (Nilsson *et al.*, 1975; Lundberg *et al.*, 1984a), and have been identified at the alveolar level in some mammalian species (Ghatei *et al.*, 1982). However, the density of substance P-containing nerve processes is greatest in the trachea and large bronchi (Nilsson *et al.*, 1975; Ghatei *et al.*, 1982; Lundberg *et al.*, 1984a,b; Sundler *et al.*, 1985; Hua *et al.*, 1985; Springall *et al.*, 1988; Hislop *et al.*, 1990); this localization probably reflects their physiological function as detailed below. Measurements of substance P content in extracts of airways confirm this distribution: substance P content is highest in the mainstem bronchi, but measurable amounts are also present in peripheral lung samples (Ghatei *et al.*, 1982; Springall *et al.*, 1988; Lilly *et al.*, 1994). Substance P-immunoreactive nerves are found in greatest density beneath and between epithelial cells. Nerve processes are often seen penetrating to the luminal surface or impinging upon blood vessels, within the airway smooth muscle layer and around parasympathetic ganglia (Lundberg *et al.*, 1984b). This distribution reflects the dual sensory and motor role of these nerves.

There are both species- and age-related variations in the amount and distribution of substance P in the airways. Substance P content per gram of airway tissue is greatest in airways from rats and guinea-pigs (Ghatei *et al.*, 1982; Springall *et al.*, 1988). It has been reported that fewer substance P-immunoreactive fibers are present in the airway epithelium of humans than in that of guinea-pigs or rats (Springall *et al.*, 1988). However, when we quantitatively measured the substance P content of human lung tissue obtained from adult autopsy specimens with high pressure liquid chromatography (a method that ensured that substance P-non-immunoreactive molecules were not measured), we found on the order of 5 pmol g^{-1} tissue (Lilly *et al.*, 1994), a level similar to that found in guinea-pig trachea (approximately 6 pmol g^{-1}; Martins *et al.*, 1991b). Examination of autopsy material from children aged 3 years or less has revealed substance P-immunoreactive nerves at all airway levels from the mainstem bronchi to alveolar ducts (Hislop *et al.*, 1990). In contrast, in older people, substance P-containing nerves are present in bronchi but are only rarely observed in smaller airways (Hislop *et al.*, 1990). Our analysis of peripheral lung samples obtained from adults at autopsy is in accord with this finding: lower levels of substance P are found per gram of peripheral lung (5 pmol) than of trachea (12 pmol; Lilly *et al.*, 1994).

Substance P content has been shown to be present at high levels in nerves subtending inflamed organs or tissues. For example, an inflammatory cytokine, interleukin 1β (IL-1β), has been found to elevate substance P content and PPT-I mRNA expression in rat superior cervical ganglion cells in culture (Jonakait *et al.*, 1991). This effect has also been documented *in vivo*; increased levels of PPT-I mRNA and substance P product have been observed in sensory nerves innervating inflamed tissue in adjuvant-induced arthritis in the rat (Colpaert *et al.*, 1983; Donnerer *et al.*, 1993; Hashimoto *et al.*, 1993; Hanesch *et al.*, 1993). A preliminary report by Fischer *et al.* (1994) indicates that levels of PPT-I mRNA are increased in the nodose ganglia of sensitized guinea-pigs 12 h after challenge with inhaled allergen and that tachykinin content is elevated in the lungs of these animals 1–2 days after challenge. Although it has been established that the amount of substance P can be upregulated in the airways of animals during induced inflammation, there is controversy concerning the amount of substance P extant in the airways of patients with asthma, a disease characterized by airway inflammation. Ollerenshaw *et al.* (1991) reported a greater density of substance P-immunoreactive nerves in airways removed from patients with asthma than in those from non-asthmatic subjects. However, Howarth *et al.* (1991) were unable to identify any substance P-immunoreactive nerves in endobronchial biopsy samples either from patients with mild asthma or from non-asthmatic individuals. We have observed that the substance P content of extracts of trachea from patients who have died of asthma is lower than that of extracts from age-matched persons dying of non-pulmonary causes, possibly because of the release and subsequent degradation of substance P during the asthmatic episode (Lilly *et al.*, 1994). Examination of extracts of peripheral lung samples from asthmatic and non-asthmatic subjects revealed no difference in substance P content. All data on the neuropeptide content of tissue must be interpreted with the understanding that the antibodies used to recognize tachykinins may not be specific and that the tachykinin content may be altered during tissue recovery and processing.

2.1.3 The Function of C-fibers and the Axon Reflex

The C-fiber neurons, which contain substance P and NkA, are predominantly sensory in function. They can be stimulated by exogenous substances, such as inhaled irritants, and by endogenous substances, such as inflammatory mediators (Fig. 7.2; Coleridge and Coleridge, 1984). Both in the airways and in the lung periphery C-fibers are also stimulated by mechanical factors, such as lung inflation or pulmonary edema (Coleridge and Coleridge 1984; Lee and Morton, 1993); responding to these stimuli constitutes the afferent function of these neurons.

In response to these activation stimuli, action potentials are conducted towards the central nervous system, where they may elicit reflex responses in respiratory pattern (e.g. rapid, shallow breathing), or alterations in pulmonary mechanics. As the action potentials, on their way to the central nervous system, pass the terminal ramifications of axon dendrites, antidromic conduction occurs and signals are propagated toward peripheral nerve terminals. Upon arrival of the action potentials at these terminals, tachykinins and other neuropeptides are released into the microenvironment surrounding the terminal. The peptides so released transduce signals at specific receptors that lead to multiple biological effects within the airways. The magnitude of these effects is limited by cleavage and inactivation of neuropeptides at or near the site of their release. Because of the multiple terminal ramifications of these nerves, a stimulus at a single locus, such as the airway surface, is spread through this "axon reflex" (Fig. 7.2), to the surrounding epithelium, smooth muscle, bronchial vasculature and secretory apparatus. The "spreading" of an irritant signal is a key feature of tachykinin-containing nerves and plays a major role in their contribution to inflammatory responses. One of the important ways that irritant signals are further amplified is through modification of the enzymatic capacity to cleave and inactivate tachykinins as they are released from nerve terminals.

2.2 AIRWAY RESPONSES TO TACHYKININS

Tachykinins transduce multiple effects in the airways, including bronchoconstriction, increased ion and water flux across the epithelium, cough, mucous secretion, increased ciliary beat frequency, increased airway blood flow and increased airway vascular permeability (Gashi *et al.*, 1986; Borson *et al.*, 1987, 1989; Kohrogi *et al.*,

1988; Shore and Drazen, 1989b; Tamaoki *et al.*, 1991; Wong *et al.*, 1991; Piedimonte *et al.*, 1992). These neurotransmitters transduce airway smooth muscle constriction both by direct action at neurokinin (Nk) receptors on smooth muscle cells and by interactions with cholinergic nerves and mast cells. For example, in rabbits, at least part of tachykinin-induced bronchoconstriction is thought to result from stimulation of parasympathetic neurons, with subsequent release of acetylcholine. The evidence for this conjecture is that atropine attenuates the contractile response of rabbit tracheal rings to substance P *in vitro* (Tanaka and Grunstein, 1984). Similarly, the addition of subthreshold amounts of substance P or NkA to ferret tracheal rings increases the contractile response to electrical field stimulation but not to exogenous acetylcholine. This observation suggests that tachykinins also stimulate cholinergic nerves in this species (Sekizawa *et al.*, 1987). In the guinea-pig, substance P-induced bronchoconstriction appears to derive predominantly from a direct effect of substance P on airway smooth muscle Nk receptors; pretreatment of guinea-pigs with anticholinergics, cyclooxygenase inhibitors or histamine receptor antagonists does not affect the changes in pulmonary resistance induced by substance P (Shore and Drazen, 1989a). In Fisher 344 rats, high concentrations of substance P or NkA increase pulmonary resistance by release of mast cell-derived amines (Joos and Pauwels, 1993). This physiological observation is in agreement with the biochemical observation that rat

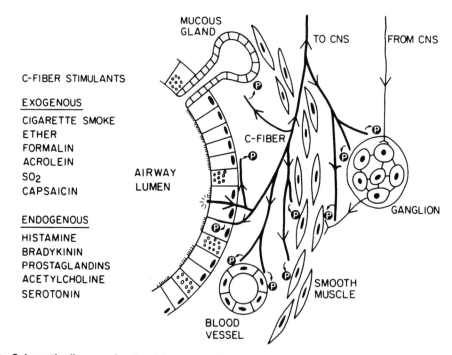

Figure 7.2 Schematic diagram showing inter-connections among axon dendrites from sensory nerves and the various components of the airway. Note that a single axon may serve C-fiber receptors, airway smooth muscle, glands and blood vessels.

airway smooth muscle does not express tachykinin receptors (Sertl et al., 1988). In Sprague–Dawley rats, lower concentrations of substance P can elicit bronchodilatation if baseline tone is raised by infusion of a cholinergic agonist (Szarek, personal communication). Relaxation of preconstricted rat tracheal rings by substance P appears to involve stimulation of Nk receptors on the airway epithelium, with the subsequent synthesis of dilator prostaglandins. The evidence for this hypothesis is that relaxation is blocked by removal of the epithelium or by cyclooxygenase inhibitors (Devillier et al., 1992). Substance P causes relaxation of isolated murine bronchus (Manzini, 1992) by the same mechanism.

In initial studies in which substance P was administered to subjects by intravenous infusion, no effect on airway caliber could be demonstrated. It is possible that potent cardiovascular side effects resulted in homeostatic reflexes with secondary effects on the airways that masked the effects of substance P (Fuller et al., 1987). However, when changes in the partial flow volume curve, a highly sensitive index of airway caliber, were used as the index of airway obstruction, inhaled NkA – and, to a lesser extent, inhaled substance P – caused airway obstruction in healthy people (Joos et al., 1987). NkA is approximately 10–100 times more potent as a bronchoconstrictor in asthmatic than in non-asthmatic subjects (Cheung et al., 1992, 1993); this ratio of potency is similar to that observed for other agonists.

2.3 DEGRADATION OF TACHYKININS

Many peptidases, including neutral endopeptidase (NEP), angiotensin-converting enzyme (ACE), aminopeptidases, dipeptidyl-peptidase IV, cathepsin G and mast cell chymase, have the capacity to cleave and inactivate substance P or NkA (Conlon and Sheehan, 1983; Turner et al., 1985; Johnson et al., 1985; Caughey et al., 1988; Skidgel et al., 1991; Ahmad et al., 1992). In normal lungs and airways, only NEP and ACE appear to be of major importance in the cleavage of substance P (Martins et al., 1991b). NEP cleaves both substance P and NkA at sites shown in Fig. 7.1; ACE cleaves substance P but not NkA. Cleavage of substance P and NkA by NEP and ACE results in fragments virtually devoid of bronchoconstrictive activity (Shore and Drazen, 1988; Shore et al., 1988). In contrast, products resulting from cleavage of substance P by aminopeptidases have enhanced biological activity compared to the native peptide. It is interesting that products resulting from cleavage of NkA by aminopeptidases do not exhibit greater bioactivity than NkA (Shore and Drazen, 1988, 1991).

The route of delivery of substance P to the lung appears to be of critical importance in determining which peptidase is likely to be responsible for degradation of the peptide. When substance P is administered intravenously to anesthetized guinea-pigs, both the NEP inhibitor thiorphan and the ACE inhibitor captopril increase the amount of peptide recovery from arterial blood, although captopril is much more effective in this respect (Shore et al., 1988). In the absence of enzyme inhibitors, when substance P is offered to isolated, tracheally perfused guinea-pig lungs, less than half of the substance P is recovered in the effluent (Martins et al., 1990). Both thiorphan and another NEP inhibitor, SCH32615, can increase the recovery of intact substance P, whereas captopril has no effect on substance P recovery from the tracheally perfused lung. These results are consistent with the known distribution of NEP and ACE in the lung. ACE is localized primarily to pulmonary vascular endothelial cells (Ryan et al., 1985), where it would be expected to come into contact with and cleave intravenously infused substance P, whereas NEP is expressed in high concentrations on the airway epithelium, in airway smooth muscle, and airway submucosa (Johnson et al., 1985; Choi et al., 1990), where it would be expected to cleave substance P delivered via the airway lumen before the activation of Nk receptors.

The relative importance of NEP and ACE in degradation of endogenously released tachykinins also appears to be dependent on the locus of release of these peptides. When capsaicin is delivered to the perfusate of guinea-pig lungs perfused via the airway lumen, NEP inhibitors but not ACE inhibitors increase the recovery of substance P and NkA from the lung effluent (Martins et al., 1991b). In contrast, both NEP and ACE appear to be important in the degradation of peptides responsible for bronchoconstriction when capsaicin is administered intravenously (Shore et al., 1993). Clearly, the administration of pharmacological agents that prevent cleavage and inactivation of substance P or NkA will enhance the peptide's bronchoconstrictive potential. This process – i.e. decreased tachykinin cleavage – can also occur in the setting of airway inflammation, as will be outlined in the section that follows.

2.4 INFLAMMATORY MODULATION OF TACHYKININ CLEAVAGE

The administration of a variety of agents capable of inducing inflammation in the airways is associated with decreased NEP activity. Airway NEP activity decreases in rodents after respiratory viral infections (Jacoby et al., 1988; Dusser, et al., 1989b; Borson et al., 1989), after exposure to ozone (Murlas et al., 1992), after inhalation of cigarette smoke (Dusser et al., 1989; Kwon, et al. 1994), and after exposure to toluene diisocyanate (Sheppard et al., 1988). The physiological importance of these changes in NEP activity is reflected in the increased magnitude of airway responses induced by substance P in animals challenged with these agents (Jacoby et al., 1988; Sheppard et al., 1988; Dusser et al., 1989a; Borson et al., 1989; Murlas et al., 1992).

Although tachykinins are important in the airway response to certain interventions, there appears to be little or no "tachykinin tone" in the absence of intervention. In non-asthmatic or stable asthmatic human airways, neither the NEP inhibitor thiorphan (Cheung *et al.*, 1992, 1993) nor the combined Nk1/Nk2 receptor antagonist FK224 (Ichinose *et al.*, 1992) has any effect on baseline specific airways resistance. Likewise, alterations in NEP activity do not contribute to the enhanced bronchoconstrictive responses to NkA observed in patients with mild asthma. Asthmatic patients respond to inhaled thiorphan with an increase in NkA-induced bronchoconstriction similar in magnitude to that documented in non-asthmatic subjects (Cheung *et al.*, 1992, 1993).

2.5 TACHYKININ RECEPTORS

2.5.1 Receptor Structure

The effects of tachykinins in the airways result from signal transduction initiated by the interaction of tachykinins with specific receptors on effector cells. Three distinct mammalian tachykinin receptors – Nk1, Nk2 and Nk3 – have now been identified and molecularly cloned (Masu *et al.*, 1987; Yokota *et al.*, 1989; Shigemoto *et al.*, 1990). In addition, subtypes of both the Nk1 and the Nk2 receptors have been proposed (Maggi, 1993). Substance P has the highest affinity for the Nk1 receptor, NkA for the Nk2 receptor, and NkB for the Nk3 receptor, although each peptide can interact with all three receptors. Even though the ligands for the Nk1 and Nk2 receptors; (i.e., substance P and NkA) are derived from transcription of the same gene (PPT-I), the genes encoding these receptors are not linked genetically. Rather, they are found on different chromosomes and are regulated independently. Specifically, the Nk1 receptor is on human chromosome 2, while the Nk2 receptor is on human chromosome 10 (Gerard *et al.*, 1990, 1991).

Structurally, the tachykinin receptors are seven transmembrane domain, G protein-linked receptors, and in that regard are similar to many other receptors mediating inflammatory events (Gerard *et al.*, 1991). The seven membrane-spanning sequences, particularly segment seven and those portions of the cytoplasmic regions close to the transmembrane regions, show a high degree of homology among the three tachykinin receptors (Nakanishi *et al.*, 1993). Data obtained by the expression of mutant or chimeric tachykinin receptors in COS cells or Chinese hamster ovary (CHO) cells indicate that the extracellular domains are critical in the binding of peptides to receptors. However, the extracellular loops and tail are not the only domains that confer the specificity of the receptor; modification of the protein sequence in the transmembrane regions also appears to be involved in signal transduction (Laneuville *et al.*, 1992; Fong *et al.*, 1993).

It has now been established by studies of numerous tissues that the interaction of all members of the Nk receptor family with appropriate ligands results in G protein-linked activation of phospholipase C and phosphoinositol turnover with increased synthesis of inositol triphosphate (IP$_3$) and consequent increases in levels of intracellular calcium (Nakanishi *et al.*, 1993). This activation occurs via a pertussis toxin-insensitive pathway and probably involves the G$_{q/11}$ subfamily of G$_\alpha$ proteins (Kwatra *et al.*, 1993). Activation of transfected tachykinin receptors in CHO cells can also result in the generation of cAMP (Nakajima *et al.*, 1992; Wiener, 1993).

Information is available concerning the regulation of Nk1 receptor gene expression. The protein coding region of the Nk1 receptor gene contains five exons and is dispersed over 45–60 kb of genomic DNA on human chromosome 2 (Krause *et al.*, 1993). The putative promoter regions of the human and rat Nk1 receptor gene are highly homologous and contain several sequence motifs displaying significant homology with previously described sequences that mediate transcriptional regulation. These elements include an AP-1 site, an AP-2 site, an epidermal growth factor, a phorbol ester and calcium site, and a cAMP site. A functional role for at least one of these sites appears to have been confirmed. Krause *et al.* (1993) showed increased levels of Nk1 receptor mRNA 4 h after treatment of U373-MG astrocytoma cells with forskolin, which also resulted in 2- to 3-fold elevations in levels of cAMP. Glucocorticoids have also been shown to down-regulate Nk1 receptor mRNA in a number of cell lines (Ihara and Nakanishi, 1990; Gerard *et al.*, 1991), although no well-defined glucocorticoid regulatory element has been identified in either rat or human genes. The organization of the Nk2 receptor gene on human chromosome 10 has been described (Gerard *et al.*, 1990), but little is known about regulation of its expression.

2.5.2 Anatomic and Functional Sites of Receptor Location

Autoradiographic studies of human and guinea-pig airways have demonstrated the presence of tachykinin receptors on vascular smooth muscle, airway epithelium and submucosal glands, with a particularly high density on airway smooth muscle from the trachea to small bronchioles (Castairs and Barnes, 1986; Hoover and Hancock, 1987). In contrast, in rat airways, there is no labelling of airway smooth muscle (Sertl *et al.*, 1988).

Both Nk1 and Nk2 receptors are present in the airway smooth muscle of guinea-pigs. Nk1 receptor selective agonists cause contraction of the isolated guinea-pig trachea or tracheally perfused lung that is inhibited by Nk1 receptor selective antagonists (Lilly *et al.*, 1994). Similarly, in these preparations, Nk2 receptor selective agonists transduce contraction that is inhibited by Nk2 receptor selective antagonists (Ellis *et al.*, 1993). It is important that substance P and NkA have the capacity to initiate signal transduction at either receptor; as a consequence, antagonists of both Nk1 and Nk2 receptors are required to inhibit the bronchoconstrictor activity of these endogenous tachykinins *in vivo* (Foulon *et al.*, 1993).

It appears that only Nk2 receptors are expressed in isolated human airway smooth muscle. Nk2 receptor selective agonists cause contraction, whereas agonists selective for the Nk1 or Nk3 receptor do not cause contraction except at very high concentrations (Naline *et al.*, 1989; Ellis *et al.*, 1993). Contractions observed at these high doses are not blocked by the Nk1 receptor selective antagonist CP96345 but are inhibited by Nk2 receptor selective antagonists (Ellis *et al.*, 1993). In isolated human airways Nk2 receptor antagonists inhibit contractions induced by exposure to either substance P or NkA (Ellis *et al.*, 1993).

Even though Nk2 receptors are the only receptors present on human airway smooth muscle, Nk1 receptors may also be involved in mediating the obstruction of human airways. It is well established that tachykinins have activities in addition to smooth muscle contraction that may contribute to airway obstruction. These peptides can elicit mucus secretion. They can enhance airway microvascular permeability and increase airway blood flow, alterations that may lead to airway edema and that, in guinea-pigs, can contribute to the changes in pulmonary resistance that follow inhalation of substance P (Lotvall *et al.*, 1990). While the action of tachykinins on airway smooth muscle is mediated by Nk2 receptors, their effects on mucus-secreting cells in human airways are mediated by Nk1 receptors (Rogers *et al.*, 1989). The tachykinin receptors mediating effects of substance P and NkA on blood vessels in human airways have not yet been characterized, but in other species the Nk1 receptor is the transduction unit involved (Lei *et al.*, 1992).

The expression of tachykinin receptors is increased in a number of inflammatory diseases, including asthma. In Crohn's disease and in ulcerative colitis, expression of Nk1 receptors in the colon is markedly up-regulated but Nk2 receptor expression is not changed (Mantyh *et al.*, 1988). In healthy colon, Nk1 receptors are primarily associated with the external circular smooth muscle. However, in resected colonic tissue from patients with inflammatory bowel diseases, there is increased Nk1 receptor expression associated with small arterioles and lymph nodes. Nk1 receptor expression is also up-regulated in the resected appendiceal tissue of patients with appendicitis (Mantyh *et al.*, 1994). Again, the increased [^{125}I]-SP binding sites are observed mainly in the germinal zone of the lymph nodes and in blood vessels. Nk1 receptor mRNA expression is increased in those portions of the rat spinal cord innervating the hind paws after the induction of inflammation of the hind paw by injection of formalin or Freund's adjuvant (McCarson and Krause, 1994).

Adcock *et al.* (1993) reported increased levels of Nk1 receptor mRNA in airway tissue obtained from asthmatic subjects compared to normal subjects. However, these authors did not evaluate the cellular distribution of receptor protein. In contrast, Bai *et al.* (1994) reported no difference between levels of Nk1 receptor mRNA in lung tissue from non-asthmatic subjects and that in lung tissue from asthmatic patients; these investigators did find that the level of Nk2 receptor mRNA was increased in the asthmatic individuals. The medications used for the treatment of asthma may explain this discrepancy; Adcock *et al.* (1993) have shown that treatment with glucocorticoids decreases Nk1 receptor expression in asthmatic airways.

2.6 INTERACTIONS OF C-FIBERS AND MAST CELLS

Histologic evidence shows a close anatomic correspondence between mast cells and C-fibers in peripheral organs (Skofitsch *et al.*, 1985; Stead *et al.*, 1987) and in the nodose ganglia (Undem and Weinrich, 1993). There is both functional and structural evidence for innervation of mast cells by C-fibers. Stimulation of the trigeminal nerve causes mast cells in the rat dura mater to degranulate. This effect, which is apparent on histologic evaluation, is abolished by treatment with neonatal capsaicin and this may be mediated by C-fibers (Dimitriadou *et al.*, 1991).

2.6.1 Activation of Mast Cells by Tachykinins

A substantial body of evidence shows that tachykinins can cause the activation and degranulation of mast cells. In rat peritoneal mast cells, substance P causes the release of histamine (H; Foreman and Jordan, 1983). This activity is related to charged residues at the amino terminus of the molecule rather than to interaction with tachykinin receptors: tachykinins such as eledoisin and SP4-11, which lack these charged moieties but can activate Nk receptors, do not mediate the release of mast cell products. In rats, both substance P and NkA, as opposed to saline, elicit enhanced recovery of serotonin and H from bronchoalveolar lavage (BAL) fluid (Joos and Pauwels, 1993). Predegranulation of mast cells by administration of compound 48/80 2 days before challenge with tachykinins prevents the release of serotonin and H; this result suggests that mast cells are indeed the source of the amines recovered. Moreover, the fact that either pretreatment with compound 48/80 or administration of a serotonin receptor antagonist prevents the changes in pulmonary resistance ordinarily induced by substance P and NkA suggests that stimulation of mast cells by tachykinins has important physiological consequences. In isolated, tracheally perfused lungs, substance P and capsaicin evoke enhanced recovery of H from perfusion fluid (Lilly *et al.*, 1994).

These effects of tachykinins appear to be mediated by Nk receptors, as both Nk1 and Nk2 receptor antagonists reduce the amounts of H/serotonin recovered.

The physiological effects of tachykinins on airway mast cells involve, in part, the release of leukotrienes. In preparations of guinea-pig bronchi that include an intact vagus nerve, stimulation of the attached nerve results in a biphasic contractile response. The first phase is cholinergically mediated, while the second phase is abolished by pretreatment with large doses of capsaicin, and thus is probably C-fiber mediated. Ellis and Undem (1991) have reported that the capsaicin-sensitive contraction is inhibited by three structurally unrelated sulfidopeptide leukotriene receptor antagonists. The addition of exogenous leukotriene D_4 (LTD_4) potentiates these second-phase responses but does not alter the response to exogenously administered substance P or NkA; this finding suggests that LTD_4 augments the release of tachykinins from C-fibers rather than the activity of the tachykinins. One potential explanation for these results is that tachykinins released from C-fibers by nerve stimulation act on mast cells in a manner that causes the release of leukotrienes. The leukotrienes then act to enhance further release of tachykinins from the C-fibers.

Activation of mast cells in the airways by allergens may sensitize C-fibers to other stimuli. Ellis and Undem (1992) reported that the addition of low concentrations of antigen – which themselves caused little contraction – to the trachea of ovalbumin-sensitized guinea-pigs markedly potentiated contractions induced by capsaicin-sensitive electrical field stimulation. They also found that H mimicked the effect of antigen, and an H_1 receptor antagonist prevented the potentiation caused by ovalbumin. Similarly, Lee and Morton (1993) reported that, in dogs, a dose of aerosolized H which evoked little change in the activity of C-fibers, increased the frequency of the action potentials observed after right atrial injection of capsaicin and prolonged the effect of capsaicin. H also increased and prolonged the effect of lung inflation on C-fiber firing frequency.

2.6.2 Stimulation of C-fibers and Tachykinin Release by Mast Cell Products

The ability of H and leukotriene C_4 (LTC_4) to increase the frequency of the action potentials recorded from single C-fibers (i.e. to activate the C-fibers) subtending canine airways was originally described by Coleridge and Coleridge (1984). Undem *et al.* (1993) have also demonstrated effects of both H and LTC_4 on electrical properties of neurons in the guinea-pig nodose ganglia subtending axons with conduction velocities consistent with those of C-fibers. These investigators reported that both autocoids caused membrane depolarization in some of these neurons. H and LTC_4 as well as prostacyclin, bradykinin and serotonin also abolished the long-lasting hyperpolarization following an action potential in some

neurons and thus increased the frequency of action potential conduction.

Products of mast cells, including H, LTC_4, LTD_4 and platelet-activating factor (PAF), have also been shown to evoke the release of substance P and NkA in the airways (Saria *et al.*, 1988; Martins *et al.*, 1991a). Martins *et al.* (1991a) detected increased amounts of substance P- and NkA-like immunoreactivity in the effluent of isolated, tracheally perfused guinea-pig lungs after the addition of H, LTC_4 or PAF to the infusion perfusate. Augmented peptide recovery after the addition of the NEP inhibitor thiorphan to the perfusate indicated that the tachykinins were released in a locus that allowed their degradation by NEP. Bloomquist and Kream reported that LTD_4 caused the release of substance P and NkA in isolated guinea-pig trachea (Bloomquist and Kream, 1990).

Tachykinins are not only released by mast cell products, but appear in some cases to be involved in mediating the contractile responses or changes in airway responsiveness induced by these products. Bronchoconstrictor responses to H decrease in anesthetized guinea-pigs previously treated with large doses of capsaicin to deplete the lungs and airways of tachykinins (Martling *et al.*, 1984; Biggs and Ladenius, 1990). In isolated perfused guinea-pig lungs, administration of the NEP inhibitor thiorphan increases the changes in airway opening pressure caused by H (Martins *et al.*, 1991a). Similarly, in isolated guinea-pig tracheal tissues, tachykinin receptor antagonists decrease and NEP inhibitors increase the contractile response to LTD_4 (Bloomquist and Kream, 1990). Perretti and Manzini (1993) reported that the airway hyperresponsiveness induced by PAF could be prevented if animals were pretreated with capsaicin to deplete the lungs and airways of tachykinins.

Interactions of C-fibers with mast cells may also contribute to airway responses in animal models of asthma. In guinea-pigs, bronchoconstriction develops after eucapnic hyperventilation. Ray *et al.* (1989) reported that the changes in pulmonary resistance induced by hyperventilation increased after the administration of the NEP inhibitor phosphoramidon, which also increases responses to substance P and NkA. In addition, treatment with large doses of capsaicin to deplete substance P and NkA in the airways substantially reduced bronchoconstriction. The same group reported that both the Nk1 receptor antagonist CP96345 and the Nk2 receptor antagonist SR48968 reduced the changes in resistance induced by hyperventilation (Solway *et al.*, 1993). Leukotriene receptor antagonists and synthesis inhibitors have also been reported to reduce bronchoconstriction in this model (Garland *et al.*, 1993). It is possible that mast cell activation resulting from cooling or drying of the airways releases leukotrienes which act to release contractile tachykinins from C-fibers. However, Yang *et al.* (1994) have demonstrated that the increase in LTC_4 measured in bile after eucapnic hyperventilation is not evident in capsaicin-pretreated guinea-pigs. These results suggest that the role of C-fibers in this model is to stimulate mast cells to release leukotrienes, which effects the contraction of airway smooth muscle.

The importance of mast cell–sensory nerve interactions in human airways remains to be established. There are, to our knowledge, no data regarding the ability of mast cell products to stimulate C-fibers in human airways. Substance P does cause the degranulation of human skin mast cells (Lawrence et al., 1987), and injection of substance P into human skin causes itching and a wheal-and-flare response (Wallengren and Hakanson, 1987). However, mast cells isolated from human lung tissue obtained at surgical resection do not release H when exposed to substance P (Lawrence et al., 1987).

2.6.3 Biological Importance of Mast Cell–Tachykinin Interactions

The amplification of physiological responses through the interactions of mast cells with C-fiber-containing neurons is an important phenomenon (Fig. 7.3). The data reviewed above demonstrate that products of mast cells stimulate C-fibers which release tachykinins from nerve terminals. In addition, tachykinins can cause mast cell activation.

The potential for amplification of airway responses is intriguing. Mast cell products released by specific antigen–IgE interactions could stimulate C-fibers to release tachykinins, and these tachykinins could then stimulate the mast cells that fell within the dendritic network of the particular nerve involved. Therefore, local antigen-induced stimulation of mast cells could spread far beyond the diffusion distance for the antigen in question. The specific relevance of this proposed amplification mechanism to human disease has not been evaluated.

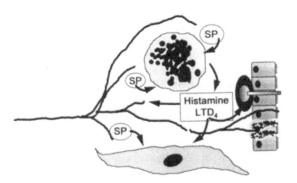

Figure 7.3 Potential mast cell interactions with C-fibers. Substance P (SP) may be released from nerve endings or from mast cells while mast cell mediators may be released by substance P. Thus, there is potential for amplification of either neural or mast cell signals by this system.

2.7 INTERACTIONS OF TACHYKININS AND THE IMMUNE SYSTEM

Important interactions of the nervous and immune systems are known to take place and to involve neuropeptides; the largest body of available evidence relates to the participation of substance P in these interactions. Most lymphoid tissues are innervated by substance P positive nerves; substance P has also been immunolocalized to immune and inflammatory cells. Cells of the immune system express receptors for substance P, and relatively low concentrations of this peptide alter the functional capabilities of these cells.

2.7.1 Substance P Innervation of Lymphoid Tissues

Immunohistochemical staining has revealed substance P in thymus (Bellinger et al., 1990), spleen (Lundberg et al., 1985; Lorton et al., 1991), lymph nodes (Nilsson et al., 1990), Peyer's patches (Stead et al., 1987), tonsils (Vanschayck and Janherwaarden, 1993) and bronchus-associated lymphoid tissue (Inoue et al., 1990; Nohr and Weihe, 1991). Nerves immunostaining for neuropeptide Y (NPY), calcitonin gene-related peptide (CGRP), VIP, cholecystokinin (CCK), somatostatin and enkephalins have also been observed in these tissues (Bellinger et al., 1992). In bronchus-associated lymphoid tissue of rats and cats, substance P-immunoreactive neurons are found under the airway epithelium in a perivascular location, intermingling with lymphoid cells in the parenchyma. Neural dendrites have also been observed in direct contact with mast cells and with cells of the macrophage/monocyte lineage (Nohra and Weihe, 1991). In most other lymphoid tissues, substance P-containing nerves are also found in close proximity to macrophages, mast cells and lymphoid cells (Stead et al., 1987; Bellinger et al., 1990; Nilsson et al., 1990; Lorton et al., 1991; Weihe and Krekel, 1991).

The innervation of the lymphoid vasculature by substance P-containing neurons and the ability of substance P to enhance lymph tissue blood flow (Lundberg et al., 1985) suggest that substance P may alter "lymphocyte traffic". Substance P induces the expression of endothelial leukocyte adhesion molecule 1 (ELAM-1) on microvascular endothelial cells (Matis et al., 1990) and may similarly affect the expression of adhesion molecules on lymphocytes. Infusion of substance P has been shown to increase lymph flow and lymphocyte traffic in sheep lymph nodes (Moore et al., 1989).

Nk receptors have been demonstrated on human and murine T and B lymphocytes, on human IM-9 cells (lymphoblasts), on human peripheral blood monocytes, and on guinea-pig peritoneal and alveolar macrophages but not on human alveolar macrophages (Payan et al., 1983; Hartung and Toyka, 1983; Hartung et al., 1986; Lotz et al., 1988; Pujol et al., 1989; Muscettola and Grasso,

1990; Bost and Pascual, 1992). In lymphocytes and lymphocytic cell lines, substance P augments mitogen-induced proliferation of B and T lymphocytes (Payan et al., 1983; Bost and Pascual, 1992), stimulates immunoglobulin synthesis (Bost and Pascual, 1992), augments the release of interleukin 2 (IL-2) from concanavalin A (Con A)-stimulated murine $CD4^+$ T cells (Rameshwar et al., 1993), and causes the release of superoxide anion as well as thromboxane A_2 (TXA_2) from guinea-pig macrophages (Hartung and Toyka, 1983; Bowden et al., 1994). Substance P also causes the release of IL-1, tumor necrosis factor 2α ($TNF\alpha$) and IL-6 from human blood monocytes (Lotz et al., 1988) and enhances production of interferon γ ($IFN\gamma$) by these cells after stimulation by staphylococcal enterotoxin A (Muscettola and Grasso, 1990). It is interesting that, in many of these assays, VIP has opposing effects (Rola-Pleszczynski et al., 1985; Wiik, 1989).

One reason to believe that substance P is an important immune effector molecule is that extraordinarily low concentrations are required for effects on lymphocytes. For example, substance P augmentation of the mitogenic effects of Con A or phytohemagglutinin M on human T cells is apparent at concentrations as low as 10^{-10} M (Payan et al., 1983). The ability of substance P to evoke the release of IL-2 from Con A-stimulated murine $CD4^+$ T cells is maximal at 10^{-13} M (Rameshwar et al., 1993). Substance P appears to bind to Nk1 receptors on murine splenocytes; the recognition of the C-terminus of the molecule by these receptors is indicated by the ability of SP4-11 to evoke the release of IL-2, whereas SP1-4 is inactive (Rameshwar et al., 1993). In addition, substance P is much more active than either NkA or NkB, and the Nk1 receptor antagonist CP96345 inhibits the release of IL-2 by substance P in a dose-dependent manner.

The tachykinin receptors on guinea-pig alveolar macrophages appear to be of the Nk2 receptor subtype. NkA is the most potent of the endogenous tachykinins in eliciting release of superoxide anion from these cells, and Nk2 selective antagonists inhibit this release (Bowden et al., 1994).

Despite the demonstrated potential for interactions of C-fibers with the immune system, the functional significance of the innervation of immune cells and tissues by substance P-containing nerves has not yet been established. It is possible that C-fibers contribute to antigen-induced responses by participating in the process of antigen sensitization, IgE production or lymphocyte recruitment. The role of C-fibers in the physiological changes that take place in animal models of allergen-induced asthma has been examined; the results obtained have been inconsistent. For example, a number of investigators (Manzini et al., 1987; Matsuse et al., 1991; Bertrand et al., 1993) have shown that chronic pretreatment of guinea-pigs with capsaicin (to deplete tachykinins) or administration of tachykinin receptor antagonists attenuates the bronchospasm, the airway hyperresponsiveness, and the increased vascular permeability that

occur following antigen exposure. In contrast, other investigators (Ingenito et al., 1991; Lai, 1991; Sakamoto et al., 1993) have been unable to demonstrate any effect of capsaicin pretreatment or of tachykinin receptor antagonists on antigen-induced airway responses. Since most of these experiments were designed to examine the participation of C-fibers or tachykinins in acute responses to inhaled allergen rather than the role of tachykinins in activation of the immune system, interpretation of the results in terms of immune function is difficult. Capsaicin treatment or Nk receptor antagonist administration may not have been undertaken at a time when it could have altered allergen sensitization or IgE production. The inconsistency of the results may reflect a complicated interplay of the experimental design and the effects of tachykinins on the immune system and/or the airways.

2.8 ROLE OF C-FIBERS IN AIRWAY INJURY

The primary locus of substance P-containing neurons in the upper airways and the ability of C-fibers to respond to many inhaled irritants suggest that these neurons play an important role in pulmonary defense. Indeed, many of the effects of C-fiber activation, including cough (Kohrogi et al., 1988), hypersecretion of mucus (Gashi et al., 1986), alterations in ventilatory pattern (Coleridge and Coleridge, 1984), increases in ion and water flux across the airway epithelium (Tamaoki et al., 1991), and increases in ciliary beat frequency (Wong et al., 1991), can be viewed as strategies to limit the ability of inhaled noxious agents or particles to reach sensitive peripheral lung tissues. Other physiological effects of neuropeptides found in C-fibers, such as augmentation of airway blood flow (Piedimonte et al., 1992), mediation of increased vascular permeability (Lundberg et al., 1984a), and stimulation of mitosis of airway epithelial cells (White et al., 1993), could be important in the repair of damaged tissue. Therefore, it is not too surprising that a protective role for C-fibers has been reported during the exposure of animals to inhaled irritants or infectious agents. Capsaicin pretreatment has been shown to augment the inflammation and/or injury induced by pulmonary exposure of rats or guinea-pigs to hydrogen sulfide (Prior et al., 1990), ozone (Sterner-Kock et al., 1994), acrolein (Turner et al., 1993), endotoxin (Long et al., 1993), SO_2 gas (Long and Shore, 1994), and Mycoplasma pulmonis (Bowden et al., 1994). It is not known whether this system can be therapeutically manipulated to the host's benefit.

3. Vasoactive Intestinal Peptide

VIP is an amidated 28 amino acid peptide that was originally discovered in extracts of pig duodenum. Both VIP and its receptor have since been found in a wide variety of tissues, including those of the digestive system, the

CNS, the respiratory system, skeletal muscle, and cells of the immune system. We limit ourselves in this chapter to a discussion of the role of VIP in the lung and a brief consideration of the importance of VIP in the immune system as it relates to the lung. One important physiological effect of VIP in the lung is the relaxation of airway smooth muscle (Altiere and Diamond, 1984; Palmer et al., 1986). In addition, VIP is a potent dilator of pulmonary (Greenberg et al., 1987) and bronchial (Palmer et al., 1986) blood vessels and affects airway mucus secretion (Peatfield and Richardson, 1983; Webber and Widdicombe, 1987). In order for VIP to influence a physiological process: (1) it must be synthesized and transported to its nerve terminals; (2) it must be released into the microenvironment, where it is (3) either processed by regulatory enzymes or (4) activates specific high affinity receptors that transduce physiological effects.

3.1 VIP SYNTHESIS AND PULMONARY TISSUE LOCALIZATION

3.1.1 VIP Gene

VIP is the product of a gene that encodes the amino acid sequence for both VIP and peptide histidine methionine (PHM-27; Linder et al., 1987; Giladi et al., 1990). There appears to be only a single copy of the gene, spanning approximately 9 kb on chromosome 6 (between 6q26 and 6q27) of the human genome (Gozes et al., 1987; Gotoh et al., 1988). The VIP gene is part of the glucagon superfamily of genes, which also includes the genes encoding glucagon, secretin, PHM-27, gastric inhibitory peptide, and growth hormone releasing factor (Bell, 1986). The gene for VIP contains seven exons, and the sequence information for the VIP peptide is contained within a single exon. Among the various mammalian species there are minor differences in the VIP gene (Giladi et al., 1990; Lamperti et al., 1991), but the 5' flanking region is highly conserved and contains a promoter sequence that is inducible by cAMP (Tsukada et al., 1987; Yamagami et al., 1988).

3.1.2 Pulmonary Distribution of VIP Gene Products

VIP is synthesized in nerve cell bodies and transported by axoplasmic flow to nerve terminals; ganglia containing VIP reactive cell bodies are found in great profusion in the walls of the trachea and major bronchi (Uddman et al., 1978; Dey et al., 1981). Nerves with VIP immunoreactivity have been found in the airways of most mammalian species studied (Springall et al., 1988; Barnes et al., 1991). This immunoreactivity is localized both to nerves and to ganglia in the airway wall; VIP-containing neurons in the airway predominantly innervate bronchial

blood vessels, seromucous glands and airway smooth muscle (Laube et al., 1992; Polak and Bloom, 1982). VIP is often, though not exclusively, co-localized with acetylcholine (Barnes et al., 1991). In the lungs of human adults, VIP-like immunoreactivity is found in the branching networks of nerve fibers that envelop airways from major bronchi down to bronchioles as small as 200 μm in diameter; however, the number of VIP-immunoreactive nerves decreases as one proceeds further "alveolarward" into the healthy lung (Hislop et al., 1990). These immunohistochemical findings have been corroborated by recent studies in which VIP was extracted from human tracheal and parenchymal samples; the VIP content of the trachea was greater than that of peripheral lung tissue (Lilly et al., 1994).

VIP-immunoreactive fibers are found within the smooth muscle and in association with vessels and mucous glands (Dey et al., 1981; Ghatei et al., 1982; Laitinen et al., 1985; Springall et al., 1988). In some mammalian species, VIP has been localized to nerves running beneath the epithelium, though these subepithelial fibers appear to be of sensory origin (Barnes et al., 1991). The presence of VIP in the airways has been confirmed by measurements of the VIP content of extracts of airway tissue (Ghatei et al., 1982, 1983).

Like those of substance P, the content and distribution of VIP in the airways differ significantly according to the animal species studied. VIP-containing nerve fibers are most abundant in cat airways (Springall et al., 1988), whose VIP content is approximately 10 times that found in rat or guinea-pig airways (Ghatei et al., 1982). In rodents and guinea-pigs VIP-containing nerves are found mostly in association with mucous glands and are less abundant in airway blood vessels and airway smooth muscle. In humans, the degree of VIP innervation in smooth muscle and in blood vessels is greater (Springall et al., 1988). There are also age-related changes in the location of VIP-containing nerves in the airways. In airways from children less than 3 years of age, VIP is localized to ganglia and to nerves from the bronchi, bronchioles, respiratory bronchioles and alveolar ducts (Hislop et al., 1990). In persons more than 10 years of age, VIP-containing nerves do not extend beyond the bronchioles. Geppetti et al. (1988) reported a decrease in VIP content (measured by both radioimmunoassay and by immunofluorescence) as a function of age in the rat.

Changes in the distribution of VIP-immunoreactive nerves have also been reported in airways (or airway biopsy specimens) from patients with airway disease. Ollerenshaw et al. (1989) reported finding few, if any, VIP-containing nerves in the airways of asthmatic patients; in contrast, such nerves were found in the airways of non-asthmatic subjects who served as controls. These data have not been widely reproduced; another group who studied the VIP content of bronchoscopically obtained airway samples from mildly asthmatic and non-asthmatic subjects was unable to detect a difference between these groups in the density of VIP-containing nerves between these (Howarth *et al.*, 1991). We measured the density of VIP-immunoreactive nerve fibers in tracheal tissue obtained at autopsy and found no significant differences in this regard between patients with fatal asthma and non-asthmatic patients. We also measured the amount of VIP in tracheal and lung parenchymal tissue extracts, using sequential high pressure liquid chromatography and immunoassay to ensure the separation of cross-reacting products from authentic VIP. All asthmatic subjects had detectable levels of VIP in lung tissue; levels of VIP and the levels detected in asthmatic lungs were not significantly different from those detected in healthy lungs (Lilly *et al.*, 1994).

Airway VIP content has also been measured in patients with the respiratory distress syndrome of the newborn. The VIP content of airways of children who died of acute respiratory distress syndrome did not differ from that of airways of age-matched control children who died of non-respiratory causes (Ghatei *et al.*, 1983).

3.1.3   Release of VIP

VIP is released from neuronal stores by appropriate neural stimulation. For example, transmural stimulation of guinea-pig tracheal segments is associated with the recovery of VIP in direct proportion to the amount of non-adrenergic, non-cholinergic (NANC) relaxation observed (Matsuzaki *et al.*, 1980; Venugopalan *et al.*, 1984). In the isolated guinea-pig trachea, VIP recovery and NANC relaxation are reduced in the presence of tetrodotoxin or neutralizing antiserum to VIP. There is increasing evidence that VIP can be released in response to capsaicin exposure (Ingenito *et al.*, 1994), as well as in response to electrical field stimulation.

3.1.4   VIP Receptors

VIP transduces intracellular signals through action at specific membrane receptors. Several groups have isolated cDNA clones encoding a 457 amino acid (52 kDa) receptor which, when transfected into COS-7 cells, confers specific and saturable VIP binding, with K_d values on the order of 1 nM (Sreedharan *et al.*, 1991, 1993a,b; Ishihara *et al.*, 1992; Couvineau *et al.*, 1994). Furthermore, when these cDNAs are transfected into COS-7 cells, binding of VIP to the transfected cells is associated with an increase in the intracellular level of cAMP; the EC_{50} for this process is also on the order of 1 nM. Analysis of the putative protein structure, based on the known cDNA sequence, reveals that the VIP receptor is a member of the secretin family of receptors. It contains seven transmembrane-spanning regions and when stimulated, results in a G protein (Gs)-coupled increase in intracellular cAMP (Kermode *et al.*, 1992; Murthy *et al.*, 1993). A second VIP receptor, known as the VIP2 receptor, has been isolated (Lutz *et al.*, 1993a,b) but appears to be expressed exclusively in the CNS.

Membranes prepared from guinea-pig, rat, mouse and human lungs have displayed specific high affinity binding of VIP (Robberecht *et al.*, 1981; Taton *et al.*, 1981; Kermode, *et al.*, 1992). VIP binding has been associated with increased levels of cAMP in rat, guinea-pig and human lung tissue (Kitamura *et al.*, 1980; Robberecht *et al.*, 1981; Taton *et al.*, 1981; Kermode *et al.*, 1992). The evidence that VIP receptors are associated with cAMP activity in lung tissue includes the observations that VIP stimulation increases the cAMP content of guinea-pig trachea; that VIP-secretin receptors in rat and human lung are coupled to adenylate cyclase (Christophe *et al.*, 1981); and that a cAMP-specific phosphodiesterase inhibitor, which would be expected to increase levels of cAMP, increases VIP-induced tracheal relaxation in the guinea-pig (Rhoden and Barnes, 1990; Shikada *et al.*, 1991).

The co-existence of parallel cAMP and cGMP pathways for VIP-induced relaxation has been demonstrated in bovine pulmonary artery (Ignarro *et al.*, 1987). In tracheally perfused lungs, VIP-induced pulmonary relaxation is augmented by the presence of methylene blue, which may act by inhibiting the degradation of cGMP (Lilly *et al.*, 1993b). This observation suggests that, in addition to the known VIP-initiated cAMP-dependent airway relaxation pathway documented in isolated airway tissues, a physiologically relevant cGMP-dependent pathway for VIP pulmonary relaxant activity exists in whole lungs. This cGMP-dependent pathway likely depends on NO· generation, which occurs in this system after tracheal injection of VIP and therefore may be an indirect consequence of VIP receptor activation.

VIP receptors, demonstrated autoradiographically as focal areas of enhanced VIP binding in tissue sections, are present on bronchial epithelium, submucosal glands, airway smooth muscle and alveolar cells (Leroux *et al.*, 1984; Barnes, 1986; Castairs and Barnes, 1986; Leys *et al.*, 1986). The presence of VIP receptors in human lungs has been confirmed by covalent cross-linking and solubilization studies (Paul and Said, 1987). VIP binding to its high affinity receptor is rapidly reversible and sensitive to GTP (Said, 1988) and to tryptic enzymes (Paul and Said, 1987). The reported distribution of the VIP receptor correlates well with functional studies demonstrating effects on airway smooth muscle, on pulmonary and bronchial blood vessels, and on airway mucus secretion. Studies of the physiological role of the VIP receptor would be greatly facilitated by the availability of a selective non-peptide VIP receptor antagonist.

3.1.5   Enzymatic Regulation of the Physiological Effects of VIP

The action of VIP at its receptor is competitive with its degradation by proteases found in the microenvironment of the receptor. VIP is known to be a favored substrate for NEP, mast cell tryptase and mast cell chymase, all of which are active at the neutral pH likely to prevail in the extracellular microenvironment. Recombinant human

NEP has been shown in isolated systems to hydrolyze VIP with kinetics that are consistent with activity under physiological conditions. The peptide bonds with the kinetics most favorable to NEP hydrolysis are Asp^3–Ala^4, Ala^4–Val^5, Lys^{21}–Tyr^{22} and Ser^{25}–Ile^{26}. VIP cleavage sites with less favorable kinetics are present at the Arg^{12}–Lys^{13} and the Gln^{16}–Met^{17} bonds (Goetzl et al., 1989; Turner et al., 1991).

Studies in isolated tracheally perfused guinea-pig lungs have demonstrated that VIP has significant pulmonary relaxant activity and that the NEP hydrolysis fragments of VIP are not physiologically active (Lilly et al., 1993a). The fact that NEP can alter the physiological activity of VIP suggests that it has the potential to function as a VIP regulatory enzyme.

Mast cell proteases have also been shown to hydrolyze VIP. Canine mastocytoma tryptase cleaves VIP with favorable kinetics at the Arg^{14}–Lys^{15} and Lys^{20}–Lys^{21} bonds (Caughey et al., 1988). The k_{cat}/K_m for hydrolysis of human VIP by canine mastocytoma tryptase at Arg^{14}–Lys^{15} and Lys^{20}–Lys^{21} is $2.2 \times 10^5 \, s^{-1} M^{-1}$ and the K_m is 3.3×10^{-3} M. These kinetic parameters compare favorably to those of other substrates screened for susceptibility to human skin and lung mast cell proteases (Tanaka et al., 1983). Chymase purified from canine mastocytoma cells has kinetics that favor the cleavage of VIP at Tyr^{22}–Leu^{23} (Caughey et al., 1988). The K_m of $3.3 \times 10^{-3} \, M^{-1}$ and the k_{cat} of $179 \, s^{-1}$ for this cleavage give a k_{cat}/K_m of $5.4 \times 10^4 \, s^{-1} M^{-1}$. While this k_{cat}/K_m is not as high as that obtained for tryptase, it is consistent with physiologically relevant inactivation of VIP by chymase. The products of hydrolysis of VIP by tryptase or chymase do not relax vascular, gastrointestinal or tracheal smooth muscle (Caughey et al., 1988); neither do they relax tracheally perfused lungs (Lilly et al., 1993a). The fact that these mast cell proteases have the capacity to alter the physiological activity of VIP makes them candidate VIP regulatory enzyme systems.

3.1.6 Autoantibodies to VIP

In addition to proteolytic enzymes, IgG autoantibodies are capable of inactivating VIP. Autoantibodies to VIP have been demonstrated in 15–20% of a limited number of asthmatic and non-asthmatic subjects (Smith et al., 1991).

3.1.7 Relevance of VIP Cleavage Sites

The physiological relevance of these cleavage sites can be inferred from studies of the effects of enzyme inhibitors on VIP responses. For example, VIP-induced pulmonary relaxation can be attenuated by exogenous mast cell tryptase or chymase in the ferret (Franconi et al., 1989) and by exogenous α-chymotrypsin or papain in the guinea-pig. Enzyme inhibitor studies have provided insight into which of these enzyme systems are involved in the limitation of VIP activity. In cat airways, a combination of inhibitors that did not include an NEP inhibitor failed

to augment VIP-induced relaxation; in contrast, in guinea-pig bronchial tissues, soybean trypsin inhibitor augmented VIP-induced tissue relaxation (Thompson et al., 1988). Inhibitors of NEP have also been reported to augment VIP-induced relaxation of guinea-pig tracheal rings (Hachisu et al., 1991), while inhibitors of ACE and aminopeptidases were ineffective in enhancing relaxation. Epithelium removal enhances VIP-induced relaxation to a degree equivalent to that seen after the treatment of epithelium-intact airways with NEP inhibitors, but not with ACE inhibitors (Farmer and Togo, 1990). The differences among the results of these studies may be due to differences in the techniques employed for handling and harvesting the tissues. Since epithelial disruption is known to affect NEP activity, the use of preparative approaches that may disrupt the airway epithelium would be expected to lead to an underestimation of the relevance of NEP; that is, NEP inhibitors would be less effective in this situation. Likewise, preparative protocols that released mast cell proteases from their inactive intracellular loci could overestimate the importance of these enzymes.

In human bronchial rings, VIP-induced relaxation is potentiated by a combination of serine protease and NEP inhibitors (Tam et al., 1990) or by epithelial removal (Hulsmann et al., 1993). Intact, tracheally perfused lungs may be less prone than the excised airway tissues to the complicating effects of epithelial disruption and mast cell activation and may offer a better model for the study of regulatory proteases. In isolated, tracheally perfused guinea-pig lungs, combinations of enzyme inhibitors augment VIP-induced pulmonary relaxation, while single agent inhibitors are ineffective (Lilly et al., 1993a). Combinations of enzyme inhibitors that exert a physiological effect also increase the amount of intact VIP recovered from lung effluent. Indeed, in light of the hydrolysis fragments obtained in the presence of various enzyme inhibitors, we deduced that NEP and a tryptic enzyme with a cleavage profile identical to that of mast cell tryptase are responsible for limiting VIP-induced pulmonary relaxation in the tracheally perfused guinea-pig lung (Lilly et al., 1993a).

3.1.8 Inflammatory Modulation of VIP Hydrolysis

Enzymatic hydrolysis regulates the airway effects of VIP, and airway inflammation induced by chronic exposure to antigens alters the activity of VIP regulatory enzymes. Whereas the effects of substance P are enhanced in lungs from antigen-exposed animals, antigen-exposed inflamed lungs are resistant to the relaxant effects of tracheally injected VIP. Furthermore, this resistance does not present a generalized inability to respond to inhibitory agonists, as isoproterenol-induced pulmonary relaxation is not altered (Lilly et al., 1993a). Instead, the data suggest that the decreased sensitivity to VIP is the result of enhanced degradation of this peptide. Evaluation of the recovery of intact [^{125}I]VIP and its hydrolysis fragments from the lungs of control guinea-pigs revealed fragments consistent with inactivation by NEP and tryptic enzyme (Lilly et al., 1993a). In contrast, no hydrolysis products consistent with NEP activity were recovered from antigen inflamed lungs while VIP1-14, a mast cell protease product, was recovered.

Alterations in activity of regulatory enzymes induced by airway inflammation provide a possible mechanism for enhanced non-specific airway responsiveness in inflamed lungs. It is known that a variety of bronchoactive mediators release tachykinins from guinea-pig lungs (Saria et al., 1988; Maggi et al., 1990; Martins et al., 1991a; Manzini and Meini, 1991). Therefore, the loss of NEP activity observed in antigen-exposed lungs could result in an enhanced response to tachykinins released by other bronchoactive mediators. This effect is in agreement with the observation that enhanced responses to methacholine (Matsuse et al., 1991) and antigen (Manzini et al., 1987) in the lungs of animals repeatedly exposed to antigen can be diminished by depletion of substance P with capsaicin pretreatment. However, studies of the impairment of VIP effects extend this mechanism by demonstrating that inflamed lungs are also resistant to the relaxant – and probably homeostatic – effects of this peptide. Therefore, chronic inflammation modulates the microenvironment in two ways: it enhances the prophlogistic effects of peptides degraded by NEP, such as substance P and NkA, while it diminishes the homeostatic effects of VIP.

Thus, through disruption of the balance between contractile and relaxant neuropeptides in the lung, inflammation can promote non-specific airway hyper-responsiveness. It is increasingly clear that regulatory enzymes affect both contractile and relaxant pathways. The implication is that the net effect of altered enzyme activity must be judged by assessing changes in the metabolism of both the relevant relaxant and the relevant contractile peptides.

In summary, VIP is a physiologically active peptide present in neurons that subtend airway smooth muscle, pulmonary and bronchial vessels, and mucosal glands. It can be released from neuronal stores by electrical stimulation and by capsaicin exposure. VIP-induced pulmonary relaxation is regulated in a complex way through hydrolytic inactivation by two physiologically competitive enzyme systems. The role of enzymes in limiting the vascular and glandular effects of VIP remains to be fully explored. The activation of the VIP receptor results in cAMP generation, but studies of this receptor have been hampered by the lack of a specific non-peptide VIP receptor antagonist.

4. Nitrogen Oxides

The discovery that mammalian cells can produce oxidized forms of nitrogen that serve as neurotransmitters has had a broad impact on our understanding of pulmonary neural responses. The biology of nitrogen oxides in the lung has recently been reviewed in general terms (Barnes and Belvisi, 1993; Gaston et al., 1994b); we will focus exclusively on the potential neural role of nitrogen oxides in the lung.

4.1 NITROGEN OXIDE CHEMISTRY IN THE LUNG

Nitric oxide (NO) is produced through the action of a class of enzymes known as nitric oxide synthases (NOSs). These enzymes utilize L-arginine, molecular oxygen and NADPH as substrates and flavoproteins, tetrahydrobiopterin and calmodulin as cofactors (Bredt et al., 1991; Nathan, 1992; Bredt and Snyder, 1994). The end products of this enzymatic reaction, NO and L-citrulline, are formed through sequential monoxidation with N^{ω}-hydroxy-L-arginine formed as an intermediate (DeMaster et al., 1989). NOSs are grouped in three broad categories: constitutively expressed neuronal or endothelial NOSs (cNOS and eNOS) and inducible (immunoactive) NOS (iNos) (Nathan, 1992). NO generated by neuronal and other NOS isoforms reacts readily in the oxygen-rich microenvironment of the lung to form an array of NO_x and related products.

4.2 IMMUNOHISTOCHEMICAL LOCALIZATION OF NOS

Neuronal (cNOS) immunostaining, near smooth muscle in the proximal airway, is uniformly evident across species lines (Kobzik et al., 1993; Fischer et al., 1993; Dey et al., 1993; Springall et al., 1994). In the guinea-pig, cNOS immunoreactivity is greatest in nerves of the major bronchi and diminishes in intensity both centrally toward the trachea and peripherally toward the terminal bronchioles (Fischer et al., 1993). In the human airway, on the other hand, cNOS immunoreactivity is greatest in the trachea and diminishes as one proceeds toward the bronchioles (Springall et al., 1994). In patients with cystic fibrosis, cNOS neuronal staining appears to be diminished overall, even in the more proximal airway (Belvisi et al., 1994). Smooth muscle is not the only tissue in the airways with cNOS immunoreactivity: in the ferret and the human, but not in the guinea-pig, cNOS immunolabels nerve cell bodies in the small ganglia of the longitudinal tracheal plexus (Dey et al., 1993; Springall et al., 1994). Fischer and co-workers have demonstrated cNOS immunoreactivity in sensory nerves of the guinea-pig subepithelial lamina propria (Fischer et al., 1993).

In the trachea of ferrets and humans, there is co-localization of immunoreactivity for cNOS and VIP (Dey et al., 1993; Springall et al., 1994). Although Fischer and co-workers were unable to demonstrate NOS immunoreactivity in the ganglia of the guinea-pig airway

(Fischer *et al.*, 1993), they did show strong immunoreactivity in the jugular and nodose ganglia and weak staining of the stellate ganglia – the putative origin of guinea-pig inhibitory NANC (iNANC) signalling. These investigators concluded that perimuscular neuronal NOS immunoreactivity in their guinea-pig tissue preparations was located primarily in vagal afferent nerve cells.

Epithelial cells from human, rat and guinea-pig airway preparations have a histochemical profile consistent with the presence of both cNOS and iNOS activity (Kobzik *et al.*, 1993; Hamid *et al.*, 1993). In addition, histochemical evidence for epithelial iNOS and cNOS is supported by biochemical data from unstimulated and cytokine-stimulated human airway cell culture lines (Asano *et al.*, 1994). The expression of airway epithelial iNOS and cNOS varies with the circumstances. Human airway specimens show reproducible iNOS staining in central cartilaginous airways but only patchy staining in bronchioles (Kobzik *et al.*, 1993). Anti-rat brain cNOS immunostains rat airway epithelium constitutively, though no such staining occurs in the corresponding human tissue. Schmidt and co-workers have shown prominent cNOS staining in rat terminal respiratory bronchioles but not in larger bronchi (Schmidt *et al.*, 1992). Thus epithelial cNOS is more commonly encountered in the distal airway, while epithelial iNOS is more intensely distributed in the proximal airway.

4.3 BIOACTIVITIES OF AIRWAY NO$_x$

The role of neuronal NOS in the airways is incompletely understood. The gene encoding the enzymes has been isolated and cloned (Bredt *et al.*, 1991), and a strain of neuronal NOS knockout mice has recently been developed (Huang *et al.*, 1993; O'Dell *et al.*, 1994). While these mice had pyloric stenosis (a characteristic that reaffirmed the important role of iNOS in gastrointestinal neurotransmission), they had no obvious abnormalities of airway structure. The relevance of these findings to the human airway remains to be determined, and the data need to be considered in light of the diverse bioactivities of NO in the lung.

NOS products of neurons, epithelial cells and other cells in the lung have both bronchodilator and inflammatory properties (Gaston *et al.*, 1994b); the specific nature of this bioactivity depends on the chemical characteristics of the functional products in the specific microenvironment under consideration. For example, NO is capable of complexing with and affecting the activity of a variety of metalloproteins and enzymes, such as guanylyl cyclase and ribonucleotide reductase (Nathan, 1992; Stamler *et al.*, 1992). NO can also complex with superoxide anion to form peroxynitrite, which has a cytotoxic immune effector role (Radi *et al.*, 1991; Stamler *et al.*, 1992). Furthermore, NO can form iron nitrosyl complexes, which are the putative intracellular macrophage products of iNOS responsible for lysis of intracellular parasites (Hibbs

et al., 1988; Stamler *et al.*, 1992). NO may also complex with free thiol groups in the microenvironment to form S-nitrosothiols – relatively stable compounds produced *in vivo* that exhibit airway smooth muscle relaxant and bacteriostatic properties (Hibbs *et al.*, 1988; Stamler *et al.*, 1992; Jansen *et al.*, 1992; Gaston *et al.*, 1993b). In line with this wide variety of potential products, there are a number of established biological roles for products of NOS in the lung. We will consider four of these roles: (1) maintenance of basal airway tone; (2) mediation of iNANC bronchodilatation; (3) modulation of bronchoconstriction; and (4) non-neuronal cell–cell signalling.

4.3.1 Maintenance of Basal Airway Tone

Intraluminal perfusion of guinea-pig tracheal preparations with inhibitors of NOS results in an increase in basal tone; this change suggests the presence of basal NOS activity and therefore the production of NO which constitutively regulates the tone of the airway (Nijkamp *et al.*, 1993). Superfusates from normal guinea-pig lungs contain NO$_x$ in concentrations sufficient to relax guinea-pig airway smooth muscle (Jansen *et al.*, 1992; Lilly *et al.*, 1993b). These findings are not limited to animals: the endogenous bronchodilator S-nitrosoglutathione is present in normal human airway lining fluid at concentrations sufficient to mediate relaxation of airway smooth muscle (Gaston *et al.*, 1994a). These findings strongly suggest a role for NO$_x$ in the regulation of airway tone in the absence of specific constrictor stimuli.

4.3.2 Inhibitory NANC Responses

Inhibitory NANC responses are mediated by both VIP and NO$_x$. NO$_x$ partially mediates the airway smooth muscle relaxant effects of VIP in the isolated perfused guinea-pig lung, as NOS inhibitors prevent relaxation induced by VIP but not that induced by isoproterenol (Lilly *et al.*, 1993b). A significant and appropriately timed increase in the NO$_x$ content of lung perfusion fluid can be measured in response to VIP (Fig. 7.4). These data support the conclusion that interactions of VIP with NO$_x$ products mediate a substantial component of iNANC transmission. In human airways, NOS stimulated as a result of iNANC neurotransmission accounts for a substantial proportion of this relaxant response. Belvisi and co-workers have shown that the α-chymotrypsin-resistant component of the human iNANC response can be abolished by treatment with NOS inhibitors (Belvisi *et al.*, 1992).

4.3.3 Modulation of Airway Constriction

The activation of NOS or the products thereof modulates airway constrictor responses as well. Atropine-sensitive, electrical field stimulation-induced bronchoconstrictor responses in the guinea-pig are enhanced by NOS inhibitors (Belvisi *et al.*, 1991). Nitrogen oxides also modulate the effects of non-cholinergic bronchoconstrictors. H stimulation of H$_1$ receptors in the guinea-pig lung leads

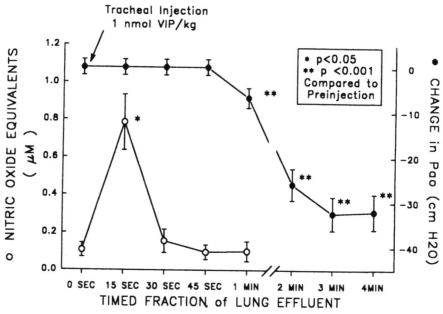

Figure 7.4 VIP-induced pulmonary relaxation is preceded by a significant increase in tracheal perfusate nitric oxide equivalents, as measured by chemiluminescence. Results are expressed as group mean and SE. P_{ao}, airway opening pressure. Reprinted with permission from Lilly *et al.* (1993b).

to NOS activation (Leurs *et al.*, 1991) both *in vivo* and *in vitro*. S-nitrosothiols have the capacity to relax guinea-pig and human airways (preconstricted with H or methacholine) at concentrations consistent those that can be recovered from the airway *in vivo* (Jansen *et al.*, 1992; Gaston *et al.*, 1993b, 1994a).

4.3.4 Non-neuronal Cell–Cell Signalling

Nitrogen oxides such as NO and S-nitrosoglutathione may serve simultaneously as a means whereby non-neuronal airway cells signal nearby smooth muscle tissues to change their contractile state and as immune effector molecules. For example, during states of immune activation or inflammation, S-nitrosoglutathione and other S-nitrosothiols with bacteriostatic properties (Morris *et al.*, 1984) are present in human airways at concentrations sufficient to cause the relaxation of precontracted human bronchial smooth muscle by more than 50% (Gaston *et al.*, 1993b). Locally produced NO may also have a pro-inflammatory role. For example, immunohistochemical analysis has established that epithelial NOSs are activated in asthma (Hamid *et al.*, 1993); this activation may explain the elevated concentrations of NO recovered in the exhaled air of asthma patients (Gaston *et al.*, 1993a; Persson *et al.*, 1994; Kharitonov *et al.*, 1994; Massaro *et al.*, 1994). The concentration of NO in exhaled air decreases as a result of treatment with corticosteroids (Massaro *et al.*, 1994; Kharitonov *et al.*, 1994; Persson *et al.*, 1994); thus it is possible that NO produced in the inflammatory microenvironment of the airway accounts for a significant component of asthmatic airway

narrowing. Whether NO_x are primarily homeostatic or inflammatory in asthma remains to be determined.

5. Summary

The complex neural network of the lung has both pro-inflammatory and anti-inflammatory activities. These activities are not simply related to the specific neurotransmitter in question. Indeed, as detailed above, each of the major classes of pulmonary neurotransmitter can have either effect, although the predominant role of substance P and NkA is prophlogistic and that of VIP is homeostatic. The predominant role of NO is not clear at this time.

6. References

Adcock, I.M., Peters, M., Gelder, C. *et al.* (1993). Increased tachykinin receptor gene expression in asthmatic lung and its modulation by steroids. J. Mol. Endocrinol. 11, 1–7.

Ahmad, S., Wang, L.H. and Ward, P.E. (1992). Dipeptidyl(amino)peptidase-IV and aminopeptidase-M metabolize circulating substance-P *in vivo*. J. Pharmacol. Exp. Ther. 260, 1257–1261.

Altiere, R.J. and Diamond, L. (1984). Comparison of vasoactive intestinal peptide and isoproterenol relaxant effects in isolated cat airways. J. Appl. Physiol. 56, 986–992.

Asano, K., Chee, C., Gaston, B. *et al.* (1994). Constitutive and inducible nitric oxide synthase gene expression, regulation and activity in human lung epithelial cells. Proc. Natl. Acad. Sci. USA 91 (1994): 10089–10093.

Bai, T.R., Zhou, D., Weir, S. *et al.* (1994). Expression of tachykinin receptors in inflammatory airway disorders. Am. J. Respir. Crit. Care Med. 149, A859 (Abstract).

Baraniuk, J.N., Lundgren, J.D., Okayama, M. *et al.* (1991). Substance P and neurokinin A in human nasal mucosa. Am. J. Respir. Cell Mol. Biol. 4, 228–236.

Barnes, P.J. Neural control of human airways in health and disease. Am. Rev. Respir. Dis. 134, 1289–1314.

Barnes, P.J. and Belvisi, M.G. (1993). Nitric oxide and lung disease. Thorax 48, 1034–1043.

Barnes, P.J., Baraniuk, J.N. and Belvisi, M.G. (1991). Neuropeptides in the respiratory tract, part I. Am. Rev. Respir. Dis. 144, 1187–1198.

Bell, G.I. (1986). The glucagon superfamily: precursor structure and gene organization. Peptides 7 Suppl 1, 27–36.

Bellinger, D.L., Lorton, D., Romano, T.D. *et al.* (1990). Neuropeptide innervation of lymphoid organs. Ann. NY Acad. Sci. 594, 17–33.

Bellinger, D.L., Lorton, D., Felten, S.Y. *et al.* (1992). Innervation of lymphoid organs and implications in development, aging, and autoimmunity. Int. J. Immunopharmacol 14, 329–344.

Belvisi, M.G., Stretton, D. and Barnes, P.J. (1991). Nitric oxide as an endogenous modulator of cholinergic neurotransmission in guinea-pig airways. Eur. J. Pharmacol. 198, 219–221.

Belvisi, M.G., Stretton, C.D., Yacoub, M. *et al.* (1992). Nitric oxide is the endogenous neurotransmitter of bronchodilator nerves in humans. Eur. J. Pharmacol. 210, 221–222.

Belvisi, M.G., Ward, J.K., Springall, D.R. *et al.* (1994). Nitrinergic innervation in the airways of patients with cystic fibrosis. Am. J. Respir. Crit. Care Med. 149, A675 (Abstract).

Bertrand, C., Geppetti, P., Baker, J. *et al.* (1993). Role of neurogenic inflammation in antigen-induced vascular extravasation in guinea-pig trachea. J. Immunol. 150, 1479–1485.

Biggs, D.F. and Ladenius, R.C. (1990). Capsaicin selectively reduces airway responses to histamine, substance P and vagal stimulation. Eur. J. Pharmacol. 175, 29–33.

Bloomquist, E.I. and Kream, R.M. (1990). Release of substance P from the guinea-pig trachea (by) leukotriene D_4. Exp. Lung Res. 16, 645–659.

Borson, D.B., Corrales, R., Varsano, S. *et al.* (1987). Enkephalinase inhibitors potentiate substance P-induced secretion of 35SO4-macromolecules from ferret trachea. Exp. Lung Res. 12, 21–36.

Borson, D.B., Brokaw, J.J., Sekizawa, K. *et al.* (1989). Neutral endopeptidase and neurogenic inflammation in rats with respiratory infections. J. Appl. Physiol. 66, 2653–2658.

Bost, K.L. and Pascual, D.W. (1992). Substance-P – a late-acting lymphocyte-B differentiation cofactor. Am. J. Physiol. 262, C537-C545.

Bowden, J.J., Baluk, P., Lefevre, P.M. *et al.* (1994). Sensory denervation by neonatal capsaicin pretreatment increases the chronic inflammatory response to *Mycoplasma pulmonis* infection in the rat respiratory tract. Am. J. Respir. Crit. Care Med. 149, A333 (Abstract).

Bredt, D.S. and Snyder, S.H. (1994). Nitric oxide: a physiologic messenger molecule. Annu. Rev. Biochem. 175–195.

Bredt, D.S., Hwang, P.M., Glatt, C.E. *et al.* (1991). Cloned and expressed nitric oxide synthase structurally resembles cytochrome P-450 reductase. Nature 351, 714–718.

Castairs, J.R. and Barnes, P.J. (1986a). Visualization of vasoactive intestinal peptide receptors in human and guinea-pig lung. J. Pharmacol. Exp. Ther. 239, 249–255.

Castairs, J.R. and Barnes, P.J. (1986b). Autoradiographic mapping of substance P receptors in lung. Eur. J. Pharmacol. 127, 295–296.

Caughey, G.H., Leidig, F., Viro, N.F. *et al.* (1988). Substance P and vasoactive intestinal peptide degradation by mast cell tryptase and chymase. J. Pharmacol. Exp. Ther. 244, 133–137.

Cheung, D., Bel, E.H., Den Hartigh, J. *et al.* (1992). The effect of an inhaled neutral endopeptidase inhibitor, thiorphan, on airway responses to neurokinin A in normal humans *in vivo*. Am. Rev. Respir. Dis. 145, 1275–1280.

Cheung, D., Timmers, M.C., Zwinderman, A.H. *et al.* (1993). Neutral endopeptidase activity and airway hyperresponsiveness to neurokinin A in asthmatic subjects *in vivo*. Am. Rev. Respir. Dis. 148, 1467–1473.

Choi, H.S., Lesser, M., Cardozo, C. *et al.* (1990) Immunohistochemical localization of endopeptidase 24.15 in rat trachea, lung tissue, and alveolar macrophages. Am. J. Respir. Cell Mol. Biol. 3, 619–624.

Christophe, J., Chatelain, P., Taton, G. *et al.* (1981). Comparison of VIP-secretin receptors in rat and human lung. Peptides 2 (Suppl 2), 253–258.

Coleridge, J.C. and Coleridge, H.M. (1984). Afferent vagal C fibre innervation of the lungs and airways and its functional significance. Rev. Physiol. Biochem. Pharmacol. 99, 1–110.

Colpaert, F.C., Donnerer, J. and Lembeck, F. (1983). Effects of capsaicin on inflammation and on the substance P content of nervous tissues in rats with adjuvant arthritis. Life Sci. 32, 1827–1834.

Conlon, J.M. and Sheehan, L. (1983). Conversion of substance P to C-terminal fragments in human plasma. Regul. Pept. 7, 335–345.

Couvineau, A., Rouyer-Fessard, C., Darmoul, D. *et al.* (1994). Human intestinal VIP receptor: cloning and functional expression of two cDNA encoding proteins with different N-terminal domains. Biochem. Biophys. Res. Commun. 200, 769–776.

DeMaster, E.G., Raij, L., Archer, S.L. *et al.* (1989). Hydroxylamine is a vasorelaxant and a possible intermediate in the oxidative conversion of L-arginine to nitric oxide. Biochem. Biophys. Res. Commun. 163, 527–533.

Devillier, P., Acker, G.M., Advenier, C. *et al.* (1992). Activation of an epithelial neurokinin Nk1 receptor induces relaxation of rat trachea through release of prostaglandin E_2. J. Pharmacol. Exp. Ther. 263, 767–772.

Dey, R.D., Shannon, W.A. Jr. and Said, S.I. (1981). Localization of VIP-immunoreactive nerves in airways and pulmonary vessels of dogs, cat, and human subjects. Cell Tissue Res. 220, 231–238.

Dey, R.D., Hoffpauir, J. and Said, S.I. (1988). Co-localization of vasoactive intestinal peptide- and substance P-containing nerves in cat bronchi. Neuroscience 24, 275–281.

Dey, R.D., Mayer, B. and Said, S.I. (1993). Colocalization of vasoactive intestinal peptide and nitric oxide synthase in neurons of the ferret trachea. Neuroscience 54, 839–843.

Dimitriadou, V., Buzzi, M.G., Moskowitz, M.A. *et al.* (1991). Trigeminal sensory fiber stimulation induces morphological changes reflecting secretion in rat dura mater mast cells. Neuroscience 44, 97–112.

Donnerer, J., Schuligoi, R., Stein, C. *et al.* (1993). Upregulation, release and axonal transport of substance P and calcitonin gene-related peptide in adjuvant inflammation and regulatory function of nerve growth factor. Regul. Pept. 46, 150–154.

Dusser, D.J., Djokic, T.D., Borson, D.B. *et al.* (1989a). Cigarette smoke induces bronchoconstrictor hyperresponsiveness to substance P and inactivates airway neutral endopeptidase in the guinea-pig. Possible role of free radicals. J. Clin. Invest. 84, 900–906.

Dusser, D.J., Jacoby, D.B., Djokic, T.D. *et al.* (1989b). Virus induces airway hyperresponsiveness to tachykinins: role of neutral endopeptidase. J. Appl. Physiol. 67, 1504–1511.

Ellis, J.L. and Undem, B.J. (1991). Role of peptidoleukotrienes in capsaicin-sensitive sensory fibre-mediated responses in guinea-pig airways. J. Physiol. Lond. 436, 469–484.

Ellis, J.L. and Undem, B.J. (1992). Antigen-induced enhancement of noncholinergic contractile responses to vagus nerve and electrical field stimulation in guinea-pig isolated trachea. J. Pharmacol. Exp. Ther. 262, 646–653.

Ellis, J.L., Undem, B.J., Kays, J.S. *et al.* (1993). Pharmacological examination of receptors mediating contractile responses to tachykinins in airways isolated from human, guinea-pig and hamster. J. Pharmacol. Exp. Ther. 267, 95–101.

Farmer, S.G. and Togo, J. (1990). Effects of epithelium removal on relaxation of airway smooth muscle induced by vasoactive intestinal peptide and electrical field stimulation. Br. J. Pharmacol. 100, 73–78.

Fischer, A., Mundel, P., Mayer, B. *et al.* (1993). Nitric oxide synthase in guinea-pig lower airway innervation. Neurosci. Lett. 149, 157–160.

Fischer, A., Philippin, B., Saria, A. *et al.* (1994). Neuronal plasticity in sensitized and challenged guinea-pigs: neuropeptides and gene expression. Am. J. Respir. Crit. Care Med. 149, A890 (Abstract).

Fong, T.M., Huang, R.R., Yu, H. *et al.* (1993). Mapping the ligand binding site of the Nk1 receptor. Regul. Pept. 46, 43–48.

Foreman, J. and Jordan, C. (1983). Histamine release and vascular changes induced by neuropeptides. Agents Actions 13, 105–116.

Foulon, D.M., Champion, E., Masson, P. *et al.* (1993). Nk(1) and Nk(2) receptors mediate tachykinin and resiniferatoxin-induced bronchospasm in guinea-pigs. Am. Rev. Respir. Dis. 148, 915–921.

Franconi, G.M., Graf, P.D., Lazarus, S.C. *et al.* (1989). Mast cell tryptase and chymase reverse airway smooth muscle relaxation induced by vasoactive intestinal peptide in the ferret. J. Pharmacol. Exp. Ther. 248, 947–951.

Fuller, R.W., Maxwell, D.L., Dixon, C.M. *et al.* (1987). Effect of substance P on cardiovascular and respiratory function in subjects. J. Appl. Physiol. 62, 1473–1479.

Garland, A., Jordan, J.E., Ray, D.W. *et al.* (1993). Role of eicosanoids in hyperpnea-induced airway responses in guinea-pigs. J. Appl. Physiol. 75, 2797–2804.

Gashi, A.A., Borson, D.B., Finkbeiner, W.E. *et al.* (1986). Neuropeptides degranulate serous cells of ferret tracheal glands. Am. J. Physiol. 251, C223–229.

Gaston, B., Drazen, J., Chee, C.B.E. *et al.* (1993a). Expired nitric oxide (NO) concentrations are elevated in patients with reactive airways disease. Endothelium 1, 87.

Gaston, B., Reilly, J., Drazen, J.M. *et al.* (1993b). Endogenous nitrogen oxides and bronchodilator S-nitrosothiols in human airways. Proc. Natl. Acad. Sci. USA 90, 10957–10961.

Gaston, B., Drazen, J.M., Jansen, A. *et al.* (1994a). Relaxation of human bronchial muscle by S-nitrosothiols *in vitro*. J. Pharmacol. Exp. Ther. 268, 978–994.

Gaston, B., Drazen, J.M., Loscalzo, J. *et al.* (1994b). The biology of nitrogen oxides in the airway. Am. J. Respir. Crit. Care Med. 149, 538–551.

Geppetti, P., De Rossi, M., Renzi, D. *et al.* (1988). Age-related changes in vasoactive intestinal polypeptide levels and distribution in the rat lung. J. Neural. Transm. 74, 1–10.

Gerard, N.P., Eddy, Jr., R.L., Shows, T.B. *et al.* (1990). The human neurokinin A (substance K) receptor. Molecular cloning of the gene, chromosome localization, and isolation of cDNA from tracheal and gastric tissues [published erratum appears in J. Biol. Chem. (1991) 266(2), 1354]. J. Biol. Chem. 265, 20455–20462.

Gerard, N.P., Garraway, L.A., Eddy, Jr., R.L. *et al.* (1991). Human substance P receptor, Nk1 organization of the gene, chromosome localization, and functional expression of cDNA clones. Biochemistry 30, 10640–10646.

Ghatei, M.A., Sheppard, M.N., O'Shaughnessy, D.J. *et al.* (1982). Regulatory peptides in the mammalian respiratory tract. Endocrinology 111, 1248–1254.

Ghatei, M.A., Sheppard, D.J., Henzen-Logman, S. *et al.* (1983). Bombesin and vasoactive intestinal polypeptide in the developing lung: marked changes in the acute respiratory distress syndrome. J. Clin. Endocrinol. Metab. 57, 1226–1232.

Giladi, E., Shani, Y. and Gozes, I. (1990). The complete structure of the rat VIP gene. Brain Res. Mol. Brain Res. 7, 261–267.

Goetzl, E.J., Sreedharan, S.P., Turck, C.W. *et al.* (1989). Preferential cleavage of amino- and carboxyl-terminal oligopeptides from vasoactive intestinal polypeptide by human recombinant enkephalinase (neutral endopeptidase, EC 3.4.24.11). Biochem. Biophys. Res. Commun. 158, 850–854.

Gotoh, E., Yamagami, T., Yamamoto, H. *et al.* (1988). Chromosomal assignment of human VIP/PHM-27 gene to 6q26---q27 region by spot blot hybridization and *in situ* hybridization. Biochem. Int. 17, 555–562.

Gozes, I., Avidor, R., Yahav, Y. *et al.* (1987). The gene encoding vasoactive intestinal peptide is located on human chromosome 6p21----6qter. Hum. Genet. 75, 41–44.

Greenberg, B., Rhoden, K. and Barnes, P.J. (1987). Relaxant effects of vasoactive intestinal peptide and peptide histidine isoleucine in human and bovine pulmonary arteries. Blood Vessels 24, 45–50.

Hachisu, M., Hiranuma, T., Tani, S. *et al.* (1991). Enzymatic degradation of helodermin and vasoactive intestinal polypeptide. J. Pharmacobiodyn. 14, 126–131.

Hamid, Q., Springall, D.R., Riverosmoreno, V. *et al.* (1993). Induction of nitric oxide synthase in asthma. Lancet 342, 1510–1513.

Hanesch, U., Pfrommer, U., Grubb, B.D. *et al.* (1993). The proportion of CGRP-immunoreactive and substance P-mRNA containing dorsal root ganglion cells is increased by a unilateral inflammation of the ankle joint of the rat. Regul. Pept. 46, 202–203.

Hartung, H.P. and Toyka, K.V. (1983). Activation of macrophages by substance P: induction of oxidative burst and thromboxane release. Eur. J. Pharmacol. 89, 301–305.

Hartung, H.P., Wolters, K. and Toyka, K.V. (1986). Substance P: binding properties and studies on cellular responses in guinea-pig macrophages. J. Immunol. 136, 3856–3863.

Hashimoto, S., Yamanaka, K., Inoue, A. et al. (1993). Preprotachykinin mRNA expression in the synovial tissue of chronic arthritis. Regul. Pept. 46, 193–194.

Hibbs, J.B., Jr., Taintor, R.R., Vavrin, Z. et al. (1988). Nitric oxide: a cytotoxic activated macrophage effector molecule [published erratum appears in Biochem. Biophys. Res. Commun. (1989) 158(2), 624]. Biochem. Biophys. Res. Commun. 157, 87–94.

Hislop, A.A., Wharton, J., Allen, K.M. et al. (1990). Immunohistochemical localization of peptide-containing nerves in human airways: age-related changes. Am. J. Respir. Cell Mol. Biol. 3, 191–198.

Hoover, D.B. and Hancock, J.C. (1987). Autoradiographic localization of substance-P binding sites in guinea-pig airways. J. Autonom. Nerv. Syst. 19, 171–174.

Howarth, P.H., Djukanovic, R., Wilson, J.W. et al. (1991). Mucosal nerves in endobronchial biopsies in asthma and non-asthma. Int. Arch. Allergy Appl. Immunol. 94, 330–333.

Hua, X.Y., Theodorsson-Noreim, E., Brodin, E. et al. (1985). Multiple tachykinins (neurokinin A, neuropeptide K and substance P) in capsaicin-sensitive sensory neurons in the guinea-pig. Regul. Pept. 13, 1–19.

Huang, P.L., Dawson, T.M., Bredt, D.S. et al. (1993). Targeted disruption of the neuronal nitric oxide synthase gene. Cell 75, 1273–1286.

Hulsmann, A.R., Jongejan, R.C., Raatgeep, H.R. et al. (1993). Epithelium removal and peptidase inhibition enhance relaxation of human airways to vasoactive intestinal peptide. Am. Rev. Respir. Dis. 147, 1483–1486.

Ichinose, M., Nakajima, N., Takahashi, T. et al. (1992). Protection against bradykinin-induced bronchoconstriction in asthmatic patients by neurokinin receptor antagonist. Lancet 340, 1248–1251.

Ignarro, L.J., Byrns, R.E., Buga, G.M. et al. (1987). Mechanisms of endothelium-dependent vascular smooth muscle relaxation elicited by bradykinin and VIP. Am. J. Physiol. 253, H1074–H1082.

Ihara, H. and Nakanishi, S. (1990). Selective inhibition of expression of the substance P receptor mRNA in pancreatic acinar AR42J cells by glucocorticoids. J. Biol. Chem. 265, 22441–22445.

Ingenito, E.I., Pliss, L.B., Martins, M.A. et al. (1991). Effects of capsaicin on mechanical, cellular, and mediator responses to antigen in sensitized guinea-pigs. Am. Rev. Respir. Dis. 143, 572–577.

Ingenito, E.P., Mark, L., Lilly, C. et al. (1994). Autonomic regulation of tissue resistance in the guinea-pig lung. J. Appl. Physiol., 78; 1382–1387.

Inoue, N., Magari, S. and Sakanaka, M. (1990). Distribution of peptidergic nerve fibers in rat bronchus-associated lymphoid tissue: light microscopic observations. Lymphology 23, 155–160.

Ishihara, T., Shigemoto, R., Mori, K. et al. (1992). Functional expression and tissue distribution of a novel receptor for vasoactive intestinal polypeptide. Neuron 8, 811–819.

Jacoby, D.B., Tamaoki, J., Borson, D.B. et al. (1988). Influenza infection causes airway hyperresponsiveness by decreasing enkephalinase. J. Appl. Physiol. 64, 2653–2658.

Jansen, A., Drazen, J., Osborne, J.A. et al. (1992). The relaxant properties in guinea-pig airways of S-nitrosothiols. J. Pharmacol. Exp. Ther. 261, 154–160.

Johnson, A.R., Coalson, J.J., Ashton, J. et al. (1985). Neutral endopeptidase in serum samples from patients with adult respiratory distress syndrome. Comparison with angiotensin-converting enzyme. Am. Rev. Respir. Dis. 132, 1262–1267.

Jonakait, G.M., Schotland, S. and Hart, R.P. (1991). Effects of lymphokines on substance P in injured ganglia of the peripheral nervous system. Ann. NY Acad. Sci. 632, 19–30.

Joos, G.F. and Pauwels, R.A. (1993). The in vivo effect of tachykinins on airway mast cells of the rat. Am. Rev. Respir. Dis. 148, 922–926.

Joos, G., Pauwels, R. and Van der Straeten, M. (1987). Effect of inhaled substance P and neurokinin A on the airways of normal and asthmatic subjects. Thorax 42, 779–783.

Kermode, J.C., Deluca, A.W., Zilberman, A. et al. (1992). Evidence for the formation of a functional complex between vasoactive intestinal peptide, its receptor, and GS in lung membranes. J. Biol. Chem. 267, 3382–3388.

Kharitonov, S.A., Yates, D., Robbins, R.A. et al. (1994). Increased nitric oxide in exhaled air of asthmatic patients. Lancet 343, 133–135.

Kitamura, S., Ishihara, Y. and Said, S.I. (1980). Effect of VIP, phenoxybenzamine and prednisolone on cyclic nucleotide content of isolated guinea-pig lung and trachea. Eur. J. Pharmacol. 67, 219–223.

Kobzik, L., Bredt, D.S., Lowenstein, C.J. et al. (1993). Nitric oxide synthase in human and rat lung – immunocytochemical and histochemical localization. Am. J. Respir. Cell Molec. Biol. 9, 371–377.

Kohrogi, H., Graf, P.D., Sekizawa, K. et al. (1988). Neutral endopeptidase inhibitors potentiate substance P- and capsaicin-induced cough in awake guinea-pigs. J. Clin. Invest. 82, 2063–2068.

Kotani, H., Hoshimaru, M., Nawa, H. et al. (1986). Structure and gene organization of bovine neuromedin K precursor. Proc. Natl. Acad. Sci. USA 83, 7074–7078.

Krause, J.E., Chirgwin, J.M., Carter, M.S. et al. (1987). Three rat preprotachykinin mRNAs encode the neuropeptides substance P and neurokinin A. Proc. Natl. Acad. Sci. USA 84, 881–885.

Krause, J.E., Bu, J.Y., Takeda, Y. et al. (1993). Structure, expression and second messenger-mediated regulation of the human and rat substance P receptors and their genes. Regul. Pept. 46, 59–66.

Kummer, W., Fischer, A., Kurkowski, R. et al. (1992). The sensory and sympathetic innervation of guinea-pig lung and trachea as studied by retrograde neuronal tracing and double-labelling immunohistochemistry. Neuroscience 49, 715–737.

Kwatra, M.M., Schwinn, D.A., Schrerurs, J. et al. (1993). The substance P receptor which couples $G_{q/11}$ is a substrate of β-adrenergic receptor kinase 1 and 2. J. Biol. Chem. 268, 9161–9164.

Kwon, O.J., Baraniuk, J.N., Kuo, H.P. *et al.* (1994). Effect of cigarette smoke and capsaicin on neutral endopeptidase. Am. J. Respir. Crit. Care Med. 145, A47 (Abstract).

Lai, Y.L. (1991). Endogenous tachykinins in antigen-induced acute bronchial responses of guinea-pigs. Exp. Lung Res. 17, 1047–1060.

Laitinen, A., Partanen, M., Hervonen, A. *et al.* (1985). VIP like immunoreactive nerves in human respiratory tract. Light and electron microscopic study. Histochemistry 82, 313–319.

Lamperti, E.D., Rosen, K.M. and Villa-Komaroff, L. (1991). Characterization of the gene and messages for vasoactive intestinal polypeptide (VIP) in rat and mouse. Brain Res. Mol. Brain Res. 9, 217–231.

Laneuville, O., Couture, R. and Paceasciak, C.R. (1992). Neurokinin A-induced contraction of guinea-pig isolated trachea – potentiation by hepoxilins. Br. J. Pharmacol. 107, 808–812.

Laube, B.L., Norman, P.S. and Adams, G.K. (1992). The effect of aerosol distribution on airway responsiveness to inhaled methacholine in patients with asthma. J. Allerg. Clin. Immunol. 89, 510–518.

Lawrence, I.D., Warner, J.A., Cohan, V.L. *et al.* (1987). Purification and characterization of human skin mast cells. Evidence for human mast cell heterogeneity. J. Immunol. 139, 3062–3069.

Lee, L.Y. and Morton, R.F. (1993). Histamine enhances vagal pulmonary C-fiber responses to capsaicin and lung inflation. Respir. Physiol. 93, 83–96.

Lei, Y.H., Barnes, P.J. and Rogers, D.F. (1992). Inhibition of neurogenic plasma exudation in guinea-pig airways by CP-96, 345, a new non-peptide Nk(1) receptor antagonist. Br. J. Pharmacol. 105, 261–262.

Leroux, P., Vaudry, H., Fournier, A. *et al.* (1984). Characterization and localization of vasoactive intestinal peptide receptors in the rat lung. Endocrinology 114, 1506–1512.

Leurs, R., Brozius, M.M., Jansen, W. *et al.* (1991). Histamine H₁-receptor-mediated cyclic GMP production in guinea-pig lung tissue is an L-arginine-dependent process. Biochem. Pharmacol. 42, 271–277.

Leys, K., Morice, A.H., Madonna, O. *et al.* (1986). Autoradiographic localisation of VIP receptors in human lung. FEBS Lett. 199, 198–202.

Lilly, C.M., Martins, M.A. and Drazen, J.M. (1993a). Peptidase modulation of vasoactive intestinal peptide pulmonary relaxation in tracheal superfused guinea-pig lungs. J. Clin. Invest. 91, 235–243.

Lilly, C.M., Stamler, J.S., Gaston, B. *et al.* (1993b). Modulation of vasoactive intestinal peptide pulmonary relaxation by NO in tracheally superfused guinea-pig lungs. Am. J. Physiol. 265, L410–L415.

Lilly, C.M., Bai, T.R., Shore, S.A. *et al.* (1994). Neuropeptide content of lungs from asthmatic and non-asthmatic patients. Am. J. Respir. Crit. Care Med., 151; 548–553.

Linder, S., Barkhem, T., Norberg, A. *et al.* (1987). Structure and expression of the gene encoding the vasoactive intestinal peptide precursor.' Proc. Natl. Acad. Sci. USA 84, 605–609.

Long, N.C. and Shore, S.A. (1994). Effect of capsaicin pretreatment on bronchoalveolar cells recovered from rats during SO₂ induction of chronic bronchitis. Am. J. Respir. Crit. Care Med. 149, A72 (Abstract).

Long, N.C., Frevert, C.W. and Shore, S.A. (1993). Capsaicin

pretreatment increases inflammatory response to intratracheal instillation of endotoxin in rats. Regul. Pept. 46, 208–210.

Lorton, D., Bellinger, D.L., Felten, S.Y. *et al.* (1991). Substance P innervation of spleen in rats: nerve fibers associated with lymphocytes and macrophages in specific compartments of the spleen. Brain Behav. Immun. 5, 29–40.

Lotvall, J.O., Lemen, R.J., Hui, K.P. *et al.* (1990). Airflow obstruction after substance P aerosol: contribution of airway and pulmonary edema. J. Appl. Physiol. 69, 1473–1478.

Lotz, M., Vaughan, J.H. and Carson, D.A. (1988). Effect of neuropeptides on production of inflammatory cytokines by human monocytes. Science 241, 1218–1221.

Lundberg, J.M., Brodin, E., Hua, X. *et al.* (1984a). Vascular permeability changes and smooth muscle contraction in relation to capsaicin-sensitive substance P afferents in the guinea-pig. Acta Physiol. Scand. 120, 217–227.

Lundberg, J.M., Hokfelt, T., Martling, C.R. *et al.* (1984b). Substance P-immunoreactive sensory neves in the lower respiratory tract of various mammals including man. Cell Tissue Res. 235, 251–261.

Lundberg, J.M., Anggard, A., Pernow, J. *et al.* (1985). Neuropeptide Y-, substance P- and VIP-immunoreactive nerves in cat spleen in relation to autonomic vascular and volume control. Cell Tissue Res. 239, 9–18.

Lutz, E.M., Sheward, W.J., West, K.M. *et al.* (1993a). The VIP(2) receptor: molecular characterization of a cDNA encoding a novel receptor for vasoactive intestinal peptide. FEBS Lett. 334, 3–8.

Lutz, E.N., Sheward, W.J., West, K.M. *et al.* (1993b). The VIP2 receptor: molecular characterisation of a cDNA encoding a novel receptor for vasoactive intestinal peptide. FEBS Lett. 334, 3–8.

MacDonald, M.R., Takeda, J., Rice, C.M. *et al.* (1989). Multiple tachykinins are produced and secreted upon post-translational processing of the three substance P precursor proteins, alpha-, beta-, and gamma-preprotachykinin. Expression of the preprotachykinins in AtT-20 cells infected with vaccinia virus recombinants. J. Biol. Chem. 264, 15578–15592.

Maggi, C.A. (1993). Tachykinin receptors and airway pathophysiology. Eur. Respir. J. 6, 735–742.

Maggi, C.A., Patacchini, R., Perretti, F. *et al.* (1990). The effect of thiorphan and epithelium removal on contractions and tachykinin release produced by activation of capsaicin-sensitive afferents in the guinea-pig isolated bronchus. Naunyn Schmiedeberas Arch. Pharmacol. 341, 74–79.

Mantyh, C.R., Gates, T.S., Zimmerman, R.P. *et al.* (1988). Receptor binding sites for substance P, but not substance K or neuromedin K, are expressed in high concentrations by arterioles, venules, and lymph nodules in surgical specimens obtained from patients with ulcerative colitis and Crohn disease. Proc. Natl. Acad. Sci. USA 85, 3235–3239.

Mantyh, C.R., Vigna, S.R. and Maggio, J.E. (1994). Receptor involvement in pathology and disease. In The Tachykinin Receptors (eds. S. Buck and E. Ottowa). Humana Press, Potana, NJ, USA.

Manzini, S. (1992). Bronchodilatation by tachykinins and capsaicin in the mouse main bronchus. Br. J. Pharmacol. 105, 968–972.

Manzini, S. and Meini, S. (1991). Involvement of capsaicin-sensitive nerves in the bronchomotor effects of arachidonic acid and melittin – A possible role for lipoxin-A₄. Br. J. Pharmacol. 103, 1027–1032.

Manzini, S., Maggi, C.A., Geppetti, P. *et al.* (1987). Capsaicin desensitization protects from antigen-induced bronchospasm in conscious guinea-pigs. Eur. J. Pharmacol. 138, 307–308.

Martins, M.A., Shore, S.A., Gerard, N.P. *et al.* (1990). Peptidase modulation of the pulmonary effects of tachykinins in tracheal superfused guinea pig lungs. J. Clin. Invest. 85, 170–176.

Martins, M.A., Shore, S.A. and Drazen, J.M. (1991a). Release of tachykinins by histamine, methacholine, PAF, LTD_4, and substance-P from guinea-pig lungs. Am. J. Physiol. 261, L449–L455.

Martins, M.A., Shore, S.A. and Drazen, J.M. (1991b). Capsaicin-induced release of tachykinins: effects of enzyme inhibitors. J. Appl. Physiol. 70, 1950–1956.

Martling, C.R., Saria, A., Andersson, P. *et al.* (1984). Capsaicin pretreatment inhibits vagal cholinergic and non-cholinergic. Naunyn Schmiedebergs Arch. Pharmacol. 325, 343–348.

Massaro, A., Gaston, B. Fanta, C.H., Stamler, J.S. *et al.* (1994). The effects of treatment with steroids on the levels of exhaled nitric oxide in patients with asthma. Am. J. Respir. Crit. Care Med., in press (1995).

Masu, Y., Nakayama, K., Tamaki, H. *et al.* (1987). cDNA cloning of bovine substance-K receptor through oocyte expression system. Nature 329, 836–838.

Matis, W.L., Lavker, R.M. and Murphy, G.F. (1990). Substance P induces the expression of an endothelial-leukocyte adhesion molecule by microvascular endothelium. J. Invest. Dermatol. 94, 492–495.

Matsuse, T., Thomson, R.J., Chen, X.R. *et al.* (1991). Capsaicin inhibits airway hyperresponsiveness but not lipoxygenase activity or eosinophilia after repeated aerosolized antigen in guinea-pigs. Am. Rev. Respir. Dis. 144, 368–372.

Matsuzaki, Y., Hamasaki, Y. and Said, S.I. (1980). Vasoactive intestinal peptide: a possible transmitter of nonadrenergic relaxation of guinea-pig airways. Science 210, 1252–1253.

McCarson, K.E. and Krause, J.E. (1994). Nk1 and Nk3 type tachykinin receptor mRNA expression in the rat spinal cord dorsal horn is increased during adjuvant- or formalin-induced nociception. J. Neurosci. 14, 712–720.

Moore, T.C., Lami, J.L. and Spruck, C.H. (1989). Substance P increases lymphocyte traffic and lymph flow through peripheral lymph nodes of sheep. Immunology 67, 109–114.

Morris, S.L., Walsh, R.C. and Hansen, J.N. (1984). Identification and characterization of some bacterial membrane sulfhydryl groups which are targets of bacteriostatic and antibiotic action. J. Biol. Chem. 259, 13590–13594.

Murlas, C.G., Lang, Z., Williams, G.J. *et al.* (1992). Aerosolized neutral endopeptidase reverses ozone-induced airway hyperreactivity to substance P. J. Appl. Physiol. 72, 1133–1141.

Murthy, K.S., Zhang, K.M., Jin, J.G. *et al.* (1993). VIP-mediated G protein-coupled Ca^{2+} influx activates a constitutive NOS in dispersed gastric muscle cells. Am. J. Physiol. 265, G660–671.

Muscettola, M. and Grasso, G. (1990). Neuropeptide modulation of interferon gamma production. Int. J. Neurosci. 51, 189–191.

Nakajima, Y., Tsuchida, K., Negishi, M. *et al.* (1992). Direct linkage of three tachykinin receptors to stimulation of both phosphatidylinositol hydrolysis and cyclic AMP cascades in

transfected Chinese hamster ovary cells. J. Biol. Chem. 267, 2437–2442.

Nakanishi, S., Nakajima, Y. and Yokota, Y. (1993). Signal transduction and ligand-binding domains of the tachykinin receptors. Regul. Pept. 46, 37–42.

Naline, E., Devillier, P., Drapeau, G. *et al.* (1989). Characterization of neurokinin effects and receptor selectivity in human isolated bronchi. Am. Rev. Respir. Dis. 140, 679–686.

Nathan, C. (1992). Nitric oxide as a secretory product of mammalian cells. FASEB J. 6, 3051–3064.

Nawa, H., Kotani, H. and Nakanishi, S. (1984). Tissue-specific generation of two preprotachykinin mRNAs from one gene by alternative RNA splicing. Nature 312, 729–734.

Nijkamp, F.P., Vanderlinde, H.J. and Folkerts, G. (1993). Nitric oxide synthesis inhibitors induce airway hyperresponsiveness in the guinea-pig *in vivo* and *in vitro* – Role of the epithelium. Am. Rev. Respir. Dis. 148, 727–734.

Nilsson, G., Dahlberg, K., Brodin, E. *et al.* (1975). Distribution and constrictor effect of substance P in guinea pig tracheobronchial tissue. In Substance P (eds U.S. von Euler and B. Pernow), pp. 75–81. Raven Press, New York.

Nilsson, G., Alving, K., Ahlstedt, S. *et al.* (1990). Peptidergic innervation of rat lymphoid tissue and lung: relation to mast cells and sensitivity to capsaicin and immunization. Cell Tissue Res. 262, 125–133.

Nohr, D. and Weihe, E. (1991). The neuroimmune link in the bronchus-associated lymphoid tissue BALT) of cat and rat: peptides and neural markers. Brain Behav. Immun. 5, 84–101.

O'Dell, T.J., Huang, P.L., Dinerman, J.L. *et al.* (1994). Endothelial NOS and the blockade of LTP by NOS inhibitors in mice lacking neuronal NOS. Science 265, 542–546.

Ollerenshaw, S., Jarvis, D., Woolcock, A. *et al.* (1989). Absence of immunoreactive vasoactive intestinal polypeptide in tissue from the lungs of patients with asthma [see comments]. N. Engl. J. Med. 320, 1244–1248.

Ollerenshaw, S.L., Jarvis, D., Sullivan, C.E. *et al.* (1991). Substance P immunoreactive nerves in airways from asthmatics and nonasthmatics. Eur. Respir. J. 4, 673–682.

Palmer, J.B., Cuss, F.M. and Barnes, P.J. (1986). VIP and PHM and their role in nonadrenergic inhibitory responses in isolated human airways. J. Appl. Physiol. 61, 1322–1328.

Paul, S. and Said, S.I. (1987). Characterization of receptors for vasoactive intestinal peptide solubilized from the lung. J. Biol. Chem. 262, 158–162.

Payan, D.G., Brewster, D.R. and Goetzl, E.J. (1983). Specific stimulation of human T lymphocytes by substance P. J. Immunol. 131, 1613–1615.

Peatfield, A.C. and Richardson, P.S. (1983). Evidence for non-cholinergic, non-adrenergic nervous control of mucus secretion into the cat trachea. J. Physiol. (Lond) 342, 335–345.

Perretti, F. and Manzini, S. (1993). Activation of capsaicin-sensitive sensory fibers modulates PAF-induced bronchial hyperresponsiveness in anesthetized guinea pigs. Am. Rev. Respir. Dis. 148, 927–931.

Persson, M.G., Zetterstrom, O., Agrenius, V. *et al.* (1994). Single-breath nitric oxide measurements in asthmatic patients and smokers. Lancet 343, 146–147.

Piedimonte, G., Hoffman, J.I.E., Husseini, W.K. *et al.* (1992). Effect of neuropeptides released from sensory nerves on blood flow in the rat airway microcirculation. J. Appl. Physiol. 72, 1563–1570.

Polak, J.M. and Bloom, S.R. (1982). Regulatory peptides and neuron-specific enolase in the respiratory tract of man and other mammals. Exp. Lung Res. 3, 313–328.

Prior, M., Green, G., Lopez, A. et al. (1990). Capsaicin pretreatment modifies hydrogen sulphide-induced pulmonary injury in rats. Toxicol. Pathol. 18, 279–288.

Pujol, J.L., Bousquet, J., Grenier, J. et al. (1989). Substance P activation of bronchoalveolar macrophages from asthmatic patients and normal subjects. Clin. Exp. Allergy 19, 625–628.

Radi, R., Beckman, J.S., Bush, K.M. et al. (1991). Peroxynitrite oxidation of sulfhydryls. The cytotoxic potential of superoxide and nitric oxide. J. Biol. Chem. 266, 4244–4250.

Rameshwar, P., Gascon, P. and Ganea, D. (1993). Stimulation of IL-2 production in murine lymphocytes by substance-P and related tachykinins. J. Immunol. 151, 2484–2496.

Ray, D.W., Hernandez, C., Leff, A.R. et al. (1989). Tachykinins mediate bronchoconstriction elicited by isocapnic hyperpnea in guinea-pigs. J. Appl. Physiol. 66, 1108–1112.

Rhoden, K.J. and Barnes, P.J. (1990). Potentiation of nonadrenergic neural relaxation in guinea pig airways by a cyclic cAMP phosphodiesterase inhibitor. J. Pharmacol. Exp. Ther. 252, 396–402.

Robberecht, P., Chatelain, P., De Neef, P. et al. (1981). Presence of vasoactive intestinal peptide receptors coupled to adenylate cyclase in rat lung membranes. Biochim. Biophys. Acta 678, 76–82.

Rogers, D.F., Aursudkij, B. and Barnes, P.J. (1989). Effects of tachykinins on mucus secretion in human bronchi in vitro. Eur. J. Pharmacol. 174, 283–286.

Rola-Pleszczynski, M., Bolduc, D. and St-Pierre, S. (1985). The effects of vasoactive intestinal peptide on human natural killer cell function. J. Immunol. 135, 2569–2573.

Ryan, U.S., Ryan, J.W. and Crutchley, D.J. (1985). The pulmonary endothelial surface. Fed. Proc. 44, 2603–2609.

Said, S.I. (1988). Vasoactive intestinal peptide in the lung. Ann. NY Acad. Sci. 527, 450–464.

Sakamoto, T., Barnes, P.J. and Chung, K.F. (1993). Effect of CP-96, 345, a non-peptide Nk1 receptor antagonist, against substance P-, bradykinin- and allergen-induced airway microvascular leakage and bronchoconstriction in the guinea-pig. Eur. J. Pharmacol. 231, 31–38.

Saria, A., Martling, C.R., Yan, Z. et al. (1988). Release of multiple tachykinins from capsaicin-sensitive sensory nerves in the lung by bradykinin, histamine, dimethylphenyl piperazinium, and vagal nerve stimulation. Am. Rev. Respir. Dis. 137, 1330–1335.

Schmidt, H.H., Gagne, G.D., Nakane, M. et al. (1992). Mapping of neural nitric oxide synthase in the rat suggests frequent co-localization with NADPH diaphorase but not with soluble guanylyl cyclase, and novel paraneural functions for nitrinergic signal transduction. J. Histochem. Cytochem. 40, 1439–1456.

Sekizawa, K., Tamaoki, J., Nadel, J.A. et al. (1987). Enkephalinase inhibitor potentiates substance P- and electrically induced contraction in ferret trachea. J. Appl. Physiol. 63, 1401–1405.

Sertl, K., Wiedermann, C.J., Kowalski, M.L. et al. (1988). Substance P: the relationship between receptor distribution in rat lung and the capacity of substance P to

stimulate vascular permeability. Am. Rev. Respir. Dis. 138, 151–159.

Sheppard, D., Thompson, J.E., Scypinski, L. et al. (1988). Toluene diisocyanate increases airway responsiveness to substance P and decreases airway neutral endopeptidase. J. Clin. Invest. 81, 1111–1115.

Shigemoto, R., Yokota, Y., Tsuchida, K. et al. (1990). Cloning and expression of a rat neuromedin K receptor cDNA. J. Biol. Chem. 265, 623–628.

Shikada, K., Yamamoto, A. and Tanaka, S. (1991). Effects of phosphodiesterase inhibitors on vasoactive intestinal peptide-induced relaxation of isolated guinea-pig trachea. Eur. J. Pharmacol. 195, 389–394.

Shore, S.A. and Drazen, J.M. (1988). Airway responses to substance P and substance P fragments in the guinea pig. Pulm. Pharmacol. 1, 113–118.

Shore, S.A. and Drazen, J.M. (1989a). Enhanced airway responses to substance P after repeated challenge in guinea-pigs. J. Appl. Physiol. 66, 955–961.

Shore, S.A. and Drazen, J.M. (1989b). Degradative enzymes modulate airway responses to intravenous neurokinins A and B. J. Appl. Physiol. 67, 2504–2511.

Shore, S.A. and Drazen, J.M. (1991). Relative bronchoconstrictor activity of neurokinin-A and neurokinin-A fragments in guinea-pigs. J. Appl. Physiol. 71, 452–457.

Shore, S.A., Stimler-Gerard, N.P., Coats, S.R. et al. (1988). Substance P-induced bronchoconstriction in the guinea-pig. Enhancement by inhibitors of neutral metalloendopeptidase and angiotensin-converting enzyme. Am. Rev. Respir. Dis. 137, 331–336.

Shore, S.A., Martins, M.A. and Drazen, J.M. (1993). Pulmonary mechanical responses to intravenously administered capsaicin in guinea-pigs: effect of peptidase inhibitors. Pulm. Pharmacol. 6, 193–199.

Skidgel, R.A., Jackman, H.L. and Erdos, E.G. (1991). Metabolism of substance P and bradykinin by human neutrophils. Biochem. Pharmacol. 41, 1335–1344.

Skofitsch, G., Savitt, J.M. and Jacobowitz, D.M. (1985). Suggestive evidence for a functional unit between mast cells and substance P fibers in the rat diaphragm and mesentery. Histochemistry 82, 5–8.

Smith, T.K., Gibson, C.L., Howlin, B.J. et al. (1991). Active transport of amino acids by gamma-glutamyl transpeptidase through Caco-2 cell monolayers. Biochem. Biophys. Res. Commun. 178, 1028–1035.

Solway, J., Kao, B.M., Jordan, J.E. et al. (1993). Tachykinin receptor antagonists inhibit hyperpnea-induced bronchoconstriction in guinea-pigs. J. Clin. Invest. 92, 315–323.

Springall, D.R., Bloom, S.R. and Polak, J.M. (1988). Distribution, nature, and origin of peptide containing nerves in mammalian airways. In "The Airways: Neural Control in Health and Disease". (eds M.A. Kaliner and P.J. Barnes), pp. 299–341. Marcel Dekker, New York.

Springall, D.R., Albeli, L., Belvisi, M. et al. (1994). Nitric oxide synthase compared with neuropeptide-immunoreactive nerves in human lung. Am. J. Respir. Crit. Care Med. 149, A859 (Abstract).

Sreedharan, S.P., Robichon, A., Peterson, K.E. et al. (1991). Cloning and expression of the human vasoactive intestinal peptide receptor. Proc. Natl. Acad. Sci. USA 88, 4986–4990.

Sreedharan, S.P., Patel, D.R., Huang, J.X. et al. (1993a).

Cloning and functional expression of a human neuroendocrine vasoactive intestinal peptide receptor. Biochem. Biophys. Res. Commun. 93, 546–553.

Sreedharan, S.P., Robichon, A., Peterson, K.E. et al. (1993b). Cloning and expression of the human vasoactive intestinal peptide receptor (Vol 88, Pg 4986, 1991) – Correction. Proc. Natl. Acad. Sci. USA 90, 9233.

Stamler, J.S., Singel, D.J. and Loscalzo, J. (1992). Biochemistry of nitric oxide and its redox-activated forms. Science 258, 1898–1902.

Stead, R.H., Tomioka, M., Quinonez, G. et al. (1987). Intestinal mucosal mast cells in normal and nematode-infected rat intestines are in intimate contact with peptidergic nerves. Proc. Natl. Acad. Sci. USA 84, 2975–2979.

Sterner-Kock, A., Green, J.F., Schelegle, E.S. et al. (1994). Protective role of afferent bronchial and pulmonary C fibers in acute O_3 induced pulmonary injury. Am. J. Respir. Crit. Care Med. 147, A485 (Abstract).

Sundler, F., Brodin, E., Ekblad, E. et al. (1985). Sensory nerve fibers: Distribution of substance P, neurokinin A and calcitonin gene-related peptide. In "Tachykinin Antagonists". (eds R. Hakason and F. Sundler). pp. 3–14, Elsevier Science, New York.

Takeda, Y., Takeda, J., Smart, B.M. et al. (1990). Regional distribution of neuropeptide gamma and other tachykinin peptides derived from the substance P gene in the rat. Regul. Pept. 28, 323–333.

Tam, E.K., Franconi, G.M., Nadel, J.A. et al. (1990). Protease inhibitors potentiate smooth muscle relaxation induced by vasoactive intestinal peptide in isolated human bronchi. Am. J. Respir. Cell Mol. Biol. 2, 449–452.

Tamaoki, J., Sakai, N., Isono, K. et al. (1991). Effect of neutral endopeptidase inhibition on substance-P-induced increase in short-circuit current of canine cultured tracheal epithelium. Int. Arch. Allergy Appl. Immunol. 95, 169–173.

Tanaka, D.T. and Grunstein, M.M. (1984). Mechanisms of substance P-induced contraction of rabbit airway smooth muscle. J. Appl. Physiol. 57, 1551–1557.

Tanaka, T., McRae, B.J., Cho, K. et al. (1983). Mammalian tissue trypsin-like enzymes. Comparative reactivities of human skin tryptase, human lung tryptase, and bovine trypsin with peptide 4-nitroanilide and thioester substrates. J. Biol. Chem. 258, 13552–13557.

Taton, G., Delhaye, M., Camus, J.C. et al. (1981). Characterization of the VIP- and secretin-stimulated adenylate cyclase system from human lung. Pflugers Arch. 391, 178–182.

Thompson, D.C., Altiere, R.J. and Diamond, L. (1988). The effects of antagonists of vasoactive intestinal peptide on nonadrenergic noncholinergic inhibitory responses in feline airways. Peptides 9, 443–447.

Tsukada, T., Fink, J.S., Mandel, G. et al. (1987). Identification of a region in the human vasoactive intestinal polypeptide gene responsible for regulation by cyclic AMP. J. Biol. Chem. 262, 8743–8747.

Turner, A.J., Matsas, R. and Kenny, A.J. (1985). Are there neuropeptide-specific peptidases? Biochem. Pharmacol. 34, 1347–1356.

Turner, C.R., Lackey, M.N., Quinlan, M.F. et al. (1991). Therapeutic intervention in a rat model of adult respiratory distress syndrome. 2. lipoxygenase pathway inhibition. Circ. Shock 34, 263–269.

Turner, C.R., Stow, R.B., Talerico, S.D. et al. (1993). Protective role for neuropeptides in acute pulmonary response to acrolein in guinea-pigs. J. Appl. Physiol. 75, 2456–2465.

Uchida, Y., Ohtsuk, M., Goto, K. et al. (1987). Neurokinin A as a potent bronchoconstrictor. Am. Rev. Respir. Dis. 136, 718–721.

Uddman, R., Alumets, J., Densert, O. et al. (1978). Occurrence and distribution of VIP nerves in the nasal mucosa and tracheobronchial wall. Acta Otolaryngol. (Stockh) 86, 443–448.

Undem, B.J. and Weinreich, D. (1993). Electrophysiological properties and chemosensitivity of guinea-pig nodose ganglion neurons in vitro. J. Autonon. Nerv. Syst. 44, 17–33.

Undem, B.J., Hubbard, W. and Weinreich, D. (1993). Immunologically induced neuromodulation of guinea-pig nodose ganglion neurons. J. Autonon. Nerv. Syst. 44, 35–44.

Vanschayck, C.P. and Vanherwaarden, C.L.A. (1993). Do bronchodilators adversely affect the prognosis of bronchial hyperresponsiveness? Thorax 48, 470–473.

Venugopalan, C.S., Said, S.I. and Drazen, J.M. (1984). Effect of vasoactive intestinal peptide on vagally mediated tracheal pouch relaxation. Respir. Physiol. 56, 205–216.

Wallengren, J. and Hakanson, R. (1987). Effects of substance P, neurokinin A and calcitonin gene-related peptide in human skin and their involvement in sensory nerve-mediated responses. Eur. J. Pharmacol. 143, 267–273.

Webber, S.E. and Widdicombe. J.G. (1987). The effect of vasoactive intestinal peptide on smooth muscle tone and mucus secretion from the ferret trachea. Br. J. Pharmacol. 91, 139–148.

Weihe, E. and Krekel, J. (1991). The neuroimmune connection in human tonsils. Brain Behav. Immun. 5, 41–54.

White, S.R., Hershenson, M.B., Sigrist, K.S. et al. (1993). Proliferation of guinea-pig tracheal epithelial cells induced by calcitonin gene-related peptide. Am. J. Respir. Cell Mol. Biol. 8, 592–596.

Wiener, C. (1993). Ventilatory management of respiratory failure in asthma. J. Am. Med. Assoc. 269, 2128–2131.

Wiik, P. (1989). Vasoactive intestinal peptide inhibits the respiratory burst in human monocytes by a cyclic AMP-mediated mechanism. Regul. Pept. 25, 187–197.

Wong, L.B., Miller, I.F. and Yeates, D.B. (1991). Pathways of substance P stimulation of canine tracheal ciliary beat frequency. J. Appl. Physiol. 70, 267–273.

Yamagami, T., Ohsawa, K., Nishizawa, M. et al. (1988). Complete nucleotide sequence of human vasoactive intestinal peptide/PHM-27 gene and its inducible promoter. Ann. NY Acad. Sci. 527, 87–102.

Yang, X.X., Powell, W.S. and Martin, J.M. (1994). Tachykinins trigger peptide-leukotriene release in hyperpnea-induced bronchoconstriction in the guinea-pig. Am. J. Respir. Crit. Care Med. 149, A892 (Abstract).

Yokota, Y., Sasai, Y., Tanaka, K. et al. (1989). Molecular characterization of a functional cDNA for rat substance P receptor. J. Biol. Chem. 264, 17649–17652.

8. The Microvasculature as a Participant in Inflammation

Jeffrey J. Bowden *and* Donald M. McDonald

1. Background

Microvascular endothelial cells play a key role in coordinating the inflammatory response of the airway mucosa. Increased endothelial permeability and adhesion of leukocytes to the luminal surface of the endothelium are characteristic features of inflammation. By governing the amount of plasma leakage and the emigration of inflammatory cells, endothelial cells can influence the magnitude of the inflammatory response. As evidence of their importance in the inflammatory response, endothelial cells are specifically targeted by many drugs used to treat inflammatory airway disease.

In this chapter we emphasize the increase in endothelial

permeability and plasma leakage that occur in the inflammatory response of the airway mucosa. We review observations made in experimental models of airway inflammation and in human disease with a particular focus on asthma. Our focus on endothelial permeability and plasma leakage is intended to complement the reviews of other aspects of endothelial cell biology, such as leukocyte adhesion and emigration, elsewhere in this series (Beynon and Haskard, 1994; Brain, 1994; Rampart, 1994; Wardlaw and Walsh, 1994).

Asthma is a respiratory illness characterized by variable and reversible airflow obstruction. Over 100 years ago, Osler, in his influential *Textbook of Medicine* (Osler, 1892), concluded that airway wall edema, bronchoconstriction and mucus plugging are responsible for the airflow obstruction in asthma. Since that time, edema of the airway mucosa has been assumed to be one of the central features of asthma, but comparatively few studies have addressed the issue directly. Therefore, there is little information about the onset, duration, magnitude, location, mechanism, consequences and management of mucosal edema in asthma.

In view of the potential clinical importance of edema of the airway mucosa, we review the evidence that vascular permeability can increase in the airways of asthmatics, assess the methods that have been used to study this change, and examine evidence that functional disturbances can result from the plasma leakage. We also review what has been learned in studies of animal models regarding the mechanism of plasma leakage and the identity of inflammatory mediators that trigger the leakage. Finally, we discuss experimental evidence showing that plasma leakage can be reversed by anti-asthma drugs.

2. Airway Microvasculature

The current framework for understanding the morphology of the vasculature of the airway mucosa is based on observations made on gross anatomical dissections (Cauldwell *et al.*, 1948; Notkovich, 1957; Daly and Hebb, 1966; Cudkowicz, 1968; Nagaishi, 1972), tissue sections examined by light and electron microscopy (Verloop, 1948, 1949; Weibel, 1960; Wanner, 1989; Laitinen and Laitinen, 1990; McDonald, 1990), angiographs of vessels filled with radio-opaque materials (Berry *et al.*, 1931), blood vessels stained with silver nitrate (McDonald, 1994b) and vascular casts (Seegal and Seegal, 1972; Verloop, 1949; Marchand *et al.*, 1950; McLaughlin *et al.*, 1961, 1966; Sobin *et al.*, 1963; Hughes, 1965; Nagaishi, 1972; Miodonski *et al.*, 1980; Magno and Fishman, 1982; McLaughlin, 1983; Laitinen *et al.*, 1989; Wanner, 1989; McDonald, 1990). Several reviews provide a historical perspective of the anatomical and histological features of the tracheal and bronchial

blood vessels (Aviado, 1965; Daly and Hebb, 1966; Cudkowicz, 1968; Nagaishi, 1972; Cudkowicz, 1979; Deffebach *et al.*, 1987; McDonald, 1990; Butler, 1991; Persson, 1991b; Widdicombe, 1993).

The airways receive their blood supply from the systemic circulation via laryngeal, tracheal and bronchial arteries. In the trachea and large bronchi of humans and other large mammals, branches of arterioles deep in the airway wall supply the airway mucosa. In some peripheral airways, mucosal vessels may be supplied by the pulmonary circulation through vascular anastomoses. The airway mucosa has a rich blood supply, with a particularly extensive vascular network located immediately beneath the epithelium. Mucosal capillaries are tributaries of an extensive system of venules. In humans, a superficial plexus of venules connects with another plexus of venules located deeper in the mucosa.

The relative simplicity of the vasculature of the rat and guinea-pig trachea and bronchi has provided the opportunity to develop a more complete understanding of the three-dimensional architecture of the airway microvasculature. It has also made it possible to determine the size and structural characteristics of mucosal arterioles, capillaries, postcapillary venules and collecting venules, analyze morphological aspects of the response of the vessels to inflammatory mediators, and examine the changes in the responsiveness of the vessels as a result of respiratory tract infections (Miodonski *et al.*, 1980; McDonald *et al.*, 1988; McDonald, 1988a, b, 1990, 1994b). The venules of the rat tracheal mucosa, in which the vessels tend to form a single plexus rather than superficial and deep layers, constitute 51% of the mucosal vessels by length and have 74% of the intravascular volume (McDonald, 1994b). By comparison, mucosal capillaries constitute 26% of the vessels by length and have only 3% of the intravascular volume.

3. Increased Vascular Permeability in Airway Inflammation

3.1 ASSESSMENT OF PLASMA LEAKAGE IN AIRWAYS OF ANIMALS AND HUMANS

In animal models of airway inflammation, plasma leakage has been assessed by measuring the amount and distribution of tracers that leak from the blood. Extravasated albumin, labelled with Evans blue (Saria and Lundberg, 1983) or radioisotopes such as ^{131}I (Erjefält *et al.*, 1985; Persson *et al.*, 1986b), can be readily quantified. Furthermore, high molecular weight dextrans, labelled with fluorescent markers such as fluorescein, can be visualized *in situ* at sites of plasma leakage (Hulstrom and Svensjö, 1979). Alternatively, the particulate tracers Monastral blue (Joris et al., 1982; McDonald, 1988a), India ink (Cotran et al., 1967), and fluorescent microspheres (McDonald, 1994b) can be used to identify sites of increased endothelial permeability in histological sections or tissue whole mounts (McDonald, 1988a). These tracers leak through endothelial gaps but are trapped by the endothelial basement membrane, thereby labelling the sites of leakage.

In human studies, where such direct measurements of plasma leakage usually are not feasible, evidence has come from bronchoscopic examinations of the bronchial mucosa after allergen challenge, measurements of plasma proteins in bronchoalveolar lavage (BAL) fluid, and histological assessments of edema in biopsies or post-mortem specimens of airway tissue (Dunnill, 1960; Dunnill et al., 1969; Ryley and Brogan, 1968; Brogan et al., 1975; Fick et al., 1987). Obviously, the methods used in humans do not give the same information on vascular permeability, plasma leakage and mucosal edema as can be obtained in experimental animals.

3.2 PLASMA LEAKAGE ASSESSED BY BRONCHIAL LAVAGE IN HUMANS

Several bronchoscopic studies have reported that direct application of allergen to the bronchial mucosa rapidly induces edema (Metzger et al., 1985, 1987; Salomonsson et al., 1992). Hunninghake and colleagues described the mucosal response to antigen in asthmatic subjects as "local blanching followed by dilation of the vessels, edema, and finally partial closure of the subsegmental bronchus" (Metzger et al., 1985).

These observations demonstrate how rapidly edema can form in the human airway mucosa, but it is unknown how the changes relate to those found in asthmatics exposed to inhaled allergens. For example, there is a report that allergens applied directly to the airway mucosa increase the amounts of protein and inflammatory cells in BAL fluid more than do inhaled allergens (Calhoun et al., 1993).

Another approach used to assess the amount of plasma leakage in the human airway mucosa is the measurement of plasma proteins in BAL fluid or sputum. For example, the amount of albumin in the sputum of subjects with acute exacerbations of asthma is greater than that in subjects with cystic fibrosis or chronic bronchitis (Ryley and Brogan, 1968; Brogan et al., 1975). Furthermore, allergen challenge can increase within minutes the amount of albumin and fibrinogen in BAL fluid of asthmatics (Lam et al., 1985; Fick et al., 1987; Liu et al., 1991; Salomonsson et al., 1992). Exposure to the allergen responsible for sensitivity to Western Red Cedar can increase the amount of albumin in BAL fluid for as long as a week (Lam et al., 1985).

3.3 HISTOPATHOLOGICAL EVIDENCE OF PLASMA LEAKAGE IN ASTHMA

In the first definitive monograph on asthma, published in 1864, Sir Henry Hyde Salter concluded that the airway obstruction in asthma results from bronchoconstriction, mucus plugging of the airway lumen, and edema of the airway wall (Salter, 1864). Salter's interpretations were probably based on clinical observations rather than on pathological examinations, because post-mortem findings were not described at that time. In fact, the initial histopathological studies of the airways of asthmatics did not recognize the presence of mucosal edema For example, in the first description of a post-mortem examination of a patient who died in status asthmaticus, Leyden concluded that the airflow obstruction resulted from mucus in the airways and that "the walls [of bronchi] ...are not essentially changed" (Leyden, 1886).

Since then many studies have described the histopathology of the airways of asthmatics who died in status asthmaticus or of related illnesses (Huber and Koessler, 1922; Kountz and Alexander, 1928; MacDonald, 1933; Craige, 1941; Unger, 1945; Walzer and Frost, 1952; Houston et al., 1953; Crepea and Harman, 1955; Dunnill, 1960; Glynn and Michaels, 1960; Messer et al., 1960; Dunnill et al., 1969; Sobonya, 1984; Saetta et al., 1991; Jeffery, 1992; Sur et al., 1993; Hogg, 1993). These studies identified the main histopathological features of the disease as obstruction of the airway lumen by mucus plugs and epithelial cells, infiltration of the bronchial wall by eosinophils, and thickening of the epithelial basement membrane. Few of the studies addressed the issue of whether mucosal edema was present or not. However, one study described an abnormal thickening of the bronchial mucosa (Kountz and Alexander, 1928), and another described a "marked degree of mucosal edema with separation of the superficial columnar epithelial cells" in asthmatics dying in status asthmaticus (Dunnill, 1960). The latter author even speculated that the extravasated plasma could separate the epithelium from its basement membrane and cause shedding of the epithelial cells into the airway lumen (Dunnill, 1960).

Ultrastructural studies have not resolved the question about how much edema is present in the airway mucosa of asthmatics. In some ultrastructural studies, a widening of intercellular spaces has been interpreted as evidence of mucosal edema (Laitinen and Laitinen, 1988; Beasley et al., 1989); however, in others no ultrastructural evidence of edema was detected in mucosal biopsies from asthmatics having a wide range of severity of disease (Laitinen et al., 1985; Jeffery et al., 1989). Furthermore, there is a report that the incidence of mucosal edema in the airways of patients with asthma or chronic bronchitis is about the same as that in the airways of normal subjects (Salvato, 1968). A possible explanation for this finding is that minor mechanical trauma, as could be caused by touching the mucosa with a bronchoscope, causes plasma leakage and mucosal edema regardless of whether the airway is normal or abnormal (Hurley, 1984).

Overall, it is evident that comparatively few histopathological studies have specifically addressed the issue of how much edema is present in the airway mucosa of asthmatics. Furthermore, the literature indicates that mucosal edema has been difficult to detect in biopsies or post-mortem specimens of human airways. One explanation is that histological changes associated with edema are relatively insensitive criteria for ascertaining whether plasma leakage occurs in the airway mucosa of asthmatics. Superimposed on such technical limitations is the possibility that the amount of plasma leakage in asthma varies during the course of the disease and also varies among different individuals.

3.4   CONSEQUENCES OF PLASMA LEAKAGE IN THE AIRWAY MUCOSA

There are several consequences of an increase in the permeability of the blood vessels of the airway mucosa. Plasma leakage could result in mucosal edema and the movement of fluid into the airway lumen, both of which could contribute to airflow obstruction. In addition, plasma-derived inflammatory mediators could form in the mucosa and airway lumen, and extravasated plasma proteins could increase the viscosity of sputum.

3.4.1   Contribution of Mucosal Edema to Airway Obstruction

Whether an increase in vascular permeability results in mucosal edema depends on the balance between the amount of leakage into the mucosa and the rate of clearance from the mucosa, either through the lymphatics or across the epithelium into the airway lumen. The increase of vascular permeability produced by inflammatory stimuli can result in the bulk flow of plasma into the airway mucosa (Renkin, 1992). The amount of plasma leakage depends upon the number of gaps that form in the endothelium of the leaky vessels, the duration of the gaps and the intravascular pressure that drives the extravasation (Clough, 1991; Taylor and Ballard, 1992). The movement of plasma proteins and other osmotically active solutes into the mucosa can increase the interstitial oncotic pressure, which favors the net movement of fluid out of vessels and further increases the amount of leakage (Taylor and Ballard, 1992).

Fluid can leave the mucosa through lymphatics or by entering the airway lumen. In this respect the airway epithelium serves as a gate across an escape route for fluid in the interstitium. This escape route has been demonstrated in the airways of the rat by Persson and colleagues (Persson, 1990, 1991; Greiff et al., 1993);

however, the relative importance of luminal versus lymphatic clearance has not been determined. On the one hand, there is a report that relatively little extravasated material enters the lymphatics of guinea-pig bronchi (Erjefält et al., 1993b). On the other hand, the plasma leakage that occurs in dog and sheep bronchi after exposure to endotoxin or smoke results in a large increase in lymph flow, but it is unknown how much fluid enters the airway lumen (Traber et al., 1992).

The effect of plasma leakage on airway resistance depends on the amount of mechanical obstruction caused by the increase in mucosal thickness, accumulation of intraluminal fluid and possible reflex bronchoconstriction. In addition, the impact on airway conductance depends on where in the tracheobronchial tree the fluid accumulates. Theoretical models predict that increased mucosal wall thickness itself may have little effect on airflow, but it could exaggerate the luminal narrowing caused by bronchoconstriction (James et al., 1989; Wiggs et al., 1990; Yager et al., 1991). Although this effect probably would be negligible in the trachea and large bronchi, it could be important in peripheral airways.

Left ventricular failure may result in airway mucosal edema, increased mucosal thickness, luminal narrowing and airflow obstruction. Indeed, pulmonary function tests show some airway obstruction in many patients with left ventricular failure, and treatment of the fluid retention associated with the heart failure tends to improve airflow in these patients (Collins et al., 1975; Light and George, 1983; Peterman et al., 1987). Left ventricular failure can also increase bronchial responsiveness to methacholine (Snashall and Chung, 1991; Rolla et al., 1992). Although this clinical state would seem to indicate that the airflow obstruction and bronchial hyperresponsiveness in patients with left ventricular failure result from mucosal edema, vascular engorgement (Laitinen et al., 1986; Corfield et al., 1991; McFadden, 1992) and reflex bronchoconstriction (Chung et al., 1983) may be contributing factors.

3.4.2   Release of Inflammatory Mediators Associated With Plasma Leakage

Plasma leakage can introduce substances into the airway mucosa that have pro-inflammatory effects (Hogg, 1981; Persson, 1986b, 1988; Persson and Erjefält, 1986a,b; Erjefält and Persson, 1991c). The plasma leakage evoked by allergen challenge in asthmatics can release kallikrein, which results in the formation of bradykinin from high molecular weight kininogen (Christiansen et al., 1987; Proud et al., 1989; Proud and Vio, 1993). Allergen challenge also results in the extravasation of fibrinogen (Fick et al., 1987), which is a rate-limiting factor in extravascular coagulation (Dvorak et al., 1985). Extravascular coagulation may in turn contribute to tissue induration and act as a focus for leukocyte migration (Keahey et al., 1989; Davidson, 1992). Moreover, the presence of fibrinogen in the air spaces can diminish surfactant function and destabilize or close gas exchange units (Seeger *et al.*, 1993). Hageman factor and components of the complement system also can participate in the inflammatory response (Williams *et al.*, 1991; Kozin and Cochrane, 1992).

3.4.3 Plasma Leakage as a "Mucosal Defense"

Persson and colleagues have proposed that increased vascular permeability, plasma leakage and the exudation of plasma into the airway lumen should be viewed as a protective mechanism or "mucosal defense" rather than as a manifestation of airway pathology (Persson, 1990, 1991b,c; Persson *et al.*, 1991). They suggest that the movement of extravasated fluid into the airway lumen is a non-injurious, fully reversible process that could serve to dilute or neutralize inflammatory mediators, inhaled irritants and allergens. This concept is supported by data from animal models (Erjefält and Persson, 1989, 1991a; Greiff *et al.*, 1991, 1993); Erjefält *et al.*, 1993a,b), but its relevance to human airway disease has just begun to be studied (Salomonsson *et al.*, 1992).

3.4.4 Effect of Plasma Leakage on Mucus Secretion

The viscosity of sputum is increased in asthma (Keal, 1971), and mucus plugging of the airways contributes to airflow obstruction (Jeffery, 1992; Hogg, 1993). As noted above, the concentration of albumin in sputum is greater in asthmatics than in normal subjects or in patients with chronic bronchitis (Ryley and Brogan, 1968; Brogan *et al.*, 1975). Studies performed *in vitro* have shown that albumin can stimulate the secretion of mucin from submucosal glands (I. P. Williams *et al.*, 1983) and can increase the viscosity of mucin (List *et al.*, 1978). Treatment of asthma with glucocorticoids or cromoglycate can reduce the amount of albumin in the sputum (Ryley and Brogan, 1968; Heilpern and Rebuck 1972).

4. Mechanism of Increased Vascular Permeability in Inflammation

Several animal models have been used to examine in detail the mechanisms and consequences of plasma leakage in the airway mucosa (Persson, 1988; McDonald, 1990, 1994a; Solway and Leff, 1991). These models have identified and characterized numerous stimuli that increase vascular permeability in the airway mucosa and have revealed many features of the plasma leakage that occurs in inflammation of the trachea and bronchi of rats, guinea-pigs and dogs (Pietra *et al.*, 1971, 1974; Pietra and Magno, 1978; Lundberg *et al.*, 1983; Lundberg and Saria, 1983; Hua *et al.*, 1985; Evans *et al.*, 1988a, 1988b; McDonald, 1988a; Sertl *et al.*, 1988; Ichinose

and Barnes, 1990; O'Donnell *et al.*, 1990; Rogers *et al.*, 1990b; Lötvall *et al.*, 1991a; Ohrui *et al.*, 1992; Bertrand *et al.*, 1993). These models have been the source of many types of data that cannot readily be obtained in humans.

4.1 EARLY PHASE INFLAMMATORY RESPONSE: GAPS IN THE ENDOTHELIUM OF POSTCAPILLARY VENULES

The amount of plasma that moves across the wall of inflamed vessels is governed by the hydrostatic and osmotic pressure gradients, the surface area of the vessels and the permeability of the endothelium. Intravascular hydrostatic pressure may change in inflammation, but endothelial permeability undergoes even more conspicuous changes. Mediators such as histamine (H), bradykinin (Bk), serotonin (5-HT), and substance P cause intercellular gaps to form in the endothelium of postcapillary venules, and as a result, plasma leaks into the surrounding tissue (Majno and Plade, 1961; Pietra and Magno, 1978; Fox *et al.*, 1980; McDonald, 1988a, 1994b). Endothelial gaps have not only been found in mucosal venules in animal models of airway inflammation, but also in biopsies of the airways of humans with asthma (Laitinen and Laitinen, 1988).

Endothelial gaps are quite small in relation to the overall size of endothelial cells. For example, the diameter of gaps that form in rat tracheal venules in the presence of neurogenic inflammation (Section 5.4) averages 1.4 μm (Baluk and McDonald, 1994; Hirata *et al.*, 1994; McDonald, 1994b). Even with an average of about 14 gaps per endothelial cell, the total area of the gaps is less than 3% of the surface area of the endothelium (McDonald, 1994b).

Endothelial gap formation is an active process mediated by a rise in intracellular calcium and effected by actin, myosin and other elements of the cytoskeleton (Schnittler *et al.*, 1990). The gaps produced by H, Bk or substance P are transient structures, so the increase in vascular permeability returns to normal within a few minutes (Horan, *et al.*, 1986; Wu and Baldwin, 1992a,b; McDonald, 1994b). Repeated exposure to these mediators results in smaller responses rather than more prolonged leakage (Brokaw and McDonald, 1988; Brokaw *et al.*, 1990; Wu and Baldwin, 1992a; Bowden *et al.*, 1994a). One of the mechanisms that can limit the amount of plasma leakage is the desensitization of receptors on the endothelial cell plasma membrane (Bowden *et al.*, 1994a). In addition, there may be mechanisms that can actively close endothelial gaps (Wu and Baldwin, 1992a).

Despite their small size, endothelial gaps have an enormous effect on endothelial permeability (Majno and Palade, 1961; Clough, 1991; Wu and Baldwin, 1992b;

McDonald, 1994b). The amount of plasma leakage produced by inflammatory mediators would be expected to be determined by the number of endothelial gaps, the duration of the gaps and the hydrostatic driving force across the endothelium. The relationship between the number of gaps and the magnitude of the increase in endothelial permeability has been examined in neurogenic inflammation (Section 5.4). The amount of plasma leakage associated with neurogenic inflammation in the rat tracheal mucosa peaks 1 min after the onset of the stimulus and decreases rapidly thereafter, with a half-life of 1.3 min (McDonald, 1994b). By comparison, the number of endothelial gaps decreases more slowly, with a half-life of 3.2 min (McDonald, 1994b). These findings suggest that vascular permeability may begin to decrease before the gaps close. Decreased leakage without gap closure could result from a decrease in the hydrostatic driving force or the sieving action of substances that accumulate within endothelial gaps or in the underlying basement membrane (Clough and Michel, 1988; Clough et al., 1988; Weinbaum et al., 1992).

4.2   LATE PHASE INFLAMMATORY RESPONSE: UNKNOWN MECHANISM OF PLASMA LEAKAGE

Some inflammatory stimuli cause plasma leakage that is more gradual in onset and longer in duration than that resulting from mediators like H, Bk and substance P (Lemanske and Kaliner, 1988). For example, exposure of sensitized animals to antigen can result in a "late phase" response that begins several hours after the immediate response and is accompanied by vasodilatation, increased vascular permeability and bronchoconstriction (Murphy, et al., 1986; Bellofiore et al., 1987; Long et al., 1990; Xu et al., 1990; Alving et al., 1991a; Walls et al., 1991; Ohrui et al., 1992; Sabirsh et al., 1993; Olivenstein et al., 1994). Exposure to platelet-activating factor (PAF), Bk, complement peptide 5a (C5a) or tumor necrosis factor (TNF) can cause a prolonged increase in vascular permeability, which in some respects mimics the late phase response to antigen (Wedmore and Williams, 1981; Williams, 1983; T. J. Williams et al., 1983; Rampart et al., 1989; Rogers et al., 1990b; Sekiya et al., 1990; O'Donnell et al., 1990).

It is uncertain which vessels leak in the late phase response. Although venules are labelled with the tracer Monastral blue in a dog model of allergen-induced late phase response (Ohrui et al., 1992), no vessels are labelled in the late phase response evoked in guinea-pig airways by PAF (O'Donnell et al., 1990). It is unclear whether this difference is due to differences in the animal models or to limitations of the methods used to identify the leaky vessels. One reason for considering that the late phase leak may not be just a longer version of the early response is evidence that prolonged inflammatory stimuli can cause plasma leakage from arterioles or capillaries as well as from venules (Cuénoud et al., 1987; Joris et al., 1990).

The late phase response may play an important role in human inflammatory airway disease (O'Byrne et al., 1987; Lemanske and Kaliner, 1988); however, little is known about the mechanism of the late phase leak or factors that can influence its magnitude. In the skin, the late phase response to C5a is neutrophil dependent (Wedmore and Williams, 1981; Williams, 1983; T. J. Williams et al., 1983). In the airways, the late phase response coincides with the influx of leukocytes, but the issue of whether the response depends on leukocytes is unresolved (Murphy et al., 1986; Hutson et al., 1988, 1990; Walls et al., 1991). In any case, the exact sites of extravasation in late phase leak have not been identified, and the mechanism by which leukocytes make vessels leak is unclear. One possibility is that the leakage occurs at sites where leukocytes penetrate the endothelium. Alternatively, leukocyte-derived mediators could induce the formation of gaps or otherwise increase endothelial permeability (Lewis and Granger, 1986; Williams et al., 1991; Yi and Ulich, 1992). There is also a question about the size of the leaky sites. Monastral blue particles (5–200 nm) can leak through the wall of vessels of the airway mucosa in the presence of antigen-induced late phase leak in dogs, and thus the leaky vessels could have endothelial gaps similar to those produced by histamine (Ohrui et al., 1992). However, in the late phase leak produced by PAF in guinea-pigs, there is apparently no leakage of colloidal carbon (10–50 nm particles), so it is unclear where the leak occurs (O'Donnell et al., 1990).

4.3   ROLE OF LEUKOCYTES IN INCREASED VASCULAR PERMEABILITY

The increase in vascular permeability associated with inflammation often is accompanied by the emigration of leukocytes out of the microvasculature. Neutrophils and eosinophils are found in the airways of guinea-pigs, rabbits, dogs and humans after allergen challenge; they are also in the airway mucosa of patients who have died from asthma (Williams et al., 1991). What then is the relationship between vascular permeability and neutrophil migration?

Neutrophil migration and endothelial gap formation are separate phenomena. In other words, neutrophils do not migrate through the endothelial gaps that are sites of plasma leakage. Endothelial gaps form in seconds and close within a few minutes (McDonald, 1994b), but neutrophil migration occurs over a much longer period. Neutrophils can adhere to the endothelium in seconds, but the cells do not begin to migrate for several minutes, and the emigration peaks several hours later (Hurley and Spector, 1961; Williams et al., 1991).

Migrating neutrophils typically do not disrupt the barrier function of the endothelium (Lewis and Granger, 1986), apparently because the membranes of leukocytes and endothelial cells are closely apposed as leukocytes pass through the vessel wall (Hurley, 1963; Meyrick et al., 1984). Vascular permeability usually is back to baseline by the time neutrophil migration is at its peak (Hurley and Spector, 1961). Furthermore, the specific regions of venules where neutrophils attach to the endothelium are not exactly the same as those where the gaps form. Studies of the rat tracheal mucosa have shown that the number of endothelial gaps is greatest in small postcapillary venules (7–20 μm diameter), whereas the number of adherent neutrophils is greatest in the largest postcapillary venules (20–40 μm diameter; McDonald, 1994b).

Despite these differences, neutrophils can play an important role in plasma leakage. For example, C5a, endotoxin, interleukin-1 (IL-1) and TNF can increase endothelial permeability through neutrophil-dependent mechanisms (Wedmore and Williams, 1981; Williams, 1983; Williams et al., 1991; Yi and Ulich, 1992). Neutrophils can release several preformed inflammatory mediators and can rapidly synthesize PAF, reactive oxygen species and arachidonic acid (AA) metabolites (Lewis and Granger, 1986). In addition, leukocyte granules contain cationic proteins that can increase vascular permeability (Gleisner, 1979). Depletion of neutrophils or inhibition of neutrophil emigration reduces the late phase response to inhaled allergen in some animal models (Murphy et al., 1986; Wegner et al., 1990).

4.4 ROLE OF MAST CELLS IN PLASMA LEAKAGE

Mast cell activation and degranulation are characteristic features of the IgE-mediated response to allergen in the airways of asthmatic subjects (Kaliner, 1989). Mast cell degranulation not only releases preformed mediators such as H and 5-HT, which can increase vascular permeability (Majno and Palade, 1961; Page and Minshall, 1993), but also releases proteases that can potentiate the effects of other mediators (Rubinstein et al., 1990; Caughey, 1991). Tryptase is a trypsin-like enzyme that is abundant in the secretory granules of human lung mast cells. Tryptase can degrade neuropeptides that have effects on blood vessels in the airway mucosa (Caughey, 1991). In addition, mast cell chymases, which are chymotrypsin-like proteases related to cathepsin G in neutrophils (Caughey, 1991), can degrade sensory neuropeptides and potentiate the increase in vascular permeability produced by H (Rubinstein et al., 1990). In some species, substances from mast cells can increase vascular permeability by triggering the release of tachykinins from sensory nerve fibers (Saria et al., 1984; Foreman, 1987; Bienenstock et al., 1987; Baraniuk et al., 1990; Alving et al., 1991b).

5. Inflammatory Mediators that Increase Vascular Permeability

Several mediators have been implicated in the plasma leakage associated with airway inflammation. Most of these mediators have relatively transient effects on vascular permeability, because of the existence of mechanisms that limit the duration of their effects. For example, the action of substance P on vascular endothelial cells is limited by the phosphorylation and internalization of neurokinin (Nk$_1$) receptors (Bowden et al., 1994a) and by the degradation of substance P by neutral endopeptidase and other enzymes (Umeno et al., 1989; Nadel, 1992; Katayama et al., 1993).

5.1 LEUKOTRIENES

Leukotrienes, formed through the action of 5-lipoxygenase on AA derived from membrane phospholipid, are likely to participate in the inflammatory response of the airway mucosa (Samuelsson, 1983; Piacentini and Kaliner, 1991; Busse and Gaddy, 1991). Leukotriene receptors may be present on endothelial cells (Evans et al., 1988b; Lam and Austen, 1992), and leukotriene C$_4$ (LTC$_4$), leukotriene D$_4$ (LTD$_4$) and leukotriene E$_4$, (LTE$_4$) can increase vascular permeability in certain model systems (Camp et al., 1980; Hua et al., 1985; Persson et al., 1986b). Inhibitors of leukotriene synthesis can decrease allergen-induced plasma leakage in the airways of guinea-pigs (Evans et al., 1988b).

Although it is unclear whether leukotrienes are involved in the airway inflammation of patients with asthma, LTC$_4$, LTD$_4$ and LTE$_4$ are present in the blood and urine of such patients (Taylor et al., 1989; Wardlaw et al., 1989). Furthermore, leukotriene synthesis inhibitors and leukotriene receptor antagonists can reduce airway narrowing under some circumstances (Israel et al., 1990; Manning et al., 1990; Busse and Gaddy, 1991), but it is not clear what effect leukotrienes have on vascular permeability in human airway disease.

5.2 PLATELET-ACTIVATING FACTOR

PAF is a phospholipid produced by leukocytes, platelets and endothelial cells through the action of phospholipase A$_2$ (PLA$_2$) on phosphoglycerides in cell membranes (Zimmerman et al., 1992). Because PAF is not stored, its release requires de novo synthesis. PAF is released from airway cells and can be detected in BAL fluid and peripheral blood of asthmatics after allergen challenge (Nakamura et al., 1987; Stenton et al., 1990). PAF released from endothelial cells may function as an intercellular messenger involved in the activation of adherent neutrophils (Zimmerman et al., 1990; Lorant et al., 1991; Carveth et al., 1992).

PAF increases vascular permeability by acting on

specific membrane receptors on endothelial cells (Evans *et al.*, 1987, 1988a, 1989; Rogers *et al.*, 1990a). The PAF-induced increase in vascular permeability in guinea-pig airways appears to be independent of sensory nerve fiber activation or Bk (Lötvall *et al.*, 1991a, 1992; Sakamoto *et al.*, 1992). However, PAF apparently does not mediate allergen-induced plasma leakage in guinea-pig airways, as the leakage is not blocked by selective PAF antagonists (Evans *et al.*, 1988a). As with leukotrienes, there is no direct evidence that PAF is a mediator of plasma leakage in human airway disease.

5.3 Bradykinin

Bk is a nine amino acid peptide formed through the action of plasma or tissue kallikrein on high molecular weight kininogen (Kozin and Cochrane, 1992; Proud and Vio, 1993). Bk can trigger plasma leakage and leukocyte adherence in blood vessels of the respiratory mucosa (Lundberg and Saria, 1983; Proud *et al.*, 1989; Kawikova *et al.*, 1993; Bowden *et al.*, 1994d). Bk acts directly on endothelial cells and indirectly through the release of mediators from mast cells and sensory nerve fibers (Lundberg and Saria, 1983; Lawrence *et al.*, 1989; Erjefält and Persson, 1991c; Sulakvelidze and McDonald, 1994). In guinea-pigs, the effects of Bk on the microvasculature are mediated by Bk_2 receptors (Ichinose and Barnes, 1990; Sakamoto *et al.*, 1992).

5.4 Tachykinins

Stimulation of sensory nerve fibers in the respiratory tract of the rat and guinea-pig results in an increase in vascular permeability and plasma leakage in the airway mucosa (Lundberg and Saria, 1983; McDonald, 1994a,c). This phenomenon, called "neurogenic inflammation" (Jancsó *et al.*, 1967, 1968), is mediated by substance P, an 11 amino acid peptide released from sensory nerve fibers (Lembeck and Holzer, 1979). A wide variety of stimuli, including tobacco smoke, gastric acid, hypertonic saline and hyperpnea, can trigger neurogenic inflammation in the respiratory tract (Lundberg and Saria, 1983; Martling and Lundberg, 1988; Umeno *et al.*, 1990; Garland *et al.*, 1991).

Substance P causes plasma leakage in the airway mucosa by binding to neurokinin-1 receptors (also known as Nk1 or substance P receptors) on the endothelial cells of postcapillary venules (Bowden *et al.*, 1994a). Consistent with the involvement of substance P from sensory nerves, the plasma leakage associated with neurogenic inflammation is abolished by pretreatment with capsaicin, which destroys or reduces the substance P content of sensory nerve fibers (Jancsó *et al.*, 1967, 1968; Lundberg and Saria, 1983). The increase in vascular permeability is thought to involve Nk1 receptors, because selective Nk1 receptor agonists can evoke plasma leakage (Abelli *et al.*, 1991a,b) and selective NK1 receptor antagonists can inhibit the

plasma leakage associated with neurogenic inflammation (Delay-Goyet and Lundberg, 1991; Delay-Goyet *et al.*, 1992; Lei *et al.*, 1992; Murai *et al.*, 1992; Sakamoto *et al.*, 1993).

Substance P-immunoreactive nerve fibers evidently do not directly supply the postcapillary venules in the mucosa that are sites of plasma leakage, although such nerves are abundant in the overlying airway epithelium and are also present around arterioles (McDonald *et al.*, 1988; Baluk *et al.*, 1992). These findings raise the question of whether substance P can act directly on the endothelial cells of postcapillary venules. This issue has been addressed by examining the distribution of Nk1 receptor immunoreactivity in the respiratory tract of the rat, using an antibody to the rat Nk1 receptor (Bowden *et al.*, 1994a; Vigna *et al.*, 1994). Nk1 receptor immunoreactivity is abundant on the endothelial cells of postcapillary venules, much less abundant on the endothelial cells of capillaries, and not detectable on the endothelial cells of arterioles (Bowden *et al.*, 1994a,b). These studies also revealed the unexpected finding that substance P stimulates a rapid internalization of Nk1 receptors, thereby increasing the number of Nk1 receptor-immunoreactive endosomes in endothelial cells (Bowden *et al.*, 1994a). Antidromic stimulation of the vagus nerve evokes neurogenic inflammation and induces the internalization of Nk1 receptors by endothelial cells, but PAF does neither (Bowden *et al.*, 1994a). We infer from these findings that the plasma leakage associated with neurogenic inflammation results when substance P, released from sensory nerves around arterioles or in the airway epithelium, diffuses to the Nk1 receptors on endothelial cells of postcapillary venules (Baluk *et al.*, 1992; Bowden *et al.*, 1994a).

Substance P may also play a role in allergen-induced plasma leakage in guinea-pig airways. Studies of this issue have had conflicting results, with some producing evidence consistent with a role of substance P in this process (Saria *et al.*, 1983; Manzini *et al.*, 1987) and others producing the opposite (Lai, 1991; Lötvall *et al.*, 1991b). However, the results of experiments using selective substance P receptor antagonists may reconcile these differences (Bertrand *et al.*, 1993). These findings suggest that substance P is not involved in the plasma leakage that occurs during the first 5 min after challenge but does participate in the leakage that occurs thereafter (Bertrand *et al.*, 1993).

To our knowledge, neurogenic inflammation, in the sense defined by Jancsó and his associates (Jancsó *et al.*, 1967, 1968), has not been demonstrated in the respiratory tract of animals other than guinea-pigs, mice and rats. Although the issue has not been examined systematically, it appears that hamsters, rabbits, dogs and pigs do not develop neurogenic plasma leakage in the airway mucosa (Matheson *et al.*, 1994; McDonald, 1994a,c). The explanation may lie in differences in the nerves, blood vessels, or both, and the lack of responsiveness in healthy animals could change in the presence of inflammatory airway diseases (McDonald, 1992, 1994a,c). Similarly, neurogenic inflammation has not been demonstrated in the human respiratory mucosa, yet there is evidence that other sensory nerve-mediated changes do occur and that these may be exaggerated in allergic rhinitis and asthma (McDonald, 1994a,c).

Regardless of the outcome of the studies of sensory nerve-mediated phenomena in humans, neurogenic inflammation in rats and guinea-pigs continues to serve as a useful experimental model for elucidating the changes in endothelial cells that occur in acute inflammation, determining the consequences of plasma leakage in the airway mucosa, and testing the anti-edema action of drugs that may be useful for treating inflammatory diseases of the nose and bronchi (McDonald, 1994a).

6. Changes in the Airway Microvasculature in Chronic Inflammation

As discussed in Section 3.3, few histopathological studies of asthma have focused specifically on the microvasculature. Nonetheless, there is evidence that neovascularization is one of the changes in the airways of asthmatics (Kuwano et al., 1991). This observation in potentially important because newly formed blood vessels may be abnormally leaky (Schoefl, 1967) or abnormally responsive to inflammatory mediators (McDonald, 1992). The plasma leakage that occurs in the airway mucosa of asthmatics could result from the heightened response of the mucosal blood vessels to inflammatory mediators.

Mycoplasma pulmonis infection in rats has been a useful model for studying changes in airway blood vessels in the presence of chronic inflammation. This infection results in extensive remodeling of the airway mucosa, with neovascularization, influx of leukocytes, mucous cell hyperplasia, and increased mucosal thickness (Lindsey et al., 1971; McDonald, 1988b; Huang et al., 1989; McDonald et al., 1991; Bowden et al., 1994c). These changes are associated with an increase in the sensitivity of the blood vessels to inflammatory mediators, which is manifested by an increase in the amount of plasma leakage produced by inflammatory stimuli, particularly stimuli that trigger neurogenic inflammation (Huang et al., 1989; McDonald, 1988b, 1992; McDonald et al., 1991). The newly formed blood vessels seem not to be abnormally leaky unless they are exposed to an inflammatory mediator (McDonald et al., 1991). The pathological changes associated with M. pulmonis infection are permanent unless the infection is treated (Bowden et al., 1994c).

Several mechanisms appear to be involved in the potentiation of neurogenic plasma leakage by M. pulmonis infection. There is an increase in the number of mucosal blood vessels that become leaky in response to inflammatory stimuli (McDonald et al., 1991). In addition, the blood vessels become abnormally sensitive to inflammatory mediators such as substance P (McDonald et al., 1991). One factor in the heightened sensitivity of the blood vessels to substance P may be an increase in the number of Nk1 receptors on endothelial cells (Baluk et al., unpublished observations). The consequences of the increased number of receptors may be augmented by a reduction in the activity of neutral endopeptidase, which degrades substance P (Borson et al., 1989). Most of the abnormalities resulting from M. pulmonis infection can be reversed by treatment with dexamethasone or oxytetracycline, which reduces the number of M. pulmonis organisms, the severity of histopathological changes, and the responsiveness of the blood vessels to substance P (Bowden et al., 1994c).

7. Effect of Anti-inflammatory Drugs on Plasma Leakage

Glucocorticoids, β_2-adrenergic agonists, methylxanthines, cromoglycate and several anti-inflammatory peptides have been shown to inhibit the increase in vascular permeability induced by inflammatory mediators. Many of these agents are widely used clinically, although usually not for their anti-edema effects. Although the anti-edema action of these drugs has considerable therapeutic potential, most information regarding this action has come from animal models of inflammation rather than from humans with airway disease.

7.1 GLUCOCORTICOIDS

Glucocorticoids are among the most potent substances that can reduce the plasma leakage induced by inflammatory stimuli. This action is not specific for a particular inflammatory stimulus (Tsurufuji et al., 1979; Ohuchi et al., 1984; Foster and McCormick, 1985; Andersson and Persson, 1988; Yarwood et al., 1988). For example, glucocorticoids can inhibit the plasma leakage induced by H, leukotrienes, PAF, Bk, C5a and N-formyl-methionyl-leucyl-phenylalanine (FMLP), presumably through a direct action on endothelial cells (Björk et al., 1985; Svensjö and Roempke, 1985b; Yarwood et al., 1993). Glucocorticoids also can decrease the release of leukotrienes, prostaglandins, H and cytokines. Furthermore, they can inhibit the influx of inflammatory cells into the airways, suppress the synthesis of substance P, and up-regulate the expression of enzymes that degrade substance P (Andersson and Persson, 1988; Ihara and Nakanishi, 1990; Borson and Gruenert, 1991; Piedimonte et al., 1991; Goldstein et al., 1992; Tobler et al., 1992; Katayama et al., 1993).

Glucocorticoids reduce plasma leakage through several mechanisms (Williams and Yarwood, 1990). Actions of glucocorticoids are typically mediated through the glucocorticoid receptor and its effect on protein synthesis (Tsurufuji *et al.*, 1979). Lipocortins and vasocortin are among the proteins that mediate the effects of glucocorticoids on endothelial cells (Williams and Yarwood, 1990; Goldstein *et al.*, 1992). Glucocorticoids can also increase intracellular cAMP (Liu, 1984; Nabishah *et al.*, 1992) and can alter the function of non-steroid receptors on endothelial cells (Davies and Lefkowitz, 1980, 1984). Changes in cAMP in endothelial cells could contribute to the small anti-edema effect of glucocorticoids that occurs too rapidly to involve protein synthesis (Inagaki *et al.*, 1992).

Dexamethasone can reduce the amount of plasma leakage associated with neurogenic inflammation in pathogen-free rats and in rats with viral or mycoplasmal infections (Huang *et al.*, 1989; Piedimonte, *et al.*, 1990; Bowden *et al.*, 1994c). The magnitude of this effect is dose dependent, ranging from a 23% to 90% reduction, and is related to the duration of the treatment (Huang *et al.*, 1989; Piedimonte *et al.*, 1990; Bowden *et al.*, 1994c). Little effect on plasma leakage is found immediately after a single dose of a glucocorticoid in this model system (Lundberg *et al.*, 1983; Norris *et al.*, 1990). In pathogen-free rats, the maximal effect is present about 2 days after the onset of treatment (Piedimonte *et al.*, 1990, 1991; Katayama *et al.*, 1993). However, in rats with mycoplasmal infections, several weeks of treatment are required to restore vascular permeability to normal (Bowden *et al.*, 1994c). Prolonged treatment is required to reverse the angiogenesis associated with the chronic inflammation of the airway mucosa in rats with mycoplasmal infections (Bowden *et al.*, 1994c). Glucocorticoids can inhibit angiogenesis by acting directly on the blood vessels (Folkman and Brem, 1992) and by decreasing the release of cytokines or growth factors that influence blood vessel growth (Goldstein *et al.*, 1992).

Depletion of endogenous glucocorticoids in experimental animals can increase the amount of plasma leakage produced by allergen challenge and other inflammatory stimuli (Boschetto *et al.*, 1992; Ohrui *et al.*, 1992). This observation may be important clinically, because the responsiveness of airway blood vessels to inflammatory mediators could increase after glucocorticoid treatment is stopped.

7.2 β_2-ADRENERGIC RECEPTOR AGONISTS

Selective β_2-adrenergic receptor agonists (β_2 agonists) are widely used in the treatment of airway diseases such as asthma because of their bronchodilating action. In addition, β_2 agonists have anti-inflammatory effects, in that they can inhibit the release of inflammatory mediators and can decrease plasma leakage (Svensjö *et al.*, 1977; Tomioka *et al.*, 1981; Erjefält and Persson, 1986; Barnes,

1993b; Persson, 1993). For example, β_2 agonists can inhibit the degranulation of mast cells and decrease the early and late cutaneous reactions to intradermal allergen (Assem and Richter, 1971; Tomioka *et al.*, 1981; Subramanian, 1986; Inagaki *et al.*, 1989; Gronneberg and Etterstrom, 1992). Furthermore, β_2 agonists can decrease the plasma leakage produced in the skin, respiratory tract and other organs by a variety of inflammatory mediators (Svensjö *et al.*, 1977; O'Donnell and Persson, 1978; Persson and Erjefält, 1986a; Ohuchi *et al.*, 1987).

Through experiments done by Svensjö, Persson, Erjefält and their colleagues, the anti-edema action of β_2 agonists has been characterized in considerable detail (Svensjö *et al.*, 1977, 1982; O'Donnell and Persson, 1978; Persson *et al.*, 1978; Arfors *et al.*, 1979; Joyner *et al.*, 1979; Persson *et al.*, 1982, 1986b; Svensjö and Roempke, 1984; Erjefält and Persson, 1985, 1991b,c; Persson and Erjefält, 1986a, 1988). For example, the β_2 agonist terbutaline can decrease the plasma leakage induced by Bk and H in the hamster cheek pouch (Svensjö *et al.*, 1977, 1982; Arfors *et al.*, 1979; Joyner *et al.*, 1979; Svensjö and Roempke 1984, 1985a), guinea-pig skin (O'Donnell and Persson, 1978; Beets and Paul, 1980), canine forelimb (Dobbins *et al.*, 1982), and human skin (Gronneberg *et al.*, 1979). Anti-edema effects have also been demonstrated for the β_2 agonists formoterol (Ida, 1981; Gronneberg and Zetterstrom, 1990; Erjefält and Persson, 1991b; Tokuyama *et al.*, 1991; Advenier *et al.*, 1992; Whelan *et al.*, 1993; Bowden *et al.*, 1994d; Sulakvelidze and McDonald, 1994), procaterol (Ohuchi *et al.*, 1990), salbutamol (Sharpe and Smith, 1979; Basran *et al.*, 1984; Ohuchi *et al.*, 1987; Inagaki *et al.*, 1989) and salmeterol (Johnson, 1990; Whelan and Johnson, 1992; Whelan *et al.*, 1993). Epinephrine and isoproterenol also have anti-edema actions (Green, 1972; Inagaki *et al.*, 1989).

Of potential relevance to their use in lung disease, β_2 agonists can reduce the plasma leakage evoked in the airways of rats, guinea-pigs and cats by a variety of inflammatory stimuli (Persson *et al.*, 1978; Persson *et al.*, 1982; Erjefält and Persson, 1985, 1991b; Persson and Erjefält, 1986a, 1988; Gronneberg and Zetterstrom, 1990; Tokuyama *et al.*, 1991; Whelan *et al.*, 1993; Sulakvelidze and McDonald 1994). Furthermore, β_2 agonists can decrease the plasma leakage in the lungs of patients with adult respiratory distress syndrome (Basran *et al.*, 1986).

Several lines of evidence indicate that the inhibitory effect of β_2 agonists on plasma leakage results from a direct effect on endothelial cells instead of the inhibition of mediator release from mast cells or sensory nerves or other indirect actions. First, β receptors are present on endothelial cells (Steinberg *et al.*, 1984; Stephenson and Summers, 1987; Aham *et al.*, 1990; Zink *et al.*, 1993). Second, the anti-edema effect of β_2 agonists is indeed mediated by β_2 receptors, as it can be reversed by the selective β_2-adrenergic antagonist ICI 118,551 (Svensjö and Roempke, 1985a; Bowden *et al.*, 1994d). Third, β_2

agonists can induce changes in endothelial cells in culture in the absence of mast cells, sensory nerves, or changes in blood flow (Killackey et al., 1986; Doukas et al., 1987; Morel et al., 1990). Fourth, the anti-edema effect does not require a change in blood flow (Joyner et al., 1979; Beets and Paul, 1980). Finally, β_2 agonists can inhibit the plasma leakage evoked by substance P and PAF, which are likely to act directly on endothelial cells (Sulakvelidze and McDonald, 1994). β_2 Agonists even can reduce the effect of the calcium ionophore A23187 (Northover, 1990).

It is likely that β agonists exert their anti-edema effect by increasing cAMP in endothelial cells (Zink et al., 1993). Agents that can increase intracellular cAMP typically reduce the effect of inflammatory mediators on the permeability of endothelial cells in culture (Carson et al., 1989; Casnocha et al., 1989). Furthermore, drugs that inhibit cAMP-degrading phosphodiesterases can decrease plasma leakage (Suttorp et al., 1993). The inhibitory effect of β_2 agonists appears to be independent of changes of intracellular calcium (Northover, 1990).

The long-standing speculation that β_2 agonists decrease plasma leakage by reducing the number of endothelial gaps that form in postcapillary venules (Svensjö et al., 1977) has recently been documented by measurements of the number of gaps per endothelial cell (Baluk and McDonald, 1994). Specifically, substance P was found to produce 70% less plasma leakage and 68% fewer endothelial gaps in venules in the tracheas of rats that had been pretreated with the β_2 agonist formoterol than in the corresponding controls (Baluk and McDonald, 1994).

7.3 SODIUM CROMOGLYCATE AND NEDOCROMIL

Although sodium cromoglycate was originally believed to exert its therapeutic effect in asthma by inhibiting the release of mediators from mast cells, it is now known to have several anti-inflammatory effects in the airways (Barnes, 1993a). One action of sodium cromoglycate that may have a non-mast cell component is the reduction of plasma leakage (Persson, 1987; Barnes, 1993c). For example, sodium cromoglycate can decrease the plasma leakage produced by H and tachykinins (Erjefält and Persson, 1991c). This effect of sodium cromoglycate does not result from a change in blood flow (Erjefält and Persson, 1991c). Sodium cromoglycate also can reduce the amount of albumin in the sputum of asthmatic subjects, suggesting that the drug has a clinically significant anti-edema effect (Heilpern and Rebuck, 1972).

Nedocromil, which is structurally different from cromoglycate, also has multiple anti-inflammatory actions (Barnes, 1993c). For example, nedocromil can decrease allergen-induced plasma leakage in guinea-pig airways (Evans et al., 1988b). This effect probably involves the inhibition of mediator release (Moqbel et al., 1988). However, an action on vascular endothelial cells

may also be involved, because nedocromil can decrease the plasma leakage produced by allergen, LTB_4, and H in the hamster cheek pouch (Dahlen et al., 1989). It is unknown whether nedocromil has an anti-edema effect in the airways of humans.

7.4 METHYLXANTHINES

Methylxanthines have been used for many years to treat airflow obstruction (Persson, 1985). Their beneficial actions in this regard include bronchodilatation, increased respiratory drive and delayed onset of respiratory muscle fatigue under conditions of increased workload (Aubier et al., 1981; Finney et al., 1985). Methylxanthines can also have anti-edema effects (Persson, 1986a; Persson and Draco, 1988). Both theophylline and enprofylline can reduce the amount of plasma leakage evoked in the guinea-pig trachea by a variety of inflammatory stimuli (Erjefält and Persson, 1986, 1991c; Persson et al., 1986b; Persson and Draco, 1988; O'Donnell et al., 1990; Raeburn and Karlsson, 1993).

Although it is unclear whether methylxanthines have anti-inflammatory actions in human airways, the exacerbation of symptoms that occurs in some patients with asthma when theophylline is withdrawn raises the possibility that the agent has clinically relevant anti-inflammatory effects which reverse when treatment is stopped (Brenner et al., 1988). Furthermore, the drugs can decrease the bronchoconstriction that occurs several hours after the inhalation of allergens, but they appear to have little effect on the immediate allergic response (Crescioli et al., 1991, 1992).

The effects of methylxanthines have been assumed to result from the increase in intracellular cAMP that follows the inhibition of phosphodiesterases (Persson, 1980, 1986a). The anti-edema effect of methylxanthines is consistent with this mechanism, because selective inhibitors of type 4 phosphodiesterase can inhibit plasma leakage (Raeburn and Karlsson, 1993). Nonetheless, theophylline's effect on phosphodiesterases occurs at concentrations far greater than can be safely achieved by systemic administration (Fredholm and Persson, 1982). Some effects of methylxanthines may involve the inhibition of adenosine receptors (Persson et al., 1981, 1986a; Fredholm and Persson, 1982). However, this action would not explain the anti-edema action of enprofylline, which is not a potent adenosine antagonist (Persson, 1982, 1986a; Persson and Draco, 1988).

7.5 ENDOGENOUS NEUROPEPTIDES

Several reports suggest that peptides released from nerve fibers in the airways can decrease the plasma leakage produced by inflammatory mediators (Raud, 1993). For example, calcitonin gene-related peptide (CGRP), which usually co-exists with substance P in sensory nerve fibers, may have an anti-edema effect in the airways (Raud et al.,

1991). In addition, neuropeptide Y, a peptide that usually co-exists with noradrenaline in sympathetic nerve fibers, may inhibit plasma leakage in the airway mucosa under some conditions (Takahashi *et al.*, 1993). Corticotropin-releasing factor, a peptide released from certain hypothalamic neurons, is also present in some sensory nerve fibers and can decrease plasma leakage in the trachea and other sites (Wei and Kiang, 1987; Gao *et al.*, 1991; Wei and Thomas, 1993). Such evidence suggests that endogenous neural mechanisms can limit the amount of plasma leakage produced in the airways by inflammatory stimuli.

8. Summary and Conclusions

The microvascular endothelium, by regulating the leakage of plasma and the emigration of cells into the airway mucosa, plays a key role in coordinating the inflammatory response of the trachea and bronchi. These processes have been studied extensively in animal models of airway inflammation, where allergens, H, Bk, 5-HT, PAF, substance P, leukotrienes and a variety of other substances have been shown to cause plasma leakage. But much less is known about the contribution of plasma leakage to the pathophysiology of human airway disease, particularly long-standing conditions. It is clear that some inflammatory stimuli can evoke plasma leakage in the human airway mucosa. Furthermore, clinically relevant stimuli such as allergens can trigger plasma leakage in the airway mucosa of asthmatics. However, the functional consequences of this leakage are unknown, and there is little information regarding the amount of plasma leakage that occurs in chronic disease in the absence of exogenous stimuli such as allergens. Another unresolved question is whether plasma leakage should be viewed as a protective mechanism or as a manifestation of airway pathology. Finally, although several drugs used in the treatment of respiratory disease can decrease plasma leakage in experimental models, additional studies in humans will be needed before it is known whether the anti-edema action occurs in humans and is clinically beneficial.

9. Acknowledgements

The authors thank Janice Wong and Amy Haskell for assisting with the references. Supported in part by NHLBI Program Project Grant HL-24136 from the National Institutes of Health (USA) and by a grant from the Royal Australasian College of Physicians.

10. References

Abelli, L., Maggi, C.A., Rovero, P., Del Bianco, E., Regoli, D., Drapeau, G. and Giachetti, A. (1991a). Effect of synthetic tachykinin analogues on airway microvascular leakage in rats and guinea-pigs: evidence for the involvement of NK-1 receptors. J. Auton. Pharmacol. 11, 267–275.

Abelli, L., Nappi, F., Maggi, C.A., Rovero, P., Astolfi, M., Regoli, D., Drapeau, G. and Giachetti, A. (1991b). NK-1 receptors and vascular permeability in rat airways. Ann. NY Acad. Sci. 632, 358–359.

Advenier, C., Qian, Y., Koune, J.D., Molimard, M., Candenas, M.L. and Naline, E. (1992). Formoterol and salbutamol inhibit bradykinin and histamine-induced airway microvascular leakage in guinea-pig. Br. J. Pharmacol. 105, 792–798.

Ahmad, S., Chretien, P., Daniel, E.E. and Shen, S.H. (1990). Characterization of beta adrenoceptors on cultured endothelial cells by radioligand binding. Life Sci. 47, 2365–2370.

Alving, K., Matran, R., Fornhem, C. and Lundberg, J.M. (1991a). Late phase bronchial and vascular responses to allergen in actively-sensitized pigs. Acta Physiol. Scand. 143, 137–138.

Alving, K., Sundström, C., Matran, R., Panula, P., Hökfelt, T. and Lundberg, J.M. (1991b). Association between histamine-containing mast cells and sensory nerves in the skin and airways of control and capsaicin-treated pigs. Cell Tissue Res. 264, 529–538.

Andersson, P.T. and Persson, C.G. (1988). Developments in anti-asthma glucocorticoids. Agents Actions Suppl. 23, 239–260.

Arfors, K.-E., Rutili, G. and Svensjö, E. (1979). Microvascular transport of macromolecules in normal and inflammatory conditions. Acta Physiol. Scand. Suppl. 463, 93–103.

Assem, E.S.K. and Richter, A.W. (1971). Comparison of *in vivo* and *in vitro* inhibition of the anaphylactic mechanism by β-adrenergic stimulants and disodium cromoglycate. Immunology 21, 729–739.

Aubier, M., de Troyer, A., Sampson, M., Macklem, P.T. and Roussos, C.M. (1981). Aminophylline improves diaphragmatic contractility. N. Engl. J. Med. 305, 249–252.

Aviado, D.M. (1965). The bronchial circulation. In "The Lung Circulation" (ed. D.M. Aviado), pp. 185–254. Pergamon Press, Oxford.

Baluk, P. and McDonald, D.M. (1994). The β₂-adrenergic receptor agonist formoterol reduces microvascular leakage by inhibiting endothelial gap formation. Am. J. Physiol. 266, L461–L468.

Baluk, P., Nadel, J.A. and McDonald, D.M. (1992). Substance P-immunoreactive sensory axons in the rat respiratory tract: a quantitative study of their distribution and role in neurogenic inflammation. J. Comp. Neurol. 319, 586–598.

Baraniuk, J.N., Kowalski, M.L. and Kaliner, M.A. (1990). Relationships between permeable vessels, nerves, and mast cells in rat cutaneous neurogenic inflammation. J. Appl. Physiol. 68, 2305–2311.

Barnes, P.J. (1993a). Anti-inflammatory therapy for asthma. Annu. Rev. Med. 44, 229–242.

Barnes, P.J. (1993b). β-adrenoceptors on smooth muscle, nerves and inflammatory cells. Life Sci. 52, 2101–2109.

Barnes, P.J. (1993c). Effect of nedocromil sodium on microvascular leakage. J. Allergy Clin. Immunol. 92, 197–199.

Basran, G.S., Paul, W. and Morley, J. (1984). β-adrenoceptor agonist responses in the skin and lungs of asthmatic subjects. Eur. J. Respir. Dis. 135 (Suppl.), 198–201.

Basran, G.S., Hardy, J.G., Woo, S.P., Ramasubramanian, R. and Byrne, A.J. (1986). β₂-adrenoceptor agonists as inhibitors of lung vascular permeability to radiolabelled transferrin in the adult respiratory distress syndrome in man. Eur. J. Nucl. Med. 12, 381–384.

Beasley, R., Roche, W.R., Roberts, J.A. and Holgate, S.T. (1989). Cellular events in the bronchi in mild asthma and after bronchial provocation. Am. Rev. Respir. Dis. 139, 806–817.

Beets, J.L. and Paul, W. (1980). Actions of locally administered adrenoceptor agonists on increased plasma protein extravasation and blood flow in guinea-pig skin. Br. J. Pharmacol. 70, 461–467.

Bellofiore, S., DiMaria, G.U. and Martin, J.G. (1987). Changes in upper and lower airway resistance after inhalation of antigen in sensitized rats. Am. Rev. Respir. Dis. 136, 363–368.

Berry, J.L., Brailsford, J.F. and Daly, I.D.B. (1931). The bronchial vascular system in the dog. Proc. R. Soc. Lond. 109, 214–228.

Bertrand, C., Geppetti, P., Baker, J., Yamawaki, I. and Nadel, J.A. (1993). Role of neurogenic inflammation in antigen-induced vascular extravasation in guinea-pig trachea. J. Immunol. 150, 1479–1485.

Beynon, H.L.C. and Haskard, D.O. (1994). Lymphocyte-endothelial cell interactions. In "Immunopharmacology of the Microcirculation" (ed. S.D. Brain), pp. 109–126. Academic Press, London.

Bienenstock, J., Tomioka, M., Matsuda, H., Stead, R.H., Quinonez, G., Simon, G.T., Coughlin, M.D. and Denburg, J.A. (1987). The role of mast cells in inflammatory processes: evidence for nerve/mast cell interactions. Int. Arch. Allergy Appl. Immunol. 82, 238–243.

Björk, J., Goldschmidt, T., Smedegard, G. and Arfors, K.E. (1985). Methylprednisolone acts at the endothelial cell level reducing inflammatory responses. Acta Physiol. Scand. 123, 221–224.

Borson, D.B. and Gruenert, D.C. (1991). Glucocorticoids induce neutral endopeptidase in transformed human tracheal epithelial cells. Am. J. Physiol. (Lung Cell. Mol. Physiol.) 260, L83–L89.

Borson, D.B., Brokaw, J.J., Sekizawa, K., McDonald, D.M. and Nadel, J.A. (1989). Neutral endopeptidase and neurogenic inflammation in rats with respiratory infections. J. Appl. Physiol. 66, 2653–2658.

Boschetto, P., Musajo, F.G., Tognetto, L., Boscaro, M., Mapp, C.E., Barnes, P.J. and Fabbri, L.M. (1992). Increase in vascular permeability produced in rat airways by PAF: potentiation by adrenalectomy. Br. J. Pharmacol. 105, 388–392.

Bowden, J.J., Garland, A., Baluk, P., Lefevre, P., Grady, E., Vigna, S.R., Bunnett, N.W. and McDonald, D.M. (1994a). Direct observation of substance P-induced internalization of neurokinin 1 (NK1) receptors at sites of inflammation. Proc. Natl. Acad. Sci. USA 91, 8964–8968.

Bowden, J.J., Lefevre, P., Vigna, S.R. and McDonald, D.M. (1994b). Internalization of substance P (NK1) receptors on endothelial cells of postcapillary venules in neurogenic inflammation in the rat respiratory tract. Am. J. Resp. Crit. Care Med. 149, A819.

Bowden, J.J., Schoeb, T., Lindsey, J.R. and McDonald, D.M. (1994c). Dexamethasone and oxytetracycline reverse the potentiation of neurogenic inflammation in airways of rats with Mycoplasma pulmonis infection. Am. J. Respir. Crit. Care Med. 150, 1391–1401.

Bowden, J.J., Sulakvelidze, I. and McDonald, D.M. (1994d). Inhibition of neutrophil and eosinophil adhesion to venules of the rat trachea by the 52-adrenergic agonist formoterol. J. Appl. Physiol. 77, 397–405.

Brain, S.D. (ed.) (1994). "Immunopharmacology of the Microcirculation". In the series "The Handbook of Immunopharmacology" (series ed. C. Page). Academic Press, London.

Brenner, M., Berkowitz, R., Marshall, N. and Strunk, R.C. (1988). Need for theophylline in severe steroid-requiring asthmatics. Clin. Allergy 18, 143–150.

Brogan, T.D., Ryley, H.C., Neale, L. and Yassa, J. (1975). Soluble proteins of bronchopulmonary secretions from patients with cystic fibrosis, asthma, and bronchitis. Thorax 30, 72–79.

Brokaw, J.J. and McDonald, D.M. (1988). The neurally-mediated increase in vascular permeability in the rat trachea: onset, duration and tachyphylaxis. Exp. Lung Res. 14, 757–767.

Brokaw, J.J., Hillenbrand, C.M., White, G.W. and McDonald, D.M. (1990). Mechanism of tachyphylaxis associated with neurogenic plasma extravasation in the rat trachea. Am. Rev. Respir. Dis. 141, 1434–1440.

Busse, W.W. and Gaddy, J.N. (1991). The role of leukotriene antagonists and inhibitors in the treatment of airway disease. Am. Rev. Respir. Dis. 143, S103–S107.

Butler, J. (ed.) (1991). "The Bronchial Circulation". In the series "Lung Biology in Health and Disease" (series ed. C. Lenfant). Marcel Dekker, New York.

Calhoun, W.J., Jarzour, N.N., Gleich, G.J., Stevens, C.A. and Busse, W.W. (1993). Increased airway inflammation with segmental versus aerosol antigen challenge. Am. Rev. Respir. Dis. 147, 1465–1471.

Camp, R.D.R., Coutts, A.A., Greaves, M.W., Kay, A.B. and Walfort, M.J. (1980). Responses of human skin to intradermal injection of leukotrienes C₄, D₄, B₄. Br. J. Pharmacol. 80, 497–502.

Carson, M.R., Shasby, S.S. and Shasby, D.M. (1989). Histamine and inositol phosphate accumulation in endothelium: cAMP and a G protein. Am. J. Physiol. 257, L259-L264.

Carveth, H.J., Shaddy, R.E., Whatley, R.E., McIntyre, T.M., Prescott, S.M. and Zimmerman, G.A. (1992). Regulation of platelet-activating factor (PAF) synthesis and PAF-mediated neutrophil adhesion to endothelial cells activated by thrombin. Semin. Thromb. Hemost. 18, 126–134.

Casnocha, S.A., Eskin, S.G., Hall, E.R. and McIntire, L.V. (1989). Permeability of human endothelial monolayers: effect of vasoactive agonists and cAMP. J. Appl. Physiol. 67, 1997–2005.

Caughey, G.H. (1991). The structure and airway biology of mast cell proteinases. Am. J. Respir. Cell. Mol. Biol. 4, 387–394.

Cauldwell, E.W., Siekert, G., Lininger, R.E. and Anson, B.J. (1948). The bronchial arteries. Surg. Gynecol. Obstet. 86, 395–412.

Christiansen, S.C., Proud, D. and Cochrane, C.G. (1987). Detection of tissue kallikrein in the bronchoalveolar lavage fluid of asthmatic subjects. J. Clin. Invest. 79, 188–197.

Chung, K.F., Keyes, S.J., Morgan, B.M., Jones, P.W. and Snashall, P.D. (1983). Mechanisms of airway narrowing in acute pulmonary oedema in dogs: influence of the vagus and lung volume. Clin. Sci. 65, 289–296.

Clough, G. (1991). Relation between microvascular permeability and ultrastructure. Prog. Biophys. Molec. Biol. 55, 47–69.

Clough, G. and Michel, C.C. (1988). The ultrastructure of frog microvessels following perfusion with the ionophore A23187. Quart. J. Exp. Physiol. 73, 123–125.

Clough, G., Michel, C.C. and Phillips, M.E. (1988). Inflammatory changes in permeability and ultrastructure of single vessels in the frog mesenteric microcirculation. J. Physiol. (Lond.) 395, 99–114.

Collins, J.V., Clark, T.J.H. and Brown, D.J. (1975). Airway function in healthy subjects and patients with left heart disease. Clin. Sci. Mol. Med. 49, 217–228.

Corfield, D.R., Hanafi, Z., Webber, S.E. and Widdicombe, J.G. (1991). Changes in tracheal mucosal thickness and blood flow in sheep. J. Appl. Physiol. 71, 1282–1288.

Cotran, R.S., Suter, E.R. and Majno, G. (1967). The use of colloidal carbon as a tracer for vascular injury. Vasc. Dis. 4, 107–125.

Craige, B. (1941). Fatal bronchial asthma. Arch. Int. Med. 67, 399–410.

Crepea, S.B. and Harman, J.W. (1955). The pathology of bronchial asthma. I. The significance of membrane changes in asthmatic and non-allergic pulmonary disease. J. Allergy 26, 453–460.

Crescioli, S., Spinazzi, A., Plebani, M., Pozzani, M., Mapp, C.E., Boschetto, P. and Fabbri, L.M. (1991). Theophylline inhibits early and late asthmatic reactions induced by allergens in asthmatic subjects. Ann. Allergy 66, 245–251.

Crescioli, S., De Marzo, N., Boschetto, P., Spinazzi, A., Plebani, M., Mapp, C.E., Fabbri, L.M. and Ciaccia, A. (1992). Theophylline inhibits late asthmatic reactions induced by toluene diisocyanate in sensitised subjects. Eur. J. Pharmacol. 228, 45–50.

Cudkowicz, L. (1968). "The Human Bronchial Circulation in Health and Disease". Williams and Wilkins Co., Baltimore.

Cudkowicz, L. (1979). Bronchial arterial circulation in man. In "Pulmonary Vascular Diseases" (ed. K.M. Moser), pp. 111–232. Marcel Dekker, New York.

Cuénoud, H.F., Joris, I., Langer, R.S. and Majno, G. (1987). Focal arteriolar insudation. A response of arterioles to chronic nonspecific irritation. Am. J. Pathol. 127, 592–604.

Dahlen, S.-E., Bjorck, T., Kumlin, M., Sydbom, A., Raud, J., Palmertz, U., Franzen, L., Gronneberg, R. and Hedqvist, P. (1989). Dual inhibitory action of nedocromil sodium on antigen-induced inflammation. Drugs 37 (Suppl.), 63–68.

Daly, I.D.B. and Hebb, C. (1966). "Pulmonary and Bronchial Vascular Systems". William and Wilkins Co., Baltimore.

Davidson, J.M. (1992). Wound repair. In "Inflammation: Basic Principles and Clinical Correlates" (eds. J.I. Gallin, I.M. Goldstein and R. Snyderman), pp. 809–819. Raven Press, New York.

Davies, A.O. and Lefkowitz, R.J. (1980). Corticosteroid-induced differential regulation of β-adrenergic receptors in circulating human polymorphonuclear leukocytes and mononuclear leukocytes. J. Clin. Endocrinol. Metab. 51, 599–605.

Davies, A.O. and Lefkowitz, R.J. (1984). Regulation of β-adrenergic receptors by steroid hormones. Annu. Rev. Physiol. 46, 119–130.

Deffebach, M.E., Charan, N.B., Lakshminarayan, S. and Butler, J. (1987). The bronchial circulation: small, but a vital attribute of the lung. Am. Rev. Respir. Dis. 135, 463–481.

Delay-Goyet, P. and Lundberg, J.M. (1991). Cigarette smoke-induced airway oedema is blocked by the NK1 antagonist, CP-96, 345. Eur. J. Pharmacol. 2, 157–158.

Delay-Goyet, P., Franco, C.A., Gonsalves, S.F., Clingan, C.A., Lowe, J. and Lundberg, J.M. (1992). CP-96,345 antagonism of NK1 receptors and smoke-induced protein extravasation in relation to its cardiovascular effects. Eur. J. Pharmacol. 222, 213–218.

Dobbins, D.E., Soika, C.Y., Premen, A.J., Grega, G.J. and Dabney, J.M. (1982). Blockage of histamine and bradykinin-induced increases in lymph flow, protein concentration, and protein transport by terbutaline in vivo. Microcirculation 2, 127–150.

Doukas, J., Shepro, D. and Hechtman, H.B. (1987). Vasoactive amines directly modify endothelial cells to affect polymorphonuclear leukocyte diapedesis in vitro. Blood 69, 1563–1569.

Dunnill, M.S. (1960). The pathology of asthma with special references to changes in the bronchial mucosa. J. Clin. Pathol. 13, 27–33.

Dunnill, M.S., Massarella, G.R. and Anderson, J.A. (1969). A comparison of the quantitative anatomy of the bronchi in normal subjects, in status asthmaticus, in chronic bronchitis, and in emphysema. Thorax 24, 176–179.

Dvorak, H.F., Senger, D.R., Dvorak, A.M., Harvey, V.S. and McDonagh, J. (1985). Regulation of extravascular coagulation by microvascular permeability. Science 227, 1059–1061.

Erjefält, I. and Persson, C.G.A. (1985). Anti-asthma drugs and capsaicine-induced microvascular effect in lower airways. Agents Actions 16, 9–10.

Erjefält, I. and Persson, C.G.A. (1986). Anti-asthma drugs attenuate inflammatory leakage of plasma into airway lumen. Acta Physiol. Scand. 128, 653–654.

Erjefält, I. and Persson, C.G.A. (1989). Inflammatory passage of plasma macromolecules into airway wall and lumen. Pulm. Pharmacol. 2, 93–102.

Erjefält, I. and Persson, C.G.A. (1991a). Allergen, bradylcinin, and capsaicin increase outward but not inward macromolecular permeability of guinea-pig tracheobronchial mucosa. Clin. Exp. Allergy 21, 217–224.

Erjefält, I. and Persson, C.G.A. (1991b). Long duration and high potency of antiexudative effects of formoterol in guinea-pig tracheobronchial airways. Am. Rev. Respir. Dis. 144, 788–791.

Erjefält, I. and Persson, C.G.A. (1991c). Pharmacologic control of plasma exudation into tracheobronchial airways. Am. Rev. Respir. Dis. 143, 1008–1014.

Erjefält, I.A., Wagner, Z.G., Strand, S.E. and Persson, C.G. (1985). A method for studies of tracheobronchial microvascular permeability fo macromolecules. J. Pharmacol. Methods 14, 275–283.

Erjefält, I., Greiff, L. and Persson, C.G.A. (1993a). Exudation versus absorption across the airway epithelium. Pharmacol. Toxicol. 72, 14–16.

Erjefält, I., Luts, A. and Persson, C.G.A. (1993b). Appearance of airway absorption and exudation tracers in guinea pig tracheobronchial lymph nodes. J. Appl. Physiol. 74, 817–824.

Evans, T.W., Chung, K.F., Rogers, D.F. and Barnes, P.J.

(1987). Effect of platelet-activating factor an airway vascular permeability: possible mechanisms. J. Appl. Physiol. 63, 479–484.

Evans, T.W., Dent, G., Rogers, D.F., Aursudkij, B., Chung, K.F. and Barnes, P.J. (1988a). Effect of a PAF antagonist, WEB 2086, on airway microvascular leakage in the guinea-pig and platelet aggregation in man. Br. J. Pharmacol. 94, 164–168.

Evans, T.W., Rogers, D.F., Aursudkij, B., Chung, K.F. and Barnes, P.J. (1988b). Inflammatory mediators involved in antigen-induced airway microvascular leakage in guinea pigs. Am. Rev. Respir. Dis. 138, 395–399.

Evans, T.W., Rogers, D.F., Aursudkij, B., Chung, K.F. and Barnes, P.J. (1989). Regional and time-dependent effects of inflammatory mediators on airway microvascular permeability in the guinea-pig. Clin. Sci. 76, 479–485.

Fick, R., Jr., Metzger, W.J., Richerson, H.B., Zavala, D.C., Moseley, P.L., Schoderbek, W.E. and Hunninghake, G.W. (1987). Increased bronchovascular permeability after allergen exposure in sensitive asthmatics. J. Appl. Physiol. 63, 1147–1155.

Finney, M.J.B., Karlsson, J.-A. and Persson, C.G.A. (1985). Effects of bronchoconstrictors and bronchodilators on a novel human small airway preparation. Br. J. Pharmacol. 85, 29–36.

Folkman, J. and Brem, H. (1992). Angiogenesis and inflammation. In "Inflammation: Basic Principles and Clinical Correlates" (eds. J.I. Gallin, I.M. Goldstein and R. Snyderman), pp. 821–839. Raven Press, New York.

Foreman, J.C. (1987). Substance P and calcitonin gene-related peptide: effects on mast cells and in human skin. Int. Arch. Allergy Appl. Immunol. 82, 366–371.

Foster, S.J. and McCormick, M.E. (1985). The mechanism of the anti-inflammatory activity of glucocorticosteroids. Agents Actions 16, 58–59.

Fox, J., Galey, F. and Wayland, H. (1980). Action of histamine on the mesenteric microvasculature. Microvasc. Res. 19, 108–126.

Fredholm, B.B. and Persson, C.G.A. (1982). Xanthine derivatives as adenosine receptor antagonists. Eur. J. Pharmacol. 81, 673–676.

Gao, G.C., Dashwood, M.R. and Wei, E.T. (1991). Corticotropin-releasing factor inhibition of substance P-induced vascular leakage in rats: possible sites of action. Peptides 12, 639–644.

Garland, A., Ray, D.W., Doerschuk, C.M., Alger, L., Eappon, S., Hernandez, C., Jackson, M. and Solway, J. (1991). Role of tachykinins in hyperpnea-induced bronchovascular hyperpermeability in guinea-pigs. J. Appl. Physiol. 70, 27–35.

Gleisner, J.M. (1979). Lysosomal factors in inflammation. In "Chemical Messengers of the Inflammatory Process" (ed. J.C. Houck), pp. 229–260. Elsevier, Amsterdam.

Glynn, A.A. and Michaels, L. (1960). Bronchial biopsy in chronic bronchitis and asthma. Thorax 15, 142–153.

Goldstein, R.A., Bowen, D.A. and Fauci, A.S. (1992). Adrenal corticosteroids. In "Inflammation: Basic Principles and Clinical Correlates" (eds J.I. Gallin, I.M. Goldstein and R. Snyderman), pp. 1061–1081. Raven Press, New York.

Green, K.L. (1972). The anti-inflammatory effect of catecholamines in the peritoneal cavity and hind paw of the mouse. Br. J. Pharmacol. 45, 322–332.

Greiff, L., Erjefält, I., Pipkorn, U. and Persson, C.G. (1991). Effects of airway luminal concentration of albumin on histamine-induced mucosal exudation of radio-iodine labelled albumin. Acta Physiol. Scand. 142, 345–349.

Greiff, L., Erjefält, I., Svensson, C., Wollmer, P., Alkner, U., Andersson, M. and Persson, C.G.A. (1993). Plasma exudation and solute absorption across the airway mucosa. Clin. Physiol. 13, 219–233.

Gronneberg, R. and Zetterstrom, O. (1990). Inhibition of anti-IgE induced skin response in normals by formoterol, a new β2-adrenoceptor agonist, and terbutaline. Allergy 45, 334–339.

Gronneberg, R. and Zetterstrom, O. (1992). Inhibitory effects of formoterol and terbutaline on the development of late phase skin reactions. Clin. Exp. Allergy 21, 257–263.

Gronneberg, R., Strandberg, K. and Hagermark, O. (1979). Effect of terbutaline, a β2-adrenergic receptor stimulating compound, on cutaneous responses to histamine, allergen, compound 48/80, and trypsin. Allergy 34, 303–309.

Heilpern, S. and Rebuck, A.S. (1972). Effect of disodium cromoglycate (Intal) on sputum protein composition. Thorax 27, 726–728.

Hirata, A., Baluk, P. and McDonald, D.M. (1994). Relationship between dose of substance P and number and size of endothelial gaps in inflamed venules. FASEB J. 8, A70.

Hogg, J.C. (1981). Bronchial mucosal permeability and its relationship to airways hyperreactivity. J. Allergy Clin. Immunol. 67, 421–425.

Hogg, J.C. (1993). Pathology of asthma. J. Allergy Clin. Immunol. 92, 1–5.

Horan, K.L., Adamski, S.W., Ayele, W., Langone, J.J. and Grega, G.J. (1986). Evidence that prolonged histamine suffusions produce transient increases in vascular permeability subsequent to the formation of venular macromolecular leakage sites. Am. J. Pathol. 123, 570–576.

Houston, J.C., De Navasquez, S. and Trounce, J.R. (1953). A clinical and pathological study of fatal cases of status asthmaticus. Thorax 8, 207–213.

Hua, X.Y., Dahlen, S.-E., Lundberg, J.M., Hammarström, S. and Hedqvist, P. (1985). Leukotrienes C4, D4 and E4 cause widespread and extensive plasma extravasation in the guinea-pig. Naunyn-Schmiedeberg's Arch. Pharmacol. 330, 136–141.

Huang, H.-T., Haskell, A. and McDonald, D.M. (1989). Changes in epithelial secretory cells and potentiation of neurogenic inflammation in the trachea of rats with respiratory tract infections. Anat. Embryol. 180, 325–341.

Huber, H.L. and Koessler, K.K. (1922). The pathology of bronchial asthma. Arch. Intern. Med. 30, 689–760.

Hughes, T. (1965). Microcirculation of the tracheobronchial tree. Nature 206, 425–426.

Hulstrom, D. and Svensjö, E. (1979). Intravital and electron microscopic study of bradykinin induced vascular permeability changes using FITC-dextran as a tracer. J. Pathol. 129, 125–133.

Hurley, J.V. (1963). An electron microscopic study of leukocytic emigration and vascular permeability in rat skin. Aust. J. Exp. Biol. 41, 171–186.

Hurley, J.V. (1984). Inflammation. In "Edema" (eds N.C. Staub and A.E. Taylor), pp. 463–488. Raven Press, New York.

Hurley, J.V. and Spector, W.G. (1961). Delayed leucocytic emigration after intradermal injections and thermal injury. J. Pathol. Bacteriol. 82, 421–429.

Hutson, P.A., Church, M.K., Clay, T.P., Miller, P. and Holgate, S.T. (1988). Early and late-phase bronchoconstriction after allergen challenge of nonanesthetized guinea pigs. I. The association of disordered airway physiology to leukocyte infiltration. Am. Rev. Respir. Dis. 137, 548–557.

Hutson, P.A., Varley, J.G., Sanjar, S., Kings, M., Holgate, S.T. and Church, M.K. (1990). Evidence that neutrophils do not participate in the late-phase airway response provoked by ovalbumin inhalation in conscious, sensitized guinea-pigs. Am. Rev. Respir. Dis. 141, 535–539.

Ichinose, M. and Barnes, P.J. (1990). Bradykinin-induced airway microvascular leakage and bronchoconstriction are mediated via a bradykin B2 receptor. Am. Rev. Respir. Dis. 142, 1104–1107.

Ida, H. (1981). Pharmacology of formoterol, (αRS)-3-formamido-4-hydroxy-α-[[[(αRS)-p-methoxy-a-methylphenethyl] amino]methyl] benzylalcohol fumarate dihydrate (BD 40A). Pharmacometrics 21, 201–210.

Ihara, H. and Nakanishi, S. (1990). Selective inhibition of expression of the substance P receptor mRNA in pancreatic acinar AR42J cells by glucocorticoids. J. Biol. Chem. 265, 22441–22445.

Inagaki, N., Miura, T., Daikoku, M., Nagai, H. and Koda, A. (1989). Inhibitory effects of β-adrenergic stimulants on increased vascular permeability caused by passive cutaneous anaphylaxis, allergic mediators, and mediator releasers in rats. Pharmacology 39, 19–27.

Inagaki, N., Miura, T., Nakajima, T., Yoshida, K., Nagai, H. and Koda, A. (1992). Studies on the anti-allergic mechanism of glucocorticoids in mice. J. Pharmacobiodyn. 15, 581–587.

Israel, E., Dermarkarian, R., Rosenberg, M., Taylor, G., Rubin, P. and Drazen, J.M. (1990). The effect of 5-lipoxygenase inhibitor on asthma induced by cold, dry air. N. Engl. J. Med. 323, 1740–1744.

James, A.L., Pare, P.D. and Hogg, J.C. (1989). The mechanics of airway narrowing in asthma. Am. Rev. Respir. Dis. 139, 242–246.

Jancsó, N., Jancsó-Gábor, A. and Szolcsányi, J. (1967). Direct evidence for neurogenic inflammation and its prevention by denervation and by pretreatment with capsaicin. Br. J. Pharmacol. Chemother. 31, 138–151.

Jancsó, N., Jancsó-Gábor, A. and Szolcsányi, J. (1968). The role of sensory nerve endings in neurogenic inflammation induced in human skin and in the eye and paw of the rat. Br. J. Pharmacol. Chemother. 33, 32–41.

Jeffery, P.K. (1992). Pathology of asthma. Br. Med. Bull. 48, 23–39.

Jeffery, P.K., Wardlaw, A.J., Nelson, F.C., V., C.J. and Kay, A.B. (1989). Bronchial biopsies in asthma: An ultrastructural, quantitative study and correlation with hyperreactivity. Am. Rev. Respir. Dis. 140, 1745–1753.

Johnson, M. (1990). The pharmacology of salmeterol. Lung 168, 115–119.

Joris, I., De Girolami, U., Wortham, K.A. and Majno, G. (1982). Vascular labeling with Monastral blue B. Stain Technol. 57, 177–183.

Joris, I., Cuenoud, H.F., Doern, G.V., Underwood, J.M. and Majno, G. (1990). Capillary leakage in inflammation. A study by vascular labeling. Am. J. Pathol. 137, 1353–1363.

Joyner, W.L., Svensjö, E. and Arfors, K.-E. (1979). Simultaneous measurements of macromolecular leakage and arteriolar blood flow as altered by PGE$_1$ and β_2-receptor stimulant in the hamster cheek pouch. Microvasc. Res. 18, 301–310.

Kaliner, M. (1989). Asthma and mast cell activation. J. Allergy Clin. Immunol. 510–520.

Katayama, M., Nadel, J.A., Piedimonte, G. and McDonald, D.M. (1993). Peptidase inhibitors reverse steroid-induced suppression of neutrophil adhesion in rat tracheal blood vessels. Am. J. Physiol. 264, L316–L322.

Kawikova, I., Arakawa, H., Lofdahl, C.G., Skoogh, B.E. and Lötvall, J. (1993). Bradykinin-induced airflow obstruction and airway plasma exudation: effects of drugs that inhibit acetylcholine, thromboxane-A(2) or leukotrienes. Br. J. Pharmacol. 110, 657–664.

Keahey, T.M., Indrisano, J. and Kaliner, M.A. (1989). Dissociation of cutaneous vascular permeability and the development of cutaneous late-phase allergic reactions. J. Allergy Clin. Immunol. 83, 669–676.

Keal, E.E. (1971). Biochemistry and rheology of sputum in asthma. Postgrad. Med. J. 47, 171–177.

Killackey, J.J.F., Johnston, M.G. and Movat, H.Z. (1986). Increased permeability of microcarrier-cultured endothelial monolayers in response to histamine and thrombin. A model for the in vitro study of increased vasopermeability. Am. J. Pathol. 122, 50–61.

Knowles, R.G., Salter, M., Brooks, S.L. and Moncada, S. (1990). Anti-inflammatory glucocorticoids inhibit the induction by endotoxin of nitric oxide synthase in the lung, liver and aorta of the rat. Biochem. Biophys. Res. Comm. 172, 1042–1048.

Kountz, W.B. and Alexander, H.L. (1928). Death from bronchial asthma. Arch. Pathol. 5, 1003–1019.

Kozin, F. and Cochrane, C.G. (1992). The contact activation system of plasma: biochemistry and pathophysiology. In "Inflammation: Basic Principles and Clinical Correlates" (eds J.I. Gallin, I.M. Goldstein and R. Snyderman), pp. 103–121. Raven Press, New York.

Kuwano, K., Bosken, C., Pare, P.D., Bai, T., Wiggs, B. and Hogg, J.C. (1991). Morphometric dimensions of small airways in asthma and chronic obstructive pulmonary disease (COPD). Am. Rev. Respir. Dis. 143, A428.

Lai, Y.L. (1991). Endogenous tachykinins in antigen-induced acute bronchial responses of guinea pigs. Exp. Lung. Res. 17, 1047–1060.

Laitinen, L.A. and Laitinen, A. (1988). Mucosal inflammation and bronchial hyperreactivity. Eur. Respir. J. 1, 488–489.

Laitinen, A. and Laitinen, L.A. (1990). Vascular beds in the airways of normal subjects and asthmatics. Eur. Respir. J. 3, 658s–662s.

Laitinen, A., Laitinen, L.A. and Widdicombe, J.G. (1989). Organisation and structure of the tracheal and bronchial blood vessels in the dog. J. Anat. 165, 133–140.

Laitinen, L.A., Heino, M., Laitinen, A. and Kava, T. (1985). Damage of the airway epithelium and bronchial reactivity in patients with asthma. Am. Rev. Respir. Dis. 131, 599–606.

Laitinen, L.A., Robinson, N.P., Laitinen, A. and Widdicombe, J.G. (1986). Relationship between tracheal mucosal thickness and vascular resistance in dogs. J. Appl. Physiol. 61, 2186–2193.

Lam, B.K. and Austen, K.F. (1992). Leukotrienes: biosynthesis, release, and actions. In "Inflammation: Basic Principles and Clinical Correlates" (eds J.I. Gallin, I.M. Goldstein and R. Snyderman), pp. 139–147. Raven Press, New York.

Lam, S., Leriche, J.C., Kijek, K. and Phillips, R.T. (1985). Effect of bronchial lavage volume on cellular and protein recovery. Chest 88, 856–859.

Lawrence, I.D., Warner, J.A., Cohan, V.L., Lichtenstein, L.M., Kagey-Sobotka, A., Vavrek, R.J., Stewart, J.M. and Proud, D. (1989). Induction of H release from human skin mast cells by bradykinin analogs. Biochem. Pharmacol. 38, 227–233.

Lei, Y.H., Barnes, P.J. and Rogers, D.F. (1992). Inhibition of neurogenic plasma exudation in guinea-pig airways by CP-96, 345, a new non-peptide NK1 receptor antagonist. Br. J. Pharmacol. 105, 261–262.

Lemanske, R. and Kaliner, M. (1988). Late-phase allergic reactions. In "Allergy, Principles and Practice" (ed. E. Middleton), pp. 224–246. Mosby, St Louis.

Lembeck, F. and Holzer, P. (1979). Substance P as neurogenic mediator of antidromic vasodilation and neurogenic plasma extravasation. Naunyn-Schmiedeberg's Arch. Pharmacol. 310, 175–183.

Lewis, R.E. and Granger, H.J. (1986). Neutrophil-dependent mediation of microvascular permeability. Fed. Proc. 45, 109–113.

Leyden, E. (1886). "Uber Bronchial-Asthma". Mittler und Sohn, Berlin.

Light, R.W. and George, R.B. (1983). Serial pulmonary function in patients with acute heart failure. Arch. Intern. Med. 143, 429–433.

Lindsey, J.R., Baker, H.J., Overcash, R.G., Cassell, G.H. and Hunt, C.E. (1971). Murine chronic respiratory disease. Significance as a research complication and experimental production with *Mycoplasma pulmonis*. Am. J. Pathol. 64, 675–708.

List, S.J., Findlay, B.P., Forstner, G.G. and Forstner, J.F. (1978). Enhancement of the viscosity of mucin by serum albumin. Biochem. J. 175, 565–571.

Liu, A.Y.C. (1984). Modulation of the function and activity of cAMP-dependent protein kinase by steroid hormones. Trends Pharmacol. Sci. 3, 106.

Liu, M.C., Hubbard, W.C., Proud, D., Stealey, B.A., Galli, S.J., Kagey-Sabotka, A., Bleecker, E.R. and Lichtenstein, L.M. (1991). Immediate and late inflammatory responses to ragweed antigen challenge of the peripheral airways in allergic asthmatics: cellular, mediator, and permeability changes. Am. Rev. Respir. Dis. 144, 51–58.

Long, W.M., Yerger, L.D., Abraham, W.M. and Lobel, C. (1990). Late-phase bronchial vascular responses in allergic sheep. J. Appl. Physiol. 69, 584–590.

Lorant, D.E., Patel, K.D., McIntyre, T.M., McEver, R.P., Prescott, S.M. and Zimmerman, G.A. (1991). Coexpression of GMP-140 and PAF by endothelium stimulated by histamine or thrombin: a juxtacrine system for adhesion and activation of neutrophils. J. Cell Biol. 115, 223–234.

Lötvall, J.O., Elwood, W., Tokuyama, K., Barnes, P.J. and Chung, K.F. (1991a). Plasma exudation into airways induced by inhaled platelet-activating factor: effect of peptidase inhibition. Clin. Sci. 80, 241–247.

Lötvall, J.O., Hui, K.P., Löfdahl, P.J., Barnes, P.J. and Chung, K.F. (1991b). Capsaicin pretreatment does not inhibit allergen-induced airway microvascular leakage in guinea-pig. Allergy 46, 105–109.

Lötvall, J., Elwood, W., Tokuyama, K., Sakamoto, T., Barnes, P.J. and Chung, K.F. (1992). A thromboxane mimetic, U-46619, produces plasma exudation in airways of the guinea-pig. J. Appl. Physiol. 72, 2415–2419.

Lundberg, J.M. and Saria, A. (1983). Capsaicin-induced desensitization of airway mucosa to cigarette smoke, mechanical and chemical irritants. Nature 302, 251–253.

Lundberg, J.M., Martling, C.R., Saria, A., Folkers, K. and Rosell, S. (1983). Cigarette smoke-induced airway oedema due to activation of capsaicin-sensitive vagal afferents and substance P release. Neuroscience 10, 1361–1368.

MacDonald, I.G. (1933). The local and constitutional pathology of bronchial asthma. Arch. Intern. Med. 6, 253–277.

Magno, M.G. and Fishman, A.P. (1982). Origin, distribution, and blood flow of bronchial circulation in anesthetized sheep. J. Appl. Physiol. 53, 272–279.

Majno, G. and Palade, G.E. (1961). Studies on inflammation. I. The effect of histamine and serotonin on vascular permeability: an electron microscopic study. J. Biophys. Biochem. Cytol. 11, 571–604.

Manning, P.J., Watson, R.M., Margolskee, D.J., Williams, V.C., Schwartz, J.I. and O'Byrne, P.M. (1990). Inhibition of exercise induced bronchoconstriction by MK-571, a potent leukotriene D4-receptor antagonist. N. Engl. J. Med. 323, 1736–1739.

Manzini, S., Maggi, C.A., Geppetti, P. and Bacciarelli, C. (1987). Capsaicin desensitization protects from antigen-induced bronchospasm in conscious guinea-pigs. Eur. J. Pharmacol. 138, 307–308.

Marchand, P., Gilroy, J.C. and Wilson, V.H. (1950). An anatomical study of the bronchial vascular system and its variations in disease. Thorax 5, 207–221.

Martling, C.R. and Lundberg, J.M. (1988). Capsaicin sensitive afferents contribute to acute airway edema following tracheal instillation of hydrochloric acid or gastric juice in the rat. Anesthesiology 68, 350–356.

Matheson, M.J., Rynell, A.-C., McClean, M.A. and Berend, N. (1994). NK-1 receptors do not mediate increases in microvascular permeability (MVP) in the trachea of the NZ white rabbit. Thoracic Society of Australia and New Zealand Annual Scientific Meeting, Hamilton Island, p. 98.

McDonald, D.M. (1988a). Neurogenic inflammation in the rat trachea. I. Changes in venules, leukocytes, and epithelial cells. J. Neurocytol. 17, 583–603.

McDonald, D.M. (1988b). Respiratory tract infections increase susceptibility to neurogenic inflammation in the rat trachea. Am. Rev. Respir. Dis. 137, 1432–1440.

McDonald, D.M. (1990). The ultrastructure and permeability of tracheobronchial blood vessels in health and disease. Eur. Respir. J. 3 (Suppl. 12), 572s–585s.

McDonald, D.M. (1992). Infections intensify neurogenic plasma extravasation in the airway mucosa. Am. Rev. Respir. Dis. 146, S40–S44.

McDonald, D.M. (1994a). The concept of neurogenic inflammation in the respiratory tract. In "Neuropeptides in the Respiratory Medicine" (eds M. Kaliner, P. Barnes, G. Kunkel and J. Baraniuk), pp. 321–349. Marcel Dekker, New York.

McDonald, D.M. (1994b). Endothelial gaps and permeability of venules in rat tracheas exposed to inflammatory stimuli. Am. J. Physiol. (Lung Cell. Mol. Physiol.) 266, L61–L83.

McDonald, D.M. (1995). Neurogenic inflammation in the airways. In "Autonomic Control of the Respiratory System" (ed. P. Barnes). Harwood Academic Publishers, London.

McDonald, D.M., Mitchell, R.A., Gabella, G. and Haskell, A. (1988). Neurogenic inflammation in the rat trachea. II. Identity and distribution of nerves mediating the increase in vascular permeability. J. Neurocytol. 17, 605–628.

McDonald, D.M., Schoeb, T.R. and Lindsey, J.R. (1991). Mycoplasma pulmonis infections cause long-lasting potentiation of neurogenic inflammation in the respiratory tract of the rat. J. Clin. Invest. 87, 787–799.

McFadden, E.R. (1992). Microvasculature and Airway Responses. Am. Rev. Respir. Dis. 145, S42–S43.

McLaughlin, R.F., Jr. (1983). Bronchial artery distribution in various mammals and in humans. Am. Rev. Respir. Dis. 128, S57–S58.

McLaughlin, R.F., Tyler, W.S. and Canada, R.O. (1961). A study of the subgross pulmonary anatomy in various mammals. Am. J. Anat. 108, 149–165.

McLaughlin, R.F., Tyler, W.S. and Canada, R.O. (1966). Subgross pulmonary anatomy of the rabbit, rat, and guinea pig, with additional notes on the human lung. Am. Rev. Respir. Dis. 94, 380–387.

Messer, J.W., Peters, G.A. and Bennett, W.A. (1960). Causes of death and pathological findings in 304 cases of bronchial asthma. Dis. Chest 38, 616–624.

Metzger, W.J., Mosely, P., Nugent, K., Richerson, H.B. and Hunninghake, G.W. (1985). Local antigen challenge and bronchoalveolar lavage of allergic asthmatic lungs. Chest 87, 155S–156S.

Metzger, W.J., Zavala, D.C., Richerson, H.B., Moseley, P.L., Iwamoto, P., Monick, M., Sjoerdsman, K. and Hunninghake, G.W. (1987). Local allergen challenge and bronchoalveolar lavage of allergic asthmatic lungs. Description of a model and local airway inflammation. Am. Rev. Respir. Dis. 135, 433–440.

Meyrick, L.C., Hoffman, L.H. and Brigham, K.L. (1984). Chemotaxis of granulocytes across bovine pulmonary artery intimal explants without endothelial cell injury. Tissue Cell 16, 1–16.

Miodonski, A., Kus, J. and Tyrankiewicz, R. (1980). Scanning electron microscopical study of tracheal vascularization in guinea pig. Arch. Otolaryngol. 106, 31–37.

Moqbel, R., Cromwell, O., Walsh, G.M., Wardlaw, A.J., Kurlak, L. and Kay, A.B. (1988). Effects of nedocromil sodium (Tilade) on the activation of human eosinophils and neutrophils and the release of H from mast cells. Allergy 43, 268–276.

Morel, N.M.L., Petruzzo, P.P., Hechtman, H.B. and Shepro, D. (1990). Inflammatory agonists that increase microvascular permeability in vivo stimulate cultured pulmonary microvessel endothelial cell contraction. Inflammation 14, 571–583.

Murai, M., Morimoto, H., Maeda, Y. and Fujii, T. (1992). Effects of the tripeptide substance P antagonist, FR113680, on airway constriction and airway edema induced by neurokinins in guinea pigs. Eur. J. Pharmacol. 217, 23–29.

Murphy, K.R., Wilson, M.C., Irvin, C.G., Glezen, L.S., Marsh, W.R., Haslett, C., Henson, P.M. and Larsen, G.L. (1986). The requirement for polymorphonuclear leukocytes in the late asthmatic response and heightened airways reactivity in an animal model. Am. Rev. Respir. Dis. 134, 62–68.

Nabishah, B.M., Khalid, B.A., Morat, P.B., Alias, A.K. and Zainuddin, M. (1992). Effects of steroid hormones on cyclic adenosine 3′,5′-monophosphate levels in the rat lung. J. Endocrinol. 134, 73–76.

Nadel, J.A. (1992). Neurogenic inflammation in airways and its modulation by peptidases. Ann. NY Acad. Sci. 664, 408–414.

Nagaishi, C. (1972). "Functional Anatomy and Histology of the Lung". University Park Press, Baltimore.

Nakamura, T., Morita, Y., Kuriyama, M., Ishihara, K., Ito, K. and Miyamoto, T. (1987). Platelet activating factor in late asthmatic responses. Int. Arch. Allergy Appl. Immunol. 82, 57–61.

Norris, A.A., Leeson, M.E., Jackson, D.M. and Holroyde, M.C. (1990). Modulation of neurogenic inflammation in the rat trachea. Pulm. Pharmacol. 3, 180–184.

Northover, A.M. (1990). Modification by some antagonists of the shape changes of venous endothelial cells in response to inflammatory agents in vitro. Agents Actions 29, 184–188.

Notkovich, H. (1957). The anatomy of the bronchial arteries of the dog. J. Thorac. Surg. 33, 242–253.

O'Byrne, P., Dolovich, J. and Hargreave, F.E. (1987). Late asthmatic responses. Am. Rev. Respir. Dis. 136, 740–751.

O'Donnell, S.R. and Persson, C.G. (1978). β₂-adrenoceptor mediated inhibition by terbutaline of histamine effects on vascular permeability. Br. J. Pharmacol. 62, 321–324.

O'Donnell, S.R., Erjefält, I. and Persson, C.G.A. (1990). Early and late tracheobronchial plasma exudation by platelet-activating factor administered to the airway mucosal surface in guinea pigs: effects of WEB 2086 and enprofylline. J. Pharmacol. Exp. Ther. 254, 65–70.

Ohrui, T., Sekizawa, K., Aikawa, T., Yamauchi, K., Sasaki, H. and Takashima, T. (1992). Vascular permeability and airway narrowing during late asthmatic response in dogs treated with metopirone. J. Allergy Clin. Immunol. 89, 933–943.

Ohuchi, K., Hirasawa, N., Takeda, H., Asano, K., Watanabe, M. and Tsurufuji, S. (1987). Mechanism of antianaphylactic action of β-agonists in allergic inflammation of air pouch type in rats. Int. Arch. Allergy Appl. Immunol. 82, 26–32.

Ohuchi, K., Watanabe, M., Hirasawa, N., Yoshizaki, S., Mue, S. and Tsurufuji, S. (1990). Suppression by adrenoceptor β-agonists of vascular permeability increase and edema formation induced by arachidonate metabolites, platelet-activating factor, and tumor-promoting phorbol ester TPA. Immunopharmacology 20, 81–88.

Ohuchi, K., Watanabe, M. and Levine, L. (1984). Arachidonate metabolites in acute and chronic allergic air pouch inflammation in rats and the anti-inflammatory effects of indomethacin and dexamethasone. Int. Arch. Allergy Appl. Immunol. 75, 157–163.

Olivenstein, R., Xu, L.J. and Martin, J.G. (1994). Microvascular leak into the airway lumen during the early and late response to antigen in rats. Am. J. Respir. Crit. Care Med. 149, A528.

Osler, W. (1892). "The Principles and Practice of Medicine, Designed for the Use of Practitioners and Students of Medicine". D. Appleton and Co., New York.

Page, C.P. and Minshall, E. (1993). Mast cells and the lung. In "Immunopharmacology of Mast Cells and Basophils" (ed. J.C. Foreman), pp. 181–195. Academic Press, London.

Persson, C.G.A. (1980). Some pharmacological aspects of xanthines in asthma. Eur. J. Respir. Dis. 109 (Suppl.), 7–16.

Persson, C.G.A. (1982). Universal adenosine receptor antagonism is neither necessary nor desirable with xanthine anti-asthmatics. Med. Hypotheses 8, 515–526.

Persson, C.G.A. (1985). On the medical history of xanthines and other remedies for asthma: a tribute to H.H. Salter. Thorax 40, 881–886.

Persson, C.G.A. (1986a). Overview of effects of theophylline. J. Allergy Clin. Immunol. 78, 780–787.

Persson, C.G.A. (1986b). The role of plasma exudation in asthmatic airways. Lancet 42, 1126–1128.

Persson, C.G.A. (1987). Cromoglycate, plasma exudation and asthma. Trends Pharmacol. Sci. 8, 202–203.

Persson, C.G.A. (1988). Plasma exudation and asthma. Lung 166, 1–23.

Persson, C.G.A. (1990). Plasma exudation in tracheobronchial and nasal airways: a mucosal defence mechanism becomes pathogenic in asthma and rhinitis. Eur. Respir. J. 3 (Suppl. 12), 652s–657s.

Persson, C.G.A. (1991a). Mucosal exudation in respiratory defence: neural or non-neural control? Int. Arch. Allergy Appl. Immunol. 94, 222–226.

Persson, C.G.A. (1991b). Plasma exudation from tracheobronchial microvessels in health and disease. In "The Bronchial Circulation" (ed. J. Butler), pp. 443–473. Marcel Dekker, New York.

Persson, C.G.A. (1991c). Plasma exudation in the airways: mechanisms and function. Eur. Respir. J. 4, 1268–1274.

Persson, C.G.A. (1993). The action of β-receptors on microvascular endothelium or: is airways plasma exudation inhibited by β-agonists? Life Sci. 52, 2111–2121.

Persson, C.G.A. and Draco, A.B. (1988). Xanthines as airway anti-inflammatory drugs. J. Allergy Clin. Immunol. 81, 615–617.

Persson, C.G.A. and Erjefält, I. (1986a). Airway microvascular permeability to large molecules. Bull. Eur. Physiopathol. Respir.: Clin. Resp. Physiol. 22 (Suppl.), 23–31.

Persson, C.G.A. and Erjefält, I. (1986b). Inflammatory leakage of macromolecules from the vascular compartment into the tracheal lumen. Acta Physiol. Scand. 126, 615–616.

Persson, C.G.A. and Erjefält, I. (1988). Non-neural and neural regulation of airway microvascular leakage of macromolecules. In "Neural Regulation of the Airways in Health and Disease" (eds M.A. Kaliner and P. Barnes), pp. 523–549. Marcel Dekker, New York.

Persson, C.G.A., Ekman, M. and Erjefält, I. (1978). Terbutaline preventing permeability effects of histamine in the lung. Acta Pharmacol. Toxicol. 42, 395–397.

Persson, C.G.A., Erjefält, I. and Karlsson, J.A. (1981). Adenosine antagonism, a less desirable characteristic of xanthine asthma drugs? Acta Pharmacol. Toxicol. (Copenh.) 49, 317–320.

Persson, C.G.A., Erjefält, I., Grega, G.J. and Svensjö, E. (1982). The role of β-receptor agonists in the inhibition of pulmonary edema. Ann. NY Acad. Sci. 384, 544–557.

Persson, C.G.A., Andersson, K.E. and Kjellin, G. (1986a). Effects of enprofylline and theophylline may show the role of adenosine. Life Sci. 38, 1057–1072.

Persson, C.G.A., Erjefält, I. and Andersson, P. (1986b). Leakage of macromolecules from guinea-pig tracheobronchial microcirculation. Effects of allergen, leukotrienes, tachykinins, and anti-asthma drugs. Acta Physiol. Scand. 127, 95–105.

Persson, C.G.A., Erjefält, I., Alkner, U., Baumgarten, C., Greiff, L., Gustafsson, B., Luts, A., Pipkorn, U., Sundler, F., Svensson, C. and Wollmer, P. (1991). Plasma exudation as a first line respiratory mucosal defence. Clin. Exp. Allergy 21, 17–24.

Peterman, W., Barth, J. and Entzian, P. (1987). Heart failure and airways obstruction. Int. J. Cardiol. 17, 207–209.

Piacentini, G.L. and Kaliner, M.A. (1991). The potential roles of leukotrienes in bronchial asthma. Am. Rev. Respir. Dis. 143, S96–S99.

Piedimonte, G., McDonald, D.M. and Nadel, J.A. (1990). Glucocorticoids inhibit neurogenic plasma extravasation and prevent virus-potentiated extravasation in the rat trachea. J. Clin. Invest. 86, 1409–1415.

Piedimonte, G., McDonald, D.M. and Nadel, J.A. (1991). Neutral endopeptidase and kininase II mediate glucocorticoid inhibition of neurogenic inflammation in the rat trachea. J. Clin. Invest. 88, 40–44.

Pietra, G.G. and Magno, M. (1978). Pharmacological factors influencing permeability of the bronchial microcirculation. Fed. Proc. 37, 2466–2470.

Pietra, G.G., Szidon, J.P., Leventhal, M.M. and Fishman, A.P. (1971). Histamine and interstitial pulmonary edema in the dog. Circ. Res. 29, 323–337.

Pietra, G.G., Szidon, J.P., Carpenter, H.A. and Fishman, A.P. (1974). Bronchial venular leakage during endotoxin shock. Am. J. Pathol. 77, 387–406.

Proud, D. and Vio, C.P. (1993). Localization of immunoreactive tissue kallikrein in human trachea. Am. J. Respir. Cell Mol. Biol. 8, 16–19.

Proud, D., Hendley, J.O., Gwaltney, J.M. and Naclerio, R.M. (1989). Recent studies on the role of kinins in inflammatory diseases of human airways. Adv. Exp. Med. Biol. 117–123.

Raeburn, D. and Karlsson, J.-A. (1993). Effects of isoenzyme-selective inhibitors of cyclic nucleotide phosphodiesterase on microvascular leak in guinea-pig airways in vivo. J. Pharmacol. Exp. Ther. 267, 1147–1152.

Rampart, M. (1994). Neutrophil-endothelial cell interactions. In "Immunopharmacology of the Microcirculation" (ed. S.D. Brain), pp. 77–107. Academic Press, London.

Rampart, M., De Smet, W., Fiers, W. and Herman, A.G. (1989). Inflammatory properties of recombinant tumor necrosis factor in rabbit skin in vivo. J. Exp. Med. 169, 2227–2232.

Raud, J. (1993). Sensory nerve activation as a potential anti-inflammatory mechanism. Pharmacol. Toxicol. 72, 30–31.

Raud, J., Lundeberg, T., Brodda-Jansen, G., Theodorsson, E. and Hedqvist, P. (1991). Potent anti-inflammatory action of calcitonin gene-related peptide. Biochem. Biophys. Res. Comm. 180, 1429–1435.

Renkin, E.M. (1992). Cellular and intercellular transport pathways in exchange vessels. Am. Rev. Respir. Dis. 146, S28–S31.

Rogers, D.F., Alton, E.W., Aursudkij, B., Boschetto, P., Dewar, A. and Barnes, P.J. (1990a). Effect of platelet activating factor on formation and composition of airway fluid in the guinea-pig trachea. J. Physiol. (Lond.) 431, 643–658.

Rogers, D.F., Dijk, S. and Barnes, P.J. (1990b). Bradykinin-induced plasma exudation in guinea-pig airways: involvement of platelet activating factor. Br. J. Pharmacol. 101, 739–745.

Rolla, G., Bucca, C., Brussino, L., Bugiani, M., Bergerone, S., Malara, D. and Morea, M. (1992). Bronchial responsiveness, oscillations of peak flow rate and symptoms in patients with mitral stenosis. Eur. Respir. J. 5, 213–218.

Rubinstein, I., Nadel, J.A., Graf, P.D. and Caughey, G.H. (1990). Mast cell chymase potentiates H-induced wheal formation in the skin of ragweed-allergic dogs. J. Clin. Invest. 86, 555–559.

Ryley, H.C. and Brogan, T.D. (1968). Variation in the composition of sputum in chronic chest disease. Br. J. Exp. Pathol. 49, 625–633.

Sabirsh, A.C., Erjefält, I.A.L. and Persson, C.G.A. (1993). Mucosal exudation of plasma following challenges restricted to the large airways. Am. Rev. Respir. Dis. 147, A558.

Saetta, M., Di Stefano, A., Rosina, C., Thiene, G. and Fabbri, L.M. (1991). Quantitative structural analysis of peripheral airways and arteries in sudden fatal asthma. Am. Rev. Respir. Dis. 143, 138–143.

Sakamoto, T., Elwood, W., Barnes, P.J. and Chung, K.F. (1992). Effect of Hoe 140, a new bradykinin receptor antagonist, on bradykinin- and platelet-activating factor-induced bronchoconstriction and airway microvascular leakage in guinea-pig. Eur. J. Pharmacol. 213, 367–373.

Sakamoto, T., Barnes, P.J. and Chung, K.F. (1993). Effect of CP-96, 345, a non-peptide NK1 receptor antagonist, against substance P-induced, bradykinin-induced and allergen-induced airway microvascular leakage and bronchoconstriction in the guinea-pig. Eur. J. Pharmacol. 231, 31–38.

Salomonsson, P., Gronneberg, R., Gilljam, H., Andersson, O., Billing, B., Enander, I., Alkner, U. and Persson, C.G.A. (1992). Bronchial exudation of bulk plasma at allergen challenge in allergic asthma. Am. Rev. Respir. Dis. 146, 1535–1542.

Salter, H.H. (1864). "On Asthma: Its Pathology and Treatment". Blanchard and Lea, Philadelphia.

Salvato, G. (1968). Some histological changes in chronic bronchitis and asthma. Thorax 23, 168–172.

Samuelsson, B. (1983). Leukotrienes: mediators of immediate hypersensitivity reactions and inflammation. Science 220, 568–575.

Saria, A. and Lundberg, J.M. (1983). Evans blue fluorescence: quantitative and morphological evaluation of vascular permeability in animal tissues. J. Neurosci. Meth. 8, 41–49.

Saria, A., Lundberg, J.M., Skofitsch, G. and Lembeck, F. (1983). Vascular protein leakage in various tissues induced by substance P, capsaicin, bradykinin, serotonin, histamine and by antigen challenge. Naunyn-Schmiedeberg's Arch. Pharmacol. 324, 212–218.

Saria, A., Hua, X., Skofitsch, G. and Lundberg, J.M. (1984). Inhibition of compound 48/80-induced vascular protein leakage by pretreatment with capsaicin and a substance P antagonist. Naunyn-Schmiedeberg's Arch. Pharmacol. 328, 9–15.

Schnittler, H.-J., Wilke, A., Gress, T., Suttorp, N. and Drenckhahn, D. (1990). Role of actin and myosin in the control of paracellular perrneability in pig, rat and human vascular endothelium. J. Physiol. (Lond.) 431, 379–401.

Schoefl, G.I. (1967). Studies on inflammation: III. Growing capillaries: Their structure and permeability. Virchows Arch. (A) 337, 97–141.

Seegal, B.C. and Seegal, D. (1927). Some observations on the lymph and blood vessels of the rabbit trachea. J. Exp. Med. 45, 203–208.

Seeger, W., Grube, C. and Gunther, A. (1993). Proteolytic cleavage of fibrinogen: amplification of its surfactant inhibitory capacity. Am. J. Respir. Cell Mol. Biol. 9, 239–247.

Sekiya, S., Yamashita, T. and Sendo, F. (1990). Suppression of late phase enhanced vascular permeability in rats by selective depletion of neutrophils with a monoclonal antibody. J. Leukocyte Biol. 48, 258–265.

Sertl, K., Kowalski, M.L., Slater, J. and Kaliner, M.A. (1988). Passive sensitization and antigen challenge increase vascular permeability in rat airways. Am. Rev. Respir. Dis. 138, 1295–1299.

Sharpe, T.J. and Smith, H. (1979). Effects of drugs on the acute inflammation following intraperitoneal injection of antigen into actively sensitized rats. Int. Arch. Allergy Appl. Immunol. 60, 216–221.

Snashall, P.D. and Chung, K.F. (1991). Airway obstruction and bronchial hyperresponsiveness in left ventricular failure and mitral stenosis. Am. Rev. Respir. Dis. 144, 945–956.

Sobin, S.S., Frasher, W.G., Tremer, H.M. and Hadley, G.G. (1963). The microcirculation of the tracheal mucosa. Angiology 14, 165–170.

Sobonya, R.E. (1984). Quantitative structural alterations in long-standing allergic asthma. Am. Rev. Respir. Dis. 130, 289–292.

Solway, J. and Leff, A.R. (1991). Sensory neuropeptides and airway function. J. Appl. Physiol. 71, 2077–2087.

Steinberg, S.F., Jaffe, E.A. and Bilezikian, J.P. (1984). Endothelial cells contain beta adrenoceptors. Naunyn-Schmiedeberg's Arch. Pharmacol. 325, 310–313.

Stenton, S.C., Court, E.N., Kingston, W.P., Goadby, P., Hendrick, D.J., Kelly, C.A. and Walters, E.H. (1990). Platelet activating factor in bronchoalveolar lavage fluid from asthmatic subjects. Eur. Resp. J. 3, 408–413.

Stephenson, J.A. and Summers, R.J. (1987). Autoradiographic analysis of receptors on vascular endothelium. Eur. J. Pharmacol. 134, 35–43.

Subramanian, N. (1986). Inhibition of immunological and non-immunological histamine release from human basophils and lung mast cells by formoterol. Arzneim-Forsch. 36, 502–505.

Sulakvelidze, I. and McDonald, D.M. (1994). Anti-edema action of formoterol in rat trachea does not depend on capsaicin-sensitive sensory nerves. Am. J. Respir. Crit. Care Med. 149, 232–238.

Sur, S., Crotty, T.B., Kephart, G.M., Hyma, B.A., Colby, T.V., Reed, C.E., Hunt, L.W. and Gleich, G.J. (1993). Sudden onset fatal asthma: a distinct entity with few eosinophils and relatively more neutrophils in the airway submucosa. Am. Rev. Respir. Dis. 148, 713–719.

Suttorp, N., Weber, U., Welsch, T. and Schudt, C. (1993). Role of phosphodiesterases in the regulation of endothelial permeability in vitro. J. Clin. Invest. 91, 1421–1428.

Svensjö, E. and Roempke, K. (1984). Microvascular aspects of oedema formation and its inhibition by $\beta_2$2-receptor stimulants and some other anti-inflammatory drugs. In "Progress in Microcirculation Research II" (eds F.C. Courtice, D.G. Garlick and M.A. Perry), pp. 449–463. Committee in Postgraduate Medical Education, Sydney, University of New South Wales.

Svensjö, E. and Roempke, K. (1985a). Dose-related antipermeability effect of terbutaline and its inhibition by a selective β_2-receptor blocking agent. Agents Actions 16, 19–20.

Svensjö, E. and Roempke, K. (1985b). Time-dependent inhibition of bradykinin- and histamine induced microvascular permeability increase by local glucocorticoid treatment. Prog. Respir. Res. 19, 173–180.

Svensjö, E., Persson, C.G.A. and Rutili, G. (1977). Inhibition of bradykinin induced macromolecular leakage from post-capillary venules by a β_2-adrenoreceptor stimulant, terbutaline. Acta Physiol. Scand. 101, 504–506.

Svensjö, E., Adamski, S.W., Su, K. and Grega, G.J. (1982). Quantitative physiological and morphological aspects of microvascular permeability changes induced by histamine and inhibited by terbutaline. Acta Physiol. Scand. 116, 265–273.

Takahashi, T., Ichinose, M., Yamauchi, H., Miura, M., Nakajima, N., Ishikawa, J., Inoue, H., Takishima, T. and Shirato, K. (1993). Neuropeptide Y inhibits neurogenic inflammation in guinea-pig airways. J. Appl. Physiol. 75, 103–107.

Taylor, A.E. and Ballard, S.T. (1992). Transvascular exchange of fluid in the airways. Am. Rev. Respir. Dis. 146, S24–S27.

Taylor, G.W., Taylor, I. and Black, P. (1989). Urinary leukotriene E4 after antigen challenge and in acute asthma and allergic rhinitis. Lancet 1, 584–588.

Tobler, A., Meier, R., Seitz, M., Dewald, B., Baggiolini, M. and Fey, M.F. (1992). Glucocorticoids downregulate gene expression of GM-CSF, NAP-1/IL-8, and IL-6, but not of M-CSF in human fibroblasts. Blood 79, 45–51.

Tokuyama, K., Lötvall, J.O., Lofdahl, C.G., Barnes, P.J. and Chung, K.F. (1991). Inhaled formoterol inhibits histamine-induced airflow obstruction and airway microvascular leakage. Eur. J. Pharmacol. 193, 35–39.

Tomioka, K., Yamada, T. and Ida, H. (1981). Anti-allergic activities of the β-adrenoceptor stimulant formoterol (BD 40A). Arch. Int. Pharmacodyn. 250, 279–292.

Traber, D.L., Lentz, C.W., Traber, L.D. and Herndon, D.N. (1992). Lymph and blood flow responses in central airways. Am. Rev. Respir. Dis. 146, S15–S18.

Tsurufuji, S., Sugio, K. and Takemasa, F. (1979). The role of glucocorticoid receptor and gene expression in the anti-inflammatory action of dexamethasone. Nature 280, 408–410.

Umeno, E., Nadel, J.A., Huang, H.-T. and McDonald, D.M. (1989). Inhibition of neutral endopeptidase potentiates neurogenic inflammation in the rat trachea. J. Appl. Physiol. 66, 2647–2652.

Umeno, E., McDonald, D.M. and Nadel, J.A. (1990). Hypertonic saline increases vascular permeability in the rat trachea by producing neurogenic inflammation. J. Clin. Invest. 85, 1905–1908.

Unger, L. (1945). Pathology of bronchial asthma with five autopsy reports. South. Med. J. 38, 513–23.

Verloop, M.C. (1948). The arteriae bronchiales and their anastomoses with the arteria pulmonalis in the human lung: a micro anatomical study. Acta Anat. 5, 171–205.

Verloop, M.C. (1949). On the arteriae bronchiales and their anastomosing with the arteria pulmonalis in some rodents: a micro-anatomical study. Acta Anat. 7, 1–32.

Vigna, S.R., Bowden, J.J., McDonald, D.M., Fisher, J., Okamoto, A., McVey, D.C., Payan, D.G. and Bunnett, N.W. (1994). Characterization of antibodies to the rat substance P (NK1 receptor and to a chimeric substance P receptor expressed in mammalian cells. J. Neurosci. 14, 834–845.

Walls, A.F., Rhee, K., Gould, D.J., Walters, C., Robinson, C., Church, M.K. and Holgate, S.T. (1991). Inflammatory mediators and cellular infiltration of the lungs in a guinea-pig model of the late asthmatic reaction. Lung 169, 227–240.

Walzer, I. and Frost, T.T. (1952). Death occurring in bronchial asthma: a report of five cases. J. Allergy 23, 204–214.

Wanner, A. (1989). Circulation of the airway mucosa. J. Appl. Physiol. 67, 917–925.

Wardlaw, A.J. and Walsh, G.M. (1994). Neutrophil adhesion receptors. In "Immunopharmacology of Neutrophils" (eds P.G. Hellewell and T.J. Williams), pp. 133–157. Academic Press, London.

Wardlaw, A.J., Hay, H., Cromwell, O., Collins, J.W. and Kay, A.B. (1989). Leukotrienes, LTC4 and LTB4 in bronchoalveolar lavage in bronchial asthma and other respiratory diseases. J. Allergy Clin. Immunol. 84, 19–26.

Wedmore, C.V. and Williams, T.J. (1981). Control of vascular permeability by polymorphonuclear leukocytes in inflammation. Nature 289, 646–650.

Wegner, C.D., Gundel, R.H., Reilly, P., Haynes, N., Letts, L.G. and Rothlein, R. (1990). Intercellular adhesion molecule-1 (ICAM) in the pathogenesis of asthma. Science 247, 456–459.

Wei, E. and Thomas, H. (1993). Anti-inflammatory peptide agonists. Annu. Rev. Pharmacol 33, 91–108.

Wei, E.T. and Kiang, J.G. (1987). Inhibition of protein exudation from the trachea by corticotropin-releasing factor. Eur. J. Pharmacol. 140, 63–67.

Weibel, E.R. (1960). Early stages in the development of collateral circulation to the lung in the rat. Circ. Res. 8, 353–376.

Weinbaum, S., Tsay, R. and Curry, F.E. (1992). A three-dimensional junction-pore-matrix model for capillary permeability. Microvasc. Res. 44, 85–111.

Whelan, C.J. and Johnson, M. (1992). Inhibition by salmeterol of increased vascular permeability and granulocyte accumulation in guinea-pig lung and skin. Br. J. Pharmacol. 105, 831–838.

Whelan, C.J., Johnson, M. and Vardey, C.J. (1993). Comparison of the anti-permeability properties of formoterol, salbutamol and salmeterol in guinea-pig skin and lung. Br. J. Pharmacol. 110, 613–618.

Widdicombe, J.G. (1993). The airway vasculature. Exp. Physiol. 78, 433–452.

Wiggs, B.R., Moreno, R., Hogg, J.C., Hilliam, C. and Pare, P.D. (1990). A model of the mechanics of airway narrowing. J. Appl. Physiol. 69, 849–860.

Williams, I.P., Rich, B. and Richardson, P.S. (1983). Action of serum on the output of secretory glycoproteins from human bronchi in vitro. Thorax 38, 682–685.

Williams, T.J. (1983). Vascular permeability changes induced by complement-derived peptides. Agents Actions 13, 451–455.

Williams, T.J. and Yarwood, H. (1990). Effect of glucocorticoids on microvascular permeability. Am. Rev. Respir. Dis. 141, S39–S43.

Williams, T.J., Das, A., von Uexkull, C. and Nourshargh, S. (1991). Neutrophils in asthma. Ann. NY Acad. Sci. 629, 73–81.

Williams, T.J., Jose, P.J., Wedmore, C.V., Peck, M.J. and Forrest, M.J. (1983). Mechanisms underlying inflammatory edema: the importance of synergism between prostaglandins, leukotrienes, and complement-derived peptides. Adv. Prostaglandin Thromboxane Leukotriene Res. 11, 33–37.

Wu, N.Z. and Baldwin, A.L. (1992a). Possible mechanism(s) for permeability recovery of venules during histamine application. Microvasc. Res. 44, 334–352.

Wu, N.Z. and Baldwin, A.L. (1992b). Transient venular

permeability increase and endothelial gap formation induced by histamine. Am. J. Physiol. 262 (Heart Circ. Physiol. 31), H1238–H1247.

Xu, L.J., Eidelman, D.H., Bates, J.H.T. and Martin, J.G. (1990). Late response of the upper airway of the rat to inhaled antigen. J. Appl. Physiol. 69, 1360–1365.

Yager, D., Shore, S. and Drazen, J.M. (1991). Airway luminal liquid. Sources and role as an amplifier of bronchoconstriction. Am. Rev. Respir. Dis. 143, S52–S54.

Yarwood, H., Nourshargh, S., Brain, S.D. and Williams, T.J. (1988). Suppression of neutrophil accumulation and neutrophil-dependent oedema by dexamethasone in rabbit skin. Br. J. Pharmacol. 95, 726.

Yarwood, H., Nourshargh, S., Brain, S. and Williams, T.J. (1993). Effect of dexamethasone on neutrophil accumulation and oedema formation in rabbit skin: an investigation of site of action. Br. J. Pharmacol. 108, 959–966.

Yi, E.S. and Ulich, T.R. (1992). Endotoxin, interleukin-l, and tumor necrosis factor cause neutrophil-dependent microvascular leakage in postcapillary venules. Am. J. Pathol. 140, 659–663.

Zimmerman, G.A., McIntyre, T.M., Mehra, M. and Prescott, S.M. (1990). Endothelial cell-associated platelet-activating factor: a novel mechanism for signaling intercellular adhesion. J. Cell Biol. 110, 529–540.

Zimmerman, G.A., Prescott, S.M. and McIntyre, T.M. (1992). Platelet activating factor. A fluid phase and cell associated mediator of inflammation. In "Inflammation: Basic Principles and Clinical Correlates" (eds J.I. Gallin, I.M. Goldstein and R. Snyderman), pp. 149–176. Raven Press, New York.

Zink, S., Rosen, P., Sackmann, B. and Lemoine, H. (1993). Regulation of endothelial permeability by β-adrenoceptor agonists: Contribution of β_1-adrenoceptor and β_2-adrenoceptors. Biochim. Biophys. Acta 1178, 286–298.

9. Regulation of Airway Smooth Muscle

Charles Twort

Immunopharmacology of the Respiratory System
ISBN 0–12–352325–7

1. Introduction

This chapter focuses on the regulation of airway smooth muscle (ASM) function. In the past it was assumed that an abnormal contractility of ASM was the underlying defect in asthma. Smooth muscle from asthmatic patients, however, does not demonstrate increased contractile responses to agonists such as histamine (H) *in vitro* (Armour *et al.*, 1984; Vincenc *et al.*, 1985; Roberts *et al*, 1985; Cerrina *et al*, 1986; Thomson, 1987), suggesting that it is not the muscle itself but the control of ASM function and of airway calibre *in vivo* which is fundamentally abnormal. Although contraction is considered to be the most significant function of the ASM cell, the less well recognized processes of relaxation and proliferation should not be ignored since they also play important roles in the pathogenesis of asthma.

The smooth muscle of the airways is morphologically similar to that of the vascular, alimentary and urogenital systems. The ASM cells are spindle shaped, with a central nucleus and prominent nucleoli, packed with longitudinally arranged myofilaments, sarcoplasmic reticulum (sr) and mitochondria. Bundles of smooth muscle cells are arranged helically around the airway, but predominantly in a circular rather than a longitudinal orientation so that contraction leads to narrowing rather than shortening of the airway (Opazo-Saez *et al.*, 1994).

In contrast to many other smooth muscle types, ASM is regarded as being multi-unit rather than single unit, which means that each smooth muscle cell is individually innervated. This is associated with a relative lack of gap junctions which normally provide a low resistance electrical pathway for cell to cell communication. Spontaneous electrical activity is rarely observed in ASM cells and action potentials only occur in exceptional circumstances, the usual electrical response to spasmogens being graded depolarization without the overall muscle bundle acting as a syncytium.

Acetylcholine is the predominant neural bronchoconstrictor of ASM. The density of cholinergic intervention is high in the central airways but reduces progressively in succeeding generations (Barnes *et al.*, 1983; Daniel *et al.*, 1986). The receptor which mediates the contractile response in ASM is the muscarinic M3 receptor which is coupled via a G-protein to phosphoinositide hydrolysis and the production of the second messenger for intracellular Ca^{2+} release – inositol 1,4,5-trisphosphate (IP_3). Another subtype of muscarinic receptor, M2, also exists in ASM. Activation of M2 receptors may support contraction by an inhibitory effect on adenylate cyclase (Yang *et al.*, 1991). A third subtype of muscarinic receptor, M_1, is not present on ASM, but excites post-ganglionic neurones in airway parasympathetic ganglia.

Adrenergic nerves do not control ASM directly, but abundant adrenoreceptors are present on ASM which respond to circulating catecholamines.

In addition to cholinergic innervation of the airways, there are neural mechanisms which act on ASM which are not blocked by cholinergic or adrenergic antagonists (Barnes, 1986). Within this non-adrenergic, non-cholinergic (NANC) system there are a number of different pathways which mediate both contraction and relaxation of ASM, and involve the release of a variety of neurotransmitters. Release is thought to occur by co-transmission from autonomic nerves rather than from a separate neural population. The peptide tachykinins, which include substance P, neurokinin A (NkA) and neurokinin B (NkB), released by the NANC systems are potent constrictors of ASM (Frossard and Barnes, 1991), while the bronchodilators include vasoactive intestinal peptide (VIP; Palmer *et al.*, 1986; Diamond *et al.*, 1991), nitric oxide (NO) and atrial naturetic peptide (ANP; Ishii and Murad 1989).

ASM is rich in receptors for many of the inflammatory mediators. It relies heavily upon pharmacomechanical coupling mechanisms for the transduction of such extracellular signals. In spite of the wide range of extracellular mediators for which the ASM cell expresses receptors, it appears that the diversity of the intracellular signalling mechanisms is more restricted, involving the release of sequestered Ca^{2+}, turnover of membrane phospholipids such as the phosphoinositides, changes in cytosolic cyclic nucleotide levels, and activation of protein kinase enzymes.

2. Contraction

2.1 CONTRACTILE PROTEINS WITHIN SMOOTH MUSCLE

Electron and light microscopy has established that, like cardiac and skeletal muscle, smooth muscle comprises actin-containing filaments and thicker myosin-containing filaments (Gerthoffer, 1991, Giembycz and Raeburn, 1992). The thin actin filaments comprise two linear polymers (MW 42 kDa) of globular actin protein wrapped together in a helical structure and in turn intertwined with another protein, tropomyosin (Fig. 9.1). The thicker myosin-containing filaments are composed of large bipolar molecules, and are arranged asymmetrically in a hexameric structure. Each myosin molecule comprises three pairs of chains, one pair of heavy chains (MW

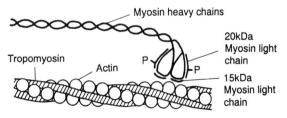

Figure 9.1   The contractile proteins of smooth muscle.

200 kDa each) and two pairs of light chains (one of 15 kDa, the other of 20 kDa). The light chains function to control binding of actin to myosin. The globular end of the myosin molecule, where the light chains are found, possesses both the site on the myosin for attachment to actin and the myosin ATPase enzymatic site, which in response to actin activation, hydrolyses ATP thereby providing the energy necessary to permit the binding of actin to myosin to occur (Adelstein and Eisenberg, 1980; Adelstein, 1983). The long tail portion of the myosin molecule imparts rigidity and length to the molecule.

The process of contraction occurs as a result of attachment of the globular myosin heads to actin. Actin then flexes in relation to myosin and the myosin head detaches and subsequently reattaches at another site further along the actin filament. This continuous attachment–detachment cycle causes the thick and thin filaments to slide over one another. This whole process is called "cross-bridge" formation and is important in the early stages of smooth muscle contraction – the rapid phasic development of tension. The myosin P light chains (MW 20 kDa) located on the head portion of the myosin macromolecule normally prevent actin–myosin interaction, and it is the phosphorylation of these myosin P light chains which commences the "cross-bridge" cycling process and the subsequent rapid development of tension.

2.2 MOLECULAR MECHANISMS OF SMOOTH MUSCLE CONTRACTION

The key biochemical trigger for initiating smooth muscle contraction is an increase in the cytosolic calcium ion (Ca^{2+}) concentration. This may be brought about by a number of mechanisms which are described later. Briefly, the rise in cytosolic Ca^{2+} may result from either an influx of extracellular Ca^{2+} across the cell membrane, or release of Ca^{2+} from intracellular stores.

The phasic component of contraction involving cross-bridge cycling occurs as a result of the interaction of elevated cytosolic Ca^{2+} combining with the Ca^{2+} receptor protein, calmodulin (CaM). The resulting Ca^{2+}/CaM complex then activates a specific Ca^{2+}/CaM-dependent enzyme, myosin light chain kinase (MLCK; Adelstein et al., 1982). This in turn leads to the phosphorylation of the myosin head P light chain (MW 20 kDa), which promotes the association between actin and myosin, and so the onset of contraction. This pathway for the initiation of contraction is described as a "thick filament-regulated" process.

There is an additional controlling pathway which is "thin filament regulated"; this is dependent upon another protein, caldesmon, which is closely associated with the actin/tropomyosin–myosin domain. Caldesmon derives its name from its intrinsic ability to bind to calmodulin with a much greater affinity than to the thin

filament domain. It functions as an inhibitor of actin/tropomyosin–myosin interaction. In the absence of the Ca^{2+}/CaM complex, caldesmon binds to actin–tropomyosin filaments (thin filaments) preventing cross-bridge formation and therefore contraction (Sobue et al., 1981, 1982). Increasing cytosolic Ca^{2+} allows Ca^{2+}/CaM to form, leading to the association of Ca^{2+}/CaM with caldesmon and the removal of the caldesmon-induced inhibition of contraction. Thus, interaction of raised cytosolic Ca^{2+} with CaM controls both the thick and thin filament-regulated mechanisms of contraction.

These mechanisms account for the phasic component of ASM contraction, and presuppose the continued elevation of intracellular Ca^{2+} levels and phosphorylation of myosin light chains during the sustained component of contraction. In fact, during the sustained phase of ASM tension intracellular Ca^{2+} concentration falls to a level just above basal (see Fig. 9.2). This anomaly of maintained contraction without maintained Ca^{2+} has led to the proposal of non-cycling or slowly cycling cross-bridges called "latch bridges" (Dillon et al., 1981; Hai and Murphy, 1988; Murphy et al, 1990), which maintain contraction at greatly reduced levels of myosin phosphorylation and energy consumption. In canine ASM latch bridges start cycling approximately 25 s after activation and cycle at one-quarter the rate of cross-bridges (Gerthoffer, 1991). One of the key characteristics of the latch bridge state is an enhanced sensitivity of the contractile machinery to Ca^{2+}.

There is evidence to suggest that an intracellular enzyme, protein kinase C (PKC), plays a role in the maintenance of ASM tension during the latch phase. It is envisaged that PKC phosphorylates myosin light chains directly, allowing actin and myosin to associate, or to facilitate the interaction of these contractile proteins at sites other than the myosin light chains. Two intracellular second messengers are produced in response to cell surface receptor activation by contractile agonists. These are IP_3, a water-soluble molecule which activates release of intracellularly stored Ca^{2+}, and diacylglycerol (DAG), which unlike IP_3 is lipid soluble and so remains in the plasmamembrane where it activates the enzyme PKC (Berridge and Irvine, 1984; Hashimoto et al., 1985).

PKC is one member of a family of serine and threonine kinases – so-called because they catalyse the phosphorylation from ATP of serine and threonine amino acid bases found in many different proteins. PKC is distinguished from other protein kinases because it is dependent on Ca^{2+} and also the phospholipid phosphatidylserine (PS) for activity. When DAG appears free in the plasma membrane, quiescent PKC "translocates" from the cytosol to the plasma membrane, binding at sites rich in PS. It is this binding of PKC to membrane PS, in the presence of DAG, which confers increased Ca^{2+} sensitivity to the enzyme and allows release of the catalytic subunit of PKC into the cytosol where target protein phosphorylation

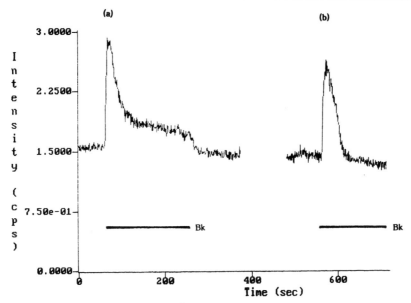

Figure 9.2   Time course of change in cytosolic Ca²⁺ in response to bradykinin in human ASM in the presence (a) and the absence (b) of extracellular Ca²⁺. Cytosolic Ca²⁺ is represented as the ratio of fluorescent intensity at 340/380 nm excitation wavelengths in Fura-2-loaded cells.

can then occur. Thus in the presence of DAG, the threshold concentration of Ca^{2+} required to activate PKC is reduced to near resting Ca^{2+} levels, allowing PKC to function at a Ca^{2+} concentration which exists during the sustained phase of agonist activation. PKC exists as a number of different isoenzymes which exert their effect by catalysing the phosphorylation of cytosolic regulatory proteins from the inactive to the active phosphorylated form (Kikkawa *et al.*, 1989).

Increases in intracellular DAG in ASM have been confirmed in response to muscarinic receptor stimulation (Takuwa *et al.*, 1986). DAG production in airway smooth muscle continues beyond the disappearance of IP_3. This raises the possibility that during continued agonist activation DAG may additionally be derived from other membrane phospholipids without IP_3 production.

Evidence for the importance of PKC activation for airway smooth muscle contraction has arisen from the experimental use of phorbol esters to stimulate PKC directly. PKC influences the sustained phase of the contractile response in a number of ways: by direct phosphorylation of actomyosin (Park and Rasmussen, 1986) and by inhibiting mechanisms of relaxation (Obanime *et al.*, 1989). Most importantly, however, PKC activation seems to increase greatly myofilament sensitivity to Ca^{2+} (Nishimura and van Breemen 1989; Gerthoffer, 1991), and this may explain the maintenance of sustained contraction during the "latch phase", despite a reduction in cytosolic Ca^{2+} to near resting levels (as in Fig. 9.2). In addition, PKC activation has been shown to act as a modulator of IP_3, both by activating its degradation via the 5-phosphatase enzyme, and by influencing its formation via phospholipase C (PLC) (Fig. 9.4).

2.3   THE ROLE OF CYTOSOLIC CALCIUM

The concentration of free calcium ions (Ca^{2+}) in the cytosol of ASM is central to the contractile response. The concentration of Ca^{2+} in the cytosol is determined by the relative activity of processes which deliver Ca^{2+} to the cytosol and which remove Ca^{2+} from it. Calcium may be delivered either by influx of extracellular Ca^{2+}, or by release of Ca^{2+} stored in intracellular organelles, both processes involving the movement of Ca^{2+} down an electrochemical gradient from pools of high concentration. Conversely, Ca^{2+} is removed from the cytosol by energy-requiring pumps and by ion-exchange mechanisms which extrude Ca^{2+} extracellularly, or which refill the intracellular stores.

2.3.1   Intracellular Calcium Concentration During Contraction

The agonist-induced rise in cytosolic Ca^{2+} associated with contraction can be measured in cultured ASM loaded with the intracellular Ca^{2+}-sensitive fluorescent dye Fura-2 (Fig. 9.2). The response is biphasic – an initial sharp rise in cytosolic Ca^{2+} occurs which peaks within the first minute following exposure to agonist. Cytosolic Ca^{2+} then falls reaching a plateau level of elevated Ca^{2+} which is sustained for the duration of exposure to agonist (Kotlikoff *et al.*, 1987; Panettieri *et al*, 1989; Senn *et al.*, 1990; Murray and Kotlikoff, 1991; Shieh *et al.*, 1991).

Contractile tension is maintained during this sustained period of only modest elevation of cytosolic Ca^{2+} by the accompanying processes involving PKC which increase myofilament sensitivity to Ca^{2+}. Calcium release from intracellular stores is responsible for the initial transient rise in cytosolic Ca^{2+}, whereas the sustained plateau elevation is provided by influx of extracellular Ca^{2+}. Following agonist activation in the absence of extracellular Ca^{2+} only the initial peak response occurs and the plateau rise cannot be sustained. Thus, Ca^{2+} released from intracellular stores activates the initial phasic contractile response to agonist and the subsequent tonic response is maintained by elevated Ca^{2+} originating from extracellular sources. The initial phasic response to agonist activation reduces the Ca^{2+} content of the intracellular store. It is unknown whether the store then refills or remains depleted in the continuing presence of agonist during the tonic phase of contraction when cytosolic Ca^{2+} is dependent on extracellular Ca^{2+}. Available evidence suggests that the store remains depleted (Twort et al., 1992).

2.3.2 Influx of Extracellular Calcium

The free Ca^{2+} concentration in the extracellular fluid surrounding airway smooth muscle is approximately 1.5 mM, while cytosolic Ca^{2+} is below the micromolar range. The large electrochemical gradient across the plasmalemmal membrane results in a continuous passive leak of Ca^{2+} into the cytosol. This leak of Ca^{2+} into the cytosol is normally compensated for by the active Ca^{2+} removal mechanisms (discussed later) which return Ca^{2+} to the extracellular space, thereby preventing a rise in the cytosolic Ca^{2+} concentration.

In addition to this passive leak, extracellular Ca^{2+} may gain access to the cytosol by passing through ion channels in the cell membrane (Fig. 9.3). As for other smooth muscle, two types of Ca^{2+} channel have been proposed

for airway smooth muscle: voltage-dependent channels (VDCs) and receptor-operated channels (ROCs; Bolton, 1979; Small and Foster, 1986; Giembycz and Rodger, 1987).

2.3.2.1 Voltage-dependent Channels

As the name suggests, the Ca^{2+} conductance of these channels is dependent on the cell membrane potential. With a reduction in the membrane potential (depolarization), as occurs with exposure to solutions containing high K^+ concentration, both the probability of individual channels being open and the duration of their open state increases (Kotlikoff, 1988), resulting in influx of extracellular Ca^{2+}. VDCs are believed to be important in cells which exhibit action potentials, such as smooth muscle of the gut or blood vessels. ASM cells, however, have a stable resting membrane potential of between -50 and -60 mV (Small and Foster, 1988), and action potentials are not normally observed. In ASM exposure to increasing concentrations of depolarizing agents such as K^+ results in a graded depolarization which is correlated with an increase in contractile force, rather than the production of action potentials. Studies in ASM demonstrate that agonist-induced contractions occur independently of changes in membrane potential (Coburn, 1979; Ahmed et al., 1984), suggesting that opening of VDCs is not involved. Drugs which characteristically block VDCs, such as verapamil, nifedipine and diltiazem, have little effect on agonist-induced contractions (Kirkpatrick, 1975; Ito and Itoh, 1984). Despite these findings, the presence of VDCs is not questioned in ASM (Kotlikoff, 1988; Marthan et al., 1989), but it seems that they have little

- **Influx of extracellular Ca^{2+}**
  - Passive Ca^{2+} leak
  - Ion channels: Voltage dependent channel (VDC)
    Receptor operated channel (ROC)

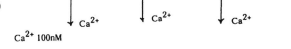

- **Release of intracellularly stored Ca^{2+}**

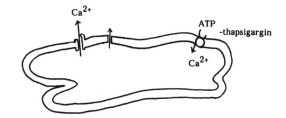

Figure 9.3   Sources of Ca^{2+} for the rise in cytosolic Ca^{2+}.

physiological importance in the contractile response to agonists.

2.3.2.2 Receptor-operated Channels

In contrast to VDCs, the permeability of ROCs to the influx of Ca^{2+} is not directly determined by changes in membrane potential (Bolton, 1979) and is governed by cell surface receptors. Unlike VDCs, ROCs are not purely selective for Ca^{2+}, frequently having a greater conductance for Na^+. They are resistant to blockage by organic inhibitors of Ca^{2+} influx which block VDCs (Murray and Kotlikoff, 1991), but can be blocked by novel compounds, H131, SKF96356 and by high concentrations of D600. Although the existence of ROCs is supported in vascular smooth muscle (Ruegg et al., 1989) there is only a little evidence, to date, to suggest that this type of Ca^{2+} influx channel exists in the ASM cell (Murray and Kotlikoff, 1991).

2.3.2.3 Passive Leak

Despite this paucity of evidence for physiologically relevant VDCs or ROCs, influx of extracellular Ca^{2+} is undoubtedly crucial for the maintenance of agonist-induced contraction of airway smooth muscle. In the absence of extracellular Ca^{2+}, tissue strips of airway smooth muscle in vitro will fail to maintain a contraction in response to agonist stimulation, and the sustained plateau of elevated cytosolic Ca^{2+} in Fura-2-loaded cells will be abolished (Fig. 9.2). One possible explanation for the net influx of extracellular Ca^{2+} in these circumstances (in the absence of VDCs or ROCs) is that, in the continuing presence of agonist, mechanisms for removing intracellular Ca^{2+} may be suppressed, thereby allowing the passive leak to provide sufficient Ca^{2+} to raise cytosolic levels.

Some evidence for this explanation is provided by studies employing two compounds – thapsigargin (see Fig. 9.3) and cyclopiazonic acid – which deplete sr Ca^{2+} by selectively blocking the sr Ca^{2+}-ATPase. These compounds have been used to study sr function in smooth muscle (Thastrup et al., 1989; Chen et al., 1992; Daniel et al., 1992; Low et al., 1992; Murray et al., 1994). In vascular smooth muscle thapsigargin abolishes Ca^{2+} uptake into the IP_3-sensitive store (Missiaen et al., 1991); it also causes a sustained rise in cytosolic Ca^{2+} in unstimulated cells indicating the importance of sr for buffering passive rises in cytosolic Ca^{2+} (Chen et al., 1992). A similar result is obtained in cultured human (Twort, 1994) and canine (Murray et al., 1994) ASM. Exposure of Fura-2-loaded human ASM cells to thapsigargin (1 μM) results in a rise in cytosolic Ca^{2+} which reaches a steady state level, similar to that achieved during the sustained tonic phase of the response to a contractile agonist. This is interpreted as resulting from the inability of the sr in the presence of thapsigargin to buffer extracellular Ca^{2+} influxing via plasmalemmal leak. Addition of the agonist bradykinin (Bk) , after this brief period of

thapsigargin exposure, results in a transient cytosolic Ca^{2+} spike due to release of that quantity of stored Ca^{2+} in the sr which has yet to leak out of the sr. The subsequent elevated plateau level of cytosolic Ca^{2+} in the presence of Bk is no different from that before application of Bk. This suggests that in the absence of the buffering function of the sr (due to thapsigargin) simple passive leak of Ca^{2+} through the plasmalemma can account for the plateau rise in Ca^{2+} which accompanies prolonged agonist activation of ASM.

The path of entry of extracellular Ca^{2+} during the sustained phase of contraction in ASM has eluded definition in the absence, to date, of the demonstration of physiologically relevant VDCs or ROCs. These data raise the possibility that following agonist activation, when the buffering capacity of the sr is abolished, the passive plasmalemmal leak can provide sufficient extracellular Ca^{2+} to sustain the agonist-induced, or "receptor-mediated" (Murray and Kotlikoff, 1991), plateau rise in cytosolic Ca^{2+} which accompanies maintained contraction.

2.3.3 The Intracellular Calcium Store

A role for intracellularly sequestered Ca^{2+} in the contraction of ASM was first suggested by Kirkpatrick (1975), who demonstrated that removal of extracellular Ca^{2+} reduced, but did not completely abolish, contractions of bovine trachealis in response to acetylcholine and H. This contrasted with depolarization-induced contractions which were entirely dependent on extracellular Ca^{2+}. Subsequent studies demonstrated that the relative contribution of Ca^{2+} from extracellular and from intracellular sources during contraction of airway smooth muscle varied (Farley and Miles, 1978; Coburn, 1979; Creese and Denborough, 1981; Foster et al., 1983a,b; Raeburn et al., 1986; Nouailhetas et al., 1988). The rapidly developing phasic component of the contractile response, occurring at the onset of agonist induced contraction, depended on release of intracellularly stored Ca^{2+}, whereas the subsequent tonic component of contraction relied on influx of extracellular Ca^{2+} (Ito and Itoh, 1984).

Removal of extracellular Ca^{2+} does not significantly impair either the initial development of tension (as opposed to the maintained contraction) or the transient peak in cytosolic Ca^{2+} in Fura-2-loaded ASM which occur in response to agonist stimulation, but it does abolish the plateau level of raised Ca^{2+} during the maintained phase (Fig. 9.2).

2.3.3.1 The Identity of the Calcium Store

The sr has been identified as the major intracellular source of activator Ca^{2+} in both smooth and striated muscle (Somlyo et al., 1981; Bond et al., 1984; Somlyo, 1985). The sr of smooth muscle is a system of membranous tubules which has components closely underlying the surface plasmalemmal membrane, as well as deeper portions contiguous with the double membrane of the nuclear envelope (Somlyo and Franzini-Armstrong, 1985). The portion of the sr network which lies close to the inner surface of the plasmalemmal membrane intermittently fuses with the plasmalemma via specialized junctions, which appear to contain structural "feet" separating the two membranes (Somlyo and Franzini-Armstrong, 1985). Calcium stored within the sr is bound to the high capacity, low affinity Ca^{2+}-binding protein calsequestrin (Wuytack et al, 1987).

Cytosolic Ca^{2+} is transferred into the store by a 105 kDa Ca^{2+}, Mg^{2+}-ATPase situated in the sr membrane. This enzyme pump undergoes Ca^{2+}-dependent phosphorylation (Sumida et al., 1984). It is also regulated through cAMP-mediated phosphorylation of phospholamban (Raeymaekers et al., 1986; Eggermont et al., 1988; Twort and van Breemen, 1989), as a result of which elevation of cytosolic cAMP lowers cytosolic Ca^{2+} by stimulating uptake into the sr.

The Ca^{2+} content of the sr is determined by a balance between the activity of the Ca^{2+}-ATPase pump and mechanisms which release Ca^{2+} from the store (Fig. 9.3). These mechanisms include the opening of ion channels in the sr membrane in response to IP_3 and passive leak of Ca^{2+} out of the sr. Both of these processes involve the movement of Ca^{2+} from the high concentration within the store, estimated to be 5 mM Ca^{2+} (Leijten and van Breemen, 1984), to the low concentration (100 nM) within the cytosol. Functional studies in vascular smooth muscle have demonstrated that the quantity of Ca^{2+} stored in the sr is sufficient to activate maximal contraction (Bond et al., 1984). An equivalent amount, 253 μmol Ca^{2+} l^{-1} of cells, is released from sr in human ASM in response to agonist activation (Twort and van Breemen, 1989).

Mitochondria are not involved in storing activator Ca^{2+} in smooth muscle (Twort and van Breemen, 1989). Despite the importance of Ca^{2+} transport systems within the mitochondria for controlling cellular metabolism, the endogenous Ca^{2+} content of mitochondria is low (Bond et al., 1984), and mitochondria in smooth muscle only accumulate Ca^{2+} at pathological (10 μM) rather than physiological (100 nM) cytosolic Ca^{2+} concentrations (Yamamoto and van Breemen, 1986).

2.3.3.2 The Store as a Source of Activator Calcium

The second messenger linking activation of surface receptors by spasmogens in smooth muscle to release of intracellular Ca^{2+} has been identified as IP_3. IP_3 is formed by activated receptor stimulation of a cell membrane enzyme, PLC, which hydrolyses a lipid constituent of the cell membrane, phosphatidylinositol bisphosphate (PIP_2; Fig. 9.4).

PIP_2 is formed in the cell membrane by the successive phosphorylation of the membrane phospholipid phosphatidylinositol (PI) to phosphatidylinositol 4-monophosphate (PIP) and then to PIP_2. PIP_2 is readily recycled by specific phosphomonoesterases to PIP and eventually back to PI. This cycle continues at rest until stimulation of surface receptors occurs. Receptor

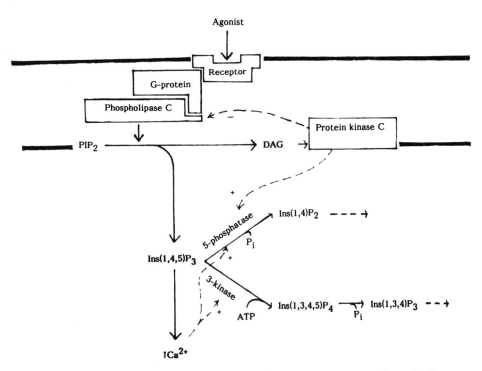

Figure 9.4 Phosphoinositide metabolism in ASM in response to agonist activation.

stimulation activates the membrane PLC thereby diverting PIP_2 from the cycle above and hydrolysing PIP_2 to form IP_3 and DAG. The step between receptor stimulation and activation of PLC in ASM is transduced by a guanine nucleotide regulatory protein (G-protein; Rodger, 1985; Chilvers and Nahorski, 1990; Birnbaumer, 1990).

This pathway is ubiquitous in many cell types and has also been implicated in the mechanism of pharmacomechanical coupling in ASM. Phosphoinositol metabolism and rapid rises in the cellular levels of IP_3 follow stimulation by a variety of agonists – carbachol, histamine, neurokinins and leukotrienes in canine, guinea-pig and bovine ASM (Hashimoto et al., 1985; Grandordy et al., 1986, 1988; Duncan et al, 1987; Kardasz et al., 1988; Chilvers et al., 1989, 1991). This rise in IP_3 precedes contraction (Duncan et al., 1987).

The initial high levels of IP_3 in response to agonist–receptor interaction are not maintained during the sustained phase of contraction (Chilvers et al, 1989; Chilvers and Nahorski, 1990). IP_3 falls to baseline levels within a minute of the onset of contraction. $I(1,4,5)P_3$ is metabolized by two pathways, both of which are activated by increases in cytosolic Ca^{2+}: hydrolysis by a 5-phosphatase enzyme to $I(1,4)P_2$, or phosphorylation to $I(1,3,4,5)P_4$ and subsequent hydrolysis to its inactive isomer $I(1,3,4)P_3$ (Chilvers and Nahorski, 1990). No physiological role for $I(1,3,4,5)P_4$ has yet been identified in airway smooth muscle, although in other cells evidence exists that $I(1,3,4,5)P_4$ may modulate plasmalemmal Ca^{2+} ion channels (Irvine and Moor, 1986). The only phosphoinositide metabolite which has been shown to release stored Ca^{2+} in ASM is $I(1,4,5)P_3$.

IP_3, released at the plasmalemma, diffuses across the cytosol and binds to an IP_3 receptor on the sr membrane. This IP_3 receptor is a Ca^{2+} channel protein (Berridge, 1993). The IP_3 receptor contains membrane spanning domains in the C-terminal region which anchor the protein in the sr membrane. Four subunits combine to form the functional IP_3-sensitive Ca^{2+} channel (Berridge, 1993). The N-terminal domain lies free in the cytosol exposing the IP_3 binding site (Mignery and Sudhof, 1990). Following IP_3 binding the receptor undergoes a conformational change which increases the probability of the open state of the channel (Ehrlich and Watras, 1988; Watras et al., 1991), resulting in Ca^{2+} efflux from sr to cytosol. It is as yet unclear whether channel opening requires sequential IP_3 binding to the four sites on the receptor (Meyer et al., 1990), or whether the binding of a single molecule of IP_3 is sufficient to result in channel opening (Meyer et al., 1990; Watras et al., 1991).

The IP_3 receptor in ASM is competitively blocked by low molecular weight heparin (Chopra et al., 1989). The sensitivity of this effect is high, half maximal inhibition of IP_3-induced Ca^{2+} release occurring at 0.8 $\mu g\,ml^{-1}$ low molecular weight heparin. The inhibitory effect is structurally specific to low molecular weight heparin and its

analogue chondroitin sulphate A. Very little inhibition is obtained with high molecular weight heparin or with De-N-sulphated heparin, suggesting that binding of heparin to the IP_3 receptor depends on the relative positions of the different sulphated residues and of the N residue on the hexosamine unit of heparin.

IP_3 binding to the receptor is also inhibited by cytosolic Ca^{2+}, half maximal inhibition occurring at 300 nM Ca^{2+} (Worley et al., 1987; Theibert et al., 1987), a cytosolic concentration present in agonist-activated ASM (Kotlikoff et al., 1987; Panettieri et al, 1989; Senn et al., 1990; Murray and Kotlikoff, 1991; Shieh et al., 1991). This effect may act as a negative feedback mechanism controlling Ca^{2+} release.

Estimates of the quantity of stored Ca^{2+} released in response to IP_3 vary depending on the preparation of ASM under study. In freshly isolated ASM from canine trachealis, maximal IP_3 releases 40% of stored Ca^{2+} (Hashimoto et al., 1985), whereas in cultured ASM cells from human bronchi a greater proportion, 84%, is released (Twort and van Breemen, 1989). This discrepancy may result from variations of the sr in ASM under conditions of tissue culture as the cells adjust from contractile to proliferative phenotype.

The intracellular Ca^{2+} store in smooth muscle will release Ca^{2+} in response to a variety of stimuli, both physiological and pharmacological. The most important effector of Ca^{2+} release is IP_3. Guanosine nucleotides have also been found to induce Ca^{2+} release in ASM, and may additionally promote translocation of Ca^{2+} between separate fractions of the overall store (Gill et al., 1988; Mullaney et al., 1988; Ghosh et al., 1989). This is not mediated by G-protein activation of IP_3 production at the plasmalemma, since neomycin, an inhibitor of PLC (Cockcroft and Gomperts, 1985; Cockcroft et al., 1987), does not affect the response (Chopra et al., 1991), indicating that the guanosine triphosphate (GTP) effect is occurring at the level of the intracellular Ca^{2+} store and not the plasmalemma. Further insight into such possible heterogeneity of the store in ASM has been provided by observation of the effects of the classical pharmacological agents which release stored Ca^{2+} in striated muscle – ryanodine (Chopra et al., 1991) and caffeine (Ito and Itoh, 1984; Nouaihletas et al., 1988; Small et al., 1988; Chopra et al., 1991).

2.3.4 Mechanisms for Removing Cytosolic Calcium

There are three main mechanisms for reducing cytosolic Ca^{2+} concentration. These mechanisms maintain the concentration gradient of Ca^{2+} across the cell or sr membranes. Activation of these processes will reduce cytosolic Ca^{2+} and will produce relaxation (Fig. 9.5).

2.3.4.1 Reuptake of Calcium into the Intracellular Store

The sr Ca^{2+} store empties during agonist activation of

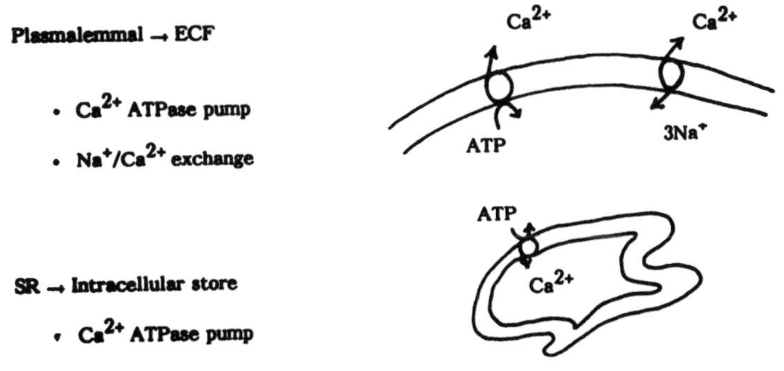

Figure 9.5   Mechanisms for lowering cytosolic Ca^{2+}. ECF, extracellular fluid; SR, sarcoplasmic reticulum.

ASM. Following removal of the agonist, relaxation occurs, cytosolic Ca^{2+} returns to resting levels and the sr refills with Ca^{2+}. Reuptake of Ca^{2+} results from the continuing activity of the sr Ca^{2+}-ATPase pump. Loading of the sr occurs since, in the absence of agonist, this pump activity exceeds efflux of Ca^{2+} from the sr via leak mechanisms or IP$_3$-gated channels. Both the rate of Ca^{2+} transport by the sr Ca^{2+}-ATPase and the steady-state Ca^{2+} content of the sr are increased in ASM by elevated cytosolic concentrations of cAMP (Twort and van Breemen, 1989). This effect of cAMP of increasing sr Ca^{2+} uptake is one of the many mechanisms by which cAMP augments relaxation of ASM (see Fig. 9.7).

2.3.4.2   Plasmalemmal Ca^{2+}-ATPase Pump
Ca^{2+} is extruded extracellularly from the cytosol by a Mg^{2+}-dependent Ca^{2+}, H$^+$-ATPase pump (Hogaboom and Fedan, 1981). This is an electrically neutral process by which Ca^{2+} is exchanged for 2H$^+$. In resting smooth muscle the Ca^{2+} pump exists in a low affinity state which is capable of extruding sufficient Ca^{2+} to maintain steady state in the face of Ca^{2+} influx across the plasmalemma down its concentration gradient via the passive leak mechanism (Carafoli, 1984).

In stimulated cells the pump adopts a high affinity state capable of extruding up to seven times more Ca^{2+}. Conversion of the pump to the high affinity state is achieved via activation by calmodulin, and by increases in cytosolic cAMP (Suematsu et al., 1984).

2.3.4.3   Sodium–Calcium Exchange
An additional method for extruding Ca^{2+} is a plasmalemmal exchange mechanism by which influxing Na$^+$ is exchanged for effluxing Ca^{2+} (Bullock et al., 1981). This process does not require ATP directly, the energy for the extrusion of Ca^{2+} being provided by the influx of Na$^+$ down its concentration gradient into the cell. However, energy is required indirectly since the Na$^+$ gradient is maintained by the Na$^+$, K$^+$-ATPase pump.

2.3.5   The Store – A Modulator of Extracellular Calcium Influx
In addition to this function of controlling the overall cytosolic Ca^{2+} concentration adjacent to the myofilaments in the deep cytosol, there is evidence that the sr also plays a role in controlling the influx of extracellular Ca^{2+} through the plasmalemmal membrane. This has led to the independent proposal of two concepts – the "superficial buffer barrier hypothesis" and the "capacitative model" – both of which attempt to link sr function to extracellular Ca^{2+} influx.

2.3.5.1   The Superficial Buffer Barrier Hypothesis
This hypothesis, proposed and subsequently reviewed by van Breemen (Loutzenhiser and van Breemen, 1983; van Breemen et al., 1986, 1988; van Breemen and Saida, 1989; Chen et al., 1992), envisages that in unstimulated resting smooth muscle cells, extracellular Ca^{2+} enters the cytosol via the leak pathway and/or via open Ca^{2+} channels. This Ca^{2+} is then largely taken up via the sr Ca$^+$-ATPase pump into superficially located sr lying only a few nanometres below the inner surface of the plasmalemma. As a result, there exists an irregular, outwardly directed Ca^{2+} gradient in the narrow region between plasmalemma and superficial sr. Influxing Ca^{2+}, having been taken up by the superficial sr in the resting cell, is then extruded extracellularly at sites of close apposition between the sr and the plasmalemma by the classical extrusion mechanisms of the plasmalemmal Ca^{2+}-ATPase and Na$^+$/Ca^{2+} exchange. In this way most of the Ca^{2+} entering the resting cell is prevented from reaching the bulk of the cytoplasm and contraction is not activated. In the presence of agonist, sr Ca^{2+} channels open and the superficial sr loses its ability to buffer Ca^{2+} influx; Ca^{2+} taken up by the superficial sr Ca^{2+}-ATPase effluxes from the superficial sr directly via the open channels and results in a rise in deep cytosolic Ca^{2+} concentration. This hypothesis proposes that the Ca^{2+} content of the sr controls the response to extracellular Ca^{2+} influx – an empty store will buffer efficiently any influxing Ca^{2+},

while a replete store will allow more influxing Ca^{2+} through to the bulk cytoplasm.

Some support exists for the superficial buffer barrier in ASM. Exposure of thin strips of rabbit trachealis to 80 mM K^+ or to carbachol (300 nM) results in a contraction within 15 s of exposure. Prior progressive depletion of the sr Ca^{2+} store by passive leak following pre-exposure to increasing durations of Ca^{2+}-free solutions results in an increasing delay, up to 115 s, in the onset of contraction following reintroduction of extracellular Ca^{2+} and agonist (Nouaihletas et al., 1988). This delay is proportional to the depletion of the sr Ca^{2+}, and is assumed to result from the requirement for the empty superficial sr to re-fill before sufficient influxing Ca^{2+} can reach the contractile apparatus. Conversely, prior exposure to 80 mM K^+ in Ca^{2+}-containing extracellular medium will augment subsequent carbachol-induced contractions in Ca^{2+}-free medium (Nouaihletas et al., 1988), indicating that the sr in ASM can take up Ca^{2+} influxing through voltage-dependent plasmalemmal channels.

2.3.5.2   *The Capacitative Model*

An additional model which aims to integrate observations that the Ca^{2+} content of the sr regulates Ca^{2+} fluxes across the plasmalemma is the "capacitative model", expounded by Putney (1986, 1990).

This model proposes that the degree of filling of the agonist-releasable Ca^{2+} store controls Ca^{2+} influx, with an empty store (as follows agonist stimulation) promoting Ca^{2+} influx by an as yet undiscovered mechanism. As in the superficial buffer barrier hypothesis, this model proposes an isolated compartment of cytosol between the plasmalemma and the superficial sr. The local Ca^{2+} concentration in this compartment modulates the Ca^{2+} permeability of the plasmalemma in that region. Ca^{2+} concentration in this compartment depends on three processes: influx across the plasmalemma, leak from the sr store and pumping into the sr store. It is proposed that agonist-mediated production of IP3 discharges the store into the deep cytosol, the store empties and Ca^{2+} concentration in the local subplasmalemmal compartment falls, reducing Ca^{2+} bound to the inner surface of the plasmalemma and increasing Ca^{2+} permeability. In this way emptying of the intracellular store by IP3 is the stimulus for extracellular Ca^{2+} influx across the plasmalemma.

3.   *Relaxation*

It would be naive to think that at any one time ASM is either in a contracted or a relaxed state. ASM *in vivo* has a degree of tone, and the extent of this tone will depend on the balance between the mechanisms producing contraction and relaxation. Relaxation, therefore, is relative, and results both from a reduction in the forces of contraction and from processes which actively produce relaxation. Relaxation will occur as stimulation of surface receptors by bronchoconstrictor agonists, and the resulting elevation of cytosolic Ca^{2+}, reduces. It will also occur as the result of stimulation of active processes of relaxation, most of which also result in a reduction of cytosolic Ca^{2+}.

3.1   PHARMACOMECHANICAL COUPLING

There are two established mechanisms for relaxation of ASM which involve pharmacomechanical coupling: stimulation of β-adrenoceptors with the production of increases in cytosolic cAMP via activation of adenylate cyclase (Rinard et al., 1983; Suematsu et al., 1984; Fig. 9.6); and activation of guanylate cyclase with resulting increases in cGMP.

3.1.1   Cyclic AMP

When adrenaline or synthetic β-adrenergic agonists (used as bronchodilators in asthmatics) bind to the β-adrenoreceptor on the cell surface of ASM, a reversible ligand–receptor complex is formed which stabilizes the receptor and allows its binding to a distinct GTP binding protein, Gs, in turn facilitating the binding of GTP. This binding of GTP promotes the dissociation of Gs from the receptor–ligand complex, allowing the GTP-bound Gs molecule (Gs-GTP) to migrate in the plane of the membrane. When the Gs-GTP complex binds to the regulatory site of the adenylate cyclase enzyme, the hydrolysis of ATP occurs to yield the second messenger cAMP. Eventually inactivation of adenylate cyclase occurs to limit the synthesis of cAMP. This is achieved by the action of an enzymatic site (GTPase) located on Gs, which hydrolyses GTP to GDP initiating the dissociation of Gs from adenylate cyclase and switching off the formation of cAMP. The β-adrenoreceptor is one of many receptors competing for the pool of Gs and adenylate cyclase molecules.

In addition to activation by β-adrenoreceptors, adenylate cyclase in ASM is activated by VIP, a neurotransmitter of the NANC nervous system (Rhoden and Barnes, 1990).

An additional mechanism controlling cAMP levels is metabolism by a specific phosphodiesterase enzyme, a cAMP-dependent phosphodiesterase. Such an enzyme exists in ASM and catalyses the degradation of cAMP to 5'-adenosine monophosphate.

The biological action of cAMP, in this case relaxation of ASM, is believed to be mediated through activation of a cAMP-dependent protein kinase, also known as protein kinase A (PKA). This enzyme, like PKC, belongs to a family of serine and threonine protein kinases, which catalyse the phosphorylation of serine and threonine amino acids in target proteins.

There are a number of different mechanisms whereby

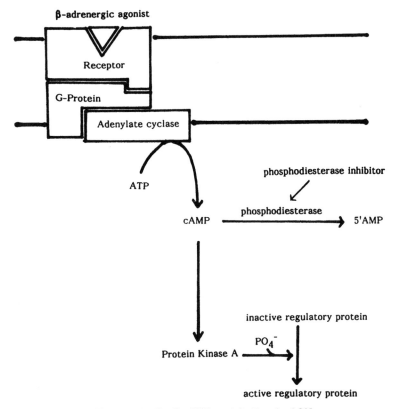

Figure 9.6   Cyclic AMP metabolism in ASM.

elevated cAMP levels are thought to mediate ASM relaxation (Fig. 9.7). Many act to reduce cytosolic Ca^{2+}; stimulation of the plasmalemmal Ca^{2+}-ATPase extrusion pump (Suematsu *et al.*, 1984), inhibition of G-protein/PLC interaction with a reduction in IP_3 and DAG production (Hall and Hill, 1988); and stimulation of the sr Ca^{2+}-ATPase pump with increased intracellular sequestration of Ca^{2+} (Twort and van Breemen, 1989).

Phosphorylation by PKA decreases the activity of MLCK, considered to be intimately involved in ASM tension development (Gerthoffer, 1991). Phosphorylation of MLCK, at certain sites on this enzyme, decreases its affinity for the Ca^{2+}/CaM complex. Ultimately this means that irrespective of the concentration of the Ca^{2+}/CaM complex, there will be fewer activated MLCK molecules, reduced myosin activation, and thus less contractile force.

3.1.2 Cyclic GMP

Much less is known about the role of guanylate cyclase stimulation in ASM relaxation. Guanylate cyclase catalyses the formation of cGMP from GTP. Guanylate cyclase differs from adenylate cyclase in that it exists in both particulate (membrane-bound) and soluble (cytosolic) forms.

Soluble guanylate cyclase is activated by NO, whereas the particulate form is activated by ANP. NO-forming compounds such as nitroprusside relax human ASM *in vitro* (Ward *et al.*, 1993), although inhalation of NO has little effect on airway function (Hulks *et al.*, 1993; Frostell *et al.*, 1993), probably because it does not penetrate the ASM due to rapid inactivation. ANP receptors have been localized to ASM (Von Schroeder *et al.*, 1985), and dose-dependent relaxation in ASM together with varying degrees of reversal of agonist-induced tone has been shown (Ishii and Murad, 1989; Amyot *et al.*, 1989).

Cyclic GMP, like AMP, can be inactivated by a specific cyclic nucleotide phosphodiesterase to 5'-GMP. Similarly, the biological effects of cGMP are believed to involve a cGMP-dependent protein kinase, also known as protein kinase G (PKG) or G-kinase. The cellular targets for G-kinase-mediated relaxation are unknown in ASM, and do not appear to involve phosphorylation of MLCK. However, in vascular smooth muscle cGMP and thus G-kinase may be involved in the extrusion of Ca^{2+} from the cytosol (Suematsu *et al.*, 1984; Popescu *et al.*, 1985) and in sr Ca^{2+} sequestration (Twort and van Breemen, 1988), ultimately leading to relaxation.

3.2 POTASSIUM CHANNELS

Potassium channels occur in the plasma membrane of a variety of different cell types throughout the body,

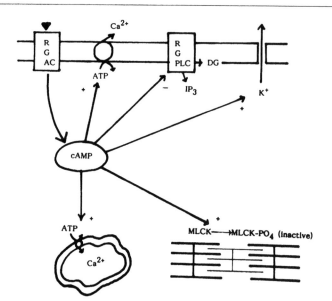

Figure 9.7   Effects of cAMP contributing to relaxation. AC, adenylate cyclase; DG, diacyl glycerol; G, guanosine nucleotide regulatory protein; IP₃, inositol trisphosphate, MLCK, myosin light chain kinase.

(1) ↑ Ca²⁺/ATPase extrusion

(2) ↓ G-Protein/PLC → IP₃ + DAG

(3) ↑ Ca²⁺ sequestration

(4) ↓ Myofilament Ca²⁺ sensitivity

(5) ↑ K⁺ channel opening

including secretory cells, nerves, skeletal and smooth muscle. These channels can exist in several different forms with one or more subtypes occurring in a single cell, and have been implicated in mechanisms of ASM relaxation (Weir and Weston, 1986; Allen *et al.*, 1986; Hamilton *et al.*, 1986; Shetty and Weiss, 1987; Quast and Cook, 1988; Black and Barnes, 1990; Bray *et al.*, 1991).

K^+ channels are associated with the recovery or repolarization of excitable cells after depolarization. In general, they function to inhibit excitatory processes. As mentioned before, ASM exhibits only a low level of electrical excitability without the development of action potentials. Instead, ASM displays slow wave activity thought to represent action potentials which are suppressed by the opening of K^+ channels, with the efflux of K^+ from the cell rectifying any tendency to depolarization (Small *et al.*, 1993). Thus, ASM is strongly rectified (i.e. more resistant to depolarization than to hyperpolarization).

These findings have heightened interest in the role of K^+ channels in ASM since it is recognized that under conditions of K^+ channel blockade, where the ASM cell has lost its ability to rectify, ASM excitability is increased and generation of action potentials is possible experimentally. This excitability is analogous to some of the electrophysiological changes recorded in ASM obtained from asthmatic airways (Akasaka *et al.*, 1975).

The subtype of K^+ channel responsible for determining the outwardly rectifying behaviour remains to be identified – possible candidates include the large conductance Ca^{2+}-dependent K^+ channel (BK_{ca}), and the delayed rectifier K^+ channel (K_v; Fleischmann *et al.*, 1994).

The opening of K^+ channels has been implicated as one of the mechanisms by which raised cytosolic cAMP relaxes ASM. Using the patch clamp technique for recording ion channels, Kume and colleagues (1989) demonstrated that extracellular application of the β-adrenergic agonist isoprenaline promotes the opening of BK_{Ca} channels; this action could be mimicked by activation of PKA. Kume *et al.* (1992) subsequently showed that isoprenaline could also promote K^+ channel opening independently of cAMP accumulation, involving purely the α-subunit of the G-protein (Gs) within the receptor/G-protein/adenylate cyclase complex.

In ASM and vascular smooth muscle relaxation occurs in response to a class of drugs, which include cromakalim and *l*-cromakalim, which are known to activate the opening of K^+ channels. These agents are thought to act on a third subtype of K^+ channel – the ATP-sensitive K^+ channel (K_{ATP}). which opens in response to a fall in cytosolic ATP (Edwards and Weston 1993).

The precise mechanism of relaxation produced by such K^+ channel openers is still unclear. Hyperpolarization brought about by K^+ channel opening is thought to close voltage-dependent Ca^{2+} channels preventing the influx of Ca^{2+} necessary for contraction. However, as described earlier, voltage-dependent Ca^{2+} influx is relatively unimportant for ASM contraction, except when induced experimentally by depolarization.

Alternatively, hyperpolarization, induced by K^+ channel opening, may prevent the maintenance of contraction by an effect on Ca^{2+} handling by the intracellular Ca^{2+} stores in ASM (Chopra *et al.*, 1992). Sarcoplasmic reticulum membranes contain K^+ selective channels. These channels facilitate the flux of K^+ into the sr from a pool of high K^+ concentration in the cytosol (Coronado *et al.*, 1980). Although the predominant mode of action of K^+ channel openers is assumed to occur by hyperpolarization of the plasmalemmal membrane, these agents undoubtedly also act on the sr Ca^{2+} store. In intact rabbit aorta cromakalim reduces both the rate and extent of store refilling with Ca^{2+} in previously Ca^{2+}-depleted tissue (Bray et al., 1991). In portal vein pinacidil actively depletes the Ca^{2+} store (Xiong et al.. 1991). Similar effects are seen in cultured ASM in which l-cromakalim reduces the loading of the IP_3-sensitive store by 26%, a response which is blocked by the K^+ channel inhibitor glibenclamide (Chopra et al, 1992). Thus an increase in the K^+ conductance decreases the ability of the sr in ASM to load Ca^{2+}, an effect comparable to the inverse effect in skeletal muscle whereby a decrease in K^+ conductance increases sr Ca^{2+} loading (Fink and Stephenson, 1987).

An additional mechanism whereby hyperpolarization resulting from K^+ channel opening may influence cytosolic Ca^{2+} is by enhancing the activity of the Na^+/Ca^{2+} exchanger leading to increased Ca^{2+} extrusion.

4. Proliferation

One of the most striking pathological features of chronic asthma is an increase in the smooth muscle mass of the airways (Dunnill et al., 1969; Heard and Hossain, 1973; Hossain, 1973; Ebina et al., 1990). Hyperplasia (increase in cell number) rather than hypertrophy (increase in cell size) predominantly produces this increase in mass, and the associated reduction in airway luminal diameter. This fixed increase in muscle mass contributes to the component of airways obstruction observed in chronic asthmatics which is irreversible with bronchodilators. It also leads to an exaggerated response to bronchoconstrictor stimuli, equivalent to the characteristic bronchial hyperresponsiveness of asthmatics (James et al., 1989; Pare et al., 1991; Pare, 1993).

4.1   GROWTH FACTORS

Compared to vascular smooth muscle, much less is known about the growth factors which stimulate proliferation of ASM and the cellular signal transduction pathways which they employ. Many of the cells present in inflamed airways synthesize growth factors, and of particular interest are the macrophage and the platelet, with contributions from other inflammatory cells such as mast cells, eosinophils and T lymphocytes. Platelet-derived growth factor (PDGF) is the most extensively studied and causes proliferation of cultured ASM obtained from a variety of species including humans and rabbits (Hirst et al., 1994), guinea-pigs (De et al., 1993) and cows (Delamere et al., 1993). Epidermal growth factors have been shown to be mitogenic in humans (Panettieri, 1991), guinea-pigs (Stewart et al., 1993) and dogs (Panettieri et al., 1992). In addition to these growth factors other products of inflammatory cells and epithelial cells, such as histamine (H), 5-hydroxytryptamine (5-HT; serotonin),

endothelin (ET) and interleukin-1 (IL-1) also cause proliferation (Panettieri et al., 1991; Noveral et al., 1992; De et al., 1993).

4.2   SIGNAL TRANSDUCTION PATHWAYS

The most important pathways for transducing receptor activation by mitogens at the cell membrane into DNA synthesis and cell division involve PKC and protein tyrosine kinase (PTK).

4.2.1   Protein Kinase C

Evidence for the link between PKC activation and cell proliferation was initially provided by the demonstration that two intracellular events associated with cell replication – a rise in cytosolic pH and the expression of the proto-oncogenes c-fos and c-myc – are controlled by PKC: PKC stimulates the membrane bound Na^+/H^+ exchange mechanism in smooth muscle which extrudes intracellular H^+ in exchange for extracellular Na^+; this leads to a rise in intracellular pH, a prerequisite for cellular DNA replication (Mitsuka and Berk, 1991). c-fos and c-myc are proto-oncogenes whose transcription to mRNA is one of the earliest markers of cell proliferation; they encode for proteins, found in the cell nucleus, which initiate the sequence of events leading to DNA synthesis (Rozengurt, 1991).

Phorbol esters, which activate PKC directly, stimulate proliferation of human (Panettieri et al., 1991), porcine (Panettieri et al., 1993) and rabbit (Hirst et al., 1993) ASM. Proliferation in response to serum in cultured rabbit ASM is inhibited by exposure to specific inhibitors of the PKC pathway (Hirst et al., 1993).

4.2.2   Protein Tyrosine Kinase

In numerous types of cell, receptor-linked activation of PTK is an important pathway in the transduction of mitogenic stimuli (Yarden, 1988). Activation of PTK by growth factors, such as PDGF and EGF, leads to phosphorylation of tyrosine residues on intracellular substrates. The identity of these substrates remains unclear, but their phosphorylation ultimately results in DNA synthesis and cell proliferation.

Evidence for involvement of PTK in the proliferation of ASM is provided by the demonstration of increased levels of tyrosine phosphorylation of proteins in cultured ASM exposed to PDGF and EGF (Delamere et al., 1993; Stewart et al., 1993), and of inhibition of proliferation by inhibition of PTK activity (Hirst et al., 1993).

The PTK and PKC pathways do not operate in isolation from each other. A high degree of cross-talk between these pathways exists – for instance, activation of PTK by EGF stimulates the production by $PLC\gamma$ of DAG from PIP_2, which in turn activates the PKC pathway.

4.3 INHIBITION OF PROLIFERATION

Future therapies aimed at preventing the proliferative response of ASM, observed in persistent asthma, will target those intracellular pathways which are believed to inhibit the proliferative response. Those pathways involving cGMP and cAMP, which produce relaxation of ASM, have been investigated as possible targets for the inhibition of proliferation.

Although activation of cGMP production inhibits vascular smooth muscle proliferation (Itoh *et al.*, 1990; Winter, 1993), no comparable effect of cGMP has been demonstrated in ASM.

However, both receptor-mediated and direct activation of adenylate cyclase by isoprenaline and forskolin respectively inhibit both phorbol ester- and EGF-induced proliferation of human ASM (Panettieri, 1991). Direct microinjection of the catalytic subunit of cAMP-dependent protein kinase inhibits proliferation of porcine ASM (Panettieri, 1993). There seems to be a marked differential sensitivity in the ability of β-adrenergic agonists, cell-permeable cAMP analogues, such as dibutryl cAMP, and forskolin to exert anti-proliferative effects on ASM stimulated via either the PKC or the PTK pathway. The increased susceptibility of the PKC pathway relative to PTK may reflect differences in intracellular cross-talk between the cAMP pathway and these pathways, and may underlie the apparent lack of anti-proliferative, or disease-modifying, effects of β-adrenergic agonists in the management of chronic asthma.

5. *References*

Adelstein, R.S. (1983). Regulation of contractile proteins by phosphorylation. J. Clin. Invest. 72, 1863–1866.

Adelstein, R.S. and Eisenberg, E. (1980). Regulation and kinetics of the actin–myosin–ATP interaction. Annu. Rev. Biochem. 49, 956–969.

Adelstein, R.S., DeLanerolle, P., Sellers, J.R., Pato, M.D. and Conti, M.A. (1982). Regulation of contractile proteins in smooth muscle and platelets by calmodulin and cyclic AMP. In "Calmodulin and Intracellular Calcium Receptors" (eds S. Kakiuchi, H. Hidaka and A.R. Means), pp. 313–331. Plenum, New York.

Ahmed, F., Foster, R.W., Small, R.C. and Weston, A.H. (1984). Some features of the spasmogenic actions of acetylcholine and histamine in guinea-pig isolated trachea. Br. J. Pharmacol. 83, 227–233.

Akasaka, K., Konno, K., Ono Y., Mue, S., Abe C., Kumagai, M. and Ise, T. (1975). Electromyographic study of bronchial smooth muscle in bronchial asthma. Tohoku J. Exp. Med. 117, 55–59.

Allen, S.L., Boyle, J.P., Cortijo, J., Foster, R.W., Morgan, G.P. and Small, R.C. (1986). Electrical and mechanical effects of BRL 34915 in guinea-pig isolatd trachealis. Br. J. Pharmacol. 89, 395–405.

Amyot, T., Lesiege, D., Michoud, M.C., Larochelle, P. and Hamet, P. (1989). Effect of atrial naturetic factor on airway sensitivity to histamine in anaesthetized dogs. Eur. Respir. J. 2 (Suppl 5), 301s.

Armour, C.L., Lazar, N.M., Schellenberg, R.R. *et al.* (1984). Comparison of *in vivo* and *in vitro* human airway reactivity to histamine. Am. Rev. Respir. Dis. 129, 907–910.

Barnes, P.J. (1986). Neural control of human airways in health and disease. Am. Rev. Respir. Dis. 134, 1289–1314.

Barnes, P.J., Basbaum, C. and Nadel J.A. (1983). Autoradiographic localization of autonomic receptors in airway smooth muscle: marked differences between large and small airways. Am. Rev. Respir. Dis. 127, 758–762.

Berridge, M.J. (1993) Inositol trisphosphate and calcium signalling. Nature 361, 315–325.

Berridge, M.J. and Irvine, R.F. (1984). Inositol phosphates and cell signalling. Nature, 341, 197–205.

Birnbaumer, L. (1990). G-proteins in signal transduction. Annu. Rev. Pharmacol. Toxicol. 30, 675–705.

Black, J.L. and Barnes, P.J. (1990). Potassium channels and airway function: new therapeutic prospects. Thorax, 45, 213–218.

Bolton, T.B. (1979). Mechanisms of action of transmitters and other substances on smooth muscle. Physiol. Rev. 59, 606–718.

Bond, M., Kitazawa, T., Somlyo, A.P. and Somlyo, A.V. (1984). Release and recycling of calcium by the sarcoplasmic reticulum in guinea-pig portal vein smooth muscle. J. Physiol. 355, 677–695.

Bray, K.M., Weston, A.H., Duty, S., Newgreen, D.T., Longmore, J., Edwards, G. and Brown, T.J. (1991). Differences between the effects of cromakalim and nifedipine on agonist-induced responses in rabbit aorta. Br. J. Pharmacol. 102, 337–344.

Bullock, C.G., Fettes, J.F. and Kirkpatrick, C.T. (1981). Tracheal smooth muscle – second thoughts on sodium-calcium exchange. J. Gen. Physiol. 318, 46.

Carafoli, E. (1984). Calmodulin-sensitive calcium-pumping ATPase of plasma membranes: Isolation, reconstitution and regulation. Fed. Proc. 43, 3005–3010.

Cerrina, J., Ladurie, M., Le, R., Labat, C., Raffestin, B., Bayol, A. and Brink, C. (1986). Comparison of human bronchial muscle responses to histamine *in vivo* with histamine and isoproterenol agonists *in vitro*. Am. Rev. Respir. Dis. 137, 57–61.

Chen, Q., Cannell, M. and Van Breemen, C. (1992). The superficial buffer barrier in vascular smooth muscle. Can. J. Physiol. Pharmacol. 70(4), 509–514.

Chilvers, E. and Nahorski, S. (1990). Phosphoinositol metabolism in airway smooth muscle. Am. Rev. Respir. Dis. 141, S137–S140.

Chilvers, E.R., Challiss, R.A., Barnes P.J. and Nahorski, S.R. (1989). Mass changes of inositol (1,4,5) trisphosphate in trachealis muscle following agonist stimulation. Eur. J. Pharmacol. 164, 587–590.

Chilvers, E.R., Giembycz, M.A., Challis, R.A.J., Barnes, P.J. and Nahorski, S.R. (1991). Lack of effect of zaprinast on methacholine-induced contraction and inositol 1,4,5 trisphosphate accumulation in bovine tracheal smooth muscle. Br. J. Pharmacol. 103, 1119–1125.

Chopra, L.C., Twort, C.H.C., Ward, J.P.T. and Cameron, I.R. (1989). Effects of heparin on inositol 1,4,5-trisphosphate and guanosine 5'-0-(3-Thio triphosphate) induced calcium release in cultured smooth muscle cells from rabbit trachea. Biochem. Biophys. Res. Commun. 163, 262–268.

Chopra, L.C., Twort, C.H.C., Cameron, I.C. and Ward, J.P.T. (1991). Inositol 1,4,5-triphosphate- and guanosine 5′-0-(3-thiotriphosphate)-induced Ca^{2+} release in cultured airway smooth muscle. Br. J. Pharmacol. 104, 901–906.

Chopra, L.C., Twort, C.H.C. and Ward, J.P.T. (1992). Direct action of BRL 38227 and glibenclamide on intracellular calcium stores in cultured airway smooth muscle of rabbit. Br. J. Pharmacol. 105, 259–260.

Coburn, R.F. (1979). Electromechanical coupling in canine trachealis muscle: acetylcholine contractions. Am. J. Physiol., 236, C177–C184.

Cockcroft, S. and Gomperts, B.D. (1985). Role of guanine nucleotide binding protein in the activation of polyphosphoinositide phosphodiesterase. Nature 314, 534–536.

Cockcroft, S., Howell, T.W. and Gomperts, B.D. (1987). Two G proteins act in series to control stimulus-secretion coupling in mast cells: use of neomycin to distinguish between G-proteins controlling polyphosphoinositide phosphodiesterase and exocytosis. J. Cell Biol., 105, 2745–2750.

Coronado, R., Rosenberg, R.L. and Miller, C. (1980). Ionic selectivity, saturation, and block in a K^+-selective channel from sarcoplasmic reticulum. J. Gen. Physiol. 76, 425–446.

Creese. B.R. and Denborough, M.A. (1981). Sources of calcium for contraction of guinea-pig isolated tracheal smooth muscle. Clin. Exp. Pharmacol. Physiol. 8, 175–182.

Daniel, E.E, Kannan, M., and Davis, C. et al., (1986). Ultrastructural studies on the neuromuscular control of human tracheal and bronchial muscle. Respir. Physiol. 63, 109–128.

Daniel, E.E., Bourreau, J.-P., Abela, A. and Jury, J. (1992). The internal calcium store in airway smooth muscle: emptying, refilling and chloride. Biochem. Pharmacol. 43, 29–37.

De, S., Zelazny, E.T. and Souhrada, M. (1993). Interleukin-l (IL-1) stimulates the proliferation of cultured ASM (ASM) cells via platelet derived growth factor (PDGF). Am. Rev. Respir. Dis. 147, A1013.

Delamere, F., Townsend, P. and Knox, A.J. (1993). Tyrosine kinases transduce serum and specific growth factor mediated mitogenesis in ASM. Am. Rev. Respir. Dis. 147, A254.

Diamond, L., Szarek, J.L., Gillespie, M.N. and Altiere, R.J. (1991). In vivo bronchodilatory activity of vasoactive intestinal peptide in the cat. Am. Rev. Respir. Dis. 128, 827–832.

Dillon, P.F., Aksoy, M.O., Driska, S.P. and Murphy, R.A. (1981). Myosin phosphorylation and the cross-bridge cycle in smooth muscle. Science (Wash). DC211, 495–497.

Duncan, R.A., Krzanowski, J., Davis, J.S., Polson, J.B., Coffey, R.G., Shimoda, T. and Szentivanyi, A. (1987). Polyphosphoinositide metabolism in canine tracheal smooth muscle (CTSM) in response to a cholinergic stimulus. Biochem. Pharmacol. 36, 307–310.

Dunnill, M.S., Masserella, G.R. and Anderson, J.A. (1969). A comparison of the quantitative anatomy of the bronchi in normal subjects, in status asthmaticus, in chronic bronchitis and in emphysema. Thorax 24, 176–179.

Ebina, M., Yaegashi, H., Chibo, R., Takahashi, T., Motomiya, M. and Tanemura, M. (1990). Hyperreactive site in the airway free of asthmatic patients revealed by thickening of bronchial muscles. Am. Rev. Respir. Dis. 141, 1327–1332.

Edwards, G. and Weston, A.H. (1993). The pharmacology of ATP-sensitive K^+-channels. Annu. Rev. Pharmacol. Toxicol. 33, 597–637.

Eggermont, J.A., Vrozlix, M., Raeymaekers, L., Wuytack, F. and Casteels, R. (1988). Ca^{2+}-transport ATPase of vascular smooth muscle. Circ. Res. 62, 266–278.

Ehrlich, B.E. and Watras, J. (1988). Inositol 1,4,5-trisphosphate activates a channel from smooth muscle sarcoplasmic reticulum. Nature 336, 583–586.

Farley, J.M. and Miles, P.R. (1978). The sources of calcium for acetylcholine-induced contractions of dog tracheal smooth muscle. J. Pharmacol. Exp. Ther. 207, 340.

Fink, R.H.A. and Stephenson, D.G. (1987). Ca^{2+} movements in muscle modulated by the state of K^+ channels in the sarcoplasmic reticulum membranes. Pflügers Arch. 409, 374–380.

Fleischmann, B.K., Hay, D.W.P. and Kotlikoff, M.I. (1994). Control of basal tone by delayed rectifier potassium channels in human airways. Am. J. Resp. Crit. Care Med. 149 (4), A1080.

Foster, R.W., Small, R.C. and Weston, A.H. (1983a). Evidence that the spasmogenic action of tetraethylammonium in guinea-pig trachealis is both direct and dependent on the cellular influx of calcium ion. Br. J. Pharmacol. 79, 255–263.

Foster. R.W., Small, R.C. and Weston, A.H. (1983b). The spasmogenic action of potassium chloride in guinea-pig trachealis. Br. J. Pharmacol. 80, 553–559.

Frossard, N. and Barnes, P.J. (1991). Effects of tachykinins on small human airways. Neuropeptides, 19, 157–162.

Frostell, C., Hogman, M., Hedenstrom, H. and Hedenstierna, G. (1993). Is nitric oxide inhalation beneficial to the asthmatic patient? Am. Rev. Respir. Dis. 147, A515.

Gerthoffer, W.T. (1991). Regulation of the contractile element of airway smooth muscle. Am. J. Physiol. 261 (Lung Cell Mol. Physiol. 5), L15–L28.

Ghosh, T.K., Mullaney, J.M., Tarazi, F.I. and Gill, D.L. (1989). GTP activated communication between inositol 1,4,5-trisphosphate-sensitive and insensitive calcium pools. Nature, 340, 236–239.

Giembycz, M.A. and Raeburn, D. (1992). Current concepts on mechanisms of force generation and maintenance in airways smooth muscle. Pulm. Pharmacol. 5, 279–297.

Giembycz, M.A. and Rodger, I.W. (1987). Electrophysiological and other aspects of excitation-contraction coupling and uncoupling in mammalian airway smooth muscle. Life Sci., 41, 111–132.

Gill, D.L., Mullaney, J.M. and Ghosh, T.K. (1988). Intracellular calcium translocation: mechanism of activation by guanine nucleotides and inositol phosphates. J. Exp. Biol. 139, 105–133.

Grandordy, B.M., Cuss, F.M., Meldrum, L., Sturon, P.G. and Barnes, P.J. (1986). Leukotriene C4 and D4 induce contraction and formation of inositol phosphates in airways and lung parenchyma (abstract). Am. Rev. Respir. Dis. 133, A239.

Grandordy, B.M., Frossard, N., Rhoden, K.J. and Barnes, P.J. (1988). Tachykinin-induced phosphoinositide breakdown in airway smooth muscle and epithelium: relationship to contraction. Mol. Pharmacol. 33, 515–519.

Hai, C. and Murphy, R.A. (1988). Cross-bridge phosphorylation and regulation of the latch state in smooth muscle. Am. J. Physiol. 254 (Cell Physiol. 23), C99–C106.

Hall, I.P. and Hill, S.J. (1988). β-Adrenoreceptor stimulation inhibits histamine-stimulated inositol phospholipid hydrolysis in bovine tracheal smooth muscle. Br. J. Pharmacol. 95, 1204–1212.

Hamilton, T.C., Weir, S.W. and Weston, A.H. (1986). Comparison of the effects of BRL 34915 and verapamil on electrical and mechanical activity in rat portal vein. Br. J. Pharmacol. 88, 103–111.

Hashimoto, T., Hirata, M. and Ito, Y. (1985). A role for inositol 1,4,5-trisphosphate in the initiation of agonist-induced contractions of dog tracheal smooth muscle. Br. J. Pharmacol. 86, 191–199.

Heard, B.E. and Hossain, S. (1973). Hyperplasia of bronchial muscle in asthma. J. Pathol. 110, 319–331.

Hirst, S.J., Barnes, P.J. and Twort, C.H.C. (1993). Protein kinase inhibitors (PKC and PTK) and phorbol ester inhibit proliferation induced by serum in cultured rabbit ASM cells. Am. Rev. Respir. Dis. 147, A253.

Hirst, S.J., Barnes, P.J. and Twort, C.H.C. (1994). Proliferation of human and rabbit airway smooth muscle in culture by platelet-derived growth factor isoforms. Am. J. Respir. and Crit. Care Med. 149 (4), A302.

Hogaboom, G.K. and Fedan, J.F. (1981). Calmodulin stimulation of calcium uptake and (Ca^{2+}-Mg^{2+})-ATPase activities in microsomes from canine tracheal smooth muscle. Biochem. Biophys. Res. Commun. 99, 737–744.

Hossain, S. (1973). Quantitative measurement of bronchial muscle in men with asthma. Am. Rev. Respir. Dis., 107, 99–109.

Hulks, G., Warren, P.M. and Douglas, N.J. (1993). The effect of inhaled nitric acid on bronchomotor tone in the normal human airway. Am. Rev. Respir. Dis. 147, A287.

Irvine, R.F. and Moor, R.M. (1986). Micro-injection of inositol 1,3,4,5-tetrakis-phosphate activates sea urchin eggs by a mechanism dependent on external Ca^{2+}. Biochem. J. 240, 917–920.

Ishii, K. and Murad, F. (1989). ANP relaxes bovine tracheal smooth muscle and increases cGMP. Am. J. Physiol. 256 (Cell Physiol 25), C495–500.

Ito, Y. and Itoh, T. (1984). The roles of stored calcium in contractions of cat tracheal smooth muscle produced by electrical stimulation, acetylcholine and high K^+. Br. J. Pharmacol. 83, 667–676.

James, A.L., Pare, P.D. and Hogg, H.C. (1989). The mechanics of airway narrowing in asthma. Am. Rev. Respir. Dis. 139, 242–246.

Kardasz, A.M., Langlands, J.M., Rodger, I.W. and Watson J. (1988). Inositol lipid turnover in isolated guinea-pig trachealis and lung parenchyma. Biochem. Soc. Trans. 15, 474.

Kikkawa, U., Kishimoto, A. and Nishizuka, U. (1989). The protein kinase C family: its heterogeneity and its implications. Annu. Rev. Biochem. 58, 31.

Kirkpatrick, C.T. (1975). Excitation and contraction in bovine tracheal smooth muscle. J. Physiol. 244, 263–281.

Kotlikoff, M.I., (1988). Calcium currents in isolated canine airway smooth muscle cells. Am. J. Physiol. 254, C793–C801.

Kotlikoff, M.I., Murray, R.K. and Reynolds, E.E. (1987). Histamine-induced calcium release and phorbol antagonism in cultured airway smooth muscle. Am. J. Physiol. 253, C561–C566.

Kume, H., Takai, A., Tokuno, H. and Tomita, T. (1989). Regulation of Ca^{2+}-dependent K^+-channel activity in tracheal myocytes by phosphorylation. Nature, 341, 152–154.

Kume, H., Graziano, M.P. and Kotlikoff, M.I. (1992). Stimulatory and inhibitory regulation of calcium-activated

potassium channels by guanine nucleotide-binding proteins. Proc. Natl. Acad. Sci. USA 89, 11051–11055.

Leijten, P.A. and van Breemen, C. (1984). The effects of caffeine on the noradrenaline-sensitive calcium store in rabbit aorta. J. Physiol. 357, 327–339.

Loutzenhiser, R. and van Breemen, C. (1983). The influence of receptor occupation on Ca^{++} influx-mediated vascular smooth muscle contraction. Circ. Res. 1983; 52 (suppl I), 97–103.

Low, A.M., Kwan, C.Y. and Daniel, E.E. (1992). Evidence for two types of internal Ca^{2+} stores in canine mesenteric artery with different refilling mechanisms. Am. J. Physiol. 262, H31–H37.

Marthan, R., Martin, C., Amedee, T. and Mironneau, J. (1989) Calcium channel currents in isolated smooth muscle cells from human bronchus. J. Appl. Physiol. 66, 1706–1714.

Meyer, T., Wensel, T. and Stryer, L. (1990). Kinetics of calcium channel opening by inositol 1,4,5-trisphosphate. Biochemistry 29, 32–37.

Mignery, G.A. and Sudhof, T.C. (1990). The ligand binding site and transduction mechanism in the inositol 1,4,5-triphosphate receptor. EMBO J. 9, 3893–3898.

Missiaen, L, de Smedt, H., Droogmans, G., Declerck, I., Plessers, L. and Casteels, R. (1991). Uptake characteristics of the IP_3-sensitive and -insensitive Ca^{2+} pools in porcine aortic smooth-muscle cells: different Ca^{2+} sensitivity of the Ca^{2+} uptake mechanism. BBRC 174(3), 1183–1188.

Mitsuka, M. and Berk, B.C. (1991). Long-term regulation of Na^+-H^+ exchange in vascular smooth muscle cells: role of protein kinase C. Am. J. Physiol., 260, C562–C569.

Mullaney, J.M., Yu, M., Ghosh, T.K. and Gill, D.L. (1988). Calcium entry into the inositol 1,4,5-trisphosphate-releasable pool is mediated by a GTP-regulatory mechanism. Proc. Natl. Acad. Sci. USA, 85, 2499–2503.

Murphy, R.A., Rembold, C.M. and Hai, C.M. (1990). Contraction in smooth muscle: What is latch? In "Frontiers in Smooth Muscle Research" (eds N. Speralakis and J.D. Wood) pp. 39–50. Liss, New York.

Murray, R.K. and Kotlikoff, M.I. (1991). Receptor-activated calcium influx in human airway smooth muscle cells. J. Physiol. 435, 123–144.

Murray, R.K., Maki, C. and Panettieri, R.A. (1994). Activation mechanisms of sustained calcium influx in human airways smooth muscle (ASM) cells. Am. J. Respir. Crit. Care Med. 149(4), A1081.

Nishimura, J. and van Breemen, C. (1989). Direct regulation of smooth muscle contractile elements by second messenger. Biochem. Biophys. Res. Comm. 163, 929–935.

Nouailhetas, V.L.A., Lodge, N.J., Twort, C.H.C. and van Breemen, C. (1988). The intracellular calcium stores in the rabbit trachealis. Eur. J. Pharmacol. 157, 165–172.

Noveral, J.P., Rosenberg, S.M., Anbar, R.A., Pawlowski, A.N. and Grunstein, M. M. (1992). Role of endothelin-l in regulating proliferation of cultured rabbit ASM cells. Am. J. Physiol. 263, (Lung Cell Mol Physiol 7), L317–24.

Obanime, A.W., Hirst, S.J. and Dale, M.M. (1989). The effects of smooth muscle relaxants on phorbol dibutyrate- and histamine-induced contraction of the lung parenchymal strip. Possible relevance for asthma. Pulm. Pharmacol. 2, 191.

Opazo-saez, A., Okazawa, M. and Pare, P.D. (1994). Airway

smooth muscle (ASM) orientation and airway stiffness favor airway narrowing rather airway shortening. Am. J. Respir. Crit. Care Med. 149(4), A583.

Palmer, J.B.D., Cuss, F.M.C and Barnes, P.J. (1986). VIP and PHM and their role in nonadrenergic inhibitory responses in isolated human airways. J. Appl. Physiol. 61, 1322–1328.

Panettieri, R.A., Murray, R.K., de Palo, L.R., Yadvish, P.A. and Kotlikoff, M.I. (1989). A human airway smooth muscle line that retains physiological responsiveness. Am. J. Physiol. 256, C329–C335.

Panettieri, R.A., Rubinstein, N.A., Feuertein, N. and Kotlikoff, M.I. (1991). Beta-adrenergic inhibition of airway smooth muscle proliferation. Am. Respir. Dis. 143, A608.

Panettieri, R.A., Eszterhas, A. and Murray, R.K. (1992). Agonist-induced proliferation of ASM is mediated by alterations in cytosolic calcium. Am. Rev. Respir. Dis. 145, A15.

Panettieri, R.A., Cohen, M.D. and Bilgen, G. (1993). ASM proliferation is inhibited by microinjection of the catalytic subunit of cAMP dependent protein kinase. Am. Rev. Respir. Dis. 147, A252.

Pare, P.D. (1993). Hyperplasia and hypertrophy of ASM in asthma: the cause of airway hyperresponsiveness. Eur. Respir. J. 6, 228s.

Pare, P.D., Wiggs, B.R., Hogg, J.C. and Busken, C. (1991). The comparative mechanics and morphology of airways in asthma and in chronic obstructive pulmonary disease. Am. Rev. Respir. Dis. 143, 1189–1193.

Park, S. and Rasmussen, H. (1986). Carbachol induced protein phosphorylation changes in bovine tracheal smooth muscle. J. Biol. Chem. 261, 15734.

Popescu, L.M., Panoiu, C., Hinescu, M. and Nutu, O. (1985). The mechanism of cGMP-induced relaxation in vascular smooth muscle. Eur. J. Pharmacol. 107, 393–394.

Putney, J.W. (1986). A model for receptor-regulated calcium entry. Cell Calcium 7, 1–12.

Putney, J.W. (1990). Capacitative calcium entry revisited. Cell Calcium 11, 611–624.

Quast, U. and Cook, N.S. (1988). Potent inhibitors of the effects of the K$^+$ channel opener BRL 34915 in vascular smooth muscle. Br. J. Pharmacol. 93, 204P.

Raeburn, D., Roberts, J.A., Rodger, I.W. and Thomson, N.C. (1986). Agonist-induced contractile responses of human bronchial muscle in vitro: effects of Ca^{2+} removal, La^{3+} and PY108068. Eur. J. Pharmacol. 121, 251–255.

Raeymaekers, L. and Jones, L.R. (1986). Evidence for the presence of phospholamban in the endoplasmic reticulum of smooth muscle. Biochem. Biophys. Acta 882, 258–265.

Rinard, G.A., Jensen, A. and Puckett, A.M. (1983). Hydrocortisone and isoproterenol effects on trachealis cAMP and relaxation. J. Appl. Physiol: Respir. Environ. Exercise Physiol. 55, 1609–1613.

Roberts, J.A., Rodger, I.W. and Thomson, N.C. (1985). Airway responsiveness to histamine in man: effect of atropine on in vivo and in vitro comparison. Thorax 40, 261–267.

Rodger, I.W. (1985). Excitation-contraction coupling and uncoupling in airway smooth muscle. Br. J. Clin. Pharmacol. 20, 155S–166S.

Rozengurt, E. (1991). Neuropeptides as cellular growth factoprs: role of multiple signalling pathways. Eur. J. Clin. Invest., 21, 123–134.

Ruegg, U.T., Wallnofer, H., Weir, S. and Cauvin. C. (1989).

Receptor-operated calcium-permeable channels in vascular smooth muscle. J. Cardiovasc. Pharmacol. 14, S49–S58.

Senn, N., Jeanclos., E. and Garay, R. (1990). Action of azelastine on intracellular Ca^{2+} in cultured airway smooth muscle. Eur. J. Pharmacol. 205, 29–34.

Shetty, S.S. and Weiss, G.B. (1987). Dissociation of actions of BRL 34915 in the rat portal vein. Eur. J. Pharmacol. 141, 485–488.

Shieh, C.C., Petrini, M.F., Dwyer, T.M., and Farley, J.M. (1991). Concentration-dependence of acetylcholine-induced changes in calcium and tension in swine trachealis. J. Pharmacol. Exp. Therap. 256, 141.

Small, R.C. and Foster, R.W. (1986). Airway smooth muscle: An overview of morphology and aspects of pharmacology: In "Asthma: Clinical Pharmacology and Therapeutic Progress" (ed A.B. Kay) pp. 101–113, Blackwell, Oxford.

Small, R.C., Foster, R.W. (1988). Electrophysiology of the airway smooth muscle cell. In "Asthma: Basic Mechanisms and clinical Management" (eds P.J. Barnes, I.W. Rodger and N.C. Thomson), pp. 35–56, Academic Press, London.

Small, R.C., Boyle, J.P., Cortijo, J., Curtis-prior, P.B., Davies, J.M., Foster, R.W. and Hofer, P. (1988).The relaxant and spasmogenic effects of some xanthine derivatives acting on guinea-pig isolated trachealis muscle. Br. J. Pharmacol. 94, 1091–1100.

Small, R.C., Berry, J.L., Cook, S.J., Foster, R.W., Green, K.A. and Murray, M.A. (1993). Potassium channels in airways. In "Lung Biology in Health and Disease. Vol. 67. Pharmacology of the Respiratory Tract: Experimental and Clinical Research" (eds K.F. Chung and P.J. Barnes), Marcel Dekker, New York.

Sobue, K., Muramoto, K., Fujita, M. and Kakiuchi, S. (1981). Purification of a calmodulin-binding protein from chicken gizzard that interacts with F-actin. Proc. Natl. Acad. Sci. USA 78, 5652–5655.

Sobue, K., Morimoto, K., Inui, M., Kanda, K. and Kakiuchi, S. (1982). Control of actin-myosin interaction of gizzard smooth muscle by calmodulin- and caldesmon-linked flip-flop mechanism. Biomed. Res. 3, 188–196.

Somlyo, A.P. (1985). Excitation-contraction coupling and the ultrastructure of smooth muscle. Circ. Res. 57(4), 497–507.

Somlyo, A.V. and Franzini-Armstrong, C. (1985). New views of smooth muscle structure using freezing, deep-etching and rotary shadowing. Experientia 41, 841–856.

Somlyo, A.V., Gonzalez-Serratos, H., Shuman, H., McClellan, G. and Somlyo, A.P. (1981). Calcium release and ionic changes in the sarcoplasmic reticulum of tetanized muscle: an electron probe study. J. Cell Biol. 90, 577–594.

Stewart, A.G., Grigoriadis, G. and Harris, T. (1993). Dissociation of epidermal growth factor induced mitogenesis and increases in intracellular calcium in cultured ASM. Am. Rev. Respir. Dis. 147, A253.

Suematsu, E., Hirata, M. and Kuriyama, H. (1984). Effects of cAMP and cGMP-dependent protein kinases, and calmodulin on Ca^{2+} uptake by highly purified sarcolemmal vesicles of vascular smooth muscle. Biochem. Biophys. Acta 773, 83–90.

Sumida, M., Okuda, H. and Hamada, M. (1984). Ca^{2+}, Mg^{2+}-ATPase of microsomal membranes from bovine aortic smooth muscle. Identification and characterization of an acid-stable phosphorylated intermediate of the Ca^{2+}, Mg^{2+}-ATPase. J. Biochem. 96, 1365–1374.

Takuwa, Y., Takuwa, N. and Rasmussen, H. (1986). Carbachol

induces a rapid and sustained hydrolysis of polyphosphoinositide in bovine tracheal smooth muscle. Measurements of the mass of polyphosphoinositides, 1,2-diacylglycerol and phosphatidic acid. J. Biol. Chem. 261, 14670–14675.

Thastrup, O., Dawson, A.P., Scharff, O., Foder, B., Cullen, P.J., Drøbak, B.K., Bjerrum, P.J., Christensen, S.B. and Hanley, M.R. (1989). Thapsigargin, a novel molecular probe for studying intracellular calcium release and storage. Agents Actions 27(1/2), 17–23.

Theibert, A.B., Suppattapone, S., Worley, P.E., Baraban, J.M., Meek, J.L. and Snyder, S.H. (1987). Demonstration of inositol 1,3,4,5 tetrakisphosphate receptor binding. Biochem. Biophys. Res. Commun. 148, 1283–1289.

Thomson, N.C. (1987). In vivo versus in vitro human airway responsiveness to different pharamologic stimuli. Am. Rev. Respir. Dis. 136, S58–S62.

Twort, C.H.C. (1994). The intracellular calcium store in airway smooth muscle. In "Airways Smooth Muscle: Biochemical Control of Contraction and Relaxation" (eds D. Raeburn and M.A. Giembycz), pp. 97–115. Virkhauser Verlag, Basel.

Twort, C.H.C. and van Breemen, C. (1988). Cyclic guanosine monophosphate-enhanced sequestration of Ca^{2+} by sarcoplasmic reticulum in vascular smooth muscle. Circ. Res. 62, 961–964.

Twort, C. and van Breemen, C. (1989). Human airway smooth muscle in cell culture: control of the intracellular calcium store. Pulm. Pharmacol. 2, 45–53.

Twort, C.H.C., Chopra, L., Lasky, R. and van Breemen, C. (1992). Human airway smooth muscle (HASM): the Ca^{2+} store during sustained agonist activation. Thorax 47, 213.

van Breemen, C. and Saida, K. (1989). Cellular mechanisms regulating $[Ca^{2+}]_i$ smooth muscle. Annu. Rev. Physiol. 51, 315–329.

van Breemen, C., Cauvin, C., Johns, A., Leijten, P. and Yamamoto, H. (1986). Ca^{2+} regulation of vascular smooth muscle. Fed. Proc. 45, 2746–2751.

van Breemen, C., Saida, K., Yamamoto, H., Hwang, K. and Twort, C. (1988). Vascular smooth muscle sarcoplasmic reticulum. Function and mechanisms of Ca^{2+} release. Ann. NY Acad. Sci. 522, 60–73.

Vincenc, K.S., Black, J.L., Yan, K., Armour, C.L., Donnelly, P.D. and Woolcock, A.J. (1983). Comparisons of in vivo and in vitro responses to histamine in human airways. Am. Rev. Respir. Dis. 128, 875–879.

von Schroeder, H.P., Nishimura, E., McIntosh, C.H.S., Buchan, A.M.J., Wilson, N. and Laidsome, J.R. (1985). Autoradiographic localisation of binding sites for atrial natriuretic factor. Can. J. Physiol. Pharmacol. 63, 1373–1377.

Ward, J.K., Belvisi, M.J., Fox, A.J., Miura, M., Tadjarimi, S., Yacoub, M.H. and Barnes, P.J. (1993). Modulation of cholingergic neural bronchoconstriction by endogenous nitric oxide and vasoactive intestinal peptide in human airways in vitro. J. Clin. Invest. 92, 736–742.

Watras, J., Bezprozvanny, I. and Ehrlich, B.E. (1991). Inositol 1,4,5-trisphosphate-gated channels in cerebellum: presence of multiple conductance states. J. Neurosci. 11, 3239–3245.

Weir, S.W. and Weston, A.H. (1986). The effects of BRL 34915 and nicorandil on electrical and mechanical activity and on ^{86}Rb efflux in rat blood vessels. Br. J. Pharmacol. 88, 121–128.

Worley, P.F., Baraban, J.M., Supattapone, S., Wilson, V.S. and Snyder, S.H. (1987). Characterization of inositol trisphosphate receptor binding in brain. J. Biol. Chem. 262, 12132–12136.

Wuytack, F., Raemaekers, L., Verbist, J., Jones, L.R. and Casteels, R. (1987). Smooth muscle endoplasmic reticulum contains a cardiac-like form of calsequestrin. Biochem. Biophys. Acta 899, 151–158.

Xiong, Z., Kajioka, S., Sakai, T., Kitamura, K. and Kuriyama, H. (1991). Pinacidil inhibits the ryanodine-sensitive outward current and glibenclamide antagonises its action in cells from the rabbit portal vein. Br. J. Pharmacol. 102, 788–790.

Yamamoto, H. and van Breemen, C. (1986). Ca^{2+} compartments in saponin-skinned cultured vascular smooth muscle cells. J. Gen. Physiol. 87, 369–389.

Yang C.M., Chow S.P. and Sung, T.C. (1991). Muscarinic receptor subtypes coupled to generation of different second messengers in isolated tracheal smooth muscle cells. Br. J. Pharmacol. 104, 613–618

Yarden, Y. (1988). Growth factor receptor tyrosine kinases. Annu. Rev. Biochem. 57, 443–478.

10. The Airway Epithelium: The Origin and Target of Inflammatory Airways Disease and Injury

Clive Robinson

1. Introduction

The airway epithelium has a central involvement in the onset and also the terminal events of some of the most important and insidious lung diseases. As the interface between the internal and external environments, the apical aspect of the bronchoalveolar epithelium acts as the initial site of interaction between foreign substances (allergens and noxious chemicals) and the immune system. That this results in mucosal inflammatory responses is now beyond doubt. This is strikingly illustrated by the local instillation of allergen into the airways of an atopic individual which results in the development of an acute inflammatory reaction that is in every way as vivid as the cutaneous allergic response in a skin prick test. A qualitative increase in mucosal blood flow and

mucus secretion may be visualized directly by fibreoptic bronchoscopy (Plate I).

2. Physiological Role of the Airway Epithelium: An Overview

It is important to recognize that the airway epithelium is much more than a sophisticated structural barrier. Although physical components of the airway defence are important in the form of interepithelial tight junctions and mucociliary transport, the protective role of the airway epithelium is more wide ranging and subtle. Current evidence suggests the existence of a delicate homeostatic balance in which the health and integrity of the airway epithelium is an essential factor that regulates not only the normal functioning of cells within the submucosa, but also the normal composition of the biomatrix, and in particular that of the epithelial basement membrane (Fig. 10.1).

Historically, most research into the physiology of the airway epithelium has focused on its barrier role, on the regulation of ion transport and on mucociliary clearance. Together, these facets constitute a fundamental element in the physical protection of the body provided by the airway epithelium. The fluid coating of the airways is approximately 10 μm thick and comprises a sol layer that bathes the cilia, beneath a gel layer comprising mucus. This mucociliary clearance system normally operates to propel materials out of the lung to the mouth. Changes in the composition of the gel layer are known to occur with disease, initially as a part of normal host defence mechanisms at the mucosal surface. However, in unresolved disease the abnormalities in mucus production are pathophysiological. Composition of the low viscosity sol layer is controlled by ion transport processes in epithelial cells (reviewed by Widdicombe, 1991). The major pathways are active absorption of Na^+ and active secretion of Cl^- into the lumen. The relative electronegativity of the lumen induces passive movement of a counter-ion, in this case Na^+, with the overall result that salt is moved from the basolateral side of the epithelium to the lumen. The osmotic gradient that this provides results in the movement of water into the lumen. Conversely, water may be absorbed under conditions where Na^+ is actively transported from the lumen and subsequently out of the epithelial cell by a $Na^+, K^+ - ATPase$ localized in the basolateral membrane. These changes in water content of epithelial lining fluid have important consequences for the aspect of mucosal protection provided by mucociliary clearance. The lower the water content of mucus, the higher its viscosity and elasticity (Lifschitz and Denning, 1970; Shih *et al.*, 1977), as is the case in cystic fibrosis. Abnormal viscoelastic properties

result in inefficient mucociliary transport and the activation of epithelial irritant receptors to evoke a reflex cough. Failure to dislodge mucus by even this reflex mechanism indicates that one major facet of mucosal defence has been compromised, with the likely result of increased epithelial injury from infectious agents or inhaled organic and inorganic particulates.

The health and integrity of ciliated columnar epithelial cells, and maintenance of their normally coordinated beating is crucial to mucosal protection. Ciliated cells are present to about the 16th generation of airways, below this level the process of epithelial clearance being handled by luminal macrophages. Ciliary beating is a Ca^{2+} and cyclic AMP-dependent process (Di Benedetto *et al.*, 1991; Lansley *et al.*, 1992) and can be influenced by the presence of inflammatory mediators or drugs. Most studies of this have employed epithelia from animal sources, and consequently the presence of species differences may be a confounding issue. Histamine has little overall effect *in vitro*, although dyskinesia is sometimes observed (Melville and Iravani, 1975). Most prostaglandins, notably excluding prostaglandin $F_{2\alpha}$ ($PGF_{2\alpha}$), increase the ciliary beat frequency (CBF; Wanner *et al.*, 1983), whereas conflicting results have been obtained with sulphidpeptide leukotrienes (Wanner *et al.*, 1986; Bisgaard and Pedersen, 1987; Weisman *et al.*, 1990).

As might be anticipated from the dependency of CBF on intracellular cAMP, β-agonists elevate CBF, although the concentrations required to do this are greater than those required for bronchodilatation (Pavia *et al.*, 1984). Evidence from primary cultures of bovine bronchial epithelium has been presented which suggests that a component of the isoprenaline-stimulated increase in CBF is dependent upon the formation of nitric oxide (NO; Jain *et al.*, 1993). The nitric oxide synthase (NOS) inhibitors L-nitromonomethyl arginine (L-NMMA) and L-nitroarginine methyl ester (L-NAME) both reduced

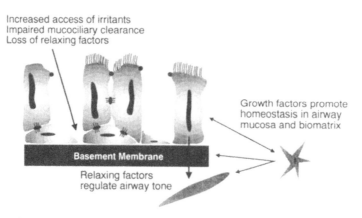

Figure 10.1   Hypothetical scheme for some of the cellular homeostatic links in the airway mucosa.

the effectiveness of isoprenaline, and similar effects have also been observed when bradykinin or substance P were used to elevate CBF (Jain *et al.*, 1993).

Corticosteroids have also been examined for their effects on CBF. Interestingly, whilst corticosteroids exert effects that are ultimately protective and regenerative in the injured airway mucosa, there is evidence from studies in nasal epithelia that they actually reduce CBF (Stafanger, 1987). Whether this occurs in the pulmonary airways, and if so its potential significance, is not known.

More recently, attention has been focused on other aspects of epithelial function that might be important in normal physiological functioning of the mucosa and its surrounding tissues. The airway mucosa is an established source of agents that might potentially regulate the tone of airway smooth muscle (Vanhoutte, 1988; Morrison and Vanhoutte, 1991). These include PGE_2 and NO. In culture, human bronchial epithelial cells continuously produce PGE_2 and this release can be augmented by activation with agents such as calcium ionophore A23187 (Robinson *et al.*, 1990). This prostaglandin has bronchorelaxant actions mediated by EP_2 receptors, although EP_1 receptor-dependent contractile action may also be observed under certain circumstances (reviewed in Robinson and Holgate, 1991). Whilst it is tempting to speculate that PGE_2 might be an endogenous bronchodilator substance capable of controlling resting tone in the airways, addition of non-steroidal anti-inflammatory drugs (NSAIDs) to isolated human airway preparations has minimal effect on their resting tension *in vitro*. Furthermore, administration of NSAIDs to healthy, non-asthmatic humans has little effect on airways calibre *in vivo*. The only exception to the evidence showing an apparently minor role for prostanoids in the normal regulation of human airway tone *in vitro* is the thromboxane antagonist BAYu3405 which produces a small relaxation of tone by an unknown mechanism (Norel *et al.*, 1991).

NO is known to reverse established bronchoconstrictor tone in humans (Högman *et al.*, 1993) and in some, but not all, studies the thiol-dependent NO donor nitroglycerin has been reported to be of benefit in asthma (Miller and Shultz, 1979; Goldstein, 1984). However, it is not clear from such phenomena whether these effects are the basis of any physiological role of NO, or whether NO is the elusive epithelium-derived relaxing factor (EDRF). In guinea-pig isolated airways, inhibition of NO formation results in an increased response to constrictor stimuli (Nijkamp *et al.*, 1993), but physicochemical evidence of NO synthesis was not apparent and the putative source of the NO was not unambiguously identified. Furthermore, caution should be exercised when extrapolating effects from animal to human airways because notable differences in the regulation of airway tone are evident between species (Robinson and Holgate, 1991).

Although its precise function remains debatable, there is little doubt that the airway epithelium does have the capacity to produce NO. Immunostaining of human lung reveals no evidence for the presence of the constitutive form of nitric oxide synthase (NOS) in epithelial cells of the bronchial or alveolar epithelium, but an enzyme that is antigenically similar to the inducible form of NOS has been localized in the epithelium of large cartilagenous bronchi (Kobzik *et al.*, 1993). Furthermore, NO production has tentatively been demonstrated in epithelial cell primary cultures and cell lines (Robbins *et al.*, 1993; Chee *et al.*, 1993). Intriguingly, epithelial NOS expression is apparently up-regulated in asthma (Springall *et al.*, 1993). As we shall see later, there is another side to the biology of NO which makes it very much a double-edged sword in mucosal biology.

Before leaving the subject of putative factors derived from the epithelium that exert protective effects on smooth muscle, it is worth noting that one important consideration when establishing such a role is that the mediator must traverse a complex and extensive vascular network before gaining access to underlying smooth muscle cells. In that context, it has been argued that this would make NO an unsuitable candidate for an epithelium-derived regulator of airway smooth muscle tone. However, when formed in low concentrations, the autoxidation kinetics of aqueous NO are probably consistent with such a role (Ford *et al.*, 1993). Thus, changes in the sensitivity of airways smooth muscle produced by epithelial injury or removal may be due to a combination of loss of a significant physical diffusion barrier and perturbations in the release of putative relaxant factors (Vanhoutte, 1988; Morrison and Vanhoutte, 1991; Sparrow and Mitchell, 1991; Omari *et al.*, 1993a,b).

However, the importance of the airway epithelium in the homeostatic control of underlying tissues may be at a rather more fundamental level than the acute changes in sensitivity of airways smooth muscle produced by acute epithelial injury or removal. The airway epithelium is now recognized to be an important source of growth factors [including interleukins (IL)-1, -4, -6, granulocyte–macrophage colony-stimulating factor (GM-CSF) granulocyte colony-stimulating factor (G-CSF), tumour necrosis factor (TNF) and interferon γ (IFNγ)] (Mattoli *et al.*, 1990; Ohtoshi *et al.*, 1991; Cox *et al.*, 1991; Nakamura *et al.*, 1991; Gauldie *et al.*, 1993; Devalia *et al.*, 1993) that have profound effects on cellular target function. In turn, the target cells can exert reciprocal control on the growth and development of the airway epithelium (Shoji *et al.*, 1990). In the midst of this intercellular control network, the basement membrane has the potential to act as a significant repository for these growth factors and other substances and provides the normal cellular differentiation signals that provide spatiotemporal regulation of cell development (Yurchenco and Schnittny, 1990). Consequently, it is possible to envisage that disruption of epithelial cells perturbs the production and storage of these growth factors that are key components in maintaining the delicate homeostatic balance in the airway mucosa, or which are responsible for activating

fibroproliferative processes. The extent of this perturbation of epithelial homeostasis could therefore be a key
event in the initiation of an inappropriate repair response
and, ultimately, impairment of airway function.

3. Structure of the Airway Mucosa

The structures of the cells lining the conducting airways
and gas exchange units have been extensively studied
(Breeze and Wheeldon, 1977; Jeffery and Reid, 1977;
Jeffery, 1983; Plopper et al., 1983; Plopper, 1983). The
larger conducting airways are mostly ciliated, columnar
and pseudo-stratified. Peripheral to the lobar bronchi the
epithelium thins to a single cell layer and the ratio of non-
ciliated to ciliated cells increases. At least eight different
types of epithelial cells have been classified as being resident in the airways mucosa, although there are some
inter-species variations in the occurrence of these cells
(Jeffery and Reid, 1977; Gail and Lenfant, 1983). Whilst
a detailed consideration of the cellular composition of the
airway mucosa is outside the scope of the present review,
brief reference will be made to some of the major cell

types that are likely to be involved in the cellular injury
and repair process.

Superficial epithelial cells in the larger airways can be
divided broadly into those that possess cilia and those
that do not. The non-ciliated cells can be further differentiated on the basis of whether they contain secretory
granules (such as Clara, mucous or serous cells; Plopper
et al., 1980) or whether they have an absence of secretory
granules (such as the intermediate or brush-type cells).
These superficial cells represent the first line in epithelial
defence. Whilst a number of these cells make direct contact with structural proteins in the basement membrane,
many of these superficial cells exert a footprint only upon
the underlying basal cells (Evans and Plopper, 1988;
Evans et al., 1989). At one time, these basal cells were
considered solely as progenitors of the highly differentiated superficial epithelial cells. However, in recent years
this view has been revised to accommodate an additional
role for basal cells as significant intermediaries in
anchorage of the columnar epithelium to the underlying
biomatrix (Evans et al., 1990; Montefort et al., 1992a).
This change in the perceived role of the basal cell has
paralleled more sophisticated studies of airway

a

b

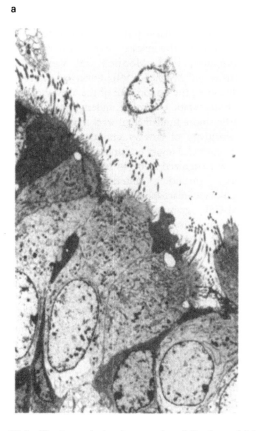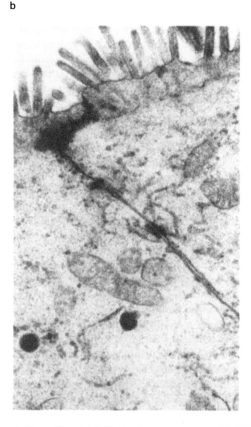

**Figure 10.2 Electron photomicrographs of the bronchial epithelium. Panel (a) illustrates columnar epithelial cells
showing the presence of cilia. Panel (b) illustrates details of intercellular junctions between adjacent columnar
cells.**

morphology and the cataloguing of the mechanisms that maintain cell–cell and cell–biomatrix adhesion in the airway mucosa (Evans *et al.*, 1989; 1990; Evans and Plopper, 1988; Montefort *et al.*, 1992a).

Superficial epithelial cells that line the airway are bonded at their apices by a continuous tight junction that encircles each cell to form an essential structural barrier to the paracellular movement of substances of large molecular radius. In addition to these structures, intercellular adhesion is maintained by a combination of intermediate junctions and desmosomes (Garrod, 1986; Fig. 10.2). Desmosomes are symmetrical structures made up from constituents from two adjacent cells and they serve as "spot welds" that vary in diameter from 0.1 to 1.5 μm. In cross section they comprise an electron-dense plaque on each side of an intercellular space divided vertically by a central disc. The central disc is attached to adjacent cell membranes by a number of short horizontal cross-bridge structures which traverse half the width of the intercellular space. The electron-dense plaques are 14–20 nm thick and lie just below the epithelial cell plasma membrane, which in the region of the desmosomes has a prominent trilaminar appearance. A number of tonofilaments fan out from this into the adjacent cytoplasm. The tonofilaments are made up of three distinct proteins, cytokeratin, vimentin and desmin (Garrod, 1986). Desmosomal structures appear to be the major structural adhesion mechanism present between superficial cells and basal cells, and for this reason they have been implicated as a factor that regulates the progress of cellular injury and detachment (Montefort *et al.*, 1992a,b). Intermediate junctions are also involved in intercellular adhesion in epithelia, and these consist of the E-cadherins which divide the lateral margins of the cell into apical and basolateral domains (McNutt and Weinstein, 1973). E-cadherin immunostaining has been localized in human airway epithelium to the intercellular contacts that are subjacent to the luminal surface (Montefort *et al.*, 1992a).

Other types of adhesion mechanism are also present in the airway epithelium (Montefort *et al.*, 1992a; Roche *et al.*, 1993). A number of integrins have been identified which appear to be the principal means of adhesion between basal cells and the underlying biomatrix, although the extent to which integrins are differentially expressed on basal as compared to superficial cells remains uncertain (Sapsford *et al.*, 1991; Damjanovich *et al.*, 1992; Sheppard, 1993; Roche *et al.*, 1993). Existing data suggest that normal human airway epithelium can express $\alpha_2\beta_1$, $\alpha_3\beta_1$ integrins, at least one α_6-containing integrin and at least one α_v-containing integrin that is not $\alpha_v\beta_3$ (Sheppard *et al.*, 1990, 1992; Sheppard, 1993). As well as providing a mechanism of anchorage of basal cells to structural proteins, there is evidence to suggest that these integrins may also play important roles in mediating the effects of biomatrix proteins on the phenotype of surrounding cells (Sheppard, 1993).

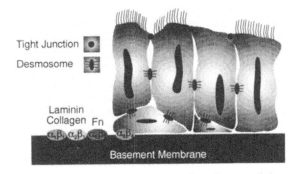

Figure 10.3 Schematic representation of some of the intercellular and cell–biomatrix adhesion mechanisms present in the airway mucosa.

Some of the adhesion molecules thought to be present in the airway epithelium are summarized in Fig. 10.3.

4. The Occurrence of Epithelial Injury

In the preceding discussion the term epithelial injury has been applied quite loosely to the broad range of abnormalities that are associated with various pathophysiological events in the airway mucosa. Any analysis of the literature in this field will reveal that what has in some cases been reported as evidence of epithelial injury is actually another pathophysiological event that is a consequence of such processes. These include altered gas exchange, pulmonary oedema, abnormal pulmonary haemodynamics and increased vascular permeability. However, not all of these have been shown to have direct correlates with the obvious feature of damage to epithelial cells and their cell–cell and cell–biomatrix attachments. To what extent disruption of epithelial morphology and cellular integrity leads or follows these other pathophysiological manifestations is not known, and so it is essential that the precise indices used to study epithelial pathophysiology are understood.

Mucosal inflammation and consequential damage of the delicate lining of the airway mucosa can be brought about by diverse factors. These include normal environmental or occupational exposure to noxious gases, vapours, fumes, aerosols, organic and inorganic particulates and complex of mixtures of these such as tobacco smoke or diesel exhaust emissions (Lippmann, 1980; Mauderly *et al.*, 1987; Churg and Green, 1988; Jones Williams, 1988; Last, 1988; Fujimaki *et al.*, 1989; Aris *et al.*, 1993). In addition to these mechanisms of injury, a number of drugs are associated with significant degrees of pulmonary toxicity which, in some cases, is manifest by direct cytotoxic effects in the airway epithelium.

Focal destruction of the epithelium and outright cellular injury is also a feature of many well-known diseases, and it seems likely that many inflammatory reactions in the airways are quite capable of resulting in epithelial injury if the inflammatory insult is allowed to develop into a chronic stimulus. Bronchial asthma is one disease state that is well associated with destruction of the epithelial architecture. In studies undertaken in the 1960s Naylor reported the presence of increased numbers of epithelial cell clusters, called creola bodies, in the expectorated sputum from asthmatics and the fact that their numbers increased further during exacerbations of the disease (Naylor, 1962). Dunnill (1960) also reported increased epithelial denudation in autopsy specimens from patients who died from status asthmaticus.

Notwithstanding the severe limitations on samples that can be taken for biopsy, more recent and detailed investigations of epithelial dysfunction in bronchial asthma have highlighted the fact that outright shedding of the epithelium is not actually an obligatory feature of the disease (Lozewicz et al., 1990), although denudation can certainly be observed in extreme cases (Laitinen et al., 1985). In lungs from patients with chronic asthma, autopsy reveals that the loss of the airway epithelium is more typically a localized rather than global process (Carroll et al., 1993), although the mere presence of the epithelial cells should not be taken as indicative of normal structure or function. From these autopsy (Carroll et al., 1993) and biopsy (Jeffery et al., 1989) studies, it would be more correct to describe the epithelium in bronchial asthma as being hyperfragile. In this hyperfragile state the superficial epithelium is especially sensitive to mechanical trauma (e.g. as artefactually induced during fibreoptic bronchoscopy; Jeffery et al., 1989) and, presumably, other forms of injury. Electron microscopy has revealed morphological evidence of disrupted intercellular adhesion in asthmatic airway mucosa that has an otherwise grossly normal appearance (Bucca et al., 1988; Ohashi et al., 1992). This disruption is typified by tight junction opening and an increase in paracellular spaces. Although subtle, a chronic manifestation of these changes may nevertheless have significant consequences for airway pathophysiology by contributing to an increasing susceptibility to injury. The fibroproliferative response that results in subepithelial fibrosis in asthma, or following chronic low level exposure to oxidant gases (Castleman et al., 1977), is most likely a major cellular response to this gradual loosening of epithelial anchorage and integrity caused by the mucosal inflammatory response. The precise sequence of events that occur to produce this epithelial hyperfragility and the factors that result in the ultimate shedding of the epithelium remain to be determined, but as discussed later there are numerous candidate mediators for these processes. Some of these putative agents may act directly on epithelial cells to produce cytotoxicity or altered expression of adhesion molecules; other mediators might act indirectly by promoting the development of subepithelial hydrostatic pressures capable of rupturing the injured mucosal barrier (Persson, 1988).

Earlier reference was made to the cellular heterogeneity of the airway mucosa and the specialized roles that different cell types have evolved to undertake. This cellular heterogeneity has possible implications for cellular injury in the airway. Differential sensitivity to injury of cells in the airway mucosa has been noted with a number of noxious gases. In dogs, cats and rats, ciliated cells in the trachea are more sensitive to injury caused by sulphur dioxide, ozone or nitrogen oxides than are non-ciliated superficial and basal cells (Boatman and Frank, 1974; Boatman et al., 1974; Cabral-Anderson et al., 1977; Man et al., 1986). These changes are frequently rapid in onset, with morphological evidence of damage observable within an hour, and have the surprising feature of focal rather than global patterns of injury despite the gaseous nature of the inciting agent (Boatman et al., 1974; Man et al., 1986). The reasons for this regional sensitivity to injury are unknown, but the presence or absence of specific cytoprotective mechanisms is an intriguing possibility and may have similar parallels in the heterogeneous pattern of epithelial injury seen in non-fatal, severe asthma (Carroll et al., 1993). In vitro studies have confirmed that the tracheal mucosa is apparently more resistant to leucocyte-mediated injury than the bronchial mucosa (Herbert et al., 1991, 1993; Kercsmar and Davis, 1993; Fig. 10.4), and within the bronchial mucosa the patterns of injury caused by these cells may be graded. For example, during in vitro studies of the interaction between eosinophils and bovine bronchial mucosa, we reported that the rapid augmentation of epithelial permeability to macromolecules was associated with an initial loss of superficial cells from the underlying basal cells and basement membrane (Herbert et al., 1991). Interestingly, some evidence for a similar pattern of injury has subsequently been claimed in bronchial asthma (Montefort et al., 1992b), although in such studies it is impossible to determine whether the basal cells observed at lesion sites are those left intact after exfoliation of superficial cells or whether they are newly arrived cells repopulating the site of injury (Keenan et al., 1982a,b,c, 1983). It has been suggested by some that the apparent resilience to injury of basal cells may be due to differences in the types of intercellular adhesion mechanisms present (Montefort et al., 1992b), but clearly the actual mediators of injury, turnover rate of the target cell, the presence or absence of cytoprotective agents (e.g. antioxidants and antiproteinases) and other regulatory mechanisms are likely to be important factors. For example, there is some evidence that basal and non-ciliated superficial epithelial cells have a greater resistance to applied osmotic stresses than ciliated cells (Spring and Ericsson, 1983; Man et al., 1986), a finding that is not inconsistent with the mitotic, and thus presumably regenerative, functional aspects of these cells.

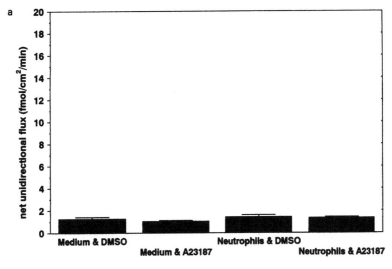

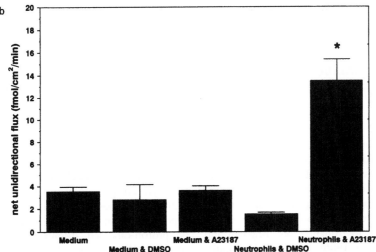

Figure 10.4 (a) Resistance of bovine tracheal mucosa to injury elicited by ionophore activated human neutrophils, and (b) susceptibility of bronchial mucosa under identical conditions. Injury, in this case cell detachment, was measured indirectly by measuring the permeability of the airway mucosa to serum albumin. Cells were applied to the basolateral surface and the vector of solute movement was basolateral–apical.

5. Cellular Events in Epithelial Injury

Only few detailed analyses of the intracellular events following epithelial injury have been performed using cells from the airways, and most mechanistic work has concentrated on pre-necrotic and necrotic processes in epithelial cells of renal origin (Trump *et al.*, 1980, 1981; Trump and Berezesky, 1984). Collectively these studies, which have employed a wide range of stimuli to evoke cellular injury (ischaemia, metabolic poisons, chaotropic agents and ionophores), suggest that similar patterns of intracellular events accompany injury evoked by the different stimuli.

An early intracellular response to epithelial injury appears to involve an increase in intracellular Ca^{2+} which coincides with the failure of the cells to regulate Na^+

balance (Blaustein, 1977). This implies that the origin of the effect is at the level of a Na^+–Ca^{2+} exchanger. Morphologically, during this phase of the response the cells exhibit vacuolated cytoplasm due to expansion of the endoplasmic reticulum, together with abnormalities of the cytoskeleton and the occurrence of blebs in the apical membrane of the cells (Trump *et al.*, 1980). These are all features that can, under appropriate conditions, be reproduced by Ca^{2+} ionophores. Derangements in the structure of mitochondria also become evident, presumably heralding the metabolic run-down through these pre-necrotic stages of injury prior to cell death (Boatman *et al.*, 1974).

Much of the intracellular response to epithelial injury remains to be defined. The mechanistic linkages between those ionic and morphological changes that have been observed remain the subject of speculation. Equally, many other intracellular events remain to be investigated. In terms of understanding the cell biology of this process, and its implications for disease, we also need to establish what level of cellular stress may be tolerated before irreversible damage or outright cell death actually occur. To date, most studies of intracellular events in epithelial injury have concentrated upon processes evoked by oxidant gases. Whilst it is not unreasonable to suspect that many of these changes are likely to be features of the intracellular response to inflammatory stimuli (of which cell-derived oxidants such as hydrogen peroxide are of course a part), initial trauma to the epithelium in asthma does not result in overwhelming cellular necrosis, but appears to result principally in loss of cellular adhesion. Thus, an important question that remains unanswered concerns the possible relationship between the established ionic changes and adhesion molecule function, and whether such a linkage might be reciprocal in nature.

6. Cell-derived Mediators Implicated in Epithelial Injury

6.1   OXIDANTS

Stimulated leucocytes are the most important sources of oxidants produced endogenously during inflammatory reactions that result in lung injury. This type of injury has most commonly been associated with oxidant production in diseases such as adult respiratory distress syndrome, following exposure to asbestos dust or in the inflammatory response to oxidant gases (Brigham and Meyrick, 1984; Mossman, 1986). However, the populations of urbanized environments are being increasingly exposed to oxidant gases and particulates in the form of industrial and vehicular exhaust emissions (American Thoracic Society, 1978; Williams et al., 1988; Hileman, 1989; Schlatter, 1994). With the exception of individuals exposed as a result of unsafe or illegal working practices, or in cases of industrial accidents, most exposure of the mass population is at a low, but nevertheless steadily increasing, level. However, even low level exposure is now considered to be a significant health hazard, particularly in individuals who may have predisposing factors for lung diseases such as asthma, or where endogenous antioxidant defences may be compromised (Bylin et al., 1988; Burney, 1993). The focus of this section will be on the role of oxidant production in airway injury and the species that may be involved as the causative agents. Because the effects of oxidants will be determined also by endogenous anti-oxidant mechanisms, readers are referred to a survey of such mechanisms in the lung (White and Repine, 1985; Heffner and Repine, 1989).

The ability of oxidant gases (e.g. sulphur dioxide, ozone, nitrogen oxides and complex mixtures of such in tobacco smoke) to produce lung inflammation and injury has been extensively studied (Freeman et al., 1966; Evans et al., 1971, 1975, 1976, 1977; Stephens et al., 1972, 1973, 1974; Bils, 1974; Cabral-Anderson et al., 1977; Castelman et al., 1977, 1980; Bils and Christie, 1980; Vai et al., 1980; Boucher et al., 1980). At the morphological level, injury produced by these agents has a number of similarities, and interestingly some of these are also found in injury processes evoked by endogenous immunological mechanisms. Initial changes include ciliary dyskinesia, ciliostasis, vacuolation, swelling of mitochondria and development of secretory cell metaplasia. An increase in paracellular permeability of the mucosal barrier (Vai et al., 1980; Gordon et al., 1983) arises as a result of tight junction failure and loss of superficial epithelial cells from the injury site. Biochemically, increased oxygen consumption, glucose utilization and changes in antioxidant enzymes ensue (Mustafa and Lee, 1976; Mustafa and Tierney, 1978). In both humans and experimental animals, these responses are associated with inflammatory cell infiltration in the airway, and this seems to be a key event in producing functional pathophysiological changes in airway responsiveness (Fabbri et al., 1984; Aris et al., 1993).

A number of oxidizing agents are candidate mediators of the mucosal injury produced by inflammatory cell activation (reviewed in Henson and Johnston, 1987; Heffner and Repine, 1989). Stimulation of leucocytes is associated with a respiratory burst as a result of activation of the NADPH oxidase/cytochrome b_{554} complex in the leucocyte cytoplasmic membrane (Babior et al., 1973). Activation of this pathway results in the formation of superoxide anion O_2^-, which readily undergoes enzyme- or proton-catalysed dismutation to form hydrogen peroxide (Weiss, 1986). Superoxide can be formed in large quantities by leucocytes, and although it is capable of producing some cellular damage it is not outstandingly cytotoxic (Fridovitch, 1986). However, in the presence of transition metal cations, a Fenton-type reaction can occur resulting in the conversion of the hydrogen peroxide to extremely toxic ·OH radicals (Halliwell and Gutteridge, 1984). Fortunately, the half-life of these radicals is short, thus limiting the extent of the damage that they can produce. Under most circumstances the concentration of free Fe^{2+} is too low to support a Fenton-type reaction, and normally the iron in haemproteins is unavailable for catalysis. However, by acidification of the microenvironment, inflammatory reactions may liberate free Fe^{2+} from the iron transporter protein ferritin and other iron-associated proteins to result in more significant participation of ·OH in cell injury (Biemond et al., 1984; Gutteridge, 1987). Hydroxyl radicals are a central component of lung injury produced by bleomycin, which produces oxidants by redox cycling (Giloni et al., 1981), and also by ionizing radiations where their formation is not dependent upon the participation of transition metals.

Superoxide anions can also be formed as a result of redox cycling induced *inter alia* by the herbicide paraquat (Sandy *et al.*, 1986; Smith, 1986). The cycling process consumes intracellular reducing agents and results in eventual exhaustion of NADPH. The action of paraquat is particularly insidious because it is selectively accumulated in the airway epithelium by an energy-dependent mechanism and exerts much of its toxic action at this site (Vijeyaratnam and Corrin, 1971; Rose *et al.*, 1976).

NO undergoes reaction with superoxide anion resulting in the formation of the peroxynitrite anion $(OONO^-)$. Peroxynitrite undergoes protonation and decomposition ultimately to yield NO_2 and hydroxyl radicals (Beckman *et al.*, 1990). Thus, a substance (NO) that in one setting is believed to be highly beneficial to mucosal function has the rather dubious property of potentially being able to generate powerful genotoxic and cytotoxic agents. However, whilst the ability of NO to participate in the killing of tumour and other cells has been established (Beckman *et al.*, 1990), its considerable potential for cytotoxic effects in the lung epithelium remains to be unambiguously confirmed.

The hydrogen peroxide produced from dismutation of superoxide anions can also undergo reaction with halides and cellular peroxidases released during leucocyte activation. In neutrophils and macrophages the major enzyme involved is a myeloperoxidase, whilst eosinophils possess their own specialized peroxidase which is described later. Neutrophil myeloperoxidase has a substrate specificity for chloride anions and transforms hydrogen peroxide into hypochlorous acid (HOCl; Weiss, 1986). This is a reactive oxidant which reacts with free amino and sulphydryl groups of proteins, forming chloramines (Thomas *et al.*, 1982; Grisham *et al.*, 1984). Formation of chloramines causes disruption of protein structure and function, and this can have profound effects on cellular physiology if membrane transport proteins such as the cellular glucose transporter and $Na^+, K^+ - ATPase$ become inactivated. Furthermore, damage to structural proteins such as fibronectin is known to occur upon reaction with chloramines (Vissers and Winterbourn, 1991), presumably providing the necessary loosening of the biomatrix to aid inflammatory cell migration and detachment of resident tissue cells. However, because the reaction of chloramines is indiscriminate they are likely to be rapidly quenched by extracellular proteins such as albumin (Halliwell, 1988).

Eosinophil peroxidase has a preferential reactivity with bromide anions (Mayeno *et al.*, 1989), although recent evidence suggests that thiocyanate might be its most effective substrate (Slungaard *et al.*, 1991). Because of its intense reactivity and evanescence (Vogel, 1970), it can only be presumed that hypobromous acid has broadly similar effects to hypochlorous acid.

However, it seems likely from experiments with stimulated leucocytes that the most important oxidant involved in cell injury and death is probably hydrogen peroxide. Plate II illustrates evidence of the ability of hydrogen peroxide to injure human bronchial epithelial cells. DNA strand breaks are produced rapidly by modest concentrations of hydrogen peroxide (Birnboim, 1982) and these lead to activation of polyadenosine diphosphate-ribose polymerase, a nuclear enzyme that hydrolyses NAD into ADP-ribose and nicotinamide. ADP-ribose polymerizes and bonds to nuclear proteins with a high turnover rate that eventually results in NAD depletion and metabolic run-down due to exhaustion of ATP reserves (Sims *et al.*, 1983; Spragg *et al.*, 1985; Schraufstatter *et al.*, 1986) and inhibition of glyceraldehyde–3-phosphate dehydrogenase. Higher concentrations of hydrogen peroxide trigger an increase in intracellular Ca^{2+} from both internal and external sources, and this is the presumed initiation signal for a number of enzyme-mediated injury promoting pathways (e.g. activation of proteinases), and disordering of the cytoskeleton (Hyslop *et al.*, 1988).

6.2 EOSINOPHIL MAJOR BASIC PROTEIN

There is no doubt that eosinophils are capable of causing epithelial injury (see e.g. Plate III and Fig. 10.5). The precise combination of mediators that contribute to this process are not known, but much attention has been paid to protein components of eosinophil granules.

Major basic protein (MBP) is a 118 amino acid, arginine-rich granule-derived protein that is initially synthesized within the cell in a pro-form that contains an acidic leader sequence (Gleich *et al.*, 1973; Wasmoen *et al.*, 1988; Barker *et al.*, 1988). In electron micrographs of eosinophils it is observable as the characteristic electron-dense core of the specific granules found within the eosinophil cytoplasm. Together with other basic proteins found within the eosinophil, it is cytotoxic towards parasites and mammalian cells (Butterworth *et al.*, 1979). A number of studies have demonstrated the susceptibility of airway epithelial cells to the damaging effects of MBP (Frigas *et al.*, 1980; Flavahan *et al.*, 1988; Gundel *et al.*, 1991), and a possible role of this protein in asthma is supported by the finding of deposits of MBP subjacent to areas of epithelial cell loss (Filley *et al.*, 1982). *In vitro* studies with airway tissue typically report that concentrations of MBP between 10 and 900 $\mu g \, ml^{-1}$ are required to produce a reduction in ciliary beat frequency, ciliostasis and the eventual loss of epithelial cellular integrity (Frigas *et al.*, 1980; Flavahan *et al.*, 1988; Motojima *et al.*, 1989; Ayars *et al.*, 1989). These concentrations are generally greater than the levels of MBP that have been observed in asthmatic sputum. For example, in 15 patients hospitalized with asthma, their sputum concentrations of MBP ranged from 0.3 to 92.9 $\mu g \, ml^{-1}$ with a geometric mean concentration of 7.1 $\mu g \, ml^{-1}$ (Frigas *et al.*, 1981). Whilst these observations concerning the amounts of MBP recoverable in biological

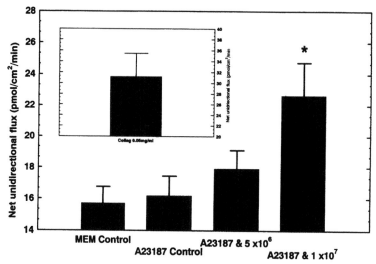

Figure 10.5 The effects of ionophore-stimulated human eosinophils on the permeability of bovine bronchial mucosa to mannitol. The inset panel in this figure illustrates the action of 0.05 mg ml^{-1} bacterial collagenase for comparison.

fluids potentially reduce the significance of MBP in producing damage to epithelial cells, several important considerations should be borne in mind. Firstly, whilst it has been possible to use MBP in experiments as one would use a conventional pharmacological agonist, its molecular mass is likely to restrict its diffusion to sites of action in complex tissues. Secondly, similar considerations apply to the possible underestimation of MBP release by its measurement in biological fluids, because the relationship between the release from eosinophils and that which finds its way into free solution is unknown. This potential problem may be particularly exacerbated for highly charged molecules such as MBP which might be expected to have a propensity to adhere to other highly charged sites such as components of the epithelial basement membrane. Thirdly, surrogates of eosinophil granule-derived proteins such as poly-L-arginine cause an increase in airway responsiveness, whereas under similar conditions, stimulated eosinophils themselves do not (Omari *et al.*, 1993a). Similar reservations ·about pharmacological potency *in vitro* and also the relationship between cellular release and concentrations detectable in biological fluids almost certainly apply to the other significant proteins of eosinophil granules.

It seems likely that MBP exerts some effects on epithelial function at much lower concentrations than those required to produce cellular dysfunction and detachment. Over a 24 h exposure period human nasal epithelium showed a 50% up-regulation of intercellular adhesion molecule-1 (ICAM-1) expression when exposed to concentrations of MBP within the range 0.1–0.2 μg ml^{-1} (Ayars *et al.*, 1991).

6.3   EOSINOPHIL CATIONIC PROTEIN

Eosinophil cationic protein (ECP) arises in the granule matrix as a pre-protein containing 160 amino acid residues, 133 of which constitute the mature secreted form of the protein (Barker *et al.*, 1989). Like MBP it is cytotoxic towards parasites and mammalian cells by promoting formation of pores within its target cells (Motojima *et al.*, 1989). ECP has numerous biological activities, some of the most notable being actions on the coagulation and fibrinolytic pathways. ECP binds to heparin to neutralize its anticoagulant activity, and it also increases the activation of plasminogen by urokinase (Dahl and Venge, 1979). Unlike major basic protein it does not stimulate histamine release from basophils (Zheutlin *et al.*, 1984). Whilst there is little doubt that application of ECP produces epithelial injury, whether these effects can be manifest over similar time scales *in vivo* is less certain. As with MBP, progress in understanding its possible contribution to epithelial injury requires studies of its chronic effects at low concentrations together with intervention in its release from endogenous sources.

6.4   EOSINOPHIL-DERIVED NEUROTOXIN

Eosinophil-derived neurotoxin (EDN) is another protein that is found in the granule matrix of eosinophils and it exhibits significant sequence homology with ECP (Barker *et al.*, 1989). As implied by its name, EDN has a powerful neurotoxic action that can damage myelinated neurones in experimental animals (Durack *et al.*, 1981). However, unlike ECP or MBP there is no significant evidence suggesting that it has a major role in destruction of the architecture of the airway mucosa, although most of the published studies with this protein have employed only tests of gross cellular injury that might not reveal evidence of more subtle participation in chronic tissue injury (Motojima et al., 1989).

6.5 EOSINOPHIL PEROXIDASE

Eosinophil peroxidase (EPO) is another distinctive component of the granule matrix of the eosinophil. Although it exhibits sequence homology with other cellular peroxidases (Ten et al., 1989), it has a distinctive absorption spectrum that demarcates it from monocyte or neutrophil myeloperoxidase (Wever et al., 1981; Olsen and Little, 1983; Bolscher et al., 1984). Each molecule of EPO consists of a heavy chain of 50–60 kDa complexed in 1:1 stoichiometry with a light chain of 10–15 kDa. EPO and hydrogen peroxide produced by the eosinophil respiratory burst act together to oxidize halides, especially bromides, to their respective hypohalous acids (Mayeno et al., 1989). As described earlier, these unstable hypohalous acids are potently cytotoxic to a wide range of cells by their interaction with cell surface proteins (Thomas et al., 1982; Grisham et al., 1984; Vissers and Winterbourn, 1991). However, recent studies have suggested that EPO interacts preferentially with thiocyanates to produce hypothiocyanous acid which is a sulphydryl-reactive oxidant (Slungaard et al., 1991). Hypothiocyanous acid is a weaker oxidant than hypobromous acid, and because thiocyanate will compete with bromide anions for conversion by EPO, there is now some doubt about whether EPO is a significant route to powerful oxidants. In addition to oxidant generation, EPO is also known to bind to rat mast cell secretory granules and this complex is capable of causing mast cell secretion (Henderson et al., 1980). It is not known whether this process has any pathophysiological significance in humans.

6.6 PROTEINASES

There is abundant evidence that a diverse array of proteinases, particularly those of macrophage or leucocyte origin, have a fundamental role in the injury processes associated with diseases such as α_1-antitrypsin deficiency, smoking-related emphysema, cystic fibrosis, bronchiectasis, bronchitis and other respiratory syndromes (Karlinsky and Snider, 1978; Stockley, 1983; Senior and Campbell, 1983; Suter et al., 1984; Janoff, 1985; Snider et al., 1991). However, there has been little exploration of their potential involvement in other airway disorders that have an inflammatory cell aetiology, such as certain types of specialized fibrotic disorders including chronic severe asthma. This is particularly true for the matrix metalloproteinases (MMPs), despite the expectation from their individual substrate specificities (Emonard and

Grimaud, 1990) that they should be of fundamental importance in diseases associated with remodelling and fibrosis of the basement membrane, epithelial injury and fibrosis.

6.6.1 Neutrophil Elastase

Neutrophil elastase is a serine proteinase and comprises the major proteinase activity released from the azurophilic granules. This enzyme, together with its endogenous inhibitor, α_1-proteinase inhibitor, has been the centre of most attention concerning the putative role of proteinases in lung diseases such as emphysema that involve the destruction of elastic structural proteins (Barrett, 1981a). Although elastin is the major substrate for the enzyme, it also has detectable activity against collagens, fibronectin, laminin and proteoglycans.

Administration of purified elastase to experimental animals certainly does produce emphysematous changes in structure of the respiratory units of the airways, but attempts to relate enzyme expression in diseased lung to manifestations of emphysema pathology have been viewed more cautiously (Fox et al., 1988; Snider et al., 1991). Elastase exposure also causes changes in lectin-binding sites on the luminal surface of Clara cells (Christensen et al., 1989) and impairs mucociliary clearance by injuring epithelial cells (Tegner et al., 1979; Smallman et al., 1984; Sykes et al., 1987). In vitro, human neutrophil elastase has been shown to cause detachment of type II alveolar cells and bronchiolar Clara cells (Bingle et al., 1990). In cystic fibrosis the airway epithelium can also be exposed to an elastase from the infective organisms Pseudomonas aeruginosa and Ps. cepacia. Unlike the neutrophil enzyme, this is a metalloenzyme, but has qualitatively similar effects to that from the neutrophil (Tournier et al., 1985; Amatini et al., 1991; Spooner et al., 1991).

6.6.2 Matrix Metalloproteinases

The association between neutrophils and airway injury in emphysema clearly establishes the potential for proteinases to play a role in injury of the airway epithelium. However, elastase is unlikely to be the only enzyme involved because neutrophils are also sources of other significant proteolytic activities. One such important group is the MMP family (Matrisian, 1990; Emonard and Grimaux, 1990). Of these, neutrophil collagenase that is found within the specific granules of the neutrophils is worthy of consideration as a mediator of mucosal injury. Instillation of interstitial-type collagenases into the airway does produce cellular injury, but unlike elastase, the wound sites are more amenable to repair and severe emphysematous changes do not usually develop (Snider et al., 1986). However, neutrophils are not the only sources of MMPs in the airway mucosa and their possible participation in other forms of mucosal injury and abnormal tissue repair in prefibrotic lesions remains an exciting field which has only recently become the focus for experimental investigation.

The ability of various MMPs to cause epithelial injury has been documented in several studies. Like exposure to activated eosinophils and neutrophils, treatment with MMPs disrupts the epithelial diffusion barrier resulting in an increase in sensitivity of underlying bronchial smooth muscle to contractile agonists introduced into the airway lumen (Herbert et al., 1993, 1994). Studies in this laboratory have demonstrated that non-neoplastic cells of the airway mucosa release a range of MMPs that have substrate activity directed towards gelatin. The major gelatinases that we have detected show characteristics (sensitivity to inhibitors, activation behaviour and cDNA probe hybridization) which indicate co-identity with the 72 kDa gelatinase known as matrix metalloproteinase-2 (MMP-2, gelatinase A) and the variably glycosylated 90–98 kDa gelatinase B (MMP-9; Emonard and Grimaux, 1990). Thus, both of these activities are similar to the well-characterized gelatinases that degrade type IV basement membrane collagen and which are released from a variety of non-pulmonary sources (Hipps et al., 1989; Emonard and Grimaud, 1990). We have found that modest amounts of catalytically competent MMP-2 and MMP-9 are released from both the apical and basolateral surfaces of freshly isolated sheets of airway mucosa, and that gelatinase activity can be increased equally by the addition of either (1) eosinophils that are primed but not actually degranulating, or (2) eosinophils that are primed and stimulated to degranulate and injure the airway epithelium. The ability of the broad spectrum antiproteinase α_2-macroglobulin to inhibit eosinophil-dependent epithelial injury supports the putative involvement of proteinases in epithelial cell detachment (Herbert et al., 1991). These observations suggest that MMP-2 and MMP-9 might increase the fragility of the airway epithelium without causing its obligatory detachment (Herbert et al., unpublished). In the presence of appropriately activated and degranulating eosinophils, the combination of augmented metalloproteinase activity and the liberation of eosinophil-derived cytotoxic proteins and oxidants might provide the additional stimuli required for cell detachment. MMP-2 and MMP-9 degrade not only type IV collagen, but also attack type V collagen, fibronectin and elastin (Collier et al., 1988; Murphy et al., 1991). These substrate profiles make them ideal putative candidates for involvement in the process of epithelial cell detachment, basement membrane remodelling and tissue proliferation that are features of bronchial asthma. Clearly, further work is needed to identify the possible contribution of these proteinases to epithelial injury and such investigations await further documentation of the cellular sources not only of the proteinases themselves but also of their endogenous inhibitors. It is not known to what extent MMPs exist physiologically in their inhibitor-free forms (Howard et al., 1991; Strongin et al., 1993), and this is likely to be dependent upon complex changes in the stoichiometry between the enzymes and their inhibitors caused by differential regulation of the expression of each (Shapiro et al., 1991).

6.6.3 Miscellaneous Cell-derived Proteinases

Many other proteinases could play contributory roles in damage to the epithelium, but at present it is difficult to establish their precise importance. Macrophages and neutrophils contain enzymes that have not been referred to thus far. These include cathepsin G, a serine proteinase from azurophilic granules of neutrophils (Barrett, 1981b; Senior and Campbell, 1983; Travis, 1988); plasminogen activator(s), a serine proteinase bound to the cytoplasmic membrane of neutrophils (Granelli-Piperno et al., 1977; Senior and Campbell, 1983; Estreicher et al., 1990); and cathepsins B and D, neutrophil- and macrophage-derived proteinases that exhibit significant proteolytic activity at low pH (Barrett and Kirschke, 1980; Senior and Campbell, 1983; Clausbruch and Tschesche, 1988). Although the content of cathepsin G in neutrophils is high (at least equivalent to that of elastase), much less is released following neutrophil stimulation (Senior and Campbell, 1983) and consequently its major role may be as an intracellular proteinase within phagosomes.

Recent studies have also highlighted the presence of the aspartic proteinases pepsinogen II and cathepsin E within the airway mucosa. Pepsinogen II has been immunolocalized to type II pneumocytes, while Clara cells have been identified as being immunoreactive for cathepsin E. In a study of 75 cases of non-neoplastic lung disease, a highly significant correlation has been shown between histopathological evidence of injury, particularly in cases of hyperplastic epithelium, and the expression of both of these proteinases (Bosi et al., 1993). The signals for the expression of these proteinases and their function remain to be established.

6.6.4 Environmental Exposure to Proteinases

The airway mucosa may also be a target of proteolytic attack from the external environment. It has been known for a number of years that asthma or other forms of reversible airways obstruction can result from occupational exposure to enzymes used in industrial processes, or fungi encountered in the environment (Pepys, 1969; Flindt, 1969; Milne and Brand, 1975). These occupational lung diseases have been considered to be due to immunological reactions to the proteins concerned, without any special regard for their enzymatic nature. However, more recent studies have revealed an interesting possibility which suggests that the enzymatic nature of certain exogenous proteins could be of considerable importance in determining their allergenicity and whether, in susceptible individuals, allergic airways disease could result.

Amino acid sequence analysis has revealed the existence

of homology between certain aeroallergens and known proteolytic enzyme classes. This homology can be expressed functionally in the limited number of cases studied in sufficient detail (Lake *et al.*, 1991; Stewart *et al.*, 1991, 1992a,b, 1993), but the significance of this to airway pathophysiology is unknown at present.

One allergen that has been studied in detail in this context is Der p 1, a major allergen from the house dust mite *Dermatophagoides pteronyssinus*. Homology searching predicted that Der p 1 could be a cysteine proteinase (Chua *et al.*, 1988), and this has been verified in standard tests for proteolytic activity. When applied to sheets of bovine bronchial mucosa and rendered catalytically competent by reducing agents, Der p 1 is capable of causing disruption of the architecture of the airway epithelium resulting in an increase in its permeability to macromolecules (Herbert *et al.*, 1990). It is not clear to what extent this process might be involved in chronic tissue injury, but these experiments suggest an exciting possibility for antigen presentation. The ability of a putative allergen to facilitate its own penetration into the airway mucosa could provide greater scope for the three-dimensional dendritic intraepithelial network of antigen-presenting cells to initiate an immunological response by detecting the antigen and shuttling it from the airway epithelium to submucosal lymphocytes and lymph nodes (Holt *et al.*, 1990; Holt, 1993). The expression by proteins of enzymatic activities to facilitate antigen presentation might therefore represent a fundamental mechanism that distinguishes between the many inert environmental antigens to which the lung is exposed, and those antigens that are immunologically significant to an individual. Equally, in a sensitized airway it is conceivable that local changes in the airway permeability induced by this enzymatic activity might facilitate access of allergen to mast cells resident in the airway submucosa. However, a number of conditions would need to be fulfilled for these actions to be of any pathophysiological relevance. Firstly, the allergen would have to be present in a suitable concentration in the airway to overcome endogenous antiproteinases. Secondly, cysteine proteinases such as Der p 1 also require reduction to achieve catalytic competence. Of these, the latter condition is more easily answered. Airway surface lining fluid contains a number of potential reducing agents, of which glutathione is likely to be the most important (Cantin *et al.*, 1987). In the studies reported in the literature, Der p 1 can be activated by concentrations of glutathione that are encountered in airway lining fluid.

6.6.5 Neutrophil-derived Toxic Proteins

The azurophil granule of the neutrophil is also a rich source of cytocidal proteins, although some of these have received relatively little attention as a mechanism of inflammatory cell-induced lung injury compared with the wealth of literature concerning oxidants and eosinophil-derived proteins. Several different cytocidal polypeptides

have been identified. These include cationic antimicrobial peptide (CAP) 57, azurocidin and the defensins (Ganz *et al.*, 1985; Selsted *et al.*, 1985; Rice *et al.*, 1987). The defensins are low molecular mass (c. 3.5–4 kDa) arginine- and cysteine-rich proteins that have been shown to be cytocidal against a number of different tumour cells including the A549 alveolar adenocarcinoma line (Lichtenstein *et al.*, 1986; Okrent *et al.*, 1990). Mechanistic studies in Madin-Darby canine kidney (MDCK) cells have demonstrated that defensins are able to reduce the integrity of an epithelial barrier without necessarily producing cytotoxicity (Nygaard *et al.*, 1993). Interestingly, these proteins also possess chemotactic activity for mononuclear cells (Territo *et al.*, 1989).

6.7 MISCELLANEOUS

The airway epithelium is also known to be injured by a variety of viral and bacterial products, a number of which have proteolytic activity themselves or which induce cellular injury as a consequence of the host immunological responses that they evoke. Episodes of airflow obstruction are known to accompany such infections and impaired mucociliary clearance is a common feature of certain infections.

It would not be surprising to find that infectious agents are sources of numerous, as yet uncharacterized, agents that produce epithelial injury. Studies with *Pseudomonas aeruginosa*, which is associated with lung lesions in cystic fibrosis (Hoiby, 1982; Fick, 1989), have demonstrated that 1-hydroxyphenazine, rhamnolipid and pyocyanin, a pigment produced by this bacterium, are capable of retarding the beating of cilia and ultimately producing ciliostasis and damage to the epithelial surface (Read *et al.*, 1992; Kanthakumar *et al.*, 1993). A decrease in the intracellular concentration of cAMP, which is required for the maintenance of ciliary beating (Di Benedetto *et al.*, 1991; Lansley *et al.*, 1992), is known to occur following pyocyanin exposure, and cells may be afforded a modest degree of protection against it by treatment with β_2-agonists (Kanthakumar *et al.*, 1994).

7. Repair of Epithelial Injury

Most studies of the repair process in the airway mucosa have concentrated on injury produced by noxious stimuli such as oxidizing gases. It is presumed, although not yet subject to experimental verification, that there will be at least some similarities with the responses to injury produced by an inflammatory cell infiltrate. Conceptually, the initial objective of "repair" is to provide continuity of the barrier properties of the airway mucosa (Gordon and Lane, 1976). Once this has been attained, the next event involves the repopulation of the airway mucosa with a normal profile of ciliated and non-ciliated cells (Keenan *et al.*, 1982a,b,c; 1983; Gordon and Lane,

1984). The initial phase is achieved by a lateral spreading of remaining superficial non-ciliated epithelial cells. The extent to which this reconciliation can occur across wide "deserts" of basement membrane from which all cells, including basal cells, have been exfoliated is not known, but in milder forms of injury this phase is known to occur within a matter of a few hours. Changes in epithelial growth factor expression are likely to be crucial in epithelial repair. Activation of fibroblasts is likely to be an essential component in providing proteolytic enzymes capable of clearing cellular debris. At the same time, stimulation of matrix production is likely to be a required component to restore normal matrix composition (and hence cellular differentiation), as well as "patch repair" zones where matrix damage renders this impossible in the short term.

tissue repair in the airway mucosa. It would be anticipated that drugs that down-regulate inflammatory cell function should act to limit mucosal damage. Bronchial biopsy studies suggest that this expectation is likely to be borne out in the case of corticosteroids, raising optimism for the novel drug candidates that are in development and designed to interfere with activation–response coupling in effector cells. However, the precise mechanism(s) of this effect of corticosteroids remains to be established. Furthermore, corticosteroids are not always successful in modulating pulmonary fibrotic responses in which it is presumed that an inappropriate repair process is mounted against a perceived source of injury, and this serves as an important reminder that the pharmacology of empirical effects is no substitute for an articulated, mechanistic understanding of the underlying cell biology.

8. Concluding Comments

Representing as it does the interface between the immune system and the exterior atmosphere, it is not surprising that the airway epithelium is a significant site of injury in lung diseases. As has been described in this review, these forms of injury are varied but their physical manifestations in resulting pathology are remarkably similar. This appears to hold remarkably true irrespective of whether the agent evoking injury acts directly on epithelial cells (as in the case of oxidant gases), or acts indirectly via the recruitment of inflammatory mechanisms (as in the case of asthma). The regiospecific occurrence of the injury process in certain parts of the tracheobronchial tree is likely to be explicable by numerous factors. Noxious gases, for example, are more likely to achieve alveolar penetration than are inhaled particulate materials such as allergens. Within a particular region of the tracheobronchial tree other, more subtle, factors may be important, such as the relative solubility of the provoking stimulus in epithelial lining fluid and cytoplasmic lipid membranes.

Many cell-derived mediators are likely to have roles in these processes of cellular injury. However, our knowledge of the precise functions of many of these candidates is currently limited. This is most outstandingly true for the MMPs and their inhibitors, where despite a vast literature on their function in other tissues, there is only a poor understanding of their possible function in the lung. Clearly, further understanding of the mechanisms of pulmonary repair and fibrosis will be necessary to gain a full appreciation of their possible roles and relevance in injury and repair.

From the pharmacological standpoint, the process of epithelial injury *per se* has received little direct attention. This situation is not difficult to understand, because our level of ignorance about many of the basic cell and molecular processes impedes a rational attempt at designing strategies for the manipulation of chronic injury and

9. Acknowledgments

Research on epithelial injury in the author's laboratory was supported by the Medical Research Council, The National Asthma Campaign, The British Lung Foundation and The Royal Society. Special thanks are due to Carolyn Herbert, Patrick Ring and Julia Carver who have been centrally involved in undertaking studies on mucosal injury in the author's laboratory. I would like to thank Kemi Olaniyan for her assistance in preparation of the manuscript.

10. References

Amatini, R., Wilson, R., Rutman, A., Read, R., Ward, C., Burnett, D., Stockley, R.A. and Cole, P.J. (1991). Effects of human neutrophil elastase and *Pseudomonas aeruginosa* proteinases on human respiratory epithelium. Am. J. Respir. Cell Mol. Biol. 4, 26–32.

American Thoracic Society (1978). "Health Effects of Air Pollution." American Lung Association, New York.

Aris, R.M., Christian, D., Hearne, P.Q., Kerr, K., Finkbeiner, W.E. and Balmes, J.R. (1993). Ozone-induced airway inflammation in human subjects as determined by airway lavage and biopsy. Am. Rev. Respir. Dis. 148, 1363–1372.

Ayars, G.H., Altman, L.C., McManus, M.M., Agosti, J.M., Baker, C., Luchtel, D.L., Loegering, D.A. and Gleich, G.J. (1989). Injurious effect of the eosinophil peroxide–hydrogen peroxide–halide system and major basic protein on human nasal epithelium *in vitro*. Am. Rev. Respir. Dis. 140, 125–131.

Ayars, G.H., Altman, L.C, Loegering, D.A. and Gleich, G.J. (1991). Eosinophil major basic protein [MBP] up regulates intercellular adhesion molecular-1 [ICAM-1] expression on human nasal epithelial cells [HNE]. J. Allergy Clin. Immunol. 87, 304.

Babior, B.M., Kipnes, R.S. and Curnutte, J.T. (1973). Biological defense mechanisms: the production by leukocytes of superoxide, a potential bactericidal agent. J. Clin. Invest. 52, 741.

Barker, R.L., Gleich, G.J. and Pease, I.R. (1988). Acidic

precursor revealed in human eosinophil granule major basic protein cDNA. J. Exp. Med. 168, 1493–1498.

Barker, R.L., Loegering, D.A., Ten, R.M., Hamann, KJ., Pease, L.R. and Gleich, G.J. (1989). Eosinophil cationic protein cDNA. Comparison with other toxic cationic proteins and ribonucleases. J. Immunol. 143, 952–955.

Barrett, A.J. (1981a). Leukocyte elastase. Methods Enzymol. 80, 581–588.

Barrett, A.J. (1981b). Cathepsin G. Methods Enzymol. 80, 561–565.

Barrett, A.J. and Kirschke, H. (1980). Cathepsin, B, cathepsin, H and cathepsin L. Methods Enzymol. 80, 535–565.

Beckman, J.S., Beckman, T.W., Chen, J., Marshall, P.A. and Freeman, B.A. (1990). Apparent hydroxyl radical production by peroxynitrite: implications for endothelial injury from nitric oxide and superoxide. Proc. Natl. Acad. Sci. USA 87, 1620–1624.

Biemond, P., van Eijk, H.G., Swask, A.J.G. and Koster, J.F. (1984). Iron mobilization from ferritin by superoxide derived from stimulated leukocytes. J. Clin. Invest. 73, 1576–1579.

Bils, R.F. (1974). Effects of nitrogen dioxide and ozone on monkey lung ultrastructure. Pneumonologie 150, 99–111.

Bils, R.F. and Christie, B.R. (1980). The experimental pathology of oxidant and air pollutant inhalation. Int. Rev. Exp. Pathol. 21, 195–293.

Bingle, L., Richards, R.J., Fox, B., Masek, L., Guz, A. and Tetley, T.D. (1990). Differential effects of neutrophil elastase and its inhibitors on alveolar and bronchial epithelial cell adherence in vitro [abstract]. Am. Rev. Respir. Dis. 141, A682.

Birnboim, H.C. (1982). DNA strand breakage in human leukocytes exposed to tumor promoter, phorbol myristate acetate. Science, 215, 1247–1249.

Bisgaard, H. and Pedersen, M. (1987). SRS-A leukotrienes decrease the activity of human respiratory cilia. Clin. Allergy 17, 95–103.

Blaustein, M.P. (1977). Sodium ions, blood pressure regulation and hypertension: a reassessment and a hypothesis. Am. J. Physiol. 232, C165–173.

Boatman, E.S. and Frank, R. (1974). Morphologic and ultrastructural changes in the lungs of animals during acute exposure to ozone. Chest 65, 9S–11S.

Boatman, E. S., Sato, S. and Frank, R. (1974). Acute effects of ozone on cat lungs. II. Structural. Am. Rev. Respir. Dis. 110, 157–169.

Bolscher, B.G.J.M., Plat, H. and Wever, R. (1984). Some properties of human eosinophil peroxidase, a comparison with other peroxidases. Biochim. Biophys. Acta 784, 177–186.

Bosi, F., Silini, E., Luisett, M., Romano, A.M., Prati, U., Silvestri, M., Tinelli, C., Samloff, M. and Fiocca, R. (1993). Aspartic proteinases in normal lung and interstitial pulmonary diseases. Am. Rev. Respir. Cell Mol. Biol. 8, 626–632.

Boucher, R.C., Johnson, J., Inoue, S., Hilbert, W. and Hogg, J.C. (1980). The effect of cigarette smoke on the permeability of guinea-pig airways. Lab. Invest. 43, 94–100.

Breeze, R.J. and Wheeldon, E.B. (1977). The cells of the pulmonary airways. Am. Rev. Respir. Dis. 116, 705–777.

Brigham, K.L. and Meyrick, B. (1984). Interaction of granulocytes with the lungs. Circ. Res. 54, 623–635.

Bucca, E.C., Rolla, G., Scappatici, E. and Cantino, D. (1988). A freeze fracture study of human bronchial epithelium in normal, bronchitic and asthmatic subjects. J. Submicrosc. Cytol. Pathol. 20, 509–517.

Burney, P. (1993). Epidemiology of asthma. Allergy 48, 17–21.

Butterworth, A.E., Wassom, D.L., Gleich, G.J., Loegering, D.A. and David, J.R. (1979). Damage to schistosomula of schistosoma mansoni induced directly by eosinophil major basic protein. J. Immunol. 122, 221–229.

Bylin, G., Hedenstirna, G., Linduall, T. and Sundin, B. (1988). Ambient nitrogen dioxide concentration increases bronchial responsiveness in subjects with mild asthma. Eur. Respir. J. 1, 606–612.

Cabral-Anderson, L.J., Evans, M.J. and Freeman, G. (1977). Effects of NO_2 on the lungs of aging rats. I. Morphology. Exp. Mol. Pathol. 27, 353–365.

Cantin, A.M., North, S.L., Hubbard, R.C. and Crystal, R.G. (1987). Normal alveolar epithelial lining fluid contains high levels of glutathione. J. Appl. Physiol. 163, 152–157.

Carroll, N., Elliot, J., Morton, A. and James, A. (1993). The structure of large and small airways in nonfatal and fatal asthma. Am. Rev. Respir. Dis. 147, 405–410.

Castleman, W.L., Tyler, W.S. and Dungworth, D.L. (1977). Lesions in respiratory bronchioles and conducting airways of monkeys exposed to ambient levels of ozone. Exp. Mol. Pathol. 26, 384–400.

Castleman, W.L, Dungworth, D.L., Schwartz, L.W. and Tyler, W.S. (1980). Acute respiratory bronchiolitis – an ultrastructural and autoradiographic study of epithelial cell injury and renewal in Rhesus monkeys exposed to ozone. Am. J. Pathol. 98, 811–840.

Chee, C., Gaston, B., Gerrard, C., Loscalzo, J. Kobzik, L., Drazen, J.M. and Stamler, J. (1993). Nitric oxide is produced by a human epithelial cell line [abstract]. Am. Rev. Respir. Dis. 147, A433.

Christensen, T.G., Breuer, R., Lucey, E.C., Stone, P.J. and Snider, G.L. (1989). Regional difference in airway epithelial response to neutrophil elastase: tracheal secretory cells discharge and recover in hamsters that develop bronchial secretory cell metaplasia. Exp. Lung Res. 15, 943–959.

Chua, K-Y., Stewart, G.A. Thomas, W.A., Simpson, R.J., Dilworth, R.J., Plozza, T.M. and Turner, K.J. (1988). Sequence analysis of cDNA coding for a major house dust mite allergen Der p I. Homology with cysteine proteases. J. Exp. Med. 167, 175–182.

Churg, A. and Green, F.H.Y. (1988). "Pathology of Occupational Lung Disease." Igaku-Shoin, New York.

Clausbruch, U.C.V. and Tschesche, H. (1988). Cathepsin D from human leukocytes. Biol. Chem. Hoppe-Seyler. 369, 683–691.

Collier, I.E., Wilhelm, S.M., Eisen, A.Z., Marmer, B.L., Grant, G.A., Seltzer, J.L., Kronberger, A., He, C., Bauer, E.A. and Goldberg, G.A. (1988). H-ras oncogene-transformed human bronchial epithelial cells (TBE-1) secrete a single metalloproteinase capable of degrading basement membrane collagen. J. Biol. Chem. 263, 6579–6587.

Cox, G., Ohtoshi, T., Vancheri, C., Denburg, J.A., Dolovich, J., Gauldie, J. and Jordana, M. (1991). Promotion of eosinophil survival by human bronchial epithelial cells and its modulation by steroids. Am. J. Respir. Cell Mol. Biol. 4, 525–531.

Dahl, R. and Venge, P. (1979). Enhancement of urokinase-induced plasminogen activation by the cationic protein of human eosinophil granulocytes. Thromb. Res. 14, 599–608.

Damjanovich, L., Albeda, S.M., Mette, S.A. and Buck, C. (1992). Distribution of integrin cell adhesion receptors in normal and malignant lung tissue. Am. J. Respir. Cell. Mol. Biol. 6, 197–206.

Devalia, J.L., Campbell, A.M., Sapsford, R.J., Rusznak, C., Quint, D., Godard, P., Bousquet, J. and Davies, R.J. (1993). Effect of nitrogen dioxide on synthesis of inflammatory cytokines expressed by human bronchial epithelial cells in vitro. Am. J. Respir. Cell Mol. Biol. 9, 271–278.

Di Benedetto, G., Marara-Shediac, F.S. and Mehta, S. (1991). Effect of cyclic AMP on ciliary activity of human respiratory epithelium. Eur. Respir. J. 4, 789–795.

Dunnill, M.S. (1960). The pathology of asthma, with specific reference to changes in the bronchial mucosa. J. Clin. Pathol. 13, 27–33.

Durack, D.T., Ackerman, S.J., Loegering, D.A. and Gleich, G.J. (1981). Purification of eosinophil-derived neurotoxin. Proc. Natl. Acad. Sci. USA 78, 5165–5169.

Emonard, H. and Grimaud, J-A. (1990). Matrix metalloproteinases: a review. Cell Mol. Biol. 4, 195–203.

Estreicher, A., Muhlhauser, J., Carpentier, J-L., Orci, L. and Vassalli, J-D. (1990). The receptor for urokinase plasminogen activator polarizes expression of the protease to the leading edge of migrating monocytes and promotes degradation of enzyme–inhibitor complexes. J. Cell Biol. 111, 783–792.

Evans, M.J. and Plopper, C.G. (1988). The role of basal cells in adhesion of columnar epithelium to airway basement membrane. Am. Rev. Respir. Dis. 138, 481–483.

Evans, M.J., Stephens, R.J. and Freeman, G. (1971). Effects of nitrogen dioxide on cell renewal in the rat lung. Arch. Intern. Med. 128, 57–60.

Evans, M.J., Cabral, L.J., Stephens, R.J. and Freeman, G. (1975). Transformation of alveolar type 2 cells to type 1 cells following exposure to NO_2. Exp. Mol. Pathol. 22, 142–150.

Evans, M.J., Johnson, L.V., Stephens, R.J. and Freeman, G. (1976). Renewal of the terminal bronchiolar epithelium in the rat following exposure to NO_2 or O_3. Lab. Invest. 35, 246–257.

Evans, M.J., Cabral-Anderson, J. and Freeman, G. (1977). Effects of NO_2 on the lungs of aging rats. II. Cell proliferation. Exp. Mol. Pathol. 27, 366–376.

Evans, M.J., Cox, R.A., Shami, S.G., Wilson, B. and Plopper, C.G. (1989). The role of the basal cells in attachment of columnar cells to the basal lamina of the trachea. Am. Rev. Respir. Cell Mol. Biol. 1, 463–469.

Evans, M.J., Cox, R.A., Shami, S.G. and Plopper, C.G. (1990). Junctional adhesion mechanisms in airway basal cells. Am. Rev. Respir. Cell Mol. Biol. 3, 341–347.

Fabbri, L.M., Aizawa, H., Alpert, S.E., Walters, E.H., Gold, R.D. and O'Byrne, P.M. (1984). Airway hyperresponsiveness and changes in cell counts in bronchoalveolar lavage after ozone exposure in dogs. Am. Rev. Respir. Dis. 129, 288–291.

Fick, R.B. (1989). Pathogenesis of the Pseudomonas lung lesion in cystic fibrosis. Chest 96, 158–164.

Filley, W.V., Holley, K.E., Kephart, G.M. and Gleich, G.J. (1982). Identification by immunofluorescence of eosinophil granule major basic protein in lung tissues of patients with bronchial asthma. Lancet ii, 11–15.

Flavahan, N.A., Slifman, N.R., Gleich, G.J. and Vanhoutte, P.M. (1988). Human eosinophil major basic protein causes hyperreactivity of respiratory smooth muscle. Am. Rev. Respir. Dis. 138, 685–688.

Flindt, M.L.H. (1969). Pulmonary disease due to inhalation of derivatives of Bacillus subtilis containing proteolytic enzymes. Lancet i, 1177–1181.

Ford, P.C., Wink, D.A. and Stanbury, D.M. (1993). Autoxidation kinetics of aqueous nitric oxide. FEBS Lett. 326, 1–3.

Fox, B., Bull, T.B., Guz, A., Harris, E. and Tetley, T.D. (1988). Is neutrophil elastase associated with elastic tissue in emphysema? J. Clin. Pathol. 41, 435–440.

Freeman, G., Funosi, N.J. and Hayden, G B. (1966). Effects of continuous exposure of 0.8 ppm NO_2 on respiration of rats. Arch. Environ. Health 13, 454–456.

Fridovitch, I. (1986). Biological effects of the superoxide radical. Arch. Biochem. Biophys. 247, 1–11.

Frigas, E., Loegering, D.A. and Gleich, G.J. (1980). Cytotoxic effects of the guinea-pig eosinophil major basic protein on tracheal epithelium. Lab. Invest. 42, 35–43.

Frigas, E., Loegering, D.A., Solley, G.O., Farrow, G.M. and Gleich, G.J. (1981). Elevated levels of eosinophil granule major basic protein in the sputum of patients with bronchial asthma. Mayo Clinic Proc. 56, 345–353.

Fujimaki, H., Kawagoe, A., Ozawa, M., Yoemoto, J. and Watanabe, N. (1989). Effects of instillation of fly ash in the lung: physicochemical properties and immune responses. Am. Rev. Respir. Dis. 140, 525–528.

Gail, D.B. and Lenfant, C. (1983). Cells of the lung: biology and clinical implications. Am. Rev. Respir. Dis. 366–383.

Ganz, T., Selsted, M.E., Harwig, S.S.L., Szklarek, D., Dahrer, K. and Lehrer, R.I. (1985). Natural peptide antibiotics of human neutrophils. J. Clin. Invest. 76, 1427–1435.

Garrod, D.R. (1986). Desmosomes, cell adhesion molecules and the adhesive properties of cells in tissues. J. Cell Sci. (Suppl.) 4, 221–237.

Gauldie, J., Jordana, M. and Cox, G. (1993). Cytokines and pulmonary fibrosis. Thorax 48, 931–935.

Giloni, L., Takeshita, M., Johnson, F., Iden, C. and Grollman, A.P. (1981). Bleomycin-induced strand scission of DNA: mechanism of deoxyribose cleavage. J. Biol. Chem. 256, 8608–8615.

Gleich, G.J., Loegering, D.A. and MacDonald, J.E. (1973). Identification of a major basic protein in guinea-pig eosinophil granules. J. Exp. Med. 137, 1459–1471.

Goldstein, J.A. (1984). Nitroglycerin therapy of asthma. Chest 85, 449.

Gordon, R.E. and Lane, B.P. (1976). Regeneration of rat tracheal epithelium after mechanical injury. Am. Rev. Respir. Dis. 113, 799–807.

Gordon, R.E. and Lane, B.P. (1984). Ciliated cell differentiation in regenerating rat tracheal epithelium. Lung 162, 233–243.

Gordon, R.E., Case, B.W. and Kleinerman, J. (1983). Acute NO_2 effects on penetration and transport of horseradish peroxidase in hamster respiratory epithelium. Am. Rev. Respir. Dis. 128, 528–533.

Granelli-Piperno, A., Vassalli, J. and Reich, E. (1977). Secretion of plasminogen activator by human polymorphonuclear leukocytes. J. Exp. Med. 146, 1693–1706.

Grisham, M.B., Jefferson, M.M., Melton, D.F. and Thomas, E.L. (1984). Chlorination of endogenous amines by isolated neutrophils. Ammonia-dependent bactericidal, cytotoxic and cytolytic activities of chloramines. J. Biol. Chem. 259, 10404–10413.

Gundel, R.H., Letts, L.G. and Gleich, G.J. (1991). Human eosinophil major basic protein induces airway constriction and airway hyperresponsiveness in primates. J. Clin. Invest. 87, 1470–1473.

Gutteridge, J.M.C. (1987). Iron promotion of the Fenton reaction and lipid peroxidation can be released from haemoglobin by peroxides. FEBS Lett. 201, 291–295.

Halliwell, B. (1988). Albumin – an important extracellular antioxidant? Biochem. Pharmacol. 37, 569–571.

Halliwell, B. and Gutteridge, J.M.C. (1984). Oxygen toxicity, oxygen radicals, transition metals and disease. Biochem. J. 219, 1–14.

Heffner, J.E. and Repine, J.E. (1989). Pulmonary strategies of antioxidant defense. Am. Rev. Respir. Dis. 140, 531–554.

Henderson, W.R., Jong, E.C. and Klebanoff, S.J. (1980). Eosinophil-mediated mammalian tumor cell cytotoxicity: role of the peroxidase system. J. Immunol. 124, 1949–1953.

Henson, P.M. and Johnston, R.B. (1987). Tissue injury in inflammation. J. Clin. Invest. 79, 669–674.

Herbert, C.A., Holgate, S.T., Robinson, C., Thompson, P.J. and Stewart, G.A. (1990). Effect of mite allergen on permeability of bronchial mucosa. Lancet 2, 1132.

Herbert, C.A., Edwards, D., Boot, J.R. and Robinson, C. (1991). In vitro modulation of the eosinophil-dependent enhancement of the permeability of the bronchial mucosa. Br. J. Pharmacol. 104, 391–398.

Herbert, C.A., Edwards, D., Boot, J.R. and Robinson, C. (1993). Stimulated eosinophils and proteinases augment the transepithelial flux of albumin in bovine bronchial mucosa. Br. J. Pharmacol. 110, 840–846.

Herbert, C.A., Omari, T.I. Sparrow, M.P. and Robinson, C. (1994). Human eosinophils and recombinant matrix metalloproteinase-2 increase the permeability of bovine and porcine airways. Br. J. Pharmacol., 112, 461P.

Hileman, B. (1989). Global warming. Chem. Eng. News. 67, 25–44.

Hipps, D.S., Hembry, R.M., McGarrity, A.M. and Reynolds, J.J. (1989). Gelatinase (type IV collagenase) immunolocalization in cells and tissues: use of an antiserum to rabbit bone gelatinase that identifies high and low Mr forms. J. Cell Sci. 92, 487–495.

Högman, M., Frostell, C.G., Hedenström and Hendenstierna, G. (1993). Inhalation of nitric oxide modulates adult bronchial tone. Am. Rev. Respir. Dis. 148, 1474–1478.

Hoiby, N. (1982). Microbiology of lung infections in cystic fibrosis. Acta Paediatr. Scand. 301(Suppl), 33–54.

Holt, P.G. (1993). Regulation of antigen-presenting cell function(s) in lung and airway tissues. Eur. Respir. J. 6, 120–129.

Holt, P.G., Schon, H.M., Oliver, J., Holt, B.J. and McMenamin, P.G. (1990). A contiguous network of dendritic antigen-presenting cells within the respiratory epithelium. Int. Arch. Allergy Appl. Immunol. 91, 155–159.

Howard, E.W., Bullen, E.C. and Banda, M.J. (1991). Preferential inhibition of 72- and 92-kDa gelatinases by tissue inhibitor of metalloproteinases-2. J. Biol. Chem. 266, 13070–13075.

Hyslop, P.A., Hinshaw, D.B., Schraufstatter, I.U. Sklar, L.A., Spragg, R.G. and Cochrane, C.G. (1986). Intracellular calcium homeostasis during H_2O_2 injury to cultured P388D1 cells. J. Cell Physiol. 129, 356–360.

Jain, B., Rubinstein, I., Robbins, R.A. Leise, K.L. and Sisson, J.H. (1993). Modulation of airway epithelial cell ciliary beat frequency by nitric oxide. Biochem. Biophys. Res. Commun. 191, 83–88.

Janoff, A. (1985). Elastases and emphysema. Am. Rev. Respir. Dis. 132, 417–433.

Jeffery, P.K. (1983). Morphologic features of airway surface epithelial cells and glands. Am. Rev. Respir. Dis. 128, S14–S20.

Jeffery, P.K. and Reid, L.M. (1977). The respiratory mucous membrane. In "Respiratory Defense Mechanisms," Part I (eds J.D. Brain, J.D., Proctor and L.M. Reid), pp. 193–238. Marcel Dekker, New York.

Jeffery, P.K, Wardlaw, A.J., Nelson, F.C., Collins, J.V. and Kay, A.B. (1989). Bronchial biopsies in asthma – an ultrastructural, quantitative study and correlation with hyperreactivity. Am. Rev. Respir. Dis. 140, 1745–1753.

Jones Williams, W. (1988). Beryllium disease. Postgrad. Med. J. 64, 511–516.

Kanthakumar, K, Taylor, G., Tsang, K.W.T., Cundell, D.R. Rutman, A., Smith, S., Jeffery, P.K., Cole, P.J. and Wilson, R. (1993). Mechanisms of action of Pseudomonas aeruginosa pyocyanin on human ciliary beat in vitro. Infect. Immun. 61, 2848–2853.

Kanthakumar, K., Cundell, D.R., Johnson, M., Wills, P.J., Taylor, G.W. and Cole, P.J. (1994). Effect of salmeterol on human nasal epithelial cell ciliary beating: inhibition of the ciliotoxin, pyocyanin. Br. J. Pharmacol. 112, 493–498.

Karlinsky, J and Snider, G.L. (1978). Animal models of emphysema. Am. Rev. Respir. Dis. 117, 1109–1133.

Keenan, K.P., Combs, J.W. and McDowell, E.M. (1982a). Regeneration of hamster tracheal epithelium after mechanical injury I. Focal lesions: quantitative morphologic study of cell proliferation. Virchows Arch. 41, 193–214.

Keenan, K, Combs, J.W. and McDowell, E.M. (1982b). Regeneration of hamster tracheal epithelium after mechanical injury II. Multifocal lesions: stathmokinetic and autoradiographic studies of cell proliferation. Virchows Arch. 41, 215–229.

Keenan, K.P., Combs, J.W. and McDowell, E.M. (1982c). Regeneration of hamster tracheal epithelium after mechanical injury III. Large and small lesions: comparative stathmokinetic and single pulse and continuous thymidine labeling autoradiographic studies. Virchows Arch. 41, 231–252.

Keenan, K.P., Wilson, T.S. and McDowell. E.M. (1983). Regeneration of hamster tracheal epithelium after mechanical injury IV. Histochemical, immunocytochemical and ultrastructural studies. Virchows Arch. 43, 213–240.

Kercsmar, C.M. and Davis, P.B. (1993). Resistance of human tracheal epithelial cells to killing by neutrophils, neutrophil elastase and Pseudomonas elastase. Am. J. Respir. Cell Mol. Biol. 8, 56–62.

Kobzik, L., Bredt, D.S., Lowenstein, C.J., Drazen, J., Gaston, B., Sugarbaker, D. and Stamler, J.S. (1993). Nitric oxide synthase in human and rat lung: immunocytochemical and histochemical localization. Am. J. Respir. Cell Mol. Biol. 9, 371–377.

Laitinen, L.A., Heino, M., Laitinen, A., Kava, T. and Haahtela, T. (1985). Damage to the airway epithelium and bronchial reactivity in patients with asthma. Am. Rev. Respir. Dis. 131, 599–606.

Lake, F.R., Ward, L.D., Simpson, R.J., Thompson, P.J. and Stewart, G.A. (1991). House dust mite derived amylase:

allergenicity and physicochemical characterization. J. Allergy Clin. Immunol. 87, 1035–1042.

Lansley, A.B., Sanderson, M.J. and Dirksen, E.R. (1992). Control of the beat cycle of respiratory tract cilia by Ca^{2+} and cAMP. Am. J. Physiol. 263, L232–L242.

Last, J.A. (1988). Biochemical and cellular interrelationships in the development of ozone-induced pulmonary fibrosis. In "Air Pollution, the Automobile, and Public Health" (eds A.Y. Watson, R.R. Bates and D. Kennedy), pp. 415–440. National Academy Press, Washington DC.

Lichtenstein, A.K, Ganz, T., Selsted, M.E. and Lehrer, R.I. (1986). In vitro tumor cell cytolysis mediated by peptide defensins of human and rabbit granulocytes. Blood 68, 1407–1410.

Lifschitz, M.I. and Denning, C.R. (1970). Am. Rev. Respir. Dis. 102, 456–458.

Lippmann, M. (1989). Health effects of ozone. A critical review. J. Air Pollut. Control Assoc. 39, 672–695.

Lozewicz, S., Wells, C., Gomez, E., Ferguson, H., Richman, P., Devalia, J. and Davies, R.J. (1990). Morphological integrity of the bronchial epithelium in mild asthma. Thorax 45, 12–15.

Man, S.F.P., Hulbert, W.C., Mok, K., Ryan, T. and Thomson, A.B.R. (1986). Effects of sulfur dioxide on pore populations of canine tracheal epithelium. J. Appl. Physiol. 60, 416–426.

Matrisian, L.M. (1990). Metalloproteinases and their inhibitors in matrix remodelling. Trends Genet. 6, 121–125.

Mattoli, S., Miante, S., Calabro, F., Mezzetti, M., Fasoli, A. and Allegra, L. (1990). Eicosanoid release from human bronchial epithelial cells exposed to isocyanates potentiate activation and proliferation of T cells. Am. J. Physiol. 259, L320–L327.

Mauderly, J.L., Jones, R.K., Griffith, W.C., Henderson, R.F. and McClellan, R.O. (1987). Diesel exhaust is a pulmonary carcinogen in rats exposed chronically. Fund. Appl. Toxicol. 9, 208–221.

Mayeno, A.N., Curran, A.J., Roberts, R.L. and Foote, C.S. (1989). Eosinophils preferentially use bromide to generate halogenating agents. J. Biol. Chem. 264, 5660–5668.

McNutt, M.S. and Winstein. R.S. (1973). Membrane ultrastructure at mammalian intercellular junction. Prog. Biophys. Mol. Biol. 26, 45–101.

Melville, G.N. and Iravani, J. (1975). Factors affecting ciliary beat frequency in the intrapulmonary airways of rats. Can. J. Physiol. Pharmacol. 53, 1122–1128.

Miller, W.C. and Schultz, T.F. (1979). Failure of nitroglycerin as a bronchodilator. Am. Rev. Respir. Dis. 120, 471.

Milne, J. and Brand, S. (1975). Occupational asthma after inhalation of dust of the proteolytic enzyme papain. Br. J. Ind. Med. 32, 302–307.

Montefort, S., Herbert, C.A., Robinson, C. and Holgate, S.T. (1992a). The bronchial epithelium as a target for inflammatory attack in asthma. Clin. Exp. Allergy 22, 511–520.

Montefort, S., Roberts, J.A., Beasley, R., Holgate, S.T. and Roche, W.R. (1992b). The site of disruption of the bronchial epithelium in asthmatic and non-asthmatic subjects. Thorax 47, 499–503.

Morrison, K.J. and Vanhoutte, P.M. (1991). Airway epithelial cells in the pathophysiology of asthma. Ann. NY Acad. Sci. 629, 82–88.

Mossman, B.T. (1986). Alteration in superoxide dismutase activity in tracheal epithelial cells by asbestos and inhibition of cytotoxicity by antioxidants. Lab. Invest. 54, 204–212.

Motojima, S., Frigas, E., Loegering, D.A. and Gleich, G.J. (1989). Toxicity of eosinophil cationic proteins for guinea-pig tracheal epithelium in vitro. Am. Rev. Respir. Dis. 138, 801–805.

Murphy, G., Cockett, M.I., Ward, R.V. and Docherty, A.J.P. (1991). Matrix metalloproteinase degradation of elastin, type IV collagen and proteoglycan. Biochem. J. 277, 277–279.

Mustafa, M.G. and Lee, S.D. (1976). Pulmonary biochemical alterations resulting from ozone exposure. Ann. Occup. Hygiene 19, 17–26.

Mustafa, M.G. and Tierney, D.F. (1978). Biochemical and metabolic changes in the lung with oxygen, ozone, and nitrogen dioxide toxicity. Am. Rev. Respir. Dis. 118, 1061–1090.

Nakamura, H., Yoshimura, K., Jaffe, H.A. and R.G. Crystal. (1991). Interleukin-8 gene expression in human bronchial epithelial cells. J. Biol. Chem. 266, 19611–19617.

Naylor, B. (1962). The shedding of the mucosa of the bronchial tree in asthma. Thorax 17, 69–72.

Nijkamp, F.P., van der Linde, H.J. and Folkerts, G. (1993). Nitric oxide synthesis inhibition induces airway hyperresponsiveness in the guinea-pig in vivo and in vitro. Am. Rev. Respir. Dis. 148, 727–734.

Norel, X., Labat, C., Gardiner, P.J. and Brink, C. (1991). Inhibitory effects of BAY u3405 on prostanoid-induced contractions in human isolated bronchial and pulmonary arterial muscle preparations. Br. J. Pharmacol. 104, 591–593.

Nygaard, S.D., Ganz, T. and Peterson, M.W. (1993). Defensins reduce the barrier integrity of a cultured epithelial monolayer without cytotoxicity. Am. J. Respir. Cell Mol. Biol. 8, 193–200.

Ohashi, Y., Motojima, S., Fukuda, T. and Makino, S. (1992). Airway hyperresponsiveness, increased intercellular spaces of bronchial epithelium and increased infiltration of eosinophils and lymphocytes in bronchial mucosa in asthma. Am. Rev. Respir. Dis. 145, 1469–1476.

Ohtoshi, T., Vancheri, C., Cox, G., Gauldie, J., Dolovich, J., Denburg, J.A. and Jordana, M. (1991). Monocyte-macrophage differentiation induced by human upper airway epithelial cells. Am. J. Respir. Cell Mol. Biol. 4, 255–263.

Okrent, D.G., Lichtenstein, A.K and Ganz, T. (1990). Direct cytotoxicity of polymorphonuclear leukocyte granule proteins to human lung-derived cells and endothelial cells. Am. Rev. Respir. Dis. 141, 179–185.

Olsen, R.L. and Little, C. (1983). Purification and some properties of myeloperoxidase and eosinophil peroxidase from human blood. Biochem. J. 209, 781–787.

Omari, T.I., Sparrow, M.P., Church, M.K, Holgate, S.T. and Robinson, C. (1993a). A comparison of the effects of polyarginine and stimulated eosinophils on the responsiveness of the bovine isovolumic bronchial segment preparation. Br. J. Pharmacol. 109, 553–561.

Omari, T.I., Sparrow, M.P. and Mitchell, H.W. (1993b). Responsiveness of human isolated bronchial segments and its relationship to epithelial loss. Br. J. Clin. Pharmacol. 35, 357–365.

Pavia, D., Agnew, J.E. Lopez-Vidriero, M.T., Newman, S.P. and Clarke, S.W. (1984). The effect of metered dose aerosols on the viscoelastic properties and clearance of bronchial secretions. In "Metered Dose Inhalers" (ed. S.W. Epstein). Astra, Mississauga.

Pepys, J. (1969). Hypersensitivity disease of the lungs due to fungi and other organic dusts. Monogr. Allergy 4, 1–145.

Persson, C.G.A. (1988). Plasma exudation and asthma. Lung 166, 1–23.

Plopper, C.G. (1983). Comparative morphologic features of bronchiolar epithelial cells. The Clara cell. Am. Rev. Respir. Dis. 128, S37–S41.

Plopper, C.G., Hill, L.H. and Mariassy, A.T. (1980). Ultrastructure of the nonciliated bronchiolar epithelial (Clara) cell of mammalian lung. III. A study of man with comparison of 15 mammalian species. Exp. Lung Res. 1, 171–180.

Plopper, C.G., Mariassy, A.T., Wilson, D.W., Alley, J.L., Nishio, S.J. and Nettesheim, P. (1983). Comparison of nonciliated tracheal epithelial cells in six mammalian species: ultrastructure and population densities. Exp. Lung Res. 5, 281–294.

Read, R.C., Roberts, P., Munro, N., Rutman, A., Hastie, A., Shryock, T., Hall, R., McDonald-Gibson, W., Lund, V., Taylor G, Cole, P.J. and Wilson, R. (1992). The effect of Pseudomonas aeruginosa rhamnolipid on guinea-pig tracheal mucociliary transport *in vivo* and human ciliary beating *in vitro*. J. Appl. Physiol. 72, 2271–2277.

Rice, W.G., Ganz, T., Kinkade, J.M. Jr, Selsted, M.E., Lehrer, R.I. and Parmly, R.T. (1987). Defensin-rich dense granules of human neutrophils. Blood, 70, 757–765.

Robbins, R.A. Hamel, F.G., Floreani, A.A., Gossman, G.L., Nelson, K.J., Blenky, S. and Rubinstein, I. (1993). Bovine bronchial epithelial cells metabolize L-arginine to L-citrulline: possible role of nitric oxide synthase. Life Sci. 52, 709–716.

Robinson, C. and Holgate S.T. (1991). In "The Lung, Scientific Foundations" (eds R.G. Crystal and J.B. West), pp. 941–951. Raven Press, New York.

Robinson, C., Campbell, A.M., Herbert, C.A., Sapsford, R., Devalia, J.L. and Davies, R.J. (1990). Calcium-dependent release of eicosanoids in human cultured bronchial epithelial cells. Br. J. Pharmacol. 100, 471P.

Roche, W.R., Montefort, S., Baker, J. and Holgate, S.T. (1993). Cell adhesion molecules and the bronchial epithelium. Am. Rev. Respir. Dis. 148(Suppl), S79–S82.

Rose, M.S., Smith, L.L. and Wyatt, I. (1976). Evidence for energy dependent accumulation of paraquat into rat lung. Nature 25, 419–423.

Sandy, M.S., Moldeus, P., Ross, D. and Smith, M.T. (1986). Role of redox cycling and lipid peroxidation in bipyridyl herbicide cytotoxicity. Studies with a compromised isolated hepatocyte model system. Biochem. Pharmacol. 35, 3095–3101.

Sapsford, R.J., Devalia, J.L., McAulay, A.E., d'Ardenne, A.J. and Davies, R.J. (1991). Expression of $\alpha 1$–6 integrin cell surface receptors in normal human bronchi biopsies and cultured bronchial epithelial cells (abstract) J. Allergy Clin. Immunol. 87S, A303.

Schlatter, C. (1994). Environmental pollution and human health. Sci. Total Environ. 143.

Schrauffstatter, I.U., Hyslop, P.A., Hinshaw, D.B., Spragg, R.G., Sklar, L.A. and Cochrane, C.G. (1986). Hydrogen peroxide-induced injury of cells and its prevention by inhibitors of poly-ADP-ribose polymerase. Proc. Natl. Acad. Sci. USA 83, 4908–4912.

Selsted, M.E., Harwig, S.S.L., Ganz, T., Schilling, J.W. and

Lehrer, R.I. (1985). Primary structures of three human neutrophil defensins. J. Clin. Invest. 76, 1436–1439.

Senior, R.M. and Campbell, E.J. (1983). Neutral proteinases from human lung inflammatory cells. Clin. Lab. Med. 3, 645–666.

Shapiro, S.D., Campbell, E.J., Kobayashi, D.K and Welgus, H.G. (1991). Immune modulation of metalloproteinase production in human macrophages. Selective pretranslational suppression of interstitial collagenase and stromelysin biosynthesis by interferon-gamma. J. Clin. Invest. 88, 1656–1662.

Sheppard, D. (1993). Identification and characterization of novel airway epithelial integrins. Am. Rev. Respir. Dis. 148(Suppl), S38–S42.

Sheppard, D., Rozzo, C., Starr, L., Quaranta, V., Erle, D.J. and Pytela, R. (1990). Complete amino acid sequence of a novel integrin beta subunit ($\beta 6$) identified in epithelial cells using the polymerase chain reaction. J. Biol. Chem. 265, 1502–1507.

Sheppard, D., Cohen, D.S., Wang, A. and Busk, M. (1992). Transforming growth factor b differentially regulates expression of integrin subunits in guinea-pig airway epithelial cells. J. Biol. Chem. 267, 17409–17414.

Shih, C.K, Litt, M., Kan, M.A. and Wolf, D.P. (1977). Effect of non-dialyzable solids concentration and viscoelasticity on ciliary transport of tracheal mucus. Am. Rev. Respir. Dis. 115, 989–995.

Shoji, S., Rickard, K.A., Takizawa, H., Ertl, R.F., Linder, J. and Rennard, S.I. (1990). Lung fibroblasts produce growth stimulatory activity for bronchial epithelial cells. Am. Rev. Respir. Dis. 141, 433–439.

Sims, J.L., Berger, S.J. and Berger, N.A. (1983). Poly-ADP-ribose polymerase inhibitors preserve NAD and ATP levels in DNA-damaged cells: mechanism of stimulation of unscheduled DNA synthesis. Biochemistry 22, 5188–5194.

Slungaard, A. and Mahoney, J.R. Jr. (1991). Thiocyanate is the major substrate for eosinophil peroxidase in physiologic fluids: implications for cytotoxicity. J. Biol. Chem. 266, 4903–4910.

Smallman, L.A., Hill, S.L. and Stockley, R.A. (1984). Reduction of ciliary beat frequency *in vitro* by sputum from patients with bronchiectasis: a serine proteinase effect. Thorax 39, 663–667.

Smith, L.L. (1986). The response of the lung to foreign compounds that produce free radicals. Annu. Rev. Physiol. 48, 681–692.

Snider, G.L., Lucey, E.C. and Stone, P.J. (1986). Animal models of emphysema. Am. Rev. Respir. Dis. 133, 149–169.

Snider, G.L., Ciccolella, D.E., Morris, S.M., Stone, P.J. and Lucey, E.C. (1991). Putative role of neutrophil elastase in the pathogenesis of emphysema. Ann. NY Acad. Sci. 624, 45–59.

Sparrow, M.P. and Mitchell, H.W. (1991). Modulation by the epithelium of the extent of bronchial narrowing produced by substances perfused through the lumen. Br. J. Pharmacol. 103, 1160–1164.

Spooner, M., Nick, H.P. and Schnebli, H.P. (1991). Different susceptibility of elastase inhibitors to inactivation by proteinases from *Staphylococcus aureus* and *Pseudomonas aeruginosa*. Biol. Chem. Hoppe Seyler 372, 963–970.

Spragg, R.G., Hinshaw, D.B., Schraufstatter, I.U. and Cochrane, C.G. (1985). Alterations in adenosine triphosphate and energy charge in endothelial and P388D1 cells after oxidant injury. J. Clin. Invest. 76, 1471–1476.

Spring, K.R. and Ericsson, A.C. (1983). Epithelial cell volume modulation and regulation. J. Membr. Biol. 69, 167–176.

Springall, D.R., Hamid, Q.A., Buttery, L.K.D., Chanea, P., Howarth, P., Bousquet, J., Holgate, S.T. and Polak, J.A. (1993). Nitric oxide synthase induction in airways of asthmatic subjects [abstract]. Am. Rev. Respir. Dis. 147, A515.

Stafanger, G. (1987). In vitro effect of beclomethasone dipropionate and flunisolide on the mobility of human nasal cilia. Allergy 42, 507–511.

Stephens, R.J., Freeman, G. and Evans, M.J. (1972). Early response of lungs to low levels of nitrogen dioxide. Arch. Environ. Health. 24, 160–179.

Stephens, R.I., Freeman, G., Satara, I.F. and Coffin, D.L. (1973). Cytologic changes in dog lungs induced by chronic exposure to ozone. Am. J. Pathol. 73, 711–726.

Stephens, R.J., Sloan, M.F., Evans, M.J. and Freeman, G. (1974). Alveolar type I cell response to exposure to 0.5 ppm O_3 for short periods. Exp. Mol. Pathol. 20, 11–23.

Stewart, G.A., Lake, F.R. and Thompson, P.J. (1991). Faecally derived hydrolytic enzymes from Dermatophagoides pteronyssinus – physicochemical characterisation of potential allergens. Int. Arch. Allergy Appl. Immunol. 95, 248–256.

Stewart, G.A., Ward, L.D. Simpson, R.J. and Thompson, P.J. (1992a). The group III allergen from house dust mite Dermatophagoides pteronyssinus is a trypsin-like enzyme. Immunology 75, 29–35.

Stewart, G.A., Bird, C.H. and Thompson, P.J. (1992b). Do the group II mite allergens correspond to lysozyme? J. Allergy Clin. Immunol. 90, 141–142.

Stewart, G.A., Boyd, S.M., Bird, C.H., Krska, KD., Kollinger, M.R. and Thompson, P.J. (1993). The immunobiology of the serine proteases from the house dust mite. Am. J. Indust. Med. 23, 105–107.

Stockley, R.A. (1983). Proteolytic enzymes, their inhibitors and lung diseases. Clin. Sci. 64, 119–126.

Strongin, A.Y., Marmer, B.L., Grant, G.A. and Goldberg, G.I. (1993). Plasma membrane dependent activation of the 72 kDa type IV collagenase is prevented by complex formation with TIMP-2. J. Biol. Chem. 268, 14033–14039.

Suter, S., Schaad, U.B., Roux, L., Nydegger, U.E. and Waldvogel, F.A. (1984). Granulocyte neutral proteases and Pseudomonas elastase as possible causes of airway damage in patients with cystic fibrosis. J. Infect. Dis. 149, 523–531.

Sykes, D.A., Wilson, R., Greenstone, M., Currie, D.C., Steinfort, C. and Cole, P.J. (1987). Deleterious effects of purulent sputum sol on human ciliary function in vitro: at least two factors identified. Thorax 42, 256–261.

Tegner, H., Ohlsson, K., Toremalm, N.G. and von Meckleburg, C. (1979). Effect of human leukocyte enzymes on tracheal mucosa and its mucociliary activity. Rhinology 17, 199–206.

Ten, R.M., Pease, L.R., McKeane, D.J., Bell, M.P. and Gleich, G.J. (1989). Molecular cloning of the human eosinophil peroxidase. Evidence for the existence of a peroxidase multigene family. J. Exp. Med. 169, 1757–1769.

Territo, M.C., Ganz, T., Selsted, M.E. and Lehrer, R.I. (1989). Monocyte chemotactic activity of defensins from human neutrophils. J. Clin. Invest. 84, 2017–2020.

Thomas, E.L., Jefferson, M.M. and Grisham, M.B. (1982). Myeloperoxidase-catalyzed incorporation of amines into proteins: role of hypochlorous acid and dichloramines. Biochemistry 21, 6299–6308.

Tournier, J.-M., Jacquot, N., Puchelle, E. and Bieth, J.G. (1985). Evidence that Pseudomonas aeruginosa elastase does not inactivate the bronchial inhibitor in the presence of leukocyte elastase. Studies with cystic fibrosis sputum and with pure proteins. Am. Rev. Respir. Dis. 132, 524–528.

Travis, J. (1988). Structure, function and control of neutrophil proteinases. Am. J. Med. 84, 37–43.

Trump, B.F. and Berezesky. I.K. (1984). Role of Sodium and calcium regulation in toxic cell injury. In "Drug Metabolism and Drug Toxicity" (eds J.R. Mitchell and M.G. Horning), pp. 261–300. Raven Press, New York.

Trump, B.F., Berezesky, L.K., Laiho, K.U., Osornio, A.R., Mergner, W.J. and Smith, M.W. (1980). The role of calcium in cell injury. A review. Scan. Electron. Microsc. 2, 1437–1462.

Trump, B.F. Berezesky, l. K. and Phelps. P.C. (1981). Sodium and calcium regulation and the role of the cytoskeleton in the pathogenesis of disease: a review and hypothesis. Scan. Electron. Microsc. 2, 435–454.

Vai, F., Fournier, M.F., Lafuma, J.C., Touaty, E. and Pariente, R. (1980). SO_2-induced bronchopathy in the rat: abnormal permeability of the bronchial epithelium in vivo and in vitro after anatomic recovery. Am. Rev. Respir. Dis. 121, 851–858.

Vanhoutte, P.M. (1988). Epithelium-derived relaxing factor(s) and bronchial reactivity. Am. Rev. Respir. Dis. 138(Suppl.), S24–S30.

Vijeyaratnam, G.S. and Corrin, B. (1971). Experimental paraquat poisoning: a histological and electron-optical study of the changes in the lung. J. Pathol. 103, 123–129.

Vissers, M.C.M. and Winterbourn, C.C. (1991). Oxidase damage to fibronectin. I. The effects of the neutrophil myeloperoxidase system and HOCI. Arch. Biochem. Biophys. 285, 53–59.

Vogel, A. I. (1970). "A Textbook of Practical Organic Chemistry, Including Qualitative Organic Analysis." Longman, London.

Wanner, A., Maurer, D., Abraham, W.M., Szepfalusi, Z. and Sielczak, M. (1983). Effects of chemical mediators of anaphylaxis on ciliary function. J. Allergy Clin. Immunol. 72, 663–667.

Wanner, A., Sielczak, M., Mella, J.F. and Abraham, W.M. (1986). Ciliary responsiveness in allergic and non-allergic airways. J. Appl. Physiol. 60, 1967–1971.

Wasmoen, T.L., Bell, M.P., Loegering, D.A., Gleich, G.J., Prendergast, F.G. and McKean, D.J. (1988). Biochemical and amino acid sequence of human eosinophil granule major basic protein. J. Biol. Chem. 263, 12559–12563.

Weisman, Z., Fink, A., Alon, A., Poliak, Z., Tabachnik, E., Priscu, L. and Bentwich, Z. (1990). Leukotriene C_4 decreases the activity of respiratory cilia in vitro. Clin. Exp. Allergy 30, 389–393.

Weiss, S.J. (1986). Oxygen, ischaemia, inflammation. Acta Phys. Scand. 548, 9–37.

Wever, R., Plat, J. and Hamers, M.N. (1981). Human eosinophil peroxidase: a novel isolation procedure, spectral properties and chlorinating activity. FEBS Lett. 123, 327–331.

White, C.W. and Repine, J.E. (1985). Pulmonary antioxidant defense mechanisms. Exp. Lung Res. 8, 81–96.

Widdicombe, J.H. (1991). Ion transport by airway epithelia. In "The Lung: Scientific Foundation" (eds R.G. Crystal and J.B. West), pp. 263–271. Raven Press, New York.

Williams, M.L., Broughton, G.J., Bower, J.S., Drury, V.J. and Lilley, K. (1988). Ambient No*x* concentrations in the UK 1976–1984: a summary. Atmos. Environ. 22, 2819–2840.

Yurchenco, P.D. and Schnittny, J.C. (1990). Molecular architecture of basement membranes. FASEB J. 4, 1577–1590.

Zheutlin, L.M., Ackerman, S.J., Gleich, G.J. and Thomas, L.L. (1984). Stimulation of basophil and rat mast cell histamine release by eosinophil granule-derived cationic proteins. J. Immunol. 133, 2180–2185.

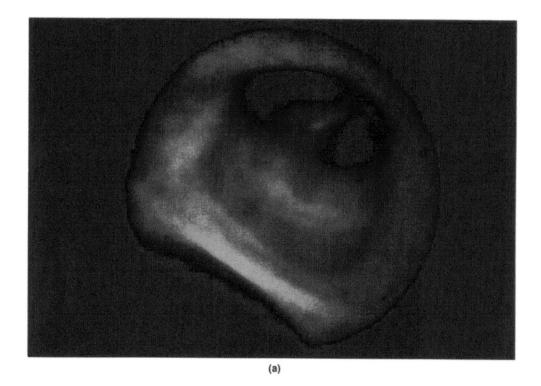

(a)

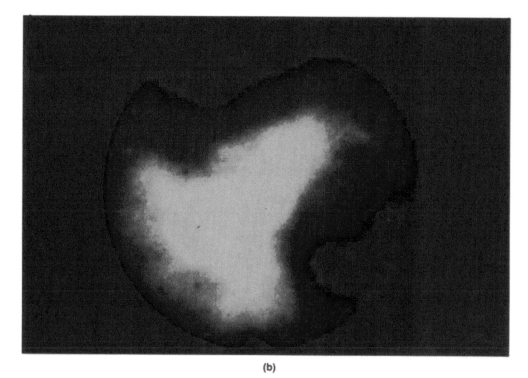

(b)

Plate I   (a) Visualization of normal human bronchus by fibreoptic bronchoscopy. (b) Corresponding image from a patient with mild allergic asthma showing evidence of mucosal inflammation after initiation of a local response to allergen.

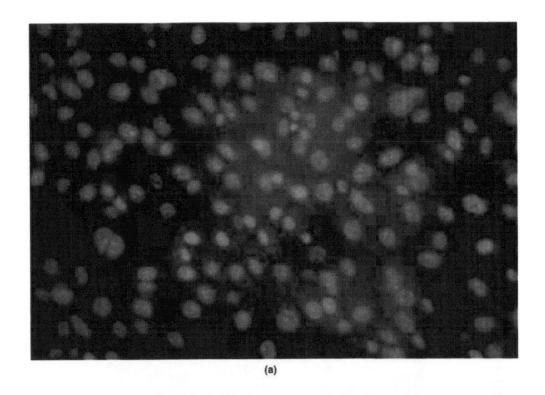

(a)

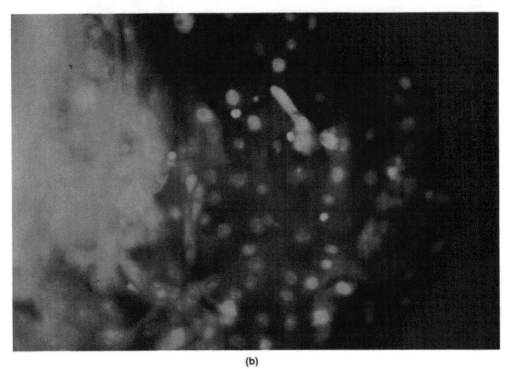

(b)

Plate II   Direct visualization of hydrogen peroxide-induced cellular injury and necrosis in cultured human bronchial epithelial cells. (a) Control. (b) After hydrogen peroxide-induced injury.

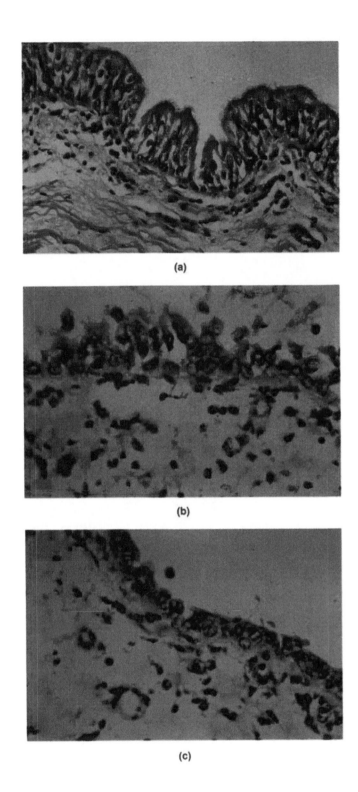

Plate III Eosinophils produce an increase in permeability of the bovine bronchial mucosa and detachment of columnar epithelial cells. Panel (a) shows bronchial mucosa exposed to unstimulated eosinophils. Panels (b) and (c) show bronchial epithelium that had been exposed to ionophore-stimulated eosinophils. Ionophore alone does not affect the airway epithelium under these conditions.

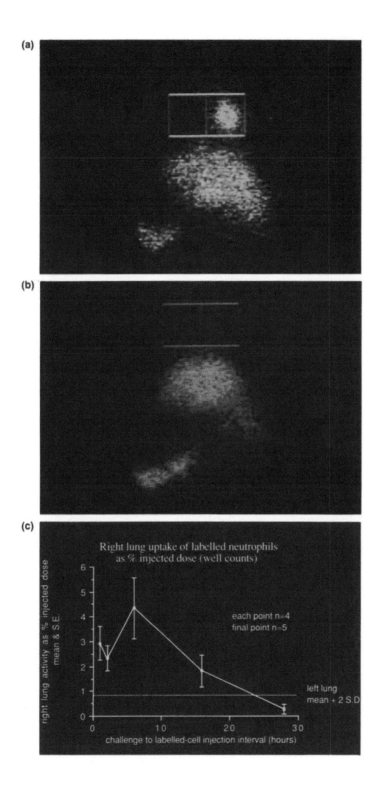

Plate IV (a) External 24 h gamma-camera scintigram of a rabbit which had received intravenous ^{111}In-labelled neutrophils 6 h after the bronchoscopic introduction of streptococcal pneumoniae into the right upper lobe. This is a major emigration of labelled cells to the lung. (b) In contrast, the 24 h scintigram of a rabbit which had received i.v. ^{111}In-neutrophils 24 h after streptococcal pneumonia instillation shows that neutrophil emigration to the lung has ceased. (c) Quantification of rates of neutrophil influx. Right lung uptake of labelled neutrophils as % injected dose (well counts).

11. Transition Between Inflammation and Fibrosis in the Lung

A. Wangoo *and* R.J. Shaw

1. Introduction

Cryptogenic fibrosing alveolitis (CFA; also called idiopathic pulmonary fibrosis) is a well-defined clinical entity of unknown aetiology which has characteristic clinical, radiological, physiological and pathological manifestations. A number of other pulmonary conditions also result in lung fibrosis, including sarcoidosis, extrinsic allergic alveolitis and many inorganic dust- and drug-induced lung diseases. The common process linking these diseases is the slow progression of chronic lung inflammation to fibrosis and irreversible scarring.

This progression can be divided into a number of phases although in reality these phases may be occurring simultaneously, or at least, may occur concurrently in nearby regions of the lung. It is believed that in all the lung fibrotic disorders there is some sort of initial injury. This initial injury is followed by an inflammatory response, which is chronic and involves a wide variety of inflammatory cells and cytokines. Intimately linked with the inflammatory response, there is the production of growth factors for fibroblasts. This is associated with activation of fibroblasts and enhanced production of collagen and other matrix proteins (Fig. 11.1). It is likely that in response to these early events, fibroblasts become activated in a manner independent of exogenous stimuli, such that they continue to cause scarring even in the absence of further exogenous pro-fibrogenic stimuli.

Immunopharmacology of the Respiratory System
ISBN 0–12–352325–7

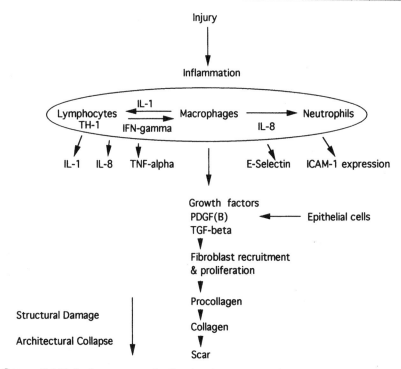

Figure 11.1 Sequential biologic processes in the development of pulmonary fibrosis: cells and cytokines involved in chronic inflammation and fibrosis.

In addition to these potentially reversible phases, structural events occur in the lung which are largely irreversible. Epithelial cell damage and intra-alveolar exudation stimulate epithelial cell regeneration on top of the organizing exudate, thus reducing alveolar size. Furthermore, damage to the normal structure of the alveolar wall causes alveolar collapse. In the setting of an active fibrotic response, this derangement of alveolar structure becomes permanent and scar tissue seals the alveolus in a crushed configuration.

If we are to improve our therapy for disorders characterized by lung fibrosis, we need to understand the link between the phases of chronic lung inflammation and fibrosis, and develop therapies which target the different phases before the irreversible changes occur.

Much of our understanding of the pathogenesis of lung fibrosis comes from studies in animal models as well as from patients with lung fibrosis.

2. *Animal Models of Fibrosis*

The mechanism linking chronic inflammation and tissue fibrosis is poorly understood. Various exposures and conditions are believed to cause pulmonary fibrosis. These include particulate agents, physical agents and infectious agents. To examine the pathogenesis of pulmonary fibrosis different experimental models in animals have been described. The most commonly used method to

induce lung fibrosis is the intratracheal instillation of bleomycin (Adamson and Bowden, 1974; Snider *et al.*, 1978; Lazo *et al.*, 1990; Hay *et al.* 1991). Other methods utilize paraquat (Smith *et al.*, 1974; Selman *et al.*, 1989), silica (Reiser *et al.*, 1982; Vuorio *et al.*, 1989), asbestos (Mossman *et al.*, 1991; Begin *et al.*, 1992), cadmium chloride (Frankel *et al.*, 1991; Driscoll *et al.*, 1992), amiodarone and desethylamiodarone (Daniels *et al.*, 1989). In bleomycin-induced fibrosis, the histological pattern of fibrosis is similar to that seen in humans, however the time course for development is much shorter (Snider *et al.*, 1978). Various animals, such as mice (Giri *et al.*, 1993), rats (Denholm and Rollins, 1993; Westergren-Thorsson *et al.*, 1993) and hamsters (Raghow *et al.*, 1989), have been used in the bleomycin model of lung fibrosis. The resultant injury is characterized by necrosis of type I pneumocytes with proliferation of type II pneumocytes. There is also evidence of interstitial oedema, sequential accumulation of inflammatory cells as well as fibroblasts, and finally collagen and fibronectin synthesis (Chandler *et al.*, 1983; Kelley *et al.*, 1985; Raghow *et al.*, 1985; Hoyt and Lazo, 1988). A common feature in any animal model of lung fibrosis is the accumulation of macrophages (Adamson and Bowden, 1974; Thrall *et al.*, 1979), which is accompanied by increases in macrophage-derived growth factors. Alveolar macrophages incubated with bleomycin have been shown to increase production of non-specific mitogenic activities (Denholm *et al.*, 1989), superoxide, tumour necrosis factor α (TNFα) and interleukin-1 (IL-1; Scheule et al., 1992), as well as transforming growth factor β (TGFβ; Khalil et al., 1989). Instillation of cadmium chloride into rat lung also results in increases in alveolar macrophage numbers and cathepsin L mRNA which is another marker of macrophage activation (Frankel et al., 1991).

In addition to an influx of macrophages, there may be an increase in the number of lymphocytes in the lung. The overwhelming majority of lymphocytes which migrate into sites of lung injury are T cells and there is an equal proportion of helper and suppressor T cells in injured lung (Thrall et al., 1982). A number of studies have used animal models to investigate T lymphocytes in the development of pulmonary interstitial fibrosis, but their precise role in these models is still controversial. T lymphocytes have been shown to have a central role in developing pulmonary interstitial fibrosis induced by bleomycin (Piguet et al., 1989) but not that induced by silica (Piguet et al., 1990). However, Kumar et al. (1990) showed that T lymphocytes participate in pulmonary interstitial fibrosis induced by silica. Recently, the immunological mechanisms that contribute to increased collagen content in the lungs of hapten immune hamsters after receiving a pulmonary challenge of the sensitizing hapten trinitrophenol have been studied. The results suggested that T lymphocytes mediate immune inflammation and regulate lung collagen deposition (Kimura et al., 1992).

Various studies have also examined animal models of pulmonary inflammation that are representative of primary eosinophil or neutrophil infiltration. Lung inflammation characterized by eosinophil influx has been used as a model of asthma and is not generally associated with lung fibrosis. After several episodes of repeated antigen challange, a subset of Ascaris suum-sensitive Cynomolgus monkeys developed a persistent eosinophilia and enhanced intercellular adhesion molecule-1 (ICAM-1) expression on pulmonary endothelial and epithelial cells when compared to control animals (Gundel et al., 1991, 1992).

By contrast, neutrophil-mediated inflammation may lead to fibrosis. Prolonged exposure to high concentrations of oxygen can lead to acute oedematous lung injury followed by pulmonary fibrosis (Clark and Lanbertsen, 1971). In a murine model of pulmonary oxygen toxicity the process is characterized by decreases in lung compliance and carbon monoxide diffusion capacity as well as by an influx of neutrophils and enhanced levels of myeloperoxidase in bronchoalveolar lavage (BAL) fluid (Wagner et al., 1992). Quartz deposition in the rat lung has been found to cause an intense persistent neutrophil alveolitis leading to parenchymal fibrosis (Kusaka et al., 1990).

The role of various cytokines has been studied in chronic lung inflammatory conditions. A mouse model of hypersensitivity pneumonitis which progresses to, fibrosis has been established by intranasal instillation of the thermophilic actinomycete Faeni rectivirgula (Denis 1992). In this model, IL-6 plays a role in regulating the cellular recruitment in the lungs during an inflammatory response and IL-4 administration partially abrogates the disease process (Ghadirian and Denis, 1992). TNFα has also been shown to play an essential role in determining hypersensitivity pneumonitis in a mouse model (Denis et al., 1991).

Altered expression of small proteoglycans, collagen and TGFβ has been shown in bleomycin-induced pulmonary fibrosis in rats (Khalil et al., 1989; Shahzeidi et al., 1993; Denholm and Rollins, 1993; Westergren-Thorsson et al., 1993). The marked alteration of biglycans and decorin during the development of fibrosis suggests that these proteoglycans have a regulating role in this process (Westergren-Thorsson et al., 1993).

Various agents have been tested for efficacy in decreasing the fibrotic response which follows lung injury in animal models. These agents include steroids, e.g. methylprednisolone which shows beneficial effects in terms of improving physiologic function or decreasing collagen accumulation (Kelley et al., 1980; Phan et al., 1981; Hesterberg and Last, 1981). Cyclooxygenase and lipoxygenase inhibition have been shown to diminish fibrosis (Thrall et al., 1979; Phan and Kunkel, 1986), proline analogues such as dehydroproline inhibit collagen production (Kelly et al., 1980) and cyclosporin (Sendelbach et al., 1985) may have beneficial effects in animal pulmonary fibrosis. Antibodies to cytokines such as TNFα (Piguet et al., 1990) and TGFβ (Giri et al., 1993) also have antifibrotic activity in models of animal pulmonary fibrosis. The availability of these animal models of interstitial pulmonary fibrosis provides the opportunity to investigate different aspects of fibrosis and novel pharmacological approaches to preventing this disease.

3. Fibrotic Lung Disease in Humans

3.1 THE INITIAL INJURY

In all lung fibrotic diseases, there is presumed to be some injurious process. This may be well recognized, as is the case when inorganic particles cause occupational lung fibrosis or recognized antigens stimulate an immune response in extrinsic allergic alveolitis. The cause of the injury is more speculative in the case of CFA, where immune complexes (Dreisin et al., 1978; Schwarz et al., 1978; Gadek et al., 1979a; Haslam et al., 1979; King et al., 1979, 1984; Cocchiara et al., 1981; Hunninghake et al., 1981; Gelb et al., 1983; Jansen et al., 1984), viruses (Vergnon et al., 1984; Ueda et al., 1992) and genetic factors (Evans, 1976; Turton et al., 1978; Varpela et al., 1979) have been suggested to play a part in initiating the disease. Similarly, no causal agent for sarcoidosis has yet been identified.

3.2 CELLS INVOLVED IN THE INFLAMMATORY PHASE

BAL studies indicate that there is a marked alveolitis in all types of lung fibrotic disorder. However, considerable heterogeneity does occur. Thus in CFA, most studies reveal a predominance of neutrophils and macrophages (Crystal *et al.*, 1976, 1981, 1984; Haslam *et al.*, 1979; Gelb *et al.*, 1983), while others identify increases in lymphocytes and mast cells (Haslam *et al.*, 1980a,b; Haslam, 1984; Davis and Gadek, 1987). The importance of inflammation is emphasized by studies in CFA correlating parenchymal inflammatory cells with the ultimate course of the disease. These studies have suggested that inflammation is a key factor in the prediction of therapeutic responsiveness (Carrington *et al.*, 1978; Dreisin *et al.*, 1978; Winterbauer *et al.*, 1978; Crystal *et al.*, 1981, 1984; Hunninghake *et al.*, 1981; Keogh *et al.*, 1983; Watters *et al.*, 1987).

Inflammatory cells including macrophages and lymphocytes are important secretory cells and have the ability to release a variety of soluble cytokines which affect other cell functions. Activated macrophages and lymphocytes have been shown to produce factors regulating fibroblast proliferation (Leibovitch and Ross, 1976; Wahl *et al.*, 1978; Korn *et al.*, 1980; Anastassiades and Wood, 1981), collagen production (Johnson and Ziff, 1976; Wahl *et al.*, 1978; Clark *et al.*, 1980), proteoglycan synthesis (Anastassiades and Wood, 1981), chemotaxis (Postlethwalte, *et al.*, 1976; Tsukamoto, 1981) and migration (Rola-Pleszczynski, 1982).

3.2.1 Macrophages

Alveolar macrophages are the most abundant nonparenchymal cell in the lung and play a critical role in pulmonary fibrosis. They exceed lymphocytes by a ratio of approximately 5–10: 1 (Hunninghake *et al.*, 1979). This population within the alveolus serves as the first line of host defence against inhaled organisms, soluble and particulate molecules. Besides phagocytic and microbicidal capacities, these cells have an extensive synthetic and secretory repertoire including lysozyme, neutral proteases, acid hydrolases and oxygen metabolites. In addition, these cell types produce a variety of pro- and anti-inflammatory cytokines which modulate lymphocyte function. Macrophages produce growth factors which have the ability to promote fibrosis by stimulating fibroblasts to migrate to the area of injury, which causes fibroblasts to proliferate, to produce matrix proteins and to contract tissues. Fibroblasts are the main effector cells in the fibrotic response, and most of the information regarding growth factors for fibroblasts relates to alveolar macrophage secretory products. These growth factors (see below) include platelet-derived growth factor

(PDGF), fibronectin, insulin-like growth factor (IGF), TGFβ and basic fibroblast growth factor (bFGF; Rennard *et al.*, 1981; Takemura and Werb, 1984; Martinet *et al.*, 1987; Nathan, 1987; Raghow *et al.*, 1987; Rom *et al.*, 1988; Kelly 1990; Nagaoka *et al.*, 1990; Shaw *et al.*, 1991). Competence factors such as PDGF render fibroblasts competent to progress around the cell cycle. Progression factors such as IGFs promote progression around the cell cycle. TGFβ has little effect on fibroblast proliferation but stimulates collagen and other matrix protein production.

3.2.2 Lymphocytes

In the inflammation of the lower respiratory tract in conditions such as fibrosing alveolitis, lymphocytes have been shown to be a prominent cellular component of the BAL fluid (Crystal *et al.*, 1984; Campbell *et al.*, 1985). Lymphocytes have also been implicated in diseases that predominantly affect the airways, such as asthma and bronchiectasis (Lapa *et al.*, 1989; Azzawi *et al.*, 1990). Lymphocytes can be broadly subdivided into T cells and B cells. T cells are responsible for a range of immune responses including delayed hypersensitivity. B cells are mainly responsible for antibody-mediated immune responses. T cells also play a central role in the early development of immune granuloma. Pulmonary immune granuloma may be initiated by infectious agents, inorganic agents and organic particulates. Early fibrotic changes, including the presence of fibroblasts and collagen, may be seen at the periphery of the granuloma (Scadding and Mitchell, 1985). Recognition of antigen by T cells is the first step in a highly complex process that results in the generation of cytokines, inflammatory cell chemotaxis and activation, as well as ultimately the release of soluble mediators responsible for inflammation. Soluble markers of T cell activation such as IL-2 and soluble CD8 have been identified in the serum of patients with fibrosing alveolitis. Expression of IL-2 receptor indicates that T cell activation is occurring within the lungs (Haslam, 1990). Activated T cells produce interferon γ (IFNγ) which is also present in lavage fluid from patients with fibrosing alveolitis (Robinson and Rose, 1990). IFNγ is an activator of macrophages and T cells, as well as stimulating the expression of ICAM-1 and HLA-DR on macrophages and endothelial cells (Kradin *et al.*, 1986). T lymphocytes have been found to influence collagen deposition in animal models of pulmonary fibrosis (Schrier and Phan, 1984). IL-1, TNF, IFNγ and IL-2 increase the adherence of T lymphocytes to fibroblasts in a dose- and time-dependent manner (Hampson *et al.*, 1989). In animal models two functional subpopulations of CD4$^+$ cells, distinguished by their patterns of cytokine production, contribute to the antigenic responses. T$_H$1 cells mediate delayed-type hypersensitivity (DTH) through secretion of IFNγ and IL-2 (Cher and Mossman, 1987) and T$_H$2 cells enhance antibody synthesis through production of IL-4, IL-5 and IL-10

(Haanen *et al.*, 1991). It is not known which type of T cell is important in human lung fibrosis although the identification of IFNγ suggests that T$_H$1 cells are involved.

Studies of lavage fluid and peripheral blood suggest that B cell stimulation, as evidenced by increased concentration of immunoglobulins, especially IgG, also has a role in the pathogenesis of fibrosing alveolitis (Weinberger *et al.*, 1978). Furthermore, lavage cells from patients with fibrosing alveolitis secrete more B cell growth factor compared to controls, suggesting a role in this disease (Emura *et al.*, 1990).

3.2.3 Neutrophils

In the normal lung, neutrophils are present but in low numbers, about 1–2% of the inflammatory cells on the epithelial surface (Hunninghake *et al.*, 1979). In contrast, in CFA neutrophils represent 5–20% of the inflammatory cells (Weinberger *et al.*, 1978; Haslam *et al.*, 1980a; Brook *et al.*, 1990). Infiltration of the lungs by neutrophils is an early event in animal models of lung injury by bleomycin and asbestos (McCullough and Collin, 1978). Neutrophils migrate to tissues after adhering to endothelial cells by expressing cell surface adhesion molecules CD18/CD11b, a ligand for ICAM-1. The cytokines IFNγ, TNFγ and IL-1 are known to enhance production of endothelial leucocyte adhesion molecule-1 (ELAM-1) which serves to bind neutrophils (Pober, 1988). Neutrophils are known to release highly reactive oxygen radicals such as hydrogen peroxide, superoxide anions and hydroxyl radicals, which can cause epithelial cell injury (Weiss, 1989; Ward, 1991). Raised concentrations of proteases and collagenases, which are probably of neutrophil origin, are seen in lavage fluid from patients with fibrosing alveolitis, indicating neutrophil involvement (Gadek *et al.*, 1979b). Increased concentrations of hyaluronic acid have been found in bleomycin-induced injury in animals, further demonstrating that neutrophil-mediated damage occurs in fibrosing alveolitis (Nettlebladt *et al.*, 1989).

3.2.4 Eosinophils

An increase in eosinophils in lung interstitium of patients with fibrosing alveolitis has been shown, and eosinophils may comprise up to 20% of the cells in lavage fluid (Allan *et al.*, 1991). Increased concentrations of eosinophil-derived eosinophil cationic protein (ECP) have been observed in CFA epithelial lining fluid and in lavage fluid samples from patients with fibrosing alveolitis, which also suggests that in some patients eosinophils may contribute to the damage of lung parenchyma (Haslam *et al.*, 1981b; Hallgren *et al.*, 1989; Hoidal, 1990).

3.2.5 Mast Cells

There is evidence for the involvement of mast cells in fibrosing alveolitis. In patients with fibrosing alveolitis mast cells are found in lung biopsy material (Haslam *et al.*, 1981a,b; Walls *et al.*, 1991). The largest population of mast cells in CFA compared with non-fibrotic patients are present in the interstitium (Fortoul and Barrios, 1990). *In vitro* studies have shown that mast cells in culture cause multiplication of fibroblasts and increased collagen production (Dayton *et al.*, 1989). Furthermore, H causes proliferation of cultured fibroblast in a dose dependent manner (Jordana *et al.*, 1988a). Similarly heparin from mast cells is mitogenic for fibroblasts (Roche, 1985). Moreover, mast cell-derived TGFβ also stimulates lung fibroblast collagen production (Fine and Goldstein, 1987). These studies all indicate that an increase in mast cells in lungs could increase inflammation and induce collagen deposition.

3.3 CHEMICAL FACTORS AND ADHESION MOLECULES

3.3.1 Nitric Oxide

Involvement of nitric oxide (NO) in the regulation of pulmonary function and diseases has been recently suggested (Barnes, 1993; Jorens *et al.*, 1993). The source of NO is not certain but it may be derived from macrophages, or alveolar or airway epithelium. Endogenous NO may have a dual role: when produced in small amounts, it may be beneficial in relaxing airway smooth muscle, but in high concentrations it may have cytotoxic effects in the airway. Inhibition of NO production significantly reduces plasma exudation and inflammation in airways (Kuo *et al.*, 1992). Inhalation of ozone stimulates NO production by alveolar and interstitial macrophages and could have a damaging effect on airway epithelial cells (Pendino *et al.*, 1992). Recently, an increase in NO in exhaled air of asthmatic patients has been observed. This increase was reduced by inhalation of the specific nitric oxide synthase (NOS) inhibitor. Asthmatic patients receiving inhaled corticosteroids had levels similar to controls (Kharitonov *et al.*, 1994). If this is the case in asthma it should be possible to use similar approaches to define the role of NO in fibrosing alveolitis.

3.3.2 Role of Adhesion Molecules in Cell–Cell Interaction

Various cell adhesion molecules have been recently identified as being important in the inflammatory response. Investigators have begun to address the role of adhesion molecules in airway inflammation in acute and chronic lung injury. The marked alterations of proteoglycans during development of fibrosis suggest that they have a regulating role in this process (Westergren-Thorsson *et al.*, 1993). In recent years, families of adhesion molecules important in different sites and situations have been discovered. These molecules can be grouped into distinct families on the basis of their molecular structure. They include integrins, members of immunoglobulin supergene family and selectins. Integrins include lympho-cyte function-associated antigen-1 (LFA-1; CD11a/CD18). T cells stimulated by an antigen have a higher LFA-1 expression than unstimulated cells and thus adhere more readily to other cells and make them suited for binding to endothelium and able to migrate to sites of inflammation (Shaw, 1990). The members of the immunoglobulin supergene family function as cell–cell adhesion molecules and are especially important in leucocyte–endothelial cell interaction. ICAM-1 belongs to this family. In areas of cellular infiltration, especially involving T lymphocytes, there is usually strong ICAM-1 expression on vascular endothelial cells (Dustin et al., 1988). The expression of ICAM-1 can be up-regulated by IFNγ, IL-1 and TNFα on human dermal fibroblasts (Pober et al., 1986; Dustin et al., 1988). Up-regulation enhances the adhesive properties of endothelial and epithelial cells to eosinophils and lymphocytes. Selectins, especially E- and P-selectin, have been shown to mediate neutrophilic inflammation in lung parenchyma in animal models of acute lung injury (Mulligan et al., 1991, 1992). Administration of antibodies to E- and P-selectins reduced neutrophil influx and lung injury, implicating their role. In the bronchial mucosa of chronic bronchitis with airway obstruction, there is an increased expression of E-selectin on vessels and of ICAM-1 on basal epithelial cells, suggesting the role of these adhesion molecules in the pathogenesis of the disease (Stefano et al., 1994).

3.3.3 Inflammatory Cytokines

The influx and activation of inflammatory cells is accompanied by the production of an array of inflamma-tory cytokines. BAL fluid from patients with CFA has been shown to contain increased concentrations of IL-8 (Carre et al., 1991) and IL-1 (Nagai et al., 1991) and alveolar macrophages encode increased amounts of IL-8 mRNA (Carre et al., 1991). Similarly, in sarcoidosis BAL fluid contains increased concentrations of IL-1 (Yamaguchi et al., 1988) and TNF (Homolka and Muller, 1993). The most compelling evidence for the importance of these cytokines comes from animal models of lung fibrosis in which the prior administration of anti-bodies to TNFα prevented the development of lung fibrosis in response to either of the two highly fibrogenic agents bleomycin or silica (Piguet et al., 1989, 1990).

3.3.3.1 Interleukin-1

There is evidence that the cytokine IL-1 is produced by alveolar macrophages derived from patients with intersti-tial lung diseases (Nagai et al., 1991). IL-1 is a major T cell growth factor, promoting the proliferation and release of lymphokines. At the site of local inflammation IL-1 is important as an early response mediator in cellular interaction leading to tissue repair. However, increased release of IL-1 results in multi organ injury to the host (Girardin et al., 1988; Shalaby et al., 1991). Investiga-tions of an IL-1 inhibitor have led to identification of an

IL-1 receptor antagonist (IL-1Ra), a 22 kDa polypeptide having 40% homology with IL-1β (Carter et al., 1990; Eisenberg et al., 1990; Hannum et al., 1990; Dinarello, 1991). Alveolar macrophages produce large amounts of IL-1Ra spontaneously in culture (Janson et al., 1993). IL-1Ra has been shown to inhibit IL-1-induced neu-trophil adhesion to endothelial cells in in vitro experi-ments (Carter et al., 1990; Arend, 1991; Dinarello, 1991). IL-1Ra has been found to attenuate the neu-trophilic alveolitis (Ulich et al., 1991) and to modulate the influence of IL-1-dependent inflammation. Produc-tion of IL-1Ra in the lungs may play a part in suppression of acute lung injury, since IL-1 is known to up-regulate the expression of ICAM-1 and enhance pathogenesis of lung disease (Arnaout, 1990; Springer, 1990). There is recent evidence of enhanced production of IL-1Ra by alveolar macrophages from non-smoking interstitial lung disease patients when compared to smoking control sub-jects and interstitial lung disease patients who smoke (Janson et al., 1993). It has been suggested that alveolar macrophages may be primed in vivo to produce aug-mented levels of this cytokine.

3.3.3.2 Tumour Necrosis Factor

TNF is another cytokine produced by macrophages in the early stages of inflammation, and its exaggerated sys-temic release can lead to organ injury and increased host morbidity and mortality (Cerami, 1992). Ingestion of particles opsonized with anti-albumin IgG markedly increases the release of TNF by alveolar macrophages (Kobzik et al., 1993). One of the central features of TNF-dependent acute inflammation in the lung appears to be the sequestration of the neutrophils and neutrophil-dependent lung injury (Sauder et al., 1984). Other func-tions include the up-regulation of ICAM-1 and E-selectin adhesion molecules (Stoolman, 1989). TNFα can have an autocrine stimulatory effect on macrophages and can also influence other inflammatory cells such as lymphocytes (Schollmeier, 1990). During the acute phases of inflammation, TNFα plays an important role and acts as a mild stimulant of fibroblast mitogenesis (Elias et al., 1987). In bleomycin-induced pulmonary fibrosis, significant increases in TNFα have been observed (Piguet et al., 1989; Phan and Kunkel, 1992). Further-more, antibodies to TNFα given prior to the injury pre-vent lung fibrosis following silica-, bleomycin- or antigen-induced allergic alveolitis in animal models (Piguet et al., 1989, 1990; Denis et al., 1991).

3.3.3.3 Interleukin-8

IL-8, an 8.0 kDa polypeptide, is a potent chemoattrac-tant and activating cytokine for neutrophils (Baggiolini et al., 1989; Matsushima and Oppenheim, 1989). In addi-tion, it causes up-regulation of neutrophil-derived β2 integrins which in turn, with ICAM-1, result in neu-trophil adhesion. The source of IL-8 is monocytes and several cellular constituents of the alveolar capillary wall including endothelial cells, fibroblasts epithelial cells, alveolar macrophages and neutrophils (Strieter et al., 1989, 1990, 1992; Standiford et al., 1990; Rolfe et al., 1991). TNFα and IL-1β are known to stimulate endothelial cells, alveolar macrophages and neutrophils to induce IL-8 gene expression (Standiford et al., 1990; Strieter et al., 1990a). The level of IL-8 mRNA and protein in alveolar macrophages was found to be significantly elevated in individuals with CFA or lung fibrosis compared to normal healthy controls (Carre et al., 1991). These data suggest that IL-8 derived from alveolar macrophages may significantly contribute to neutrophil involvement in the pathogenesis of CFA. Elevated concentrations of IL-8 in the BAL fluid of patients with CFA and pulmonary sarcoidosis are thought to contribute to the influx of neutrophils into the pulmonary alveolus and interstitium (Car et al., 1994). Recently, a supergene family of chemotactic and activating cytokines has been described which includes macrophage inflammatory protein-1α (MIP-1α) MIP-1α is expressed in increased amounts within the airspace and interstitium of patients with sarcoidosis and CFA, and this cytokine may be an important mediator of the macrophage activation and recruitment that characterizes these disease states (Standiford et al., 1993).

3.3.4 Growth Factors

Two main growth factors have been the source of most study in the setting of lung fibrosis. PDGF containing the B subunit is an exquisitely potent mitogen for fibroblasts and is produced by macrophages. This probably occurs as circulating monocytes enter inflamed tissue and differentiate into macrophages (Shaw et al., 1990). TGFβ, rather than stimulating fibroblast proliferation, causes fibroblasts to produce collagen and other matrix proteins. TGFβ is a product of many cells including macrophages. Other growth factors include IGF-1 and bFGF.

3.3.4.1 Platelet-derived Growth Factor-B

Human PDGF, a potent mitogen and chemoattractant for connective tissue cells, was originally identified and purified from human platelets (Raines et al., 1990). It is a 32 kDa peptide comprising two subunits joined by disulphide bonds. It exists as a homo- or heterodimer of two chains, PDGF-A or -B. PDGF-B is encoded by the c-sis proto-oncogene localized on chromosome 22 and PDGF-A is encoded by a gene on chromosome 7 (Antoniades et al., 1990). PDGF-B is thought to be more important with respect to lung fibrosis (Bonner et al., 1991). The main role played by PDGF is as a proliferation factor, particularly for fibroblasts and smooth muscle cells (Larson et al., 1989; Marinelli et al., 1991). This potent mitogenic cytokine is produced mainly by alveolar macrophages in the lungs. Other cells producing PDGF-B include activated fibroblasts, and smooth muscle endothelial and epithelial cells (Sariban et al., 1988; Fabisiak et al., 1990; Fabisiak and Kelly, 1992). PDGF-B

mRNA is expressed in human alveolar macrophages, macrophages derived from cultured monocytes (Mornex et al., 1986; Shaw et al., 1990) and macrophages derived from differentiation of human monocytic cells (Pantazis et al., 1986). There is increasing evidence for the importance of PDGF-B in human lung fibrosis. In patients with pulmonary fibrosis, there is an increased abundance of PDGF-B mRNA (Nagoaka et al., 1990; Shaw et al., 1991), increased transcription (Nagoaka et al., 1990) and spontaneous release of PDGF protein (Martinet et al., 1987) in alveolar macrophages. The use of immunohistochemistry and in situ hybridization on lung sections showed that in established CFA, the alveolar macrophages and epithelial cells were the main source of PDGF-B (Antoniades et al., 1990). Furthermore, PDGF-B was shown to be prominent in interstitial macrophages in lung fibrosis even in areas with little fibrosis, suggesting PDGF accumulation might precede fibrosis (Vignaud et al., 1991).

The role of PDGF-B is modulated by other cells present in lungs. Lymphocytes are one of the other dominant cell types in chronic inflammatory diseases and may influence PDGF-B production by macrophages (see below). Physical stimuli such as adherence are also known to cause gene activation including an increase in PDGF-B mRNA in monocytes (Shaw et al., 1990). Similar events may occur while in contact with large particles such as asbestos fibres or silica, since they are known to induce PDGF-B secretion by macrophages (Schapira et al., 1991; Bonner et al., 1991). Immune complexes are also known to increase production of PDGF-B protein by alveolar macrophages (Martinet et al., 1987), which may contribute to the initiation of fibrotic responses in certain autoimmune disorders.

3.3.4.2 Transforming Growth Factor β

TGFβ is a 25 kDa homodimer that is a multifunctional regulator of cell growth and differentiation (Roberts and Sporn, 1990). It is produced in an inactive form as part of a larger molecule which requires either acid treatment or enzymatic action to cleave the active TGFβ. It is present in a wide variety of tissues and interacts with specific cell membrane receptors (Massague and Like 1985). The major storage sites for TGFβ are platelets (Assoian et al., 1983) and these platelets release the peptide at sites of injury (Assoian and Sporn, 1986). Activated macrophages also contain TGFβ mRNA and secrete TGFβ protein (Assoian et al., 1987). Like several other growth factors TGFβ has been found to exist in multiple isoforms which are approximately 64–82% homologous to each other (Roberts and Sporn, 1990). There are five known subtypes but only three are known to be present in mammalian cells and tissues, and of these TGFβ_1 is the most prominent. TGFβ has several actions that potentially have a role in the repair process of fibrosis. It can stimulate fibroblasts to synthesize collagen, fibronectin and other proteoglycans, and directly affects the gene expression of extracellular matrix molecules in stromal cells so as to induce collagen synthesis (Ignotz and Massague, 1986; Roberts and Sporn, 1990; Kahari et al., 1991). TGFβ is a chemoattractant for monocytes and macrophages and induces transcription of IL-1, PDGF, FGF, TNFα and TGFβ itself (McCartney-Francis et al., 1990; Wahl et al., 1987). TGFβ can also establish an apparent state of autocrine stimulation in fibroblasts, resulting in chronic activation, such as occurs in progressive tissue fibrosis (Pelton and Moses, 1990; Roberts and Sporn, 1990).

TGFβ has been detected in a number of animal models. In the mouse bleomycin model, there is an increase in total lung TGFβ mRNA (Hoyt and Lazo, 1988; Phan and Kunkel, 1992; Westergren-Thorsson et al., 1993) and TGFβ production occurs (Khalil et al., 1989) prior to the increase in collagen I and III gene expression and protein production. Raghow et al. (1989), using hamsters, showed a significant increase in the steady-state expression of TGFβ about 1 week after intratracheal injection of bleomycin. In vitro experiments with endothelial cells and fibroblasts have shown bleomycin to induce TGFβ mRNA and protein synthesis in these cells (Phan et al., 1991; Breen et al., 1992).

TGFβ has been detected in human idiopathic pulmonary fibrosis tissue and fluids during the active phase of the disease. In lung biopsies from patients with idiopathic pulmonary fibrosis intracellular TGFβ was seen predominantly in bronchiolar epithelial cells as well as in epithelial cells of honeycomb cysts and in hyperplastic type II pneumocytes with a lower abundance in macrophages (Khalil et al., 1991). In areas of dense fibrosis, extracellular TGFβ was localized in the lamina propria of bronchioles and in subepithelial regions of honeycomb cysts. The relationship between TGFβ expression and collagen gene activation in tissues from CFA has been shown by combining immunochemistry for TGFβ expression and in situ hybridization for collagen gene expression (Broekelmann et al., 1991). In the areas of fibrotic tissue involvement, these two activities were co-expressed suggesting a role for TGFβ in promoting collagen synthesis. In sarcoidosis, a chronic inflammatory disease of unknown aetiology, enhanced tissue localization of TGFβ$_1$ and related extracellular matrix proteins has also been described (Limper et al., 1994). This may modulate the fibrotic repair process accompanying granuloma healing in sarcoidosis. In our study measuring TGFβ mRNA in alveolar macrophages, constitutive expression was found in patients with a range of interstitial lung diseases as well as in normal subjects (Shaw et al., 1991). Thus it is currently unclear whether the increase in TGFβ in human pulmonary fibrosis is mainly of macrophage or epithelial origin. The factors which result in increased TGFβ production are currently not known.

3.3.4.3 Insulin-like Growth Factor-1

The IGFs are peptide growth factors which are structurally related, to pro-insulin. Macrophages produce IGF-1, which was originally described as alveolar macrophage-derived growth factor. Alveolar macrophages express IGF-1 mRNA and spontaneously release molecules with IGF-1-like growth factor activity (Rom et al., 1988). There is evidence that IGF-1 production by macrophages may be increased during embryo development. There is also evidence of increased IGF-1 mRNA in skin wounds (Rappolee et al., 1988) and in alveolar macrophages in fibrotic lung disorders such as cryptogenic fibrosing alveolitis, systemic sclerosis and asbestosis (Bitterman et al., 1983; Rossi et al., 1985; Rom et al., 1988). IGF-1 also promotes fibroblast proliferation and may also contribute to the stimulus for macrophage replication in the lungs (Rom and Paakko, 1991).

3.3.4.4 Fibronectin

Fibronectin is an attachment protein found in extracellular matrix. It has a MW of approximately 500 000 and binds to a variety of molecules including collagens, proteoglycans and fibrin. The main source of fibronectin in adults is the liver, which synthesizes plasma fibronectin. Fibronectin can also be produced by alveolar macrophages. Fibronectin acts as a stimulant and as a chemotactic agent for fibroblasts (Bitterman et al., 1983). In the healthy lungs, interstitial fibronectin is likely to be of plasma origin, whereas during injury, fibronectin is increased in basal lamina, interstitium and smaller air spaces (Torikata et al., 1985). In patients with serious bronchial inflammation levels of plasma fibronectin are elevated (Nagy et al., 1988). In animal models of acute lung injury caused by instillation of agents such as bleomycin, there are increased levels of mRNA for fibronectin, as well as accumulation of fibronectin, as observed by immunocytochemistry, accompanying the fibroproliferative response (Raghow et al., 1985; Bray et al., 1986). Alveolar macrophages from lungs of rats with bleomycin-induced fibrosis release a monocyte chemotactic factor which is inactivated by anti-fibronectin antibodies (Denholm et al., 1989). In experimental animals exposed to silica or titanium oxide, there is a relationship between alveolar macrophage production of fibronectin and development of fibrosis (Driscoll et al., 1992). There is evidence that fibronectins are found in alveolar lining fluid in patients with fibrotic lung disease (Rennard and Crystal, 1982) and sarcoidosis (O'Connor et al., 1988). These studies suggest that the increase in fibronectin in lungs can at least in part be attributed to alveolar macrophages and phagocytic cells which are activated by injury and respond by secreting increased amounts of fibronectin and other cytokines.

3.4   THE DEVELOPMENT OF THE FIBROTIC PROCESS

3.4.1   Link Between Chronic Inflammation and Growth Factor Production

How is chronic inflammation linked to the production of growth factors? Many pro-inflammatory stimuli may stimulate the production of growth factors. Using the clinical model of tuberculosis we and others have provided data suggesting that lymphocytes may influence macrophage production of PDGF-B. IFNγ is thought to be a prominent product of the T lymphocytes of the T$_H$1 subtype (Robinson and Rose, 1990). In patients with pulmonary tuberculosis we and others have shown that IFNγ mRNA is present in CD4$^+$ cells (Barnes et al., 1993; Robinson et al., 1994), and IFNγ mRNA has been documented to be present in skin biopsies of positive tuberculin heaf tests (Tsicopolous et al., 1992). IFNγ is known to increase the mRNA abundance of the potent fibroblast mitogen PDGF-B in alveolar macrophages (Shaw et al., 1991) and we have recently shown that IFNγ derived in vitro from purified protein derivative (PPD)-stimulated lymphocytes also increases PDGF-B mRNA in alveolar macrophages (Wangoo et al., 1993). Thus, this study suggests that persisting antigen causes T$_H$1 lymphocyte activation and release of IFNγ. This in turn stimulates macrophages to produce PDGF-B and thus initiates a fibrotic response. There is also evidence in the setting of acute lung injury in the adult respiratory distress syndrome and obliterative bronchiolitis post-transplantation, that molecules with PDGF-B-like activity can be found in BAL (Snyder et al., 1991; Hertz et al., 1992). To dissect further the relationship between chronic inflammation and fibrosis we have studied profibrotic growth factor production and new collagen production in a human in vivo model of DTH which progresses to fibrosis. We used the tuberculin Heaf test in individuals previously inoculated with BCG and obtained biopsies at different time points after the test. Within 5 days of the test there was a large influx of lymphocytes and inflammatory cells. On day 5 following the tuberculin Heaf test, there was a marked increase in TGFβ staining in the tissue matrix (Wangoo et al., 1994). Immunohistochemical analysis of skin biopsies showed extensive type I procollagen staining in the biopsies as early as 5 days, which was maximal in biopsies obtained on Day 13 (Wangoo et al., 1994). There was also an increased abundance of pro-collagen mRNA located in the area of cellular infiltration in these biopsies. This study thus provided in vivo evidence of a link between chronic inflammation as seen in a delayed hypersensitivity response, growth factors and new collagen production, the first stage of a fibrotic response.

3.4.2   Fibroblast Activation and Collagens

Fibroblasts are the effector cells of the fibrotic response. They migrate to the site of injury, proliferate and produce collagen as well as other matrix proteins. Interstitial collagens are vital in the maintenance of the structural properties of lung tissue while allowing gas exchange to proceed freely across alveoli. Collagens account for 15–20% of the dry weight of normal lung. The most abundant are type I and III collagens. These are widely distributed in both airways and parenchymal structures. Fibrosis associated with CFA is characterized by excessive deposition of type I and III collagens (Madri and Furthmayr, 1980; Kirk et al., 1984, 1986). In bleomycin-induced fibrosis using in vitro and in vivo techniques, several groups have reported increased rates of collagen synthesis in the lung (Clark et al., 1980; Laurent and McAnulty, 1983). Recently, enhanced type III collagen gene expression during bleomycin-induced lung fibrosis has been reported (Shahzeidi et al., 1993). However, determining the collagen content of the lung is not a reliable guide to disease activity because lung collagen may accumulate slowly or rapidly, and the balance between collagen production and degradation is critical to the ultimate abundance of collagen (Bateman et al., 1983; Kirk et al., 1986).

Newly synthesized collagen can be identified by immunostaining using antibodies directed at the N or C terminal peptides of procollagen, since these peptides are cleaved and degraded when procollagen converts to mature tissue collagen (McDonald et al., 1986). Similarly, in situ hybridization for mRNA of type I procollagen can be used to identify cells actively synthesizing collagen (Peltonen et al., 1991). There is evidence in patients with CFA of increased type I procollagen in biopsies and increased type III procollagen in BAL fluid and serum (Bateman et al., 1981, 1983; Kirk et al., 1983, 1984, 1986; Low et al., 1983, 1992; McDonald et al., 1986; Cantin, 1988; Bjermer et al., 1989; Last et al., 1990). This also indicates that there is persistent fibroblast activation in established disease. Certain matrix molecules such as fibronectin may also stimulate fibroblasts to make collagen in CFA (Rennard et al., 1981; Bitterman et al., 1983). Increased amounts of mRNA coding for procollagen I and III have been reported in experimental models of bleomycin-induced interstitial pulmonary fibrosis (Raghow et al., 1985; Hoyt and Lazo, 1988). This suggests that there is enhanced expression of procollagen genes in lung parenchymal cells.

There is also some evidence that, once a sufficient stimulus has been provided for fibroblast proliferation and activation, this will continue even if the stimulus has been removed. Thus, dermal fibroblasts from keloid scars exhibit resistance to exogenous down-regulation (Russell et al., 1978, 1989; Myles et al., 1992), skin fibroblasts from scleroderma patients proliferate more rapidly than control fibroblasts even following several generations in vitro (Bordin et al., 1984), and cloned fibroblasts from patients with CFA exhibit an increased proliferation rate *in vitro* (Jordana *et al.*, 1988b).

3.4.3 Structural Damage

Immunocytochemistry studies examining the fate of the basement membrane in lung samples from patients with CFA have revealed that in areas of fibrosis, the alveolar architecture has been irreversibly deranged. It has been postulated (Kuhn *et al.*, 1989) that initial injury causes alveolar epithelial damage. This results in an exudate forming over the denuded basement membrane. Inflammatory cells and fibroblasts migrate into this exudate. Type II cells in the alveoli proliferate and generate a new epithelium but on the luminal side of the organizing exudate, thus leading to irreversibly thickened alveolar walls. In addition to this process, damage to the alveolus causes it to physically collapse. The fibrotic process then seals the alveolus in this collapsed state (Burkhardt, 1989). This process of structural collapse is probably exacerbated by alteration in the surface tension characteristics of the alveolar lining fluid, since there are also well recognized changes in the surfactant produced by type II cells in CFA (Low, 1989; McCormack *et al.*, 1991).

4.   *Therapeutic Approaches*

Examining each phase of the fibrotic response allows possible therapeutic interventions to be assessed. The first approach is to remove the inflammatory cells from the lungs. Corticosteroids are likely to exert their major action via this process, and indeed patients with greater degrees of inflammation have a better clinical response to corticosteroids (Fulmer *et al.*, 1979; Watters *et al.*, 1987). Similarly cyclophosphamide, which has been shown to be useful in the treatment of some patients with CFA (Johnson *et al.*, 1989), has also been shown to cause a reduction in BAL neutrophils (O'Donell *et al.*, 1987). Preventing inflammatory cells entering areas of inflammation may prove to be possible. Antibodies to the adhesion molecule CD11, even if given after the injury, prevented lung fibrosis in bleomycin-treated mice (Piguet *et al.*, 1993).

Selective inhibition of inflammatory cytokines is also an attractive approach. In animal models, antibodies to TNFα prevent lung fibrosis when given prior to injury from silica, bleomycin or antigen-induced extrinsic allergic alveolitis (Piguet *et al.*, 1989, 1990; Denis *et al.*, 1991). This approach may be an option in humans since TNFα antibodies have been used successfully as a therapy in patients with rheumatoid arthritis (Elliot *et al.*, 1993).

Inhibition of increased release of IL-1 may be equally useful in preventing lung damage (Girardin *et al.*, 1988; Shalaby *et al.*, 1991). The identification of IL-1Ra (Eisenberg *et al.*, 1990; Dinarello, 1991) raises the possibility that this may have therapeutic benefit.

Administration of cytokines which antagonize a T$_H$1-mediated inflammatory process may be beneficial. IL-4 administration partially abrogates the disease process in murine hypersensitivity pneumonitis. IL-4 partially blocked the appearance of the fibrosis and also decreased cell numbers in BAL (Ghadirian and Denise *et al.*, 1992). Such an approach has not been tested in humans.

Relatively non-specific T cell immunosuppression has been successful in animal models. Treatment of hypersensitivity pneumonitis in mice with cyclosporin A led to improvement of the disease as seen by an abrogation of the increase in lung index, lack of IL-1 and TNFα release in BAL (Denis *et al.*, 1992). However, this drug has not proven to be useful in clinical lung fibrosis in humans.

To date, few studies have been carried out specifically to inhibit growth factors. Of interest, drugs with anti-inflammatory actions such as corticosteroids and colchicine increase rather than decrease the abundance of PDGF-B mRNA in alveolar macrophages (Haynes and Shaw, 1992; Wangoo *et al.*, 1992). Furthermore, alveolar macrophage secretion of TGFβ is not inhibited by the presence of high concentrations of corticosteroids (Khalil *et al.*, 1993). Suramin, a polyanionic antitrypanosomal and anti-filarial drug, binds to and inhibits PDGF as well as TGFβ and other cationic fibrogenic cytokines (Hosang, 1985). However, clinical benefit in pulmonary fibrosis has yet to be confirmed.

Decorin is a natural inhibitor of TGFβ (Border *et al.*, 1992) which is present in reduced amounts in areas of fibrosis (Westergren-Thorsson *et al.*, 1993). This natural inhibition of TGFβ has been shown to protect animals against scarring in experimental kidney disease (Border *et al.*, 1992). The therapeutic application of this molecule in lung fibrosis has yet to be determined. Recently antibodies to TGFβ in bleomycin-induced accumulation of lung collagen in mice showed that neutralization of TGFβ was possible following systemic administration, suggesting that this may be useful in treating lung fibrosis (Giri *et al.*, 1993).

Other suggested approaches to prevent the progression of lung inflammation to fibrosis have included the interferon inducer, bropirimine, which has been found to reduce the bleomycin-induced accumulation of collagen in the lung as measured by hydroxyproline content (Zia *et al.*, 1992). The angiotensin-converting enzyme inhibitior captopril also reduces collagen and mast cell accumulation in the pathogenesis of radiation fibrosis (Ward *et al.*, 1990).

The protective effects of liposome-entrapped superoxide dismutase and catalase have also been studied on bleomycin-induced collagen deposition in the lungs of mice (Ledwozyn, 1991). It has been suggested that liposomes might be good vectors for drugs in the treatment of bleomycin-induced fibrosis.

Oral *N*-acetylcysteine has been employed in the treatment of acute lung injury and was found to reduce bleomycin-induced collagen deposition in the lungs of mice (Shahzeidi *et al.*, 1991).

Finally, a direct approach to inhibit fibroblasts has been considered. However, clinical trials with agents such as colchicine and penicillamine, which have some ability to suppress fibroblast collagen production, have not been successful (Cegla *et al.*, 1975; Goodman and Turner-Warwick, 1978; Meier-Sydow *et al.*, 1979; Liebetrau *et al.*, 1982).

5. Conclusion

By the time the patient presents with lung fibrosis, the disease is already established. Many areas of the lung have already been irreversibly damaged by the inflammatory and fibrotic processes. We need to develop means of identifying these patients early in the natural history before inflammation has progressed to fibrosis and irreversible structural damage. We now have many insights into the components of both the inflammatory and fibrotic processes and the link between them. The challenge is to develop new therapies which break the link so that inflammation can subside without the development of lung fibrosis.

6. References

Adamson, I.Y. and Bowden, D.H. (1974). The pathogenesis of bleomycin induced pulmonary fibrosis in mice. Am. J. Pathol. 77, 185–197.

Allan, J.N, Davies, W.B. and Pacht, E.R. (1991). Diagnostic significance of increased bronchoalveolar lavage fluid eosinophils. Am. Rev. Respir. Dis. 42, 642–647.

Anastassiades, T.P. and Wood, A. (1981). Effect of soluble products from lectin-stimulated lymphocytes on the growth, adhesiveness and glycosaminoglycan synthesis of cultured synovial fibroblast cells. J. Clin. Invest. 68, 792–802.

Antoniades, H.N., Bravo, M.A., Avila, R.E., Galanopoulos, T., Neville-Golden, J., Maxwell, M. and Selman, M. (1990). Platelet-derived growth factor in idiopathic pulmonary fibrosis. J. Clin. Invest. 86, 1055–1064.

Arend, W.P. (1991) Interleukin-1 receptor antagonist a new member of the Interleukin-1 family. J. Clin. Invest. 88, 1445–1451.

Arnaout, M.A. (1990). Structure and function of the leukocyte adhesion molecule CD11/CD18. Blood 75, 1037–1050.

Assoian, R.K. and Sporn, M.B. (1986). Type B transforming growth factor in human platelets: release during platelet degranulation and action on vascular smooth muscle walls. J. Cell Biol. 102, 1217–1223.

Assoian, R.K., Komoriya, A., Meyers, C.A., Miller, D.M. and Sporn, M.B. (1983). Transforming growth factor β in human platelets. J. Biol. Chem. 258, 7155–7171.

Assoian, R.K., Fleurdelys, B.E, Stevenson, H.C., Miller, P.J., Madtes, D.K., Raines, E.W., Ross, R. and Sporn, M.B. (1987). Expression and secretion of type β transforming growth factor by activated human macrophages. Proc. Natl. Acad. Sci. 84, 6020–6024.

Azzawi, M., Bradley, B., Jeffery, P.K., Frew, A.J. *et al*

(1990). Identification of activated T lymphocytes and eosinophils in bronchial biopsies in stable atopic asthma. Am. Rev. Respir. Dis. 142, 1407–1413.

Baggiolini, M., Walz, A. and Kunkel, S.L. (1989). Neutrophil activating peptide-1/ interleukin 8, a novel cytokine that activates neutrophils. J. Clin. Invest. 84, 1045–1049.

Barnes, P.J. (1993). Nitric oxide and airways. Eur. Respir. J. 6, 163–165.

Barnes, P.F., Lu, S., Abrams, J.S. *et al.* (1993). Cytokine production at the site of disease in human tuberculosis. Infect. Immun. 61, 3482–3489.

Bateman, E.D., Turner-Warwick, M. and Adelmann-Grill, B.C. (1981). Immunohistochemical study of collagen types in human foetal and fibrotic lung disease. Thorax 36, 645–653.

Bateman, E.D., Turner-Warwick, M. Haslam, P.L. *et al.* (1983). Cryptogenic fibrosing alveolitis: prediction of fibrogenic activity from immunohistochemical studies of collagen types in lung biopsy specimens. Thorax 38, 93–101.

Begin, R., Masse, S. and Bureau, M.A. (1982). Morphologic features and function of the airways in early asbestosis in the sheep model. Am. Rev. Respir. Dis. 126, 870–876.

Bitterman, P.B., Rennard, S.I., Hunninghake, C.W. and Crystal, R.G. (1982). Human alveolar macrophage growth factor for fibroblasts: regulation and partial characterization. J. Clin. Invest. 70, 1801–1813.

Bitterman, P.B., Rennard, S.I., Adelberg, S. and Crystal, R.G. (1983). Role of fibronectin as a growth factor for fibroblasts. J. Cell Biol. 97, 1925–1931.

Bjermer, L., Lundgren, R. and Haellgren, R. (1989). Hyaluronan and type III procollagen peptide concentrations in bronchoalveolar lavage fluid in idiopathic pulmonary fibrosis. Thorax 44, 126–131.

Bonner, J.C., Osorino-Vergas, A.R., Badgett, A. and Brody, A.R. (1991). Differential proliferation of rat lung fibroblasts induced by the platelet-derived growth factor AA, -AB, and -BB isoforms secreted by rat alveolar macrophages. Am. J. Respir. Cell. Mol. Biol. 5, 539–547.

Border, W.A., Noble, N.A., Yamamoto, T., Harper, J.R., Yamaguchi, Y.U., Pierschbacher, M.D. and Ruoslahti, E. (1992). Natural inhibitor of transforming growth factor β protects against scarring in experimental kidney disease. Nature 360, 361–364.

Bordin, S., Page, R.C. and Narayanan, A.S. (1984). Heterogeneity of normal human diploid fibroblasts: isolation and characterization of one phenotype. Science 223, 171–173.

Bray, B.A., Osman, M., Ashtyani, H., Mandl, I. and Turino, G.M. (1986). The fibronectin content of canine lungs is increased in bleomycin induced fibrosis. Exp. Mol. Pathol. 44, 353–363.

Breen, E., Absher, M., Kelly, J., Phan, S. and Cutroneo, K.R. (1992). Bleomycin regulation of TGF-B mRNA in rat lung fibroblasts. Am. J. Respir. Cell. Mol. Biol. 6, 146–152.

Broekelmann, T.J., Limper, A.H., Colby, T.V. and McDonald, J.A. (1991). Transforming growth factor β 1 is present at sites of extracellular matrix gene expression in human pulmonary fibrosis. Proc. Natl. Acad. Sci. USA 88, 6642–6646.

Brook, Z., Trapnell, B.C. and Crystal, R.G. (1990). Neutrophils and the pathogenesis of idiopathic pulmonary fibrosis. In "Pathophysiology of Pulmonary Cells: Neutrophils and Lymphocytes" (ed. M. Baggiolini *et al.*). Masson Italia, Milan.

Burkhardt, A. (1989). Alveolitis and collapse in the pathogenesis of pulmonary fibrosis. Am. Rev. Respir. Dis. 140, 513–524.

Campbell, D.A., Poulter, L.W., Janossy, G. and du Bois, R.M. (1985). Immunohistological analysis of lung tissue from patients with cryptogenic fibrosing alveolitis suggesting local expression of immune hypersensitivity. Thorax 40, 405–411.

Cantin, A.M., Boileau, R. and Begin, R. (1988). Increased procollagen III aminoterminal peptide-related antigens and fibroblast growth signals in the lung of patients with idiopathic pulmonary fibrosis. Am. Rev. Respir. Dis. 137, 572–578.

Car, B.D., Meloni, F., Luisetti, M. et al. (1994). Elevated IL-8 and MCP-1 in the bronchoalveolar lavage fluid of patients with idiopathic pulmonary fibrosis and pulmonary sarcoidosis. Am. J. Respir. Crit. Care Med. 149, 655–659.

Carre, P.C., Mortenson, R.L., King, T.E.J.R., Noble, P.W., Sable, C.L. and Riches, D.W.H. (1991). Increased expression of the interleukin-8 gene by alveolar macrophages in idiopathic pulmonary fibrosis. J. Clin. Invest. 88, 1802–1810.

Carrington, C.B., Gaensler, E.A., Coutu, R.E. et al. (1978). Natural history and treated course of usual and desquamative interstitial pneumonia. N. Engl. J. Med. 298, 801–809.

Carter, D.B., Deibel, M.R., Dunn, C.J. et al. (1990). Purification, cloning, expression and biological characterization of an interleukin-1 receptor antagonist protein. Nature 334, 633–638.

Cegla, U.H., Kroidl, R.F., Meier-Sydow, J. et al. (1975). Therapy of the idiopathic fibrosis of the lung. Experiences with three therapeutic principles corticosteroids in combination with azathioprine, D-penicillamine, and para-aminobenzoate. Pneumonologie 152, 75–92.

Cerami, A. (1992). Inflammatory cytokines. Clin. Immunol. Immunopathol. 62, (S)3–10.

Chandler, D.B., Hyde, D.M. and Giri, S.N. (1983). Morphometric estimates of infiltrative cellular changes during the development of bleomycin induced pulmonary fibrosis in hamsters. Am. J. Pathol. 112, 170–177.

Cher, D.J. and Mosmann, T.R. (1987). Two types of murine helper T cell clone. ll. Delayed type hypersensitivity is mediated by T_H1 clones. J. Immunol. 138, 3688–3694.

Clark, J.G., Overton, J,E., Marino, B.A., Vitto, J. and Starcher, B.C. (1980). Collagen biosynthesis in bleomycininduced pulmonary fibrosis in hamster. J. Lab. Clin. Med. 96, 943–953.

Clark, J.M. and Lanbertsen, C.J. (1971). Pulmonary oxygen toxicity: a review. Pharmacol. Rev. 23, 37–133.

Cocchiara, R., Giallongo, A., Amoroso, S. et al. (1981). Circulating immune complexes in patients with usual interstitial pulmonary fibrosis: partial characterization and relationship with Thermoactinomyces vulgaris. Immunology 44, 817–825.

Crystal, R.G., Fulmer, J. D., Roberts, W.C. et al. (1976). Idiopathic pulmonary fibrosis: clinical, histologic, radiographic, physiologic, scintigraphic, cytologic and biochemical aspects. Ann. Intern. Med. 85, 769–788.

Crystal, R.G., Gadek, J.E., Ferrans, V.J. et al. (1981). Interstitial lung disease: current concepts of pathogenesis, staging, and therapy. Am. J. Med. 70, 542–568.

Crystal, R.G., Bitterman, P.B., Rennard, S.I. et al. (1984). Interstitial lung diseases of unknown cause. Disorders characterized by chronic inflammation of the lower respiratory tract. N. Engl. J. Med. 310, 154–166.

Daniels, J.M., Brien, J.F. and Massey, T.E. (1989). Pulmonary fibrosis induced in the hamster by amiodarone and desethylamiodarone. Toxicol. Appl. Pharmacol. 100, 350–359.

Davis, W.B. and Gadek, J.E. (1987). Detection of pulmonary lymphoma by bronchoalveolar lavage. Chest, 91, 787–790.

Dayton, E.T., Caulfield, J.P., Hein, A., Austen, K.F. and Stevens, R.L. (1989). Regulation of the growth rate of mouse bone marrow fibroblasts by IL-3 activated mouse bone marrow-derived mast cells. J. Immunol. 142, 4307–4313.

Denholm, E.M. and Rollins, S.M. (1993). Expression and secretion of transforming growth factor β by bleomycin-stimulated rat alveolar macrophages . Am. J. Physiol. 264 (Lung. Cell. Mol. Physiol. 8), L36–L42.

Denholm, E.M., Wolber, F.M. and Phan, S.H. (1989). Secretion of monocyte chemotactic activity by alveolar macrophages. Am. J. Pathol. 135, 571–580.

Denis, M. (1992). Interleukin-6 in mouse hypersensitivity pneumonitis: changes in lung free cells following depletion of endogenous IL-6 or direct administration of IL-6. J. Leuk. Biol. 52, 197–201.

Denis, M., Cornier, Y., Fournier, M., Tardif, J. and Laviolette, M. (1991). Tumor necrosis factor plays an essential role in determining hypersensitivity pneumonitis in a mouse model. Am. J. Respir. Cell. Mol. Biol. 5, 477–483.

Denis, M., Comier, Y. and Laviolette, M. (1992). Murine hypersensitivity pneumonitis: a study of cellular infiltrates and cytokine production and its modulation by cyclosporin A. Am. J. Respir. Cell. Mol. Biol. 6, 68–74.

Dinarello, C.A. (1991). Interleukin-1 and interleukin-1 antagonism. Blood 77, 1627–1652.

Dreisin, R.B., Schwarz, M.I., Theofilopoulos, A.N. et al. (1978). Circulating immune complexes in the idiopathic interstitial pneumonias. N. Engl. J. Med. 298, 353–357.

Driscoll, K.E., Maurer, J.K., Poynter, J., Higgins, J., Asquith, T. and Miller, N.S. (1992). Stimulation of rat alveolar macrophage fibronectin release in a cadmium chloride model of lung injury and fibrosis. Toxicol. Appl. Pharmacol. 116, 30–37.

Dustin, M.L., Singer, K.H., Tuck, D.T. and Springer, T.A. (1988). Adhesion of T lymphoblasts to epidermal keratinocytes is regulated by IFNγ and is mediated by ICAM-1. J. Exp. Med. 167, 1323–1340.

Eisenberg, S.P., Evans, R.J., Arend, W.P. et al. (1990). Primary structure and functional expression from complimentary DNA of a human interleukin-1 receptor antagonist. Nature 343, 343–346.

Elias, J.A., Gustilo, K., Basder, W. and Freundlich, B. (1987). Synergistic stimulation of fibroblast prostaglandin production by recombinant interleukin-1 and tumor necrosis factor. J. Immunol. 138, 3821–3826.

Elliot, M.J., Maini, R.N., Feldman, M. et al. (1993). Treatment of rheumatoid arthritis with chimeric monoclonal antibodies to tumor necrosis factor α. Arthritis Rheum. 36, 1681–1690.

Emura, M., Nagai, S., Takouchi, M., Kitaichi, M. and Izumi, T. (1990). In vitro production of B-cell growth factor and B-cell differentiation factor by peripheral blood mononuclear cells and bronchoalveolar lavage T lymphocytes from patients with idiopathic pulmonary fibrosis. Clin. Exp. Immunol. 82, 133–139.

Evans, C.C. (1976). HLA antigens in diffuse fibrosing alveolitis. Thorax 31, 483.

Fabisiak, J.P. and Kelly, J. (1992). Platelet derived growth factor. In "Cytokines of the Lung" (ed. J. Kelly), pp. 3–40. Marcel Dekker, New York.

Fabisiak, J.P., Absher, M.P. and Kelly, J. (1990). Production of platelet derived growth factor like cytokines by rat lung fibroblasts *in vitro*. Am. Rev. Respir. Dis. 141(Supp.), 915.

Fine, A. and Goldstein, R.H. (1987). The effect of transforming growth factor β on cell proliferation and collagen formation by lung fibroblasts. J. Biol. Chem. 262, 3897–3902.

Fortoul, T.I. and Barrios, R. (1990). Mast cells and idiopathic lung fibrosis. Arch. Invest. Med. 21, 5–10.

Frankel, F.R., Steeger, J.R., Damiano, V.V., Sohn, M., Oppenheim, D. and Weinbaum, G. (1991). Induction of unilateral pulmonary fibrosis in the rat by cadmium chloride. Am. J. Respir. Cell. Mol. Biol. 5, 385–394.

Fulmer, J.D., Roberts, W.C., von Gal, E.R. *et al.* (1979). Morphologic-physiologic correlates of the severity of fibrosis and degree of cellularity in idiopathic pulmonary fibrosis. J. Clin. Invest. 63, 665–676.

Gadek, J., Hunninghake, G., Zimmerman, R. *et al.* (1979a). Pathogenetic studies in idiopathic pulmonary fibrosis. Control of neutrophil migration by immune complexes. Chest 2, 264–265.

Gadek, J.E., Kelman, J.A., Fells, G.A. *et al.* (1979b). Collagenase in the lower respiratory tract of patients with idiopathic pulmonary fibrosis. N. Engl. J. Med. 301, 737–742.

Gelb, A.F., Dreisin, R.B., Epstein, J.D. *et al.* (1983). Immune complexes, gallium lung scans, and bronchoalveolar lavage in idiopathic interstitial pneumonitis-fibrosis a structure–function clinical study. Chest 84, 148–153.

Ghadirian, E. and Denis, M. (1992). Murine hypersensitivity pneumonitis: interleukin-4 administration partially abrogates the disease process. Microb. Pathog. 12, 377–382.

Girardin, E., Grau, G.E., Dayer, J.M. *et al.* (1988). Tumor necrosis factor and interleukin-1 in the serum of children with severe infectious purpura. N. Engl. J. Med. 319, 397–400.

Giri, S.N., Hyde, D.M. and Hollinger, M.A. (1993). Effect of antibody to transforming growth factor β on bleomycin induced accumulation of lung collagen in mice. Thorax 48, 959–966.

Goodman, M. and Turner-Warwick, M. (1978). Pilot study of penicillamine therapy in corticosteroid failure patients with widespread pulmonary fibrosis. Chest 74, 338.

Gundel, R.H., Wegner, C.D., Torcellini, C.D., Clark, C.C, Haynes, N., Rothlein, R., Smith, C.W. and Lenst, L.G. (1991). Endothelial leukocyte adhesion molecule-1 mediates antigen-induced acute airway inflammation and late-phase obstruction in monkeys. J. Clin. Invest. 88, 1407–1411.

Gundel, R.H., Wegner, C.D. and Letts, L.G. (1992). Antigen-induced acute and late-phase responses in primates. Am. Rev. Respir. Dis. 145, 369–373.

Haanen, J.B.A., de Wall Malefijt, R., Res. P.C.M., *et al.* (1991). Selection of human T helper type I. like T cell subset of Mycobacteria, J. Exp. Med. 174, 583–592.

Hallgren, R., Bjermer, L., Lundgren, R. and Venge, P. (1989). Tho eosinophil component of the alveolitis in idiopathic pulmonary fibrosis. Signs of eosinophil activation in the lung are related to impaired lung functions. Am. Rev. Respir. Dis. 139, 373–377.

Hampson, F., Monick, M., Peterson, M.W. and Hunninghake, G.W. (1989). Immune mediators increase adherence

of T-lymphocytes to human lung fibroblasts. Am. J. Physiol. 256, C336–340.

Hannum, C.H., Wilcox, C.J., Arend, W.P. *et al.* (1990). Interleukin-1 receptor antagonist activity of a human interleukin-1 inhibitor. Nature 343, 336–340.

Haslam, P. (1984). Bronchoalveolar lavage. Semin. Respir. Med. 6, 55–70.

Haslam, P.L. (1990). Evaluation of alveolitis by studies of lung biopsies. Lung 168 (Suppl), 984–992.

Haslam, P.L., Thompson, B., Mohammed, I. *et al.* (1979). Circulating immune complexes in patients with cryptogenic fibrosing alveolitis. Clin. Exp. Immunol. 37, 381–390.

Haslam, P.L., Turton, C.W.G., Heard, B. *et al.* (1980a). Bronchoalveolar lavage in pulmonary fibrosis: comparison of cells obtained with lung biopsy and clinical features. Thorax 35, 9–18.

Haslam, P.L., Turton, C.W.G., Lukoszek, A. *et al.* (1980b). Bronchoalveolar lavage fluid cell counts in cryptogenic fibrosing alveolitis and their relation to therapy. Thorax 35, 328–339.

Haslam, P.L., Cromwell, O., Dewar, A. and Turner-Warwick, M. (1981a). Evidence of increased histamine levels in lung lavage fluids from patients with cryptogenic fibrosing alveolitis. Clin. Exp. Immunol. 44, 587–593.

Haslam, P.L., Dewar, A. and Turner-Warwick, M. (1981b). Lavage eosinophils and histamine. In "Cellular Biology of the Lung" (eds G. Cumming and G., Bonsignore), pp. 77–94. Plenum, New York.

Hay, J., Shahazeidi, S. and Laurent, G.J. (1991). Mechanisms of bleomycin-induced lung damage. Arch. Toxicol. 65, 81–94.

Haynes, A.R. and Shaw, R.J. (1992). Dexamethasone-induced increase in platelet-derived growth factor (B) mRNA in human alveolar macrophages and myelomonocytic HL60 macrophage-like cells. Am. J. Respir. Cell. Mol. Biol. 7, 198–206.

Hertz, M.I., Henke, C.A., Nakhleh, R.E. *et al.* (1992). Obliterative bronchiolitis after lung transplantation: A fibroproliferative disorder associated with platelet derived growth factor. Proc. Natl. Acad. Sci. USA 89, 10385–10389.

Hesterberg, T.W. and Last, J. A. (1981). Ozone iduced acute pulmonary fibrosis in rats. Prevention of increased rates of collagen synthesis by methylprednisolone. Am. Rev. Respir. Dis. 123, 47–52.

Hoidal, J.R. (1990). The eosinophil and lung injury. Am. Rev. Respir. Dis. 142, 1245–1246.

Homolka, J. and Muller Q.M. (1993). Increased Interleukin 6 production by bronchoalveolar lavage in patients with active sarcoidosis. Lung 171, 173–183.

Hosang, M. (1985). Suramin binds to platelet derived growth factor and inhibits its biological activity. Cell. Biochem. 29, 265–273.

Hoyt, D.G. and Lazo, J.S. (1988). Alteration of pulmonary mRNA encoding procollagens, fibronectin and transforming growth factor b precede bleomycin-induced pulmonary fibrosis in mice. J. Pharmacol. Exp. Ther. 246, 765–771.

Hunninghake, G.W., Gadek, J.E., Kawanami, O., Ferrany, V.J. and Crystal, R.G. (1979). Inflammatory and immune processes in the human lung in health and disease: evaluation of bronchoalveolar lavage. Am. J. Pathol. 97, 149–206.

Hunninghake, G.W., Gadek, J.E., Lawley, T.J. *et al.* (1981). Mechanisms of neutrophil accumulation in the lungs of patients with idiopathic pulmonary fibrosis. J. Clin. Invest. 68, 259–269.

Ignotz, R.A. and Massague, J. (1986). Transforming growth factor -B stimulates the expression of fibronectin and collagen and their incorporation in to the extracellular matrix. J. Biol. Chem. 261, 4337–4345.

Jansen, J.M., Schutte, A.J.H., I.D. Elema, I.D. et al. (1984). Local immune complexes and inflammatory response in patients with chronic interstitial pulmonary disorders associated with collagen vascular diseases. Clin. Exp. Immunol. 56, 311–320.

Janson, R.W., King, T.E. Jr., Hance, K.R. and Arend, W.P. (1993). Enhanced production of IL-1 receptor antagonist by alveolar macrophages from patients with interstitial lung disease. Am. Rev. Respir. Dis. 148, 495–503.

Johnson, M.A., Kwan, S., Snell, N.J.C. et al. (1989). Randomized controlled trial comparing prednisolone alone with cyclophosphamide and low dose prednisolone in combination in cryptogenic fibrosing alveolitis. Thorax 44, 280–288.

Johnson, R.L. and Ziff, M. (1976). Lymphokine stimulation of collagen accumulation. J. Clin. Invest. 58, 240–252.

Jordana, M., Befus, A.D., Newhouse, M.T., Bienenstock, J. and Gauldie, J. (1988a). Effect of histamine on proliferation of normal human adult lung fibroblasts. Thorax 43, 552–558.

Jordana, M., Schulman, J., Mc Sharry, C., Irving, L.B., Newhouse, M.T., Jordana, G. and Gauldie, J. (1988b). Heterogeneous proliferative characteristics of human adult lung fibroblast lines and clonally derived fibroblasts from control and fibrotic tissue. Am. Rev. Respir. Dis. 137, 579–584.

Jorens, P.G., Vermeire, P.A. and Herman, A.G. (1993). L-Arginine dependent nitric oxide synthase: a new metabolic pathway in lung and airways. Eur. Respir. J. 6, 258–266.

Kahari, V.M., Larjava, H. and Uitto, J. (1991). Differential regulation of extracellular matrix proteoglycan (PG) gene expression. Transforming growth factor β 1 upregulates biglycan (PGI), and versican (large fibroblast PG) but down regulates decorin (PGII) mRNA levels in human fibroblasts in culture. J. Biol. Chem. 266, 10608–10615.

Kelley, J. (1990). Cytokines of the lung. Am. Rev. Respir. Dis. 141, 765–788.

Kelley, J., Newman, R.A. and Evans, J.N. (1980). Bleomycin induced pulmonary fibrosis in the rat. Prevention with an inhibitor of collagen synthesis. J. Lab. Clin. Med. 96, 954–964.

Kelley, J., Chrin, L., Shull, S., Rowe, D.W. and Cutroneo, F.R. (1985). Bleomycin selectively elevates mRNA levels for procollagen and fibronectin following acute lung injury. Biochem. Biophys. Res. Commun. 131, 836–843.

Keogh, B.A., Bernardo, J., Hunninghake, G. W. et al. (1983). Effect of intermittent high dose parenteral corticosteroids on the alveolitis of idiopathic pulmonary fibrosis. Am. Rev. Respir. Dis. 127, 18–22.

Khalil, N., Bereznay, O.H., Sporn, M. and Greenberg, A.H. (1989). Macrophage production of transforming growth factor β and fibroblast collagen synthesis in chronic pulmonary inflammation. J. Exp. Med. 170, 727–737.

Khalil, N., O'Connor, R.N., Unruh, H.W., Warren, P.W., Flanders, K.C., Kemp, A., Bereznay, O.H. and Greenberg, A.H. (1991). Increased production and immunohistochemical localization of transforming growth factor β in idiopathic pulmonary fibrosis. Am. J. Res. Cell. Mol. Biol. 5, 155–162.

Khalil, N., Whitman, C., Zuo, L., Danielpour, D. and Greenberg, A. (1993). Regulation of alveolar macrophage trans-forming growth factor β secretion by corticosteroids in bleomycin-induced pulmonary inflammation in the rat. J. Clin. Invest. 92, 1812–1818.

Kharitonov, S.A, Yates, D., Robbins, R.A., Logan sinclair, R., Shinebourne, E.A. and Bames, P.J. (1994). Increased nitric oxide in exhaled air of asthmatic patients. Lancet 343, 133–135.

Kimura, R., Hu, H. and Stein-Streilein, J. (1992). Delayed type hypersensitivity responses regulate collagen deposition in the lung. Immunology 77, 550–555.

King, T.E., Schwarz, M.I., Dreisin, R.B. et al. (1979). Circulating immune complexes in pulmonary eosinophilic granuloma. Ann. Intern. Med. 91, 397–399.

King, T.E., Christopher, K.L., Zeballos, J. et al. (1984). Bronchoalveolar lavage, gallium-67 citrate lung scanning and circulating immune complexes in the staging of idiopathic pulmonary fibrosis: Correlation with physiologic and morphologic features. Clin. Res. 32, 63A.

Kirk, J.M.E., Laurent, G.J., Bateman, E. et al. (1983). The measurement of serum procollagen peptide levels in cryptogenic fibrosing alveolitis. Am. Rev. Respir. Dis. 127 (Supp), 272.

Kirk, J.M.E., Heard, B.E., Kerr, I. et al. (1984). Quantitation of types I and III collagen in biopsy lung samples from patients with cryptogenic fibrosing alveolitis. Collagen. Rel. Res. 4, 169–182.

Kirk, J.M.E., DaCosta, P.E., Turner-Warwick, M., et al. (1986). Biochemical evidence for an increased and progressive deposition of collagen in lungs of patients with pulmonary fibrosis. Clin. Sci. 70, 39–45.

Kobzik, L., Huang, S., Paulauskis, J.D. and Godleski, J.J. (1993). Particle opsonization and lung macrophage cytokine response. J. Immunol. 151, 2753–2759.

Korn, J.H., Halushka, P.V. and Le Roy, E.C. (1980). Mononuclear cell modulation of connective tissue function: suppression of fibroblast growth by stimulation of endogenous prostaglandin production. J. Clin. Invest. 65, 543–554.

Kradin, R.L., Divertie, M.B. and Colvin, R.B. (1986). Usual interstitial pneumonitis is a T cell alveolitis. Immunol. Immunopathol. 40, 224–235.

Kuhn, C., Boldt, J., King, T.E.Jr. et al. (1989). An immunohistochemical study of architectural remodeling and connective tissue synthesis in pulmonary fibrosis. Am. Rev. Respir. Dis. 140, 1693–1703.

Kumar, R.K., Li, W. and O'Grady, R. (1990). Activation of lymphocytes in the pulmonary inflammatory response to silica. Immunol. Invest. 19, 363–372.

Kuo, H.P., Liu, S. and Barnes, P.J. (1992). The effect of endogenous nitric oxide on neurogenic plasma exudation in guinea pig airways. Eur. J. Pharmacol. 221, 385–388.

Kusaka, Y., Cullen, R.T. and Donaldson, K. (1990). Immunomodulation in mineral dust exposed lungs: stimulatory effect and interleukin-1 release by neutrophils from quartz elicited alveolitis. Clin. Exp. Immunol. 80, 293–298.

Lapa e Silva, J.R., Guerreiro, D., Noble, B., Poulter, L.W. and Cole, P.G. (1989). Immunopathology of experimental bronchiectasis. Am. J. Respir. Cell. Mol. Biol. 1, 297–304.

Larson, O., Latham, C., Zickert, P. and Zetterberg, A. (1989). Cell cycle regulation of human diploid fibroblasts: possible mechanisms of platelet derived growth factor. J. Cell. Physiol. 139, 477–483.

Last, J.A., King, T.E.Jr., Nerlich, A. M. *et al.* (1990). Collagen crosslinking in adult patients with acute and chronic fibrotic lung disease: molecular markers for fibrotic collagen. Am. Rev. Respir. Dis. 141, 307–313.

Laurent, G.J. and McAnulty, R.J. (1983). Protein metabolism during bleomycin-induced pulmonary fibrosis in rabbits. *In vivo* evidence for collagen accumulation because of increased synthesis and decreased degradation of newly synthesised collagen. Am. Rev. Respir. Dis. 128, 82–88.

Lazo, J.S., Hoyt, D.G., Sebti, S.M. and Pitt, B.R. (1990). Bleomycin: a pharmacologic tool in the study of the pathogenesis of interstitial pulmonary fibrosis. Pharmacol. Ther. 47, 347–358.

Ledwozyn, A. (1991). Protective effect of liposome entrapped superoxide dismutase and catalase on bleomycin induced lung injury in rats. I. Antioxidant enzyme activities and lipid peroxidation. Acta. Vet. Hung. 39, 215–224.

Leibovitch, S.J. and Ross, R. (1976). A macrophage-dependent factor that stimulates the proliferation of fibroblasts *in vitro*. Am. J. Pathol. 84, 501–513.

Liebetrau, G., Pielesch, W., Ganguin, H.G. *et al.* (1982). Therapy of pulmonary fibrosis with D-penicillamine. Z. Gesamte. Inn. Med. 37, 263–266.

Limper, A.H., Colby, T.V., Sanders, M.S. *et al.* (1994). Immunohistochemical localization of Transforming Growth Factor β1 in the nonnecrotizing granulomas of pulmonary sarcoidosis. Am. J. Respir. Crit. Care. Med. 149, 197–204.

Low, R.B. (1989). Bronchoalveolar lavage lipids in idiopathic pulmonary fibrosis. Chest 95, 3–5.

Low, R.B., Cutroneo, K.R., Davis, G.S. *et al.* (1983). Lavage type III procollagen N-terminal peptides in human pulmonary fibrosis and sarcoidosis. Lab. Invest. 48, 755–759.

Low, R.B., Giancola, M.S., King, T.E.Jr. *et al.* (1992). Serum and bronchoalveolar lavage N-terminal type III procollagen peptides in idiopathic pulmonary fibrosis. Am. Rev. Respir. Dis. 146, 701–706.

Madri, J.A. and Furthmayr, H. (1980). Collagen polymorphism in the lung. An immunohistochemical study of pulmonary fibrosis. Hum. Pathol. 11, 355–365.

Marinelli, W.A., Polunovsky, V.A., Harmon, K.R. and Bitterman, P.B. (1991). Role of platelet-derived growth factor in pulmonary fibrosis. Am. J. Respir. Cell. Mol. Biol. 5, 503–504.

Martinet, Y., Rom, W.N., Grotendorst, G.R., Martin, G.R. and Crystal, R.G. (1987). Exaggerated spontaneous release of platelet-derived growth factor by alveolar macrophages from patients with idiopathic pulmonary fibrosis. N. Engl. J. Med. 317, 202–209.

Massague, J. and Like, B. (1985). Cellular receptors for type-B transforming growth factor. J. Biol. Chem. 260, 2636–2645.

Matsushima, K. and Oppenheim, J.J. (1989). Interleukin 8 and MCAF: novel inflammatory cytokines inducible by IL-1 and TNF. Cytokine 1, 2–13.

McCartney-Francis, N., Mizel, D., Wong, H., Wahl, L. and Wahl, S. (1990). TGF-beta regulates production of growth factors and TGF-beta by human peripheral blood monocytes. Growth Factors 4, 27–35.

McCormack, F.X., King T.E. Jr., Voelker, D.R. *et al.* (1991). Idiopathic pulmonary fibrosis: Abnormalities in the bronchoalveolar lavage content of surfactant protein A. Am. Rev. Respir. Dis. 144, 160–166.

McCullough, B. and Collins, J.F. (1978). Bleomycin-induced

diffuse interstitial pulmonary fibrosis in baboons. J. Clin. Invest. 61, 79–88.

McDonald, J.A., Broekelmann, T.J., Matheke, M.L. *et al.* (1986). A monoclonal antibody to the carboxyterminal domain of procollagen type I visualizes collage-synthesizing fibroblasts. Detection of an altered fibroblast phenotype in lungs of patients with pulmonary fibrosis. J. Clin. Invest. 78, 1237–1244.

Meier-Sydow, J., Rust, M., Kronenberger, H. *et al.* (1979). Long-term follow-up of lung function parameters in patients with idiopathic pulmonary fibrosis treated with prednisone and azathioprine or D-penicillamine. Prax. Pneumol. 33, 680–688.

Mornex, J.F., Martinent, Y., Yamaguchi, K. *et al.* (1986). Spontaneous expression of the c-*sis* gene and release of a platelet derived growth factor like molecule by human alveolar macrophages. J. Clin. Invest. 78, 61–66.

Mossman, B.T., Janssen, Y.M., Marsh, J.P. *et al.* (1991). Development and characterization of a rapid onset rodent inhalation model of asbestosis for disease prevention. Toxicol. Pathol. 19, 412–418.

Mulligan, M.S., Verani, J. and Dame, M.K. (1991). Role of endothelial leucocyte adhesion molecule-1 in neutrophil mediated lung injury in rats. J. Clin Invest. 88, 1396–1406.

Mulligan, M.S., Polley, M.J., Bayer, R.J. *et al.* (1992). Neutrophil dependent acute lung injury: requirement for P-sellectin (Gmp-140). J. Clin. Invest. 90, 1600–1607.

Myles, M.E., Russell, J.D., Trupin, J.S., Smith, J.C., and Russell, S.B. (1992). Keloid fibroblasts are refractory to inhibition of DNA synthesis by phorbol esters. Altered response is accompanied by reduced sensitivity to prostaglandin E2 and altered down-regulation of phorbol ester binding sites. J. Biol. Chem. 267, 9014–9020.

Nagai, S., Aung, H., Takoudhi, M., Kusume, K. and Izumi, T. (1991). IL-1 and IL-1 inhibitory activity in the culture supematants of alveolar macrophages from patients with interstitial lung diseases. Chest 99, 674–680.

Nagaoka, I., Trapnell, B.C. and Crystal, R.G. (1990). Upregulation of platelet-derived growth factor-A and -B gene expression in alveolar macrophages of individuals with idiopathic pulmonary fibrosis. J. Clin. Invest. 85, 2023–2027.

Nagy, B., Katona, E., Erdei, J. *et al.* (1988). Fibronectin in bronchoalveolar lavage fluid and plasma from children with chronic inflammation of lungs. Acta. Paediatr. Scand. 77, 727–733.

Nathan, C.F. (1987). Secretory products of macrophages. J. Clin. Invest. 79, 319–326.

Nettlebladt, O., Bergh, J., Schenhom, M., Tengblad, A. and Hallgren, R. (1989). Accumulation of hyaluronic acid in the alveolar interstitial tissue in bleomycin-induced alveolitis. Am. Rev. Respir. Dis. 139, 759–762.

O'Connor ,C., Odlum, C., Van Breda, A., Power, C. and htzgerald, M.X. (1988). Collagenase and fibronectin in bronchoalveolar lavage fluid in patients with sarcoidosis. Thorax 43, 393–400.

O'Donell, K., Keogh, B., Cantin, A. and Crystal, R.G. (1987). Pharmacological suppression of the neutrophil component of the alveolitis in idiopathic pulmonary fibrosis. Am. Rev. Respir. Dis. 136, 288–292.

Pantazis, P., Sariban, E., Kufe, D. and Antoniades, H.N. (1986). Induction of c-*sis* gene expression and synthesis of platelet derived growth factor in human myeloid leukemia cells during monocytic differentiation. Proc. Natl. Acad. Sci. USA 83, 6455–6459.

Pelton, R.W. and Moses, H.L. (1990). The beta type transforming growth factor. Mediators of cell regulation in the lung. Am. Rev. Resp. Dis. 142, S31–35.

Peltonen, J., Hsiao, L.L., Jaakkola, S. et al. (1991). Activation of collagen gene expression in keloids: colocalization of Type I and VI collagen and transforming growth factor β 1 mRNA. J. Invest. Dermatol. 97, 240–248.

Pendino, K., Punjabi, C. and Lavnikova, N. (1992). Inhalation of ozone stimulates nitric acid production by pulmonary and interstitial macrophages. Am. Rev. Respir. Dis. 145, A650.

Phan, S.H. and Kunkel, S.L. (1986). Inhibition of bleomycin induced pulmonary fibrosis by nordihydroguaiaretic acid. Am. J. Pathol. 124, 343–352.

Phan, S.H. and Kunkel, S.L. (1992). Lung cytokine production in bleomycin induced pulmonary fibrosis. Exp. Lung. Res. 18, 29–43.

Phan, S.H., Thrall, R.S. and Williams, C. (1981). Bleomycin induced pulmonary fibrosis. Effect of steroid on lung collagen metabolism. Am. Rev. Respir. Dis. 124, 428–434.

Phan, S.H., Gharaee-Kermani, M., Wolber, F. and Ryan, U.S. (1991). Stimulation of rat endothelial cell transforming growth factor production by bleomycin. J. Clin. Invest. 87, 148–154.

Piguet, P.F., Collart, M.A., Grau, G.E., Kapanci, Y. and Vassalli, P. (1989). Tumour necrosis factor/cachectin plays a key role in bleomycin-induced pneumopathy and fibrosis. J. Exp. Med. 170, 655–663.

Piguet, P.F., Collart, M.A., Grau, G.E., Sappino, A.P. and Vassalli, P. (1990). Requirement of tumour necrosis factor for development of silica-induced pulmonary fibrosis. Nature. 344, 245–247.

Piguet, P.F., Rosen, H., Vesin, C. and Grau, G.E. (1993). Effective treatment of the pulmonary fibrosis elicited in mice by bleomycin or silica with anti-CD-11 Antibodies. Am. Rev. Respir. Dis. 147, 435–441.

Pober, J.S. (1988). Cytokine-mediated activation of vascular endothelium. Am. J. Pathol. 133, 426–433.

Pober, J.S., Gimbrone, M.A.Jr., Lopiene, L.A. et al. (1986). Overlapping patterns of activation human endothelial cells by IL-1, TNF and IFNγ. J. Immunol. 137, 1893–1896.

Postlethwaite, A.E., Snyderman, R. and Kang, A.H. (1976). The chemotactic attraction of human fibroblasts to a lymphocyte derived factor. J. Exp. Med. 144, 1188–1203.

Raghow, R., Lurie, S., Seyer, J.M. and Kang, A.H. (1985). Profiles of steady state levels of messenger RNAs coding for type 1 procollagen, elastin, and fibronectin in hamster lungs undergoing bleomycin-induced interstitial pulmonary fibrosis. J. Clin. Invest. 76, 1733–1739.

Raghow, R., Portiethwaite, A.E., Keski-Oja, J., Moses, H. and Kang, A.H. (1987). Transforming growth factor β increases steady state levels of type 1 procollagen and fibronectin mRNAs post transcriptionally in cultured human dermal fibroblasts. J. Clin. Invest. 79, 1285–1288.

Raghow, R., Irish, P. and Kang, A.H. (1989). Coordinate regulation of transforming growth factor gene expression and cell proliferation in hamster lungs undergoing bleomycin induced pulmonary fibrosis. J. Clin. Invest. 84, 1836–1842.

Raines, E.W., Bowen, D.F. and Ross, R. (1990). Platelet-derived growth factor. In: "Handbook of Experimental Pharmacology, Vol. 95. Peptide Growth Factors and their Receptors". (eds M.B. Sporn and A.B. Roberts), pp. 173–262. Springer-Verlag, Heidelberg.

Rappolee, D.A., Mark, D., Banda, M.J. and Werb, Z. (1988). Wound macrophages express TGFα and other growth factors in vivo: analysis by mRNA phenotyping. Science 241, 708–712.

Reiser, K.M., Hesterberg, T.W., Haschek, W.M. and Last, J.A. (1982). Experimental silicosis. I. Acute effects of intratracheally instilled quartz on collagen metabolism and morphologic characteristics of rat lungs. Am. J. Pathol. 107, 176–185.

Rennard, S.I. and Crystal, R.G. (1982). Fibronectin in human bronchopulmonary lavage fluid. Elevation in patients with interstitial lung disease. J. Clin. Invest. 69, 113–122.

Rennard, S.I., Hunninghake, G.W., Bitterman, P.B. and Crystal, R.G. (1981). Production of fibronectin by the human alveolar macrophage: mechanism for the recruitment of fibroblasts to sites of tissue injury in interstitial lung diseases. Proc. Natl. Acad. Sci. USA 78, 7147–7151.

Roberts, A.B. and Sporn, M.B. (1990). The transforming growth factor betas. In "Handbook of Experimental Pharmacology, Vol. 95. Peptide Growth Factors and their Receptors", pp. 419–473, Springer-Verlag, Heidelberg.

Robinson, B.W. and Rose, A.H. (1990). Pulmonary gamma interferon production in patients with fibrosing alveolitis. Thorax 45, 105–108.

Robinson, D.S., Ying, S., Taylor, I.K., Wangoo, A., Mitchell, D.M., Kay, A.B., Hamid, Q. and Shaw R.J. (1994). Evidence for a Th-1 bronchoalveolar T cell subset and predominance of interferon gamma gene activation in pulmonary tuberculosis. Am. J. Respir. Crit. Care. Med. 149, 989–993.

Roche, W.R. (1985). Mast cells and tumours. The specific enhanchment of tumor proliferation in vitro. Am. J. Pathol. 119, 57–64.

Rola-Pleszczynski, M., Liew, H. and Lemaire, I. (1982). Stimulated human lymphocytes produce a soluble factor which inhibits fibroblast migration. Cell. Immunol. 74, 104–110.

Rolfe, M.W., Kunkel, S.L., Standiford, T.J., Chensue, S.W., Allen, R.M., Evanoff, H.L. et al. (1991). Pulmonary fibroblast expression of interleukin-8: a model for alveolar macrophage-derived cytokine networking. Am. J. Respir. Cell. Mol. Biol. 5, 493–501.

Rom, W.N. and Paakko, P. (1991). Activated alveolar macrophages express the insulin like growth factor-1 receptor. Am. J. Respir. Cell. Mol. Biol. 4, 432–439.

Rom, W.N., Basset, P., Fells, G.A., Nukiwa, T. and Crystal, R.G. (1988). Alveolar macrophages releases an insulin like growth factor 1-type molecule. J. Clin. Invest. 82, 1685–1693.

Rossi G.A., Bitterman, P.B., Rennard, S.I., Ferrans, V.J. and Crystal, R.G. (1985). Evidence for chronic inflammation as a component of the interstitial lung disease associated with progressive systemic sclerosis. Am. Rev. Respir. Dis. 131, 612–617.

Russell, J.D., Russell, S.B. and Trupin, K.M. (1978). Differential effects of hydrocortisone on both growth and collagen metabolism of human fibroblasts from normal and keloid tissue. J. Cell. Physiol. 97, 221–223.

Russell, S.B., Trupin, J.S., Myers, J.C., Broquist, A.H., Smith, J.C., Myles, M.E. and Russell, J.D. (1989). Differential glucocorticoid regulation of collagen mRNAs in human dermal fibroblasts. J. Biol. Chem. 264, 13730–13735.

Sariban, E., Sitaras, N.M., Antoniades, H.N., Kufe, D.W. and Pantazis, P. (1988). Expression of platelet-derived growth factor (PDGF)-related transcripts and synthesis of biologically active PDGF-like proteins by human malignant epithelial cell lines. J. Clin. Invest. 82, 1157–1164.

Sauder, D.N., Mounessa, N.L., Kary, S.l., Dinarello, C.A. and Gallin, J.I. (1984). Chemotactic cytokines: the role of leukocyte pyrogen and epidermal cell thymocyte activating factor in neutrophi chemotaxis. J. Immunol. 132, 828–837.

Scadding, J.G. and Mitchell, D.N. (1985). "Sarcoidosis", 2nd edn. Chapman and Hall, London.

Schapira, R.M., Osornio Vargas, A.R. and Brody, A.R. (1991). Inorganic particles induce secretion of a macrophage homologue of platelet derived growth factor in a density and time dependent manner in vitro. Expt. Lung Res., 17, 1011–1024.

Scheule, R.K., Perkins, R.C., Hamilton, R. and Holian, A. (1992). Bleomycin stimulation of cytokine secretion by the human alveolar macrophage. Am. J. Physiol. 262, 386–391.

Schollmeier, K. (1990). Immunologic and pathophysiologic role of tumour necrosis factor. Am. J. Respir. Cell. Mol. Biol. 3, 11–12.

Schrier, D.J. and Phan, S.H. (1984). Modulation of bleomycin-induced pulmonary fibrosis in the BALB/c mouse by cyclophosphamide-sensitive T cells. Am. J. Pathol. 116, 270–278.

Schwarz, M.I., Dreisin, R.B., Pratt, D.S. et al. (1978). Immunofluorescent patterns in the idiopathic interstitial pneumonias. J. Lab. Clin. Med. 91, 929–938.

Selman, M., Montano, M., Ramos, C. et al. (1989). Experimental pulmonary fibrosis induced by paraquat plus oxygen in rats: a morphologic and biochemical sequential study. Exp. Mol. Pathol. 50, 147–166.

Sendelbach, L.E., Lindenschmidt, R.C. and Witschi, H.P. (1985). The effect of cyclosporine A on pulmonary fibrosis induced by butylated hydroxytoluene, bleomycin and beryllium sulfate. Toxicol. Lett. 26, 169–173.

Shahzeidi, S., Sarnstrand, B., Jeffery, P.K., McAnulty, R.J. and Laurent, G.J. (1991). Oral N-acetylcysteine reduces bleomycin induced collagen production in the lungs of mice. Eur. Respir. J. 4, 845–852.

Shahzeidi, S., Mulier, B., de Crombrugghe, B., Jeffery, P.K., McAnulty, R.J. and Laurent, G.J. (1993). Enhanced type III collagen gene expression during bleomycin induced lung fibrosis. Thorax 48, 622–628.

Shalaby, M.R., Halgunset, J., Haugen, O.A. et al. (1991). Cytokine associated tissue injury and lethality in mice: a comparative study. Clin. Immunol. Immunopathol. 61, 69–82.

Shaw, S. (1990). Leucocyte adhesion molecules: normal function and clinical relevence. Postgrad. Educ. Course. Syllabus. A.A.A.I Wisconsin. 132–137.

Shaw, R.J., Doherty, D.E., Ritter, A., Benedict, S.H. and Clark, R.A.F. (1990). Adherence-dependent human monocyte PDGF(B) gene activation is associated with increases in c-fos, c-jun and EGR2 mRNA. J. Cell. Biol. 111, 2139–2148.

Shaw, R.J., Benedict, S.H., Clark, R.A. and King, T.E. Jr. (1991). Pathogenesis of pulmonary fibrosis in interstitial lung disease. Alveolar macrophage PDGF(B) gene activation and up-regulation by interferon gamma. Am. Rev. Respir. Dis. 143, 167–173.

Smith, P., Heath, D. and Kay, J.M. (1974). The pathogenesis and structure of paraquat-induced pulmonary fibrosis in rats. J. Pathol. 114, 57–67.

Snider, G.L., Bartolosne, R.C., Ronald, H.G., O'Brien, J.J. and Lucey, E.C. (1978). Chronic interstitial pulmonary fibrosis produced in hamsters by endotracheal bleomycin. Am. Rev. Respir. Dis. 117, 289–297.

Snyder, L.S., Hertz, M.I., Peterson, M.S. et al. (1991). Acute lung injury: pathogenesis of intra-alveolar fibrosis. J. Clin. Invest. 88, 663–673.

Springer, T.A. (1990). Adhesion receptors of the immune system. Nature 346, 425–434.

Standiford, T.J., Kunkel, S.L., Basha, M.A., Chensue, S.W., Lynch, J.P., Toews, G.B. et al. (1990). Interleukin-8 gene expression by a pulmonary epithelial cell line: a model for cytokine networks in the lung. J. Clin. Invest. 86, 1945–1953.

Standiford, T.J., Rolfe, M.W., Kunkel, S.L. et al. (1993). Macrophage inflammatory protein-1 alpha expression in interstitial lung disease. J. Immunol. 151, 2852–2863.

Stefano, A.D., Maestrelli, P., Roggeri, A. et al. (1994). Upregulation of adhesion molecules in the bronchial mucosa of subjects with chronic obstructive bronchitis. Am. J. Respir. Crit. Care Med. 149, 803–810.

Stoolman, L.M. (1989). Adhesion molecules controlling lymphocyte migration. Cell. 56, 907–910.

Strieter, R.M., Kunkel, S.L., Showell, H.J., Remick, D.G., Phan, S.H., Ward, P.A. et al. (1989). Endothelial cell gene expression of a neutrophil chemotactic factor by TNFα, LPS, and IL-1 beta. Science. 243, 1467–1469.

Strieter, R.M., Chensue, S.W., Basha, M.A., Standiford, T.J., Lynch, J.P., Baggiolini, M. et al. (1990a). Human alveolar macrophage gene expression of interleukin-8 by tumor necrosis factor α, lipopolysaccharide, and interleukin-1β. Am. J. Respir. Cell. Mol. Biol. 2, 321–326.

Strieter, R.M., Kasahara, K., Allen, R., Showell, H.J., Standiford, T.J. and Kunkel, S.L. (1990b). Human neutrophils exhibit disparate chemotactic factor gene expression. Biochem. Biophys. Res. Commun. 173, 725–730.

Strieter, R.M., Kasahara, K., Allen, R., Stadiford, T.J., Rolfe, M.W., Becker, F.S. et al. (1992). Cytokine-induced neutrophil-derived interleukin-8. Am. J. Pathol. 141, 397–407.

Takemura, R. and Werb, Z. (1984). Secretory products of macrophages and their physiological functions. Am. J. Physiol. (Cell. Physiol.) 246, C1–C9.

Thrall, R.S., McCormic, J.R., Jack, R.M., McReynolds, R.A. and Ward, P.A. (1979). Bleomycin induced pulmonary fibrosis in the rat. Am. J. Pathol. 95, 117–130.

Thrall, R.S., Barton, R.W., D'Amato, D.A. and Sulavik, S.B. (1982). Differential cellular analysis of bronchoalveolar lavage fluid obtained at various stages during the development of bleomycin induced pulmonary fibrosis in the rat. Am. Rev. Respir. Dis. 126, 488–492.

Torikata, C., Villiger, B., Kuhn, C. and McDonald, J.A. (1985). Ultrastructural distribution of fibronectin in normal and fibrotic human lung. Lab. Invest. 52, 399–408.

Tsicopoulos, A., Hamid, Q., Varney, V. et al. (1992). Preferential messenger RNA expression of Th-1 type cells (IFN-gamma +, IL-2 +) in classical delayed type delayed type (tuberculin) hypersensitivity reactions in human skin. J. Immunol. 148, 2058–2061.

Tsukamoto, Y., Helsel, W.E. and Wahl, S.M. (1981). Macrophage production of fibronectin, a chemoattractant for fibroblasts. J. Immunol. 127, 673–678.

Turton, C.W.G., Morris, L.M., Lawler, S.D. *et al.* (1978). HLA in cryptogenic fibrosing alveolitis. Lancet 1(8062), 507–508.

Ueda, T., Ohta, K., Suzuki, N. *et al.* (1992). Idiopathic pulmonary fibrosis and high prevalence of serum antibodies to hepatitis C virus. Am. Rev. Respir. Dis. 146, 266–268.

Ulich, T.R, Yin, S., Guo, K. *et al.* (1991). The intratracheal administration of endotoxin and cytokines: III. The interleukin-1 receptor antagonist inhibits endotoxin and IL-1 induced acute inflammation. Am. J. Pathol. 138, 521–524.

Varpela, E., Tiilikainen, A., Varpela, M. and Tukiainen P. (1979). High prevalences of HLA-BI5 and HLA-Dw6 in patients with cryptogenic fibrosing alveolitis. Tissue Antigens 14, 68–71.

Vergnon, J.M., Vincent, M., DeThe, G. *et al.* (1984). Cryptogenic fibrosing alveolitis and Epstein–Barr virus: an association? Lancet 2, 768–771.

Vignaud, J.M., Allam, M., Martinet, N., Pech, M., Plenat, F. and Martinet, Y. (1991). Presence of platelet-derived growth factor in normal and fibrotic lung is specifically associated with interstitial macrophages, while both interstitial macrophages and alveolar epithelial cells express the c-*sis* proto-oncogene. Am. J. Respir. Cell. Mol. Biol. 5, 531–538.

Vuorio, E.I., Makela, J.K., Vuorio, T.K., Poole, A. and Wagner, J.C. (1989). Characterization of excessive collagen production during development of pulmonary fibrosis induced by chronic silica inhalation in rats. Br. J. Exp. Pathol. 70, 305–315.

Wagner, C.D., Wolyniec, W.W., LePlante, A.M., Marschman, K., Lubbe, K., Haynes, N., Rothlein, R. and Letts, L.G. (1992). Intercellular adhesion molecule-1 contributes to pulmonary oxygen toxicity in mice. Lung 170, 267–279.

Wahl, S.M., Wahl, L.M. and McCarthy, J.B. (1978). Lymphocyte-mediated activation of fibroblast proliferation and collagen production. J. Immunol. 121, 942–946.

Wahl, S.M., Hunt, D.A., Wakefield, L.M. *et al.* (1987). Transforming growth factor type beta induces monocyte chemotaxis and growth factor production. Proc. Natl. Acad. Sci. USA 84, 5788–5792.

Walls, A.F., Bennen, A.R., Godfrey, R.C., Holgate, S.T. and Church, M.K. (1991). Mast cell tryptase and histamine concentrations in bronchoalveolar lavage fluid from patients with interstitial lung disease. Clin. Sci. 81, 183–188.

Wangoo, A., Haynes, A.R., Sutcliffe, S.P., Sorooshian, M. and Shaw, R.J. (1992). Modulation of PDGF(B) mRNA abundance in macrophages by colchicine and dibutyril cAMP. Mol. Pharmacol. 42, 584–589.

Wangoo, A., Taylor, I.K., Haynes, A.R. and Shaw, R.J. (1993). Upregulation of alveolar macrophage PDGF(B) mRNA by interferon-γ from M. tuberculosis antigen (PPD) stimulated lymphocytes. Clin. Exp. Immunol. 94, 43–50.

Wangoo, A., Cook, H.T., Taylor, G.M. and Shaw, R.J. (1994). Enhanced expression of type 1 Procollagen and Transforming growth factor β in tuberculin induced delayed type hypersensitivity. Am. J. Respir. Crit. Care Med. 149, 682.

Ward, P.A. (1991). Overview of the processes of cellular injury in interstitial lung disease. Chest 100, 230–232.

Ward, W.F., Molteni, A., Ts'ao, C.H. and Hinz, J.M. (1990). Captopril reduces collagen and mast cell accumulation in irradiated rat lung. Int. J. Radiat. Oncol. Biol. Phys. 19, 1405–409.

Watters, L.C. (1986). Genetic aspects of idiopathic pulmonary fibrosis and hypersensitivity pneumonitis. Semin. Respir. Med. 7, 317–325.

Watters, L.C., Schwarz, M.I., Cherniack, R.M. *et al.* (1987). Idiopathic pulmonary fibrosis. Pretreament bronchoalveolar lavage cellular constituents and their relationships with lung histopathology and clinical response to therapy. Am. Rev. Respir. Dis. 135, 696–704.

Weinberger, S.E., Kelman, J.A., ELson, N.A. *et al.* (1978). Bronchoalveolar lavage in interstitial lung disease. Ann. Intern. Med. 89, 459–466.

Weiss, S.J. (1989). Tissue destruction by neutrophils. N. Engl. J. Med. 320, 365–376.

Westergren-Thorsson, G., Hemas, J., Samstrand, B., Oldberg, A., Heinegard, D. and Malmstrom, A. (1993). Altered expression of small proteoglycans, collagen, and transforming growth factor -beta 1 in developing bleomycin-induced pulmonary fibrosis in rats. J. Clin. Invest. 92, 632–637.

Winterbauer, R.H., Hammar, S.P., Hallman, K.O. *et al.* (1978). Diffuse interstitial pneumonitis. Clinicopathologic correlations in 20 patients treated with prednizone/azathioprine. Am. J. Med. 65, 661–672.

Yamaguchi, E., Okazaki, N., Tsuneta, Y., Abe, S., Terai, T. and Kawakami, Y. (1988). Interleukins in plumonary sarcoidosis. Dissociative correlations of lung interleukins 1 and 2 with the intensity of alveolitis. Am. Rev. Respir. Dis. 138, 645–651.

Zia, S., Hyde, D.M. and Giri, S.N. (1992). Effect of an interferon inducer bropirimine on bleomycin induced lung fibrosis in hamsters. Pharmacol. Toxicol. 71, 11–18.

12. The Cell Biology of the Resolution of Inflammation

C. Haslett

1. Introduction

In recent years it has become clear that the inflammatory response is centrally involved in the pathogenesis of a wide range of diseases afflicting developed societies. Many of these are associated with a heavy burden of morbidity and untimely deaths. Important inflammatory lung diseases include asthma, chronic bronchitis and emphysema, respiratory distress syndromes of the adult and the neonate, the pneumoconioses and a variety of other chronic inflammatory/scarring conditions. Extensive lists of analogous conditions can be drawn up for other organs. Most of these diseases are characterized by the persistent accumulation of inflammatory cells, associated with chronic tissue injury and the development of a scarring response, which in organs with delicate

exchange membranes such as the lung and kidney, can result in catastrophic deterioration of organ function.

However, the central paradox in our consideration of inflammatory disease is that the inflammatory response evolved as a highly effective component of the innate immune response of the body to infection or injury. Indeed, until the last two or three decades, inflammation was perceived as an entirely beneficial host response to injury or infection. Elias Metchnikoff, the father of modern inflammatory cell biology, emphasized this concept in his work. Neutrophil and eosinophil granulocytes play key defensive roles in infections such as lobar streptococcal pneumonia and in parasitic infestations such as schistosomiasis. The acute inflammatory response in lobar streptococcal pneumonia represents one of the most dramatic examples of the effectiveness of a rapidly mounted inflammatory response in host antibacterial defences. In the pre-antibiotic era, streptococcal infection was highly prevalent and responsible for more than 90% of pneumonias, yet the inflammatory response was effective enough to save the lives of most of the affected individuals. It is also remarkable that despite what we now know of the potential of neutrophils and activated macrophages to injure tissue and to promote scarring, the massive accumulation of inflammatory cells in lobar pneumonia cleared completely in more than 95% of cases with less than 2.5% progressing to fibrosis (Robertson and Uhley, 1938).

By contrast with initiation and amplification mechanisms, little research has been directed at the processes responsible for termination of inflammation. Yet, it is just as reasonable to suppose that understanding how inflammation normally resolves will not only provide important insights into the circumstances leading to the persistent inflammation which characterizes most inflammatory diseases but will also suggest novel therapies directed at promoting resolution mechanisms rather than those favouring amplification and persistence of inflammation. In his treatise on acute inflammation, Hurley (1983) considered that the acute inflammatory response might terminate by development of chronic inflammation, suppuration, scarring or by resolution. All the alternatives to resolution are potentially detrimental and could contribute to disease processes, particularly in organs whose function depends on the integrity of delicate exchange membranes. However, until recently there has been little information about the cellular and molecular mechanisms underlying the normal resolution processes of inflammation. The remainder of this chapter therefore represents somewhat incomplete and hypothetical consideration of some of the processes which are likely to be necessary for inflammation to resolve, and speculation concerning how a better understanding of these mechanisms will help elucidate the pathogenesis of inflammatory disease and suggest novel anti-inflammatory therapy.

In order for tissues to return to normal during the resolution of inflammation, all of the processes occurring during the establishment of the inflamed state must be reversed. Thus in the simplest model of a self-limited inflammatory response, such as might occur in response to the instillation of bacteria into the alveolar airspace, these would include: removal of the inciting stimulus and dissipation of the mediators so generated; cessation of granulocyte emigration from blood vessels; restoration of normal microvascular permeability; limitation of granulocyte secretion of potentially histotoxic and pro-inflammatory agents; cessation of the emigration of monocytes from blood vessels and their maturation into inflammatory macrophages; and, finally, removal of extravasated fluid, proteins, bacterial and cellular debris,

granulocytes and macrophages. Whilst *in vitro* experiments demonstrate that neutrophils and monocytes are able to emigrate between endothelial cell and epithelial cell monolayers without necessarily causing injury to these "barrier cells", it is clear that even at sites of "beneficial" inflammation such as streptococcal pneumonia there may be quite extensive endothelial and epithelial injury, but the capacity of streptococcal pneumonia to resolve implies that this injury must not be sufficient in degree or extent to prevent effective repair mechanisms. With the completion of resolution and repair events, the stage should be set for full recovery of normal tissue architecture and function.

Each of these events will be considered in the following discussion, but factors relevant to the behaviour of neutrophil and eosinophil granulocytes in the resolution of inflammation will receive most attention. The neutrophil is the archetypal acute inflammatory cell. It is essential for host defence, but it is also implicated in the pathogenesis of a wide range of inflammatory diseases (Malech and Gallin, 1988). It is usually the first cell to arrive at the scene of tissue perturbation, and a number of key inflammatory events including monocyte emigration (Doherty et al., 1988) and the generation of inflammatory oedema (Wedmore and Williams, 1981) appear to depend upon the initial accumulation of neutrophils. Neutrophils contain a variety of agents with the capacity not only to injure tissues (Weiss, 1989), but also to cleave matrix proteins into chemotactic fragments (Vartio et al., 1981) with the potential to amplify inflammation by attracting more cells, and they have recently been shown to contain a granule component CAP37 (Spitznagel, 1990) which is a specific monocyte chemotaxin. Eosinophils also play an important role in host defence, particularly against parasites, yet paradoxically they are also specifically implicated in the pathogenesis of allergic diseases such as bronchial asthma, and the presence of large numbers of eosinophils in tissue is often associated with a local fibrogenic response. Termination of granulocyte emigration from blood vessels and their subsequent clearance from inflamed sites are obvious prerequisites for inflammation to resolve, and are important events to consider in the control of inflammatory tissue injury generally. Moreover, understanding the mechanisms controlling these events may suggest new therapeutic opportunities for manipulating inflammation and promoting mechanisms which favour resolution rather than the persistence of inflammation.

2. Mediator Dissipation

During the resolution of inflammation the powerful mediators initiating the response must somehow be removed, inactivated or otherwise rendered impotent. This aspect of mediator biology has received much less attention than mechanisms involved in their initial generation, and it is likely that different mechanisms may be utilized for different mediators. For example thromboxane A_2 (TXA_2) and endothelial-derived relaxing factor (nitric oxide) are labile factors which are spontaneously unstable. Platelet-activating factor (PAF) and C5a are inhibited in vitro by an inactivating enzyme (Berenberg and Ward, 1973), and some chemotactic cytokines such as interleukin-8 (IL-8) are thought to become inactivated by binding to other cells, e.g. erythrocytes. Reduction of mediator efficacy might occur by local reduction of their concentration consequent upon dilution during the generation of inflammatory oedema. Mediator efficacy may also be reduced by attenuation of target cell responsiveness, for example in the downregulation of receptors which occurs during desensitization of neutrophils to high concentrations of a variety of inflammatory mediators (Henson et al., 1981). Locally generated factors which can exert opposing influences must also be considered; for example neutrophil immobilizing factor would tend to counteract the chemotactic effects of locally generated chemotactic peptides. In cytokine biology, much attention has been paid to agents which initiate or amplify inflammation, but, by analogy with the proteins involved in the blood coagulation cascade, the whole system must be kept under close control by very effective inhibitors and other negative influences. Some such agents have been discovered, e.g. the IL-1 receptor antagonist (IL-1Ra), yet the inhibitory "partners" of the most newly described cytokines and chemotactic peptides have yet to be described. The final requirement for the success of most of the above mechanisms is that the production of mediators at the site must cease.

It is thus likely that control of a single, complex function such as neutrophil chemotaxis in response to a chemotactic peptide, e.g. C5a or IL-8, is influenced at a number of points and by a number of factors, including the concentration of mediators, the concentration of their inhibitors or inactivators, possible desensitizing mechanisms, and the effects of other locally generated agents with negative influences on chemotaxis. The redundancy of the inflammatory response in vivo must also be taken into consideration. Not only may single mediators exert multiple effects under different circumstances, but important events may be provoked by agents from different mediator families. For example C5a, leukotriene B_4 (LTB_4), IL-8, ENA-78, and probably many more factors are likely to exert neutrophil chemotactic effects in vivo. In order to gain a dynamic perspective of the resolution of inflammation, it will therefore be necessary to consider how a variety of important mediators may act in concert at the inflamed site and seek to appreciate the integrated impact of negative and positive stimuli on dynamic events in situ. Thus, the overall propensity for inflammation to persist would be expected to cease when the balance of mediator effects tips towards the inhibitory rather than the stimulatory, presumably as

a result of the combination of at least some of the possible mechanisms considered above.

3. Cessation of Granulocyte and Monocyte Emigration

Until quite recently it was considered that the differential rate of emigration of granulocytes and monocytes at the inflamed site was mainly due to a slower responsiveness of monocytes to "common" chemotactic factors, e.g. C5a. However, in the light of new discoveries in chemokine and adhesive molecule biology, it is likely that the emigration through microvascular endothelium of specific leucocytes in different pathological circumstances is caused by the combined effects of the local release of cell-specific chemokines and the utilization of different components of the adhesive molecule repertoire that control inflammatory cell endothelial cell adhesion. For example, IL-8 is a specific neutrophil chemotaxin, and transcapillary neutrophil migration is likely to be mediated by the adhesive interaction between the leucocyte integrins on the neutrophil surface and adhesive molecules such as intercellular adhesion molecule-1 (ICAM-1) on the endothelial surface; whereas specific chemotactic peptides such as monocyte chemotactic protein-1 (MCP-1) and the use of an alternative adhesive molecule interaction between e.g. very late activation antigen 4 (VLA_4) on the monocyte surface and vascular cell adhesion molecule-1 (VCAM-1) on the endothelium may be utilized to achieve specific monocyte emigration. Since VCAM-1 tends to be expressed later than E-selectin by stimulated endothelial cells, sequential emigration of leucocytes may also be influenced by the time course of endothelial adhesin molecular expression. Experiments in vitro suggest that eosinophils may also have the capacity to use the VCAM-1/VLA_4 adhesive axis but PAF, RANTES and IL-5 are important in their initial attraction and stimulation.

The factors controlling cessation of inflammatory cell emigration remain obscure. The evolution and resolution of inflammation are dynamic processes, and simple histological techniques do not represent these events. Because poorly understood factors such as cell removal rates may also exert major influences on the number of cells observed in "static" histological sections, the study of neutrophil emigration kinetics requires the careful monitoring of labelled populations of cells. When intravenous pulses of radiolabelled neutrophils were used to define the emigration profiles of neutrophils from blood into acutely inflamed skin (Colditz and Movat, 1984), joints (Haslett et al., 1989a) or lung (Clark et al., 1989), it was found that neutrophil influx ceased remarkably early, by contrast with the greatly prolonged influx which occurred in an inflammatory model which progressed to chronic tissue injury and scarring (Haslett et al., 1989b). Indeed in experimental streptococcal pneumonia, in which the lung tissues are massively infiltrated by neutrophils and inflammatory macrophages for several days before resolution occurs, neutrophil emigration to the site ceases within 24 h of the initiation of pneumonia (Plate IV). These data are supported by human studies in lobar pneumonia in which intravenously delivered radiolabelled neutrophils fail to emigrate to the pneumonic lung by the time the patients are admitted to hospital (Saverymuttu *et al.*, 1985). Cessation of granulocyte emigration occurring so soon in the evolution of acute inflammation may therefore represent one of the earliest resolution events, and a number of hypothetical mechanisms could be responsible.

(1) Locally generated chemotactic factor inhibitors could inactivate neutrophil chemotactic factors. Agents with the capacity to inactive C5a activity *in vitro* have been isolated in plasma (Berenberg and Ward, 1973), but these factors have not been characterized and quantified at inflamed sites, and a plasma-derived inactivator is unlikely to account for cessation of neutrophil emigration in situations where extravascular protein leakage is minimal or absent.

(2) "Deactivation" or desensitization of neutrophils to high concentrations of inflammatory mediators may lead to extravasated neutrophils becoming unresponsive to further chemotactic factor stimulation (Ward and Becker, 1967). This might be expected to occur at the centre of an inflamed site where the concentration of chemotaxins would be expected to be highest, but it seems unlikely that this mechanism could be involved in the cessation of neutrophils entering the site.

(3) A negative feedback loop might operate whereby neutrophils which have already accumulated exert an influence that prevents more neutrophils entering from the bloodstream.

(4) Cessation of neutrophil emigration may simply occur as a result of dissipation or removal of chemotactic factors from the inflamed site.

(5) The layers of endothelial and epithelial cells which normally permit neutrophils to emigrate during the initiation of inflammation could alter to form a "barrier" to further neutrophil emigration. It is well recognized that neutrophils can migrate between endothelial cells (Hurley, 1963) and epithelial cells (Milks and Cramer, 1984) without causing obvious injury. This process, known as "diapedesis" or "transmigration", involves complex intercellular adhesive mechanisms together with the opening of intercellular endothelial and endothelial junctions by mechanisms which are presently obscure.

Which of these hypothetical events are important *in vivo* is by no means clear. In a skin model of inflammation it appeared that desensitization (Colditz and Movat, 1984) was important operating in some forms of human disease involving persistent inflammation there have been suggestions that chemotactic factor inhibitory agents may be defective. However, in experimental arthritis we found no evidence for a desensitization mechanism or for a chemotactic factor inhibitory mechanism (Haslett *et al.*, 1989a). Cessation of neutrophil emigration into the joint coincided with loss of chemoattractants from the joint space. Loss of chemoattractants was not dependent upon cellular accumulation at the site, an observation providing evidence against a simple negative feedback mechanism (Haslett *et al.*, 1989a). Although the mechanism responsible for the loss of chemotaxin was not identified, these observations suggest that the local generation and removal of chemoattractants are likely to be centrally important in the persistence and cessation of neutrophil emigration.

Neutrophil surface adhesive molecules become rapidly up-regulated upon neutrophil exposure to chemotaxins such as C5a and IL-8. It is now thought that L-selectin on the surface of the neutrophil is important in the initial interaction with endothelial cells under the conditions of shear stress which exist *in vivo*, whereas the leucocyte integrins, e.g. CD11b/CD18 (Mac-1), are particularly important in the second phase of "tight" adhesion necessary for capillary transmigration. Neutrophil adhesive molecules must then uncouple to permit the next stage of migration to proceed. Molecular mechanisms controlling the "turn-on" and "turn-off" signals of the integrins and other surface molecules are now the subject of detailed scrutiny. The endothelium plays an active role in these events. Neutrophil adhesins interact with counter-receptors on the endothelium, e.g. E-selectin, ICAM-1, ICAM-2 and P-selectin. It appears that endothelial P-selectin and E-selectin are involved in the initial neutrophil adhesion (P-selectin and E-selectin), whereas the link between ICAM-1 on the endothelium and MAC-1 (CD11b/CD18) integrin on the neutrophil surface is likely to be important in the second stage of adhesion and transmigration. Endothelial adhesive molecules are markedly up-regulated by factors such as IL-1 and tumour necrosis factor (TNF) which are generated by local cells, particularly macrophages, during the initiation of inflammation. There has been no detailed *in vivo* research on changes of adhesion molecule expression during the termination of neutrophil emigration, but in experimental arthritis it is clear that the inflamed site will permit a further wave of neutrophil emigration in response to a second inflammatory stimulus (Haslett *et al.*, 1989a). Therefore any "barriers" to cell adhesion or transmigration existing at the time of cessation of neutrophil emigration must be readily reversible, presumably by the further action of newly generated inflammatory cytokines which in time induce renewed expression/activation of endothelial surface adhesive molecules and exert parallel effects on neutrophil locomotion and the expression/activation of neutrophil surface adhesive molecules. Thus it is clear that the detailed identification of mechanisms controlling the local generation and dissipation of agents which promote chemotaxis and up-regulation and activation of adhesive molecules is essential for our understanding of the processes of termination or persistence of neutrophil emigration at inflamed sites.

Much less is understood of the control of eosinophil and monocyte emigration *in situ*, although similar principles would be applicable to the identification of mechanisms involved in the cessation of their emigration.

4. Restoration of Normal Microvascular Permeability

In some experimental models of lung inflammation there may be no detectable leak of plasma proteins from the microvessels, and from classical ultrastructural studies it is apparent that neutrophil migration to inflamed sites is not necessarily associated with overt endothelial or epithelial injury (Majno and Palade, 1961). Nevertheless, in "real" acute inflammation, such as experimental pneumococcal pneumonia (Larsen *et al.*, 1980), there is clear morphological evidence of endothelial injury ranging from cytoplasmic vacuolation to areas of complete denudation and fluid leakage into alveolar spaces. However, the sheets of endothelial and epithelial cells must retain the capacity for complete repair as the pneumonia resolves. Since many inflammatory diseases, e.g. the adult respiratory distress syndrome (ARDS), are characterized by severe and persistent endothelial and epithelial injury and there is evidence of at least a degree of inevitable endothelial and epithelial injury in examples of "beneficial inflammation", this may represent a pivotal point at which the loss of the normal controls of tissue injury and repair might represent a major mechanism in the development of inflammatory disease. Although the underlying processes are poorly understood, repair is likely to occur by a combination of local cell proliferation to bridge gaps and the recovery of some cells from sublethal injury. Little is known of how endothelial cells recover from sublethal injury, but epithelial cells (Parsons *et al.*, 1987) *in vitro* appear to be able to recover from hydrogen peroxide-induced injury by a mechanism which requires new protein synthesis. Such cytoprotective mechanisms have received little study. Similarly it is known that endothelial monolayers, deliberately "wounded" *in vitro*, have a remarkable capacity to reform, yet little is known of the underlying mechanisms (Haudenschild and Schwartz, 1979).

5. Control of Inflammatory Cell Secretion

Tight control, and ultimately cessation of neutrophil secretion of granule enzymes is likely to be important in the limitation of inflammatory tissue injury and is necessary for resolution to occur. Although there has been much recent study of the initiation and up-regulation of phagocyte secretion *in vitro* (Henson *et al.*, 1988), little is understood of how secretion is down-regulated or how these processes are controlled *in vivo*. Phagocyte secretion *in situ* is likely to be modulated by the balance between stimulatory and inhibitory mediators. The simplest mechanism for termination of secretion, i.e. the cell exhausting its secretory potential, is unlikely since cells removed from inflamed sites retain significant residual capacity for further secretion upon stimulation *ex vivo* (Zimmerli *et al.*, 1986). Other factors which may contribute to down-regulation or termination of secretion are the exhaustion of internal energy supplies, receptor down-regulation, dissipation of stimuli and finally death or removal of the cell itself.

In a short-lived, terminally differentiated cell like the neutrophil granulocyte, which normally has a blood half-life of about 6 h, the ultimate demise of the cell could itself represent an important mechanism in the irreversible down-regulation of its secretory function. We have recently observed that ageing neutrophils and eosinophils undergo programmed cell death or apoptosis (see below). During apoptosis the neutrophil retains its granule enzyme and membrane function, including the ability to exclude vital dyes, but loses the ability to secrete granule contents in response to external stimulation with inflammatory mediators (Whyte *et al.*, 1993a). The apoptotic neutrophil undergoes surface changes by which it becomes recognized as "senescent self" by inflammatory macrophages which phagocytose the intact senescent cell. Apoptosis therefore provides a mechanism which renders the neutrophil inert and functionally isolated from inflammatory mediators in its microenvironment, thus greatly limiting the destructive potential of the neutrophil before it is removed by local phagocytes.

6. The Clearance Phase of Inflammation

Once extravasated inflammatory cells have completed their defence tasks for the host, and inciting agents, e.g. bacteria, have been removed, the site must then be cleared of fluid, proteins, antibodies and debris. Finally, the key cellular players of inflammation – granulocytes and inflammatory macrophages – must be removed before the tissues can return to normality.

6.1 CLEARANCE OF FLUID, PROTEINS AND DEBRIS

Most fluid is probably removed via the lymphatic vessels, although reconstitution of normal haemodynamics may contribute by restoring the balance of hydrostatic and osmotic forces in favour of net fluid absorption at the venous end of the capillary. Proteolytic enzymes in plasma exudate and inflammatory cell secretions are likely to break down any fibrin clot at the inflamed site, and products of this digestion are likely to be drained by the lymphatics which become widely distended as the removal of fluid and proteins increases.

The macrophage may also play a role in this phase. It can remove fluids (which might contain a variety of proteins) by pinocytosis. In activated inflammatory macrophages, pinocytosis can occur at a rate such that 25% of the cell surface is reused each minute (Steinmann et al., 1976). Inflammatory macrophages also possess a greatly increased phagocytic potential. They can recognize opsonized and non-opsonized particles and they express cell surface receptors for a wide variety of altered and damaged cells and proteins. The critical role of macrophages in the clearance phase of inflammation was first recognized by Metchnikoff more than a century ago, and we are now just beginning to elucidate the molecular mechanisms of some of his seminal light microscopical observations.

6.2 THE CLEARANCE OF EXTRAVASATED GRANULOCYTES

Although for some time we have been aware of the histotoxic potential of a wide variety of neutrophil contents, the fate of this cell *in situ* has not received much attention until very recently. There is no evidence that extravasated neutrophils return to the bloodstream or that lymphatic drainage provides an important disposal route, and it is generally agreed that the bulk of neutrophils meet their fate at the inflamed site. It was widely assumed that the majority of neutrophils as a rule disintegrate at the inflamed site before their fragments are removed by local macrophages (Hurley, 1983). However, if this was the rule, healthy tissues would inevitably be exposed to large quantities of potentially damaging neutrophil contents. Although a number of pathological descriptions have favoured neutrophil necrosis as a major mechanism operating in inflammation, many of these examples have derived from histological observations on diseased tissues rather than from examples of more "benign", self-limited inflammation. Furthermore, since the classical observations of Elias Metchnikoff (1891) there has been evidence for over a century of an alternative fate for extravasated neutrophils. Metchnikoff was the first to catalogue the cellular events of the evolution and resolution of acute inflammation in vital preparations. Rather than neutrophil necrosis as the major mechanism, he described an alternative process whereby intact senescent neutrophils were removed by local macrophages.

Over the ensuing decades there have been a number of sporadic reports, in both health and disease, of macrophage phagocytosis of neutrophils, and of particular relevance to the resolution of inflammation is the clinical phenomenon of "Reiter's cells" – neutrophil-containing macrophages which have been described in cytology of synovial fluid from the inflamed joints of patients with Reiter's disease and other forms of acute arthritis (Pekin et al., 1967; Spriggs et al., 1978). In experimental peritonitis, where it is possible to sample the inflammatory exudate with ease, it appears that macrophage ingestion of apparently intact neutrophils is the dominant mode of neutrophil removal from the inflamed site (Chapes and Haskill, 1983). The mechanisms underlying these *in vivo* observations have only recently been addressed *in vitro*. Newman et al. (1982) showed that human neutrophils harvested from peripheral blood and "aged" overnight were recognized and ingested by inflammatory macrophages (but not by monocytes) whereas freshly isolated neutrophils were not ingested. This suggested that, during ageing, a time-related process must have been associated with changes in the neutrophil surface leading to its recognition as "non-self" or "senescent self". The development of improved methods for harvesting and culturing human neutrophils with minimal activation and avoiding cell losses caused by aggregation and clumping allowed us to study in detail the changes occurring in cultured neutrophils. We have found that ageing granulocytes constitutively undergo apoptosis (programmed cell death) and that this process is responsible for the recognition of intact senescent neutrophils by macrophages (Savill et al., 1989a).

This is an appropriate point at which to consider what is known of the processes of apoptosis and necrosis in other cellular systems, how this terminology evolved, and the possible relevance of these alternative neutrophil fates for our understanding of the control of inflammation.

6.3 NECROSIS VERSUS APOPTOSIS

From the work of Wyllie and his colleagues it is now recognized that the death of nucleated cells can be classified into at least two distinct types: necrosis or accidental death, and apoptosis (programmed cell death; Kerr et al., 1972; Wyllie et al., 1980).

Necrosis can be observed where tissues are exposed to gross insults such as high concentrations of toxins or hypoxia. It is characterized by rapid loss of membrane function and abnormal permeability of the cell membrane which can be recognized by failure to exclude vital dyes such as trypan blue. There is early disruption of organelles including liposomal disintegration and irreversible damage to mitochondria. The stimuli inducing necrosis usually affect large numbers of contiguous cells and the resultant widespread release of liposomal contents may obviously be associated with local tissue injury and the initiation or amplification of a local inflammatory response. By contrast, apoptosis occurs in situations where death is controlled, or physiological, such as the removal of unwanted cells during embryological remodelling, or where cell turnover is rapid, e.g. crypt cells in the gut epithelium. Recognizing the widespread importance of this process in tissue kinetics, Wyllie and his colleagues named it "apoptosis", meaning the "the falling off, as of leaves from a tree" in ancient Greek. This had an appealing analogy with leaf fall during Autumn; a carefully programmed and regulated event in which the loss of individual leaves occurs in a random fashion and where the overall process is not detrimental to the host.

In the many physiological and pathophysiological situations where apoptosis is now recognized, the process occurs with remarkably reproducible structural changes, implying a common underlying series of molecular mechanisms (Wyllie, 1981). During apoptosis, cells shrink and there are major changes in the cell surface, which becomes featureless with the loss of microvilli and with the development of deep invaginations in the surface. However, the membrane remains intact and continues to exclude vital dyes, and organelles such as mitochondria and lysosomes remain intact until very late in the process, although the endoplasmic reticulum appears to undergo characteristic, marked dilatation which, on light microscopy, may give the appearance of vacuoles in the cytoplasm. The ultrastructural changes in

the nucleus are most characteristic, with condensation of chromatin into dense crescent-shaped aggregates and prominence of the nucleolus (Fig. 12.1). Apoptotic cells are very swiftly ingested by phagocytes *in vivo* such that in tissue sections of the remodelling embryo apoptotic cells are usually seen contained within other cells. Usually macrophages are responsible for their ingestion, but other "semi-professional" phagocytes, e.g. epithelia and fibroblast-like cells, can also participate. The speed and efficiency of clearance of cells undergoing apoptosis together with the fact that it occurs in cells at random within the population renders this mode of cell death much less conspicuous than necrosis in histological sections. It is remarkable that in embryological remodelling or during thymus involution whole tracts of tissue can be removed by this process over a few hours without causing local tissue injury or inciting an inflammatory response.

A prominent biochemical feature of apoptosis is internucleosomal cleavage of chromatin in a pattern indicative of endogenous endonuclease activation (Wyllie, 1981). This creates low molecular-sized fragments of chromatin which are integers of the 180 base pairs of DNA associated with a nucleosome. When DNA extracted from apoptotic cells is subjected to agarose gel elec-

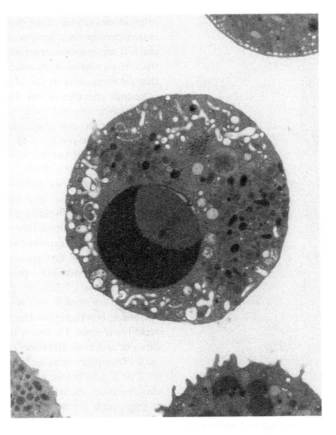

Figure 12.1 Electron micrograph of an apoptotic human neutrophil showing the characteristic chromatin aggregation, prominent nucleolus and dilated cytoplasmic vacuoles. Note that the cell membrane is intact and the granule structure appears normal. (EM taken by Jan Henson). (× 11 000).

trophoresis this results in a characteristic "ladder" pattern of DNA fragments (Fig. 12.2). Over the past decade a number of laboratories have been pursuing the endonuclease(s) responsible for this particular feature of apoptosis. Activities have been isolated which cleave DNA in a characteristic fashion. Final characterization has remained elusive and while a number of candidates have been put forward (Shi *et al.*, 1992; Pietsch *et al.*, 1993), there is as yet no agreement on the molecular nature of the endonuclease concerned. Recent work in our laboratory (Nunn *et al.*, 1995) suggests that the neutrophil possesses a single endonuclease which has the biochemical characteristics of DNAase II.

It is now recognized that the process of apoptosis can be influenced by external mediators which are to some extent cell lineage restricted, but can also be modulated by internal genetic influences. However, the biochemical processes integrating the external influences, the genetic influences and how they relate to the nuclease responsible for chromatin cleavage and the cell surface changes

responsible for cell removal represent major "black boxes".

6.4   APOPTOSIS IN AGEING GRANULOCYTES LEADS TO THEIR PHAGOCYTOSIS BY MACROPHAGES

Neutrophils harvested from blood or from acutely inflamed human joints remain intact and continue to exclude trypan blue for up to 24 h "ageing" in culture (Savill *et al.*, 1989a). With time in culture there is a progressive increase in the proportion of cells exhibiting the light microscopical features of apoptosis, confirmed by electron microscopy and by the chromatin cleavage ladder pattern indicative of endogenous endonuclease activation (Savill *et al.*, 1989a). Macrophage recognition was directly related to the apoptotic cells in the ageing neutrophil population (Savill *et al.*, 1989a). Neutrophil apoptosis is obvious at 2–3 h in culture and by 10 h up to 50% of the cells would be expected to show apoptotic morphology. By 24 h in culture the majority of neutrophils display features of apoptosis, but there is minimal (usually less than 3%) evidence of cell necrosis or spontaneous release of the granule enzyme markers such as myeloperoxidase (Savill *et al.*, 1989a). Apoptotic neutrophils are not indestructible. Beyond 24 h in culture there is a progressive increase in the percentage of cells that fail to exclude trypan blue and spontaneous release of granule enzymes occurs. However, when neutrophils are cultured beyond 24 h in the presence of macrophages, the removal of apoptotic cells is so effective that no trypan blue-positive cells are seen and there is no release of granule enzyme markers into the surrounding medium (Kar *et al.*, 1995).

We have recently carried out similar experiments with human eosinophilic granulocytes purified from the peripheral blood of healthy individuals. In the absence of external stimuli, cultured eosinophils undergo apoptosis at a much slower apparent constitutive rate than neutrophils, with the first apoptotic cells being obvious at about 72 h and maximum percentage apoptosis by about 96 h (Stern *et al.*, 1992). Up to 96 h there is minimal evidence of eosinophil necrosis assessed by trypan blue exclusion, but thereafter the percentage of necrotic cells steadily increases. However, the time difference between onset of apoptosis and onset of necrosis suggests that the bulk of apoptotic eosinophils can exist in culture for up to 2 or 3 days before undergoing necrosis and ultimate disintegration. As in the neutrophil, eosinophil apoptosis is responsible for macrophage recognition of the intact senescent eosinophil (Stern *et al.*, 1992).

The speed with which monocyte-derived macrophages recognize, ingest and destroy apoptotic neutrophils *in vitro* is quite remarkable. Individual macrophages can ingest several neutrophils (see Fig. 12.3), and once

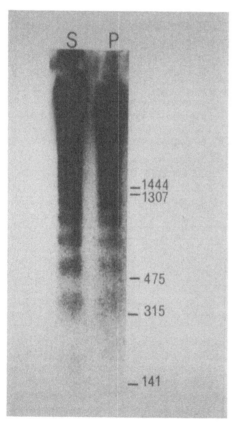

Figure 12.2   The characteristic "ladder" pattern on agarose gel electophoresis of DNA extracted from apoptotic neutrophils. This cleavage of fragments which are integers of 180 kb is indicative of internucleosomal cleavage.

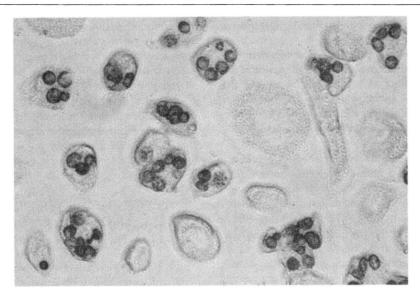

Figure 12.3   Macrophages which *in vitro* have ingested large numbers of neutrophils.

ingested there is extremely rapid degradation of the neutrophils such that in electron microscopic studies it is necessary to fix macrophages within minutes of the initial interaction between apoptotic cells and the macrophage in order to demonstrate recognizable neutrophils within the phagocytes. Thereafter, ingested cells are no longer recognizable. This may represent part of the explanation why the dynamic contribution of this process to cell and tissue kinetics has not been fully appreciated until recently. However, there are now several examples demonstrating clear histological evidence of a role for apoptosis in the *in vivo* removal of granulocytes in acute inflammation. These include acute arthritides (Savill *et al.*, 1989a), neonatal acute lung injury (Grigg *et al.*, 1991) and experimental pneumococcal pneumonia during its resolution phase (Fig. 12.4). Histological evidence of eosinophil apoptosis and ingestion by local macrophages has also been described in dexamethasone-

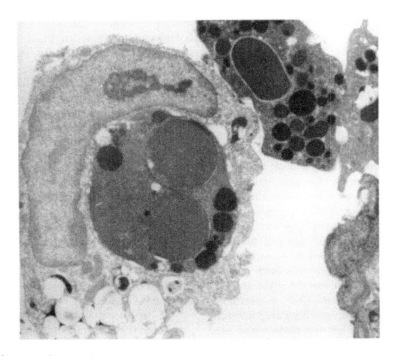

Figure 12.4   An electron micrograph of resolving experimental streptococcal pneumonia showing a macrophage which contains an apoptotic neutrophil.

treated experimental eosinophilic jejunitis (Meagher *et al.*, 1992).

Given the pro-inflammatory potential of neutrophils and their contents, there are now several lines of *in vitro* experimental evidence to support the hypothesis that apoptosis provides an injury-limiting neutrophil clearance mechanism in tissues which would tend to promote resolution rather than persistence of inflammation.

(1) During the process of neutrophil apoptosis there is marked loss of a number of neutrophil functions, including chemotaxis, superoxide production and stimulated granule enzyme secretion. These data suggest that apoptosis may lead to "shutting off" of neutrophil functions resulting in it becoming functionally isolated from external stimuli which would otherwise trigger responses which could damage tissue (Whyte *et al.*, 1993a). This mechanism could be important if fully mature and competent phagocytes are not immediately available in the vicinity of the neutrophil undergoing apoptosis.

(2) Neutrophils undergoing apoptosis are very rapidly phagocytosed by macrophages, and in these model systems apoptotic neutrophils retain their enzyme contents and are ingested while still intact, thus preventing the leakage of granule enzymes which would occur should the cell disintegrate before or during uptake by macrophage. This is emphasized by preliminary *in vitro* data of a simple model in which macrophages and neutrophils are co-cultured. If macrophage uptake of apoptotic neutrophils is blocked (with colchicine for example; Kar *et al.*, 1995), rather than being ingested the apoptotic cells then disintegrate and release toxic contents such as myeloperoxidase and elastase before their cellular fragments are taken up by macrophages.

(3) The usual response of macrophages to the ingestion of particles *in vitro* is to release pro-inflammatory mediators such as thromboxane, enzymes and pro-inflammatory cytokines. However, it has been found that even maximal uptake of apoptotic neutrophils failed to stimulate the release of pro-inflammatory mediators (Meagher *et al.*, 1992). This was not simply the result of a toxic or inhibitory effect of the apoptotic neutrophil on the macrophage since phagocytes which had ingested apoptotic cells were able to generate maximum release of potential mediators when subsequently stimulated by opsonized zymozan. Moreover, when apoptotic neutrophils were deliberately opsonized prior to ingestion, macrophages did respond by the release of thromboxane (Meagher *et al.*, 1992). Furthermore, when granulocytes are cultured beyond apoptosis to a point at which they fail to exclude trypan blue their ingestion by macrophages induces the release of pro-inflammatory mediators. From these experiments it was concluded that it is recognition of the senescent granulocyte in the apoptotic morphology rather than the necrotic morphology which determines the lack of macrophage pro-inflammatory response, and secondly, this lack of macrophage response is not a function of the apoptotic particle itself, but relates to the mechanism by which the apoptotic cell is normally ingested. These observations provided considerable impetus for our work on the molecular mechanisms by which macrophages recognize and ingest apoptotic cells.

6.5 MECHANISMS WHEREBY MACROPHAGES RECOGNIZE APOPTOTIC NEUTROPHILS

Early work by Duval *et al.* (1985) using various sugars to inhibit the interactions between macrophages had suggested that phagocytes possess a lectin mechanism capable of recognizing sugar residues on the apoptotic thymocyte surface exposed by loss of sialic acid. This mechanism does not appear to be involved in the macrophage recognition of apoptotic granulocytes but these findings stimulated our early work demonstrating that recognition of apoptotic neutrophils occurred by a "charge-sensitive" mechanism, inhibitable by cationic molecules such as amino sugars and amino acids and directly influenced by minor changes of pH, in a fashion suggesting involvement of negatively charged residues on the apoptotic neutrophil surface (Savill *et al.*, 1989b). As well as drawing attention to analogous recognition systems inhibited by amino sugars (which ultimately turned out to be of relevance to macrophage/apoptotic cell recognition), these data implied that low pH and the presence of cationic molecules might be expected to adversely influence the clearance of apoptotic cells at inflamed sites. This was of particular interest, since a number of granulocyte-derived products such as elastase, myeloperoxidase and eosinophil-derived major basic protein from eosinophils are known to be highly cationic and have been detected in significant amounts in the tissues. Furthermore, in situations where inflammation is chronic or where there is abscess formation, interstitial pH may be very low.

The amino sugar inhibition pattern suggested two new lines of inquiry which led to the definition of macrophage cell surface molecules involved in the clearance of apoptotic neutrophils.

(1) Since amino sugars are known to inhibit the functions of certain members of the integrin family, this led to a detailed series of investigations using a range of monoclonal antibodies directed against candidate integrins in the β_2 and β_3 family leading to the implication of macrophage surface α v β_3 in the recognition of apoptotic neutrophils and also apoptotic lymphocytes (Savill *et al.*, 1990).

(2) This amino sugar inhibition pattern was previously described in platelet–platelet interactions occurring via thrombospondin and thrombospondin receptors on their surfaces. This led to work which now implicates CD36 on the macrophage surface (Savill *et al.*, 1992a), and it is thought that thrombospondin itself may serve as an intracellular bridging molecule between the macrophage and the apoptotic cell surface. The present model of this recognition mechanism so far is depicted in Fig. 12.5. The moiety on the surface of the apoptotic cell which is responsible for apoptotic cell recognition in this system has not yet been identified.

Recent studies from others have focused on the putative ionic sites on the apoptotic cell surface that might be involved in macrophage recognition. Initially working in a system utilizing murine macrophage recognition of murine thymocytes which had been induced to undergo apoptosis with glucocorticoids, Fadok *et al.* (1992a) showed that (as yet uncharacterized) receptors on macrophages can recognize exposed phosphatidlyserine residues on the surface of apoptotic cells. These normally reside on the inner leaflet of the membrane lipid bilayer, but during apoptosis it is hypothesized that there is "flipping" of this layer in a fashion which may be analogous to that occurring during the sickling of erythrocytes. It appears that the main difference between these two recognition systems relates to the utilization of alternative recognition mechanisms by different subpopulations of macrophages (Fadok *et al.*, 1992b). The *in vivo* significance of these observations is as yet uncertain.

The definition of cell surface molecules involved in macrophage uptake of apoptotic neutrophils suggests mechanisms by which this function might be regulated. We have recently found that agents which modulate cAMP can greatly influence the rate of apoptotic neutrophil ingestion by macrophages, possibly through

clustering of $\alpha \ v \ \beta_3$. Moreover, a number of cytokines, including granulocyte–macrophage colony-stimulating factor (GM-CSF), IL-1β, TNFα, and interferon γ (IFNγ), promote macrophage uptake of apoptotic neutrophils (Ren and Savill, 1995).

6.6 CLEARANCE OF APOPTOTIC GRANULOCYTES BY CELLS OTHER THAN MACROPHAGES

In embryonic remodelling and in thymus involution apoptotic cells are usually taken up by local macrophages, but they may also be seen within epithelial cells or fibroblast-like cells. We therefore compared the ability of monolayers of fibroblasts, endothelial cells and epithelial cells from a variety of sources to recognize apoptotic neutrophils *in vitro*. In these experiments only the fibroblast appeared to recognize and ingest apoptotic neutrophils (Hall *et al.*, 1990). The fibroblast has long been recognized as a "semi-professional" phagocyte capable of ingesting latex beads, dye particles and mast cell granules. The significance of fibroblast phagocytosis of senescent neutrophils is uncertain, but the fibroblast appears to employ recognition mechanisms differing from the macrophage in that fibroblasts appear to utilize a sugar–lectin recognition mechanism in addition to the integrin mechanism described in the macrophage–neutrophil system (Hall *et al.*, 1994). More recently, Savill *et al.* (1992b) have shown that renal mesangial cells, also recognized as "semi-professional" phagocytes, have the capacity to take up large numbers of apoptotic neutrophils.

The significance of these observations is uncertain. It is possible that uptake of apoptotic neutrophils by resident cells, including fibroblasts, serves as a clearance mechanism before extravasated monocytes have fully matured into inflammatory macrophages capable of recognizing and ingesting apoptotic cells. Alternatively, it may

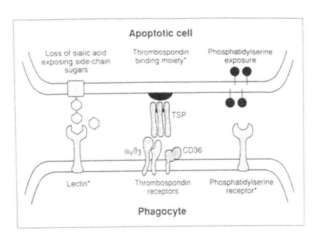

Figure 12.5 A model of possible surface mechanisms by which macrophages recognize apoptotic cells. (Reproduced from Savill et al., 1993.)

represent a "back up" mechanism should the macrophage disposal mechanism be overwhelmed by waves of neutrophil apoptosis. However, since the fibroblast is responsible for scar tissue matrix protein secretion, it is possible that this clearance route is an "undesirable" alternative, particularly if the uptake of apoptotic neutrophils should cause fibroblast replication and secretion of collagen.

6.7   REGULATION OF GRANULOCYTE APOPTOSIS BY EXTERNAL MEDIATORS – A CONTROL POINT FOR GRANULOCYTE TISSUE LONGEVITY?

Recent histological observations of resolving pulmonary inflammation suggested that extravasated neutrophils undergo apoptosis at a slower rate than neutrophils derived from peripheral blood. In experimental pneumonia, at 48 h large numbers of neutrophils without any significant evidence of apoptosis are seen. The use of radiolabelled pulses of neutrophils delivered intravenously during the evolution of this model suggested that neutrophil emigration from the blood to the inflamed lung had largely ceased by 16 h (Plate IV). This implied that the bulk of neutrophils observed at 48 h had been present for at least 24 h, yet the half-life of neutrophils in blood is about 4–5 h. These observations suggested that factors present at the inflamed site might have delayed the constitutive rate of neutrophil apoptosis. We have now shown that the rate neutrophil apoptosis *in vitro* is inhibited by a variety of inflammatory mediators (Lee *et al.*, 1993) including endotoxic lipopolysaccharide, C5a and GM-CSF (which is particularly potent at inhibiting the rate of neutrophil apoptosis). If, as seems likely, apoptosis, by leading to macrophage removal of unwanted cells, controls the tissue longevity of neutrophils these might represent important mechanisms controlling the "tissue load" of inflammatory cells *in situ*. Experiments with eosinophils *in vitro* show that GM-CSF inhibits eosinophil apoptosis, but IL-5 is also extremely potent in this regard, whereas it has no effect on neutrophil longevity.

Intracellular mechanisms governing apoptosis are as yet poorly understood. However, there are indications that internal controls in granulocytes may differ from lymphoid cells. In thymocytes, elevation of intracellular calcium concentration ($[Ca^{2+}]_i$) by calcium ionophores induces apoptosis, and apoptosis induced by other stimuli, such as glucocorticoids, is associated with rises in $[Ca^{2+}]_i$ (McConkey *et al.*, 1989). However, in neutrophils spontaneously undergoing apoptosis, there were no such rises in $[Ca^{2+}]_i$, and agents increasing $[Ca^{2+}]_i$ caused dramatic slowing of neutrophil apoptosis without

inducing necrosis (Whyte *et al.*, 1993b). Furthermore, treatment of ageing neutrophils with MAPTAM and BAPTA, which bind intracellular calcium, is associated with an increase in the rate of neutrophil apoptosis (Whyte *et al.*, 1993b). Rises in intracellular calcium are known to occur when neutrophils are primed or activated *in vitro*, and this is likely to represent at least part of the mechanism underlying the retardation of neutrophil apoptosis observed after external stimulation with inflammatory mediators. Further, when neutrophils are aged in the presence of inhibitors of protein synthesis, e.g. cycloheximide, there is an acceleration of the constitutive rate of apoptosis. This is in marked contrast to corticosteroid-induced thymocyte apoptosis which is inhibited by cycloheximide. These observations suggest the existence of a protein synthesis-dependent, apoptosis inhibitory factor in neutrophils. It is hypothesized that external inflammatory mediators inhibit the rate of neutrophil apoptosis by causing a rise in intracellular calcium which subsequently acts on downstream processes, including mRNA and protein synthesis of this putative inhibitory factor(s).

It is now clear that apoptosis in a variety of cell types can be influenced by proto-oncogene expression. The best worked out system is *bcl*-2 (Vaux *et al.*, 1992; Fanidi *et al.*, 1992), the expression of which is associated with inhibition of apoptosis in a variety of cell types, but its significance in the neutrophil is uncertain. *c-myc* expression in fibroblasts and lymphoid cells appears to induce apoptosis (Evan *et al.*, 1992), and more recently p53 and Rb-1 have also been implicated. It remains to be established whether the products of these genes are relevant for the control of inflammatory cell apoptosis, and it is presently unclear how the genetic controls are linked with second messenger controls and the final effector events including endonuclease activation, down-regulation of cell function and cell surface changes responsible for phagocyte recognition.

Clearly, we are only just beginning to scratch at the surface of the internal mechanisms of apoptosis, but ultimately it may be possible to regulate granulocyte apoptosis for therapeutic benefit. It is intriguing that two such closely related cells as the neutrophil and eosinophil granulocyte should appear to possess different inherent rates of apoptosis which are further influenced by different external mediators. Furthermore, recent experiments show that apoptosis in neutrophils is influenced very differently by corticosteroids, as compared with eosinophils (Meagher *et al.*, 1995). These observations suggest that in the future it may be possible to specifically induce eosinophil apoptosis without influencing neutrophil apoptosis. This would presumably result in the tissue clearance of eosinophils by the non-phlogistic mechanisms "which nature intended" and yet leave other inflammatory cells available for host defence purposes.

6.8   A ROLE FOR GRANULOCYTE APOPTOSIS IN THE CONTROL OF INFLAMMATION?

Granulocyte apoptosis could play at least two important roles in the control of inflammation. By controlling the functional longevity and tissue removal of unwanted granulocytes and providing a pivotal point at which inflammatory cytokines and growth factors exert their controls on inflammatory cell longevity, it is likely that (together with neutrophil influx) neutrophil apoptosis is a critical determinant of the overall tissue load of inflammatory cells (see Fig. 12.6). Secondly, it is reasonable to suggest that whether neutrophil fate occurs by apoptosis or by necrosis is an important factor in the control of inflammation. Inflammatory disease is widely considered to result from a quantitative imbalance between potentially damaging inflammatory influences and their tissue protective mechanisms, as exemplified by the proteinase/antiproteinase theory of emphysema. Whether neutrophils meet their fate by a mechanism which involves removal of the whole cell or whether they meet their fate via a mechanism which results in disintegration and disgorgement of their potentially histotoxic and pro-inflammatory contents is likely to impinge on such an equation. We have given detailed consideration to the observations that apoptosis may serve to keep potentially injurious granule contents within the cell membrane while at the same time the cell becomes unable to respond by degranulation in response to external stimuli and finally, the intact cell is removed by a novel phagocytic recognition mechanism which determines that macrophages fail to release pro-inflammatory mediators during macrophage recognition and phagocytosis. This is not to suggest that apoptosis is the only mechanism for removal of neutrophils at an inflamed site. While in all the spontaneously resolving examples of inflammation we have examined in humans and in experimental models, the removal of whole granulocytes by apoptosis appears to be a major mechanism, examples of neutrophil necrosis are also seen, particularly in some inflammatory diseases such as systemic vasculitis, where light microscopical features of neutrophil necrosis and disintegration have been described in tissues close to inflamed vessels and called "leucocytoclastic vasculitis". It is possible therefore that the balance between the degree of neutrophil apoptosis and the degree of neutrophil necrosis at an inflamed site may represent a pivotal point in the control of tissue injury and in the propensity of an inflamed site to resolve or to progress.

At the present time the proper *in vivo* validation of this hypothesis presents a number of problems. It is difficult

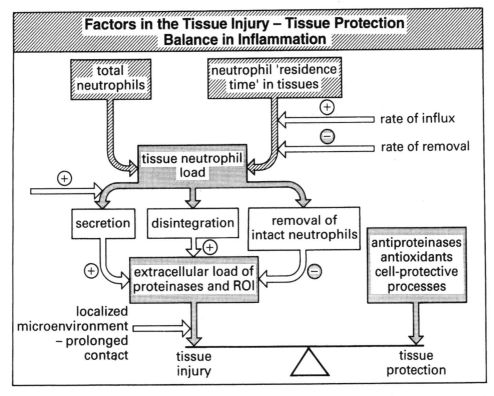

Figure 12.6   The "balance" between the capacity of inflammatory cells to injure tissues and tissue defence mechanisms including pivotal points at which factors involved in the emigration and clearance of granulocytes may impinge on the equation.

to make even semi-quantitative estimates of the degree of apoptosis and necrosis at inflamed sites, since we do not have specific surface markers to enable us to develop tools for quantifying with confidence the rates of these processes *in situ*. This difficulty is compounded by the rapidity with which apoptotic cells are known to be recognized and ingested *in vitro* and degraded to the point at which the cell of origin is no longer easily recognizable. Moreover, at the present time we can only speculate what the actual consequences of neutrophil disintegration *in situ* would be. Attention has been drawn to the enormous histotoxic potential of a wide variety of granulocyte contents which would inevitably by disgorged during granulocyte necrosis. It is also recognized that there are tissue defence mechanisms including antioxidants and antiproteinases which normally serve to protect tissues from the effects of these toxic agents. However, there is also evidence that these tissue protective mechanisms can be overcome, e.g. in the intercellular microenvironment between cells adherent to matrix or other cells, wherein very high localized concentrations of the damaging agent can be generated. Furthermore, potent antiproteinases can be rendered ineffective in the presence of oxidants generated locally by inflammatory cells. Moreover, a variety of toxic neutrophil and eosinophil contents do not have obvious tissue inhibitors or regulators, such as the non-enzymatic, cationic antimicrobial proteins of neutrophils and the major basic protein and eosinophil cationic protein content of eosinophils. Neutrophil enzymes can cleave chemotactic fragments from complement and from matrix proteins such as fibronectin, while the granule protein CAP37 also has monocyte-specific chemoattractant properties. Thus neutrophil contents have the capacity to recruit more leucocytes to an inflamed site, amplifying and prolonging the response.

Therefore, it is reasonable to hypothesize that granulocyte necrosis at inflamed sites can be regarded as a deleterious and undesirable mode of clearance which, by contrast with apoptosis, is likely to favour persistence of inflammatory tissue injury rather than resolution of inflammation.

7.   The Fate of Macrophages

Although monocytes appear to undergo apoptosis spontaneously *in vitro* (Mangan and Wahl, 1991), as they mature into macrophages they have a very low constitutive rate of apoptosis (Bellingan *et al.*, 1994), but can be induced to undergo apoptosis with a number of toxic stimuli (Bellingan *et al.*, 1995a) including cycloheximide (although higher concentrations of cycloheximide are required to induce macrophage apoptosis compared with those which promote neutrophil apoptosis). Furthermore, during the resolution of experimental streptococcal pneumonia, whilst there is clear evidence of

neutrophil apoptosis and ingestion by macrophages, there is no histological clue as to the local fate of inflammatory macrophages (Haslett, unpublished observations). These observations suggest that macrophages might leave the inflamed tissue and meet their fate at a distal site. We have recently carried out studies of the fate of inflammatory macrophages during the resolution of experimental murine peritonitis and have found that inflammatory macrophages have a short tissue resistance time, compared with resident macrophages, and that they exit the inflamed peritoneum, to be cleared in the draining lymph nodes (Bellingan *et al.*, 1995b). Whether their final demise involves apoptosis and clearance by specialized local phagocytes is as yet uncertain.

7.1   RESOLUTION MECHANISMS AND THE CONTROL OF INFLAMMATION

On preliminary evidence a speculative scheme can be proposed whereby a stereotyped sequence of inflammatory events results in the resolution of inflammation. Injury to endothelial and epithelial cell "barriers" is minimized, but is associated with leak of fluid and protein; reconstitution of normal microvascular permeability occurs by the re-forming of cell junctions and the regeneration of cell layers; neutrophil influx ceases early in the evolution of acute inflammation and relates to cessation of local chemoattractant generation and their dissipation; monocytes mature into inflammatory macrophages to remove proteins and other debris; neutrophil secretion is restricted and the aged cell undergoes apoptosis which controls neutrophil longevity and determines the macrophage removal of the intact senescent cell without stimulating macrophage release of pro-inflammatory mediators; excess tissue macrophages leave the inflamed site and migrate to the local lymph nodes where they are removed by unknown processes.

There are several steps in the above scheme which could go awry and lead to circumstances favouring the development of persistent inflammation. For example, if the macrophage fails to develop the appropriate receptors for removing apoptotic cells, neutrophils would eventually become necrotic and disgorge their damaging contents, or alternatively they may be taken up by local fibroblasts, possibly with a pro-fibrotic response. Once a chronic inflammatory state begins to develop there is evidence that the local pH falls. This would tend to inhibit macrophage recognition of apoptotic neutrophils, as would the continued accumulation of inflammatory cell cationic products such as elastase, major basic protein etc., which would contribute to the inhibition of macrophage clearance of apoptotic neutrophils.

Hurley recognized that there were several mechanisms other than resolution whereby acute inflammation could terminate. These include chronic inflammation, scarring and abscess formation. By comparison with resolution, all of these termination events clearly must be regarded as detrimental to organ function. As we begin to learn more about the mechanisms involved in the resolution of inflammation it may be possible to "divert" inflammatory processes down resolution pathways rather than one of the other less desirable pathways. More specifically, with increasing knowledge of apoptosis and its internal mechanisms, it may be possible to use this "controlled" process of cell suicide or programmed cell death to remove specific inflammatory cells at particular pathogenetic stages when they are critical to the disease process. In this regard it is interesting that the two very closely related cells, the neutrophil and eosinophil granulocytes, appear to have different constitutive rates of apoptosis and be under the control of different external mediators. In the future it may be possible to utilize these differences to specifically induce eosinophil apoptosis, for example, thereby clearing these cells from tissues by the "mechanism" which nature intended to remove cells in a fashion which would neither injure local tissues nor promote inflammatory mechanisms.

8. Acknowledgments

This work was supported by the Medical Research Council of Great Britain in the form of a programme grant and a number of preceeding project grants and fellowships. Further support was obtained from the National Asthma Campaign and the British Lung Foundation. Thanks are due to close colleagues who were responsible for much of the work and who were a constant source of stimulation. These include John Savill in particular, Moira Whyte, Laura Meagher and Ian Dransfield. The manuscript was typed by Mrs J. McMahon.

9. References

Bellingan, G.J., Dransfield, I. and Haslett, C. (1994). Characterisation of apoptosis in the human macrophage. Clin. Sci.

Bellingan, G.J., Dransfield, I. and Haslett, C. (1995a). Resistance of the human macrophage to the induction of apoptosis; the role of the oncogenes bcl-2 and c-myc. In preparation.

Bellingan, G.J., Caldwell, H., Howie, S.E.M., Dransfield, I. and Haslett, C. (1995b). In vivo fate of the inflammatory macrophage during the resolution of inflammation: inflammatory macrophages do not die locally but emigrate to the draining lymph nodes. Submitted.

Berenberg, J.L. and Ward, P.A. (1973). Chemotactic factor of inactivator in normal human serum. J. Clin. Invest. 52, 1200–1207.

Chapes, S.K. and Haskill, S. (1983). Evidence for granulocyte-mediated macrophage activation after C. parvum immunization. Cell Immunol. 75, 367–377.

Clark, R.J., Jones, H.A., Rhodes, C.G. and Haslett, C. (1989). Non-invasive assessment in self-limited pulmonary inflammation by external scintigraphy of [111]Indium-labelled neutrophil influx and by measurement of the local metabolic response with positron emission tomography. Am. Rev. Respir. Dis. 139, A58.

Colditz, I.G. and Movat, H.Z. (1984). Desensitisation of acute inflammatory lesions to chemotaxins and endotoxin. J. Immunol. 133, 2163–2168.

Doherty, D.E., Downey, G.P., Worthen, G.S., Haslett, C. and Henson, P.M. (1988). Monocyte retention and migration in pulmonary inflammation. Lab. Invest. 59, 200–213.

Duvall, E., Wyllie, A.H. and Morris, R.G. (1985). Macrophage recognition of cells undergoing programmed cell death. Immunology 56, 351–358.

Evan, G.I., Wyllie, A.H., Gilbert, G.S., Littlewood, T.D., Land, H., Brooks, M., Waters, C.M., Penn, L.Z. and Hancock, D.C. (1992). Induction of apoptosis in fibroblasts by c-myc protein. Cell 69, 119–128.

Fadok, V.A., Voelker, D.R., Campbell, P.A., Cohen, J.J., Bratton, D.L. and Henson, P.M. (1992a). Exposure of phosphatidyserine on the surface of apoptotic lymphocytes triggers specific recognition and removal by macrophages. J. Immunol. 148, 2207–2216.

Fadok, V., Savill, J.S., Haslett, C., Bratton, D.L., Doherty, D.E., Campbell, P.A. and Henson, P.M. (1992b). Different populations of macrophages use either the vitronectin receptor or the phosphatidylserine receptor to recognise and remove apoptotic cells. J. Immunol. 149, 4029–4035.

Fanidi, A., Harrington, E.A. and Evan, G.I. (1992). Cooperative interaction between c-myc and bcl-2 proto-oncogenes. Nature 359, 554–556.

Grigg, J.M., Savill, J.S., Sarraf, C., Haslett, C. and Silverman, M. Neutrophil apoptosis and clearance from neonatal lungs. Lancet 338, 720–722.

Hall, S.E., Savill, J.S. and Haslett, C. (1990). Fibroblast recognition of aged neutrophils is mediated by the RGD adhesion signal and is modulated by charged particles [Abstract]. Clin. Sci. 78, 17p.

Hall, S.E., Savill, J.S., Henson, P.M. and Haslett, C. (1994). Apoptotic neutrophils are phagocytosed by fibroblasts with participation of the fibroblast vitronectin receptor and involvement of a mannose/fucose-specific lectin. J. Immunol. 153, 3218–3227.

Haslett, C., Jose, P.J., Giclas, P.C., Williams, T.J. and Henson, P.M. (1989a). Cessation of neutrophil influx in C5a-induced acute experimental arthritis is associated with loss of chemoattractant activity from joint spaces. J. Immunol. 142, 3510–3517.

Haslett, C., Shen, A.S., Feldsien, D.C., Allen, D., Henson, P.M. and Cherniack, R.M. (1989b). [111]Indium-labelled neutrophil flux into the lungs of bleomycin-treated rabbits assessed non-invasively by external scintigraphy. Am. Rev. Respir. Dis. 140, 756–763.

Haudenschild, C.L. and Schwartz, S.M. (1979). Endothelial regeneration. II. Restitution of endothelial continuity. Lab. Invest. 41, 407–418.

Henson, P.M., Schwartzmann, N.A. and Zanolari, B. (1981). Intracellular control of human neutrophil secretion. II. Stimulus specificity of desensitisation induced by six different soluble and particulate stimuli. J. Immunol. 127, 754–759.

Henson, P.M., Henson, J.E., Fittschen, C., Kimani, G., Bratton, D.L. and Riches, D.H.W. (1988). Phagocytic cells: degranulation and secretion. In "Inflammation, Basic

Principles and Clinical Correlates" (ed. J.I. Gallin), pp. 363–390. Raven Press, New York.

Hurley, J.V. (1963). An electron microscopic study of leukocyte emigration and vascular permeability in rat skin. Aust. J. Exp. Biol. Med. Sci. 41, 171–179.

Hurley, J.V. (1983). Termination of acute inflammation. I. Resolution. In "Acute Inflammation," 2nd edn., pp. 109–117. Churchill Livingstone, London.

Kar, S., Ren, Y., Savill, J.S. and Haslett, C. (1995). Inhibition of macrophage phagocytosis in vitro of aged neutrophils increases release of neutrophil contents [Abstract]. Clin. Sci., in press.

Kerr, J.F.R., Wyllie, A.H. and Currie, A.R. (1972). Apoptosis: a basic biological phenomenon with wide-ranging implications in tissue kinetics. Br. J. Cancer 26, 239–257.

Larsen, G.L., McCarthy, K., Webster, R.O., Henson, J.E. and Henson, P.M. (1980). A differential effect of C5a and C5a des arg in the induction of pulmonary inflammation. Am. J. Pathol. 100, 179–192.

Lee, A., Whyte, M.B.K. and Haslett, C. (1993). Prolonged in vitro lifespan and functional longevity of neutrophils induced by inflammatory mediators acting through inhibition of apoptosis. J. Leuk. Biol. 54, 283–288.

Majno, E. and Palade, G.E. (1961). Studies on inflammation. I. The effect of histamine and serotonin on vascular permeability, an electron microscopic study. J. Biol. Phys. Biochem. Cytol. 11, 571–605.

Malech, H.D. and Gallin, J.I. (1988). Neutrophils in human diseases. N. Engl. Med. J. 37, 687–694.

Mangan, D.F. and Wahl, S.M. (1991). Differential regulation of human monocytes in programmed cell death (apoptosis) by chemotactic factors and pro-inflammatory cytokines. J. Immunol. 147, 3408–3412.

McConkey, D.J., Nicotera, P., Hartzell, P., Bellomo, G., Wyllie, A.H. and Orrenius, S. (1989). Glucocorticoids activate a suicide process in thymocytes through an elevation of cytosolic Ca^{2+} concentration. Arch. Biochem. Biophys. 269, 365–370.

McCutcheon, J., Haslett, C. and Dransfield, I. (1995). Regulation of macrophage recognition of apoptotic cells by activation of protein kinases, redistribution of the integrin α v β_3. Submitted.

Meagher, L.C., Savill, J.S., Baker, A., Fuller, R.W. and Haslett, C. (1992). Phagocytosis of apoptotic neutrophils does not induce macrophage release of thromboxane B_2. J. Leuk. Biol. 52, 269–273.

Meagher, L., Seckl, J.R. and Haslett, C. (1995). Opposing effects of glucocorticoids on neutrophil and eosinophil apoptosis. Submitted.

Metchnikoff, E. (1891). Lectures on the comparative pathology of inflammation. Lecture VII. Delivered at the Pasteur Institute in 1891 (transl. Starling, F.A. and Starling, E.H. (1968). Dover, New York).

Milks, L. and Cramer, E. (1984). Transepithelial electrical resistance studies during in vitro neutrophil migration. Fed. Proc. 43, 477.

Newman, S.L., Henson, J.E. and Henson, P.M. (1982). Phagocytosis of senescent neutrophils by human monocyte-derived macrophages and rabbit inflammatory macrophages. J. Exp. Med. 156, 430–442.

Nunn, A.V.W., Turner, N., Savill, J.S., Wyllie, A.H. and Haslett, C. (1995). A single acid nuclease in neutrophils with similarities to DNase II; a candidate apoptotic endonuclease. Submitted.

Parsons, P.E., Sugahara, K., Cott, G.R., Mason, R.J. and Henson, P.M. (1987). The effect of neutrophil migration and prolonged neutrophil contact on epithelial permeability. Am. J. Pathol. 129, 302–312.

Pekin, T., Malinin, T.I. and Zwaifler, R. (1967). Unusual synovial fluid findings in Reiter's syndrome. Ann. Intern. Med. 66, 677–684.

Pietsch, M.C., Polzar, B., Stephan, H., Crompton, T., MacDonald, H.R., Mannherz, H.G. and Tschopp, J. (1993). Characterisation of the endogenous deoxyribonuclease involved in nuclear DNA degradation during apoptosis (programmed cell death). EMBO J. 12, 371–377.

Ren, Y. and Savill, J.S. (1995). Increased macrophage secretion of thrombospondin is a mechanism by which cytokines potentiate phagocytosis of neutrophils undergoing apoptosis. In Press.

Robertson, O.H. and Uhley, C.G. (1938). Changes occurring in the macrophage system of the lungs in pneumococcus lobar pneumonia. J. Clin. Invest. 15, 115–130.

Saverymuttu, S.H., Phillips, G., Peters, A.M. and Lavender, J.P. (1985). Indium-III autologous leucocyte scanning in lobar pneumonia and lung abscesses. Thorax 40, 925–930.

Savill, J.S., Wyllie, A.H., Henson, J.E., Henson, P.M. and Haslett, C. (1989a). Macrophage phagocytosis of aging neutrophils in inflammation – programmed cell death leads to its recognition by macrophages. J. Clin. Invest. 83, 865–875.

Savill, J.S., Henson, P.M. and Haslett, C. (1989b). Phagocytosis of aged human neutrophils by macrophages is mediated by a novel "charge sensitive" recognition mechanism. J. Clin. Invest. 84, 1518–1527.

Savill, J.S., Dransfield, I., Hogg, N. and Haslett, C. (1990). Macrophage recognition of "senescent self". The vitronectin receptor mediates phagocytosis of cells undergoing apoptosis. Nature 342, 170–173.

Savill, J.S., Hogg, N. and Haslett, C. (1992a). Thrombospondin co-operates with CD36 and the vitronectin receptor macrophage in recognition of aged neutrophils. J. Clin. Invest. 90, 1513–1529.

Savill, J.S., Smith, J., Sarraf, C., Ren, Y., Abbott, F. and Rees, A. (1992b). Glomerular mesangial cells and inflammatory macrophages ingest neutrophils undergoing apoptosis. Kidney Int. 42, 924–936.

Savill, J. et al. (1993). Immunol. Today 14, 131–136.

Shi, Y., Glynn, J.M., Guilbert, L.J., Cotter, T.G., Bissonnette, R.P. and Green, D.R. (1992). Role for c-myc in activation-induced cell death in T cell hybridomas. Science 257, 212–215.

Spitznagel, J.K. (1990). Antibiotic proteins of neutrophils. J. Clin. Invest. 86, 1851–1854.

Spriggs, R.S., Boddington, M.M. and Mowat, A.G. (1978). Joint fluid cytology in Reiter's syndrome. Ann. Rheum. Dis. 37, 557–560.

Steinmann, R.M., Brodie, S.E. and Cohn, Z.A. (1976). Membrane flow during pinocytosis – a sterological analysis. J. Cell. Biol. 68, 665–687.

Stern, M., Meagher, L., Savill, J. and Haslett, C. (1992). Apoptosis in human eosinophils. Programmed cell death in the eosinophil leads to phagocytosis by macrophages and is modulated by IL-5. J. Immunol. 148, 3543–3549.

Vartio, T., Seppa, H. and Vaheri, A. (1981). Susceptibility of soluble and matrix fibronectin to degradation by tissue proteinases, mast cell chymase and cathepsin G. J. Biol. Chem. 256, 471–477.

Vaux, D.L., Cory, S. and Adams, J.M. (1992). *Bcl-2* gene promotes haemopoitic cell survival and cooperates with c-*myc* to immortalise pre-B cells. Nature 335, 440–442.

Ward, P.A. and Becker, E.L. (1967). The deactivation of rabbit neutrophils by chemotactic factor and the nature of the activatable esterase. J. Exp. Med. 127, 693–709.

Wedmore, C.V. and Williams, T.J. (1989). Control of vascular permeability by polymorphonuclear leukocytes in inflammation. Nature 289, 646–650.

Weiss, S.J. (1989). Tissue destruction by neutrophils. N. Engl. Med. J. 320, 365–376.

Whyte, M.K.B., Meagher, L.C., MacDermott, J. and Haslett, C. (1993a). Down-regulation of neutrophil function by apoptosis: A mechanism for functional isolation of neutrophils from inflammatory mediator stimulation. J. Immunol. 150, 5124–5134.

Whyte, M.K.B., Meagher, L.C., Hardwick, S.J., Savill, J.S. and Haslett, C. (1993b). Transient elevations of cytosolic free calcium retard subsequent apoptosis in neutrophils *in vitro*. J. Clin. Invest. 92, 446–455.

Wyllie, A.H. (1981). Glucocorticoid-induced thymocyte apoptosis is associated with endogenous endonuclease activation. Nature 284, 555–558.

Wyllie, A.H., Kerr, J.F.R. and Currie, A.R. (1980). Cell death: the significance of apoptosis. Int. Rev. Cytol. 68, 251.

Zimmerli, W., Seligmann, B. and Gallin, J.I. (1986). Exudation primes human and guinea-pig neutrophils for subsequent responsiveness to the chemotactic peptide N-formyl methionyl leucyl phenylamine and increases complement C3bi receptor expression. J. Clin. Invest. 77, 925–933.

Glossary

Note: This glossary is up to date for the current volume only and will be supplemented with each subsequent volume.

α_1, α_2 receptors Adrenoceptor subtypes
α_1-ACT α_1-Antichymotrypsin
α_1-AP α_1-antiproteinase *also known as* α_1-antitrypsin and α_1-proteinase inhibitor
α_1-AT α_1-Antitrypsin inhibitor *also known as* α_1-antiproteinase and α_1-proteinase inhibitor
α_1-PI α_1-Proteinase inhibitor *also known as* α_1-antitrypsin and α_1-antiproteinase
α_2-M α_2-macroglobulin
A Absorbance
AI, AII Angiotensin I, II
Å Angstrom
AA Arachidonic acid
aa Amino acids
AAb Autoantibody
ABAP 2',2'-azobis-2-amidino propane
Ab Antibody
Ab1 Idiotype antibody
Ab2 Anti-idiotype antibody
Ab2α Anti-idiotype antibody which binds outside the antigen binding region
Ab2β Anti-idiotype antibody which binds to the antigen binding region
Ab3 Anti-anti-idiotype antibody
Abcc Antibody dependent cellular cytotoxicity
ABA-L-GAT Arsanilic acid conjugated with the synthetic polypeptide L-GAT
AC Adenylate cyclase
ACAT Acyl-co-enzyme-A acyltransferase
ACAID Anterior chamber-associated immune deviation
ACE Angiotensin-converting enzyme
ACh Acetylcholine
ACTH Adrenocorticotrophin hormone
ADH Alcohol dehydrogenase
Ado Adenosine
ADP Adenosine diphosphate
ADPRT Adenosine diphosphate ribosyl transferase
AES Anti-eosinophil serum

Ag Antigen
AGE Advanced glycosylation end-product
AGEPC 1-*O*-alkyl-2-acetyl-*sn*-glyceryl-3-phosphocholine; *also known as* PAF and APRL
AH Acetylhydrolase
AID Autoimmune disease
AIDS Acquired immune deficiency syndrome
A/J A Jackson inbred mouse strain
ALP Anti-leukoprotease
ALS Amyotrophic lateral sclerosis
cAMP Cyclic adenosine monophosphate *also known as* adenosine 3', 5'-phosphate
AM Alveolar macrophage
AML Acute myelogenous leukaemia
AMP Adenosine monophosphate
AMVN 2,2'-azobis (2,4-dimethylvaleronitrile)
ANAb Anti-nuclear antibodies
ANCA Anti-neutrophil cytoplasmic auto antibodies
cANCA Cytoplasmic ANCA
pANCA Perinuclear ANCA
AND Anaphylactic degranulation
ANF Atrial natriuretic factor
ANP Atrial natriuretic peptide
Anti-I-A, Anti-I-E Antibody against class II MHC molecule encoded by I-A locus, I-E locus
anti-Ig Antibody against an immunoglobulin
anti-RTE Anti-tubular epithelium
AP-1 Activator protein-1
APA B-azaprostanoic acid
APAS Antiplatelet antiserum
APC Antigen-presenting cell
APD Action potential duration
apo-B Apolipoprotein B
APRL Anti-hypertensive polar renal lipid *also known as* PAF
APUD Amine precursor uptake and decarboxylation
AR Aldose reductase
AR-CGD Autosomal recessive form of chronic granulomatous disease
ARDS Adult respiratory distress syndrome

AS Ankylosing spondylitis
ASA Acetylsalicylic acid *also known as* aspirin
4-ASA, 5-ASA 4-, 5-aminosalicylic acid
ATHERO-ELAM A monocyte adhesion molecule
ATL Adult T cell leukaemia
ATP Adenosine triphosphate
ATPase Adenosine triphosphatase
ATPγs Adenosine 3' thiotriphosphate
AITP Autoimmune thrombocytopenic purpura
AUC Area under curve
AVP Arginine vasopressin

β_1, β_2 receptors Adrenoceptor subtypes
β_2 (CD18) A leucocyte integrin
β_2M β_2-Microglobulin
β-TG β-Thromboglobulin
B$_7$/BB$_1$ *Known to be* expressed on B cell blasts and immunostimulatory dendritic cells
BAF Basophil-activating factor
BAL Bronchoalveolar lavage
BALF Bronchoalveolar lavage fluid
BALT Bronchus-associated lymphoid tissue
B cell Bone marrow-derived lymphocyte
BCF Basophil chemotactic factor
B-CFC Basophil colony-forming cell
BCG Bacillus Calmette-Guérin
BCNU 1,3-bis(2-chloroethyl)-1-nitrosourea
bFGF Basic fibroblast growth factor
Bg Birbeck granules
BHR Bronchial hyperresponsiveness
BHT Butylated hydroxytoluene
b.i.d. *Bis in die* (twice a day)
Bk Bradykinin
Bk$_1$, Bk$_2$ receptors Bradykinin receptor subtypes *also known as* B$_1$ and B$_2$ receptors
Bk$_2$ receptor Bradykinin receptor subtype
Bl-CFC Blast colony-forming cells

B-lymphocyte Bursa-derived lymphocyte
BM Bone marrow
BMCMC Bone marrow cultured mast cell
BMMC Bone marrow mast cell
BOC-FMLP Butoxycarbonyl-FMLP
bp Base pair
BPB Para-bromophenacyl bromide
BPI Bacterial permeability-increasing protein
BSA Bovine serum albumin
BSS Bernard-Soulier Syndrome

^{51}**Cr** Chromium^{51}
C1, C2…C9 The 9 main components of complement
C1 inhibitor A serine protease inhibitor which inactivates C1r/C1s
C1q Complement fragment 1q
C1qR Receptor for C1w; facilitates attachment of immune complexes to mononuclear leucocytes and endothelium
C3a Complement fragment 3a (anaphylatoxin)
C3a$_{72-77}$ A synthetic carboxyterminal peptide C3a analogue
C3aR Receptor for anaphylatoxins, C3a, C4a, C5a
C3b Complement fragment 3b (anaphylatoxin)
C3bi Inactivated form of C3b fragment of complement
C4b Complement fragment 4b (anaphylatoxin)
C4BP C4 binding protein; plasma protein which acts as co-factor to factor I inactivate C3 convertase
C5a Complement fragment 5a (anaphylatoxin)
C5aR Receptor for anaphylatoxins C3a, C4a and C5a
C5b Complement fragment 5b (anaphylatoxin)
C$_\epsilon$2, C$_\epsilon$3, C$_\epsilon$4 Heavy chain of immunoglobulin E: domains 2, 3 and 4
Ca *The chemical symbol for* calcium
[Ca^{2+}]$_i$ Intracellular free calcium concentration
CAH Chronic active hepatitis
CALLA Common lymphoblastic leukaemia antigen
CALT Conjunctival associated lymphoid tissue
CaM Calmodulin
cAMP Cyclic adenosine monophosphate *also known as* adenosine 3′, 5′-phosphate
CAM Cell adhesion molecule
CAP57 Cationic protein from neutrophils
CAT Catalase
CatG Cathepsin G

CB Cytochalasin B
CBH Cutaneous basophil hypersensitivity
CBP Cromolyn-binding protein
CCK Cholecystokinin
CCR Creatinine clearance rate
CD Cluster of differentiation (a system of nomenclature for surface molecules on cells of the immune system); cluster determinant
CD1 Cluster of differentiation 1 *also known as* MHC class I-like surface glycoprotein
CD1a Isoform a *also known as* non-classical MHC class I-like surface antigen; present on thymocytes and dendritic cells
CD1b *Known to be* present on thymocytes and dendritic cells
CD1c Isoform c *also known as* non-classical MHC class I-like surface antigen; present on thymocytes
CD2 Defines T cells involved in antigen non-specific cell activation
CD3 *Also known as* T cell receptor-associated surface glycoprotein on T cells
CD4 Defines MHC class II-restricted T cell subsets
CD5 *Known to be* present on T cells and a subset of B cells; *also known as* Lyt 1 in mouse
CD7 Cluster of differentiation 7; present on most T cells and NK cells
CD8 Defines MHC class I-restricted T cell subset; present on NK cells
CD10 *Known to be* common acute leukaemia antigen
CD11a *Known to be* an α chain of LFA-1 (leucocyte function antigen-1) present on several types of leucocyte and which mediates adhesion
CD11c *Known to be* a complement receptor 4 α chain.
CD13 Aminopeptidase N; present on myeloid cells
CD14 *Known to be* a lipid-anchored glycoprotein; present on monocytes
CD15 *Known to be* Lewis X, fucosyl-N-acetyllactosamine
CD16 *Known to be* Fcγ receptor III
CD16-1, CD16-2 Isoforms of CD16
CD19 Recognizes B cells and follicular dendritic cells
CD20 *Known to be* a pan B cell
CD21 C3d receptor
CD23 Low affinity FcεR
CD25 Low affinity receptor for interleukin-2
CD27 Present on T cells and plasma cells
CD28 Present on resting and activated T cells and plasma cells

CD30 Present on activated B and T cells
CD31 *Known to be* on platelets, monocytes, macrophages, granulocytes, B-cells and endothelial cells; *also known as* PECAM
CD32 Fcγ receptor II
CD33$^+$ *Known to be* a monocyte and stem cell marker
CD34$^-$ *Known to be* a stem cell marker
CD35 C3b receptor
CD36 *Known to be* a macrophage thrombospondin receptor
CD40 Present on B cells and follicular dendritic cells
CD41 *Known to be* a platelet glycoprotein
CD44 *Known to be* a leucocyte adhesion molecule; *also known as* hyaluronic acid cell adhesion molecule (H-CAM), Hermes antigen, extracellular matrix receptor III (ECMIII); present on polymorphonuclear leucocytes
CD45 *Known to be* a pan leucocyte marker
CD45RO *Known to be* the isoform of leukosialin present on memory T cells
CD46 *Known to be* a membrane cofactor protein
CD49 Cluster of differentiation 49
CD51 *Known to be* vitronectin receptor alpha chain
CD54 *Known to be* Intercellular adhesion molecule-1 *also known as* ICAM-1
CD57 Present on T cells and NK subsets
CD58 A leucocyte function-associated antigen-3, *also known to be* as a member of the β-2 integrin family of cell adhesion molecules
CD59 *Known to be* a low molecular weight HRf present to many haematopoetic and non-haematopoetic cells
CD62 *Known to be present on* activated platelets and endothelial cells; *also known as* P-selectin
CD64 *Known to be* Fcγ receptor I
CD65 *Known to be* fucoganglioside
CD68 Present on macrophages
CD69 *Known to be* an activation inducer molecule; present on activated lymphocytes
CD72 Present on B-lineage cells
CD74 An invariant chain of class II B cells
CDC Complement-dependent cytotoxicity
cDNA Complementary DNA
CDP Choline diphosphate
CDR Complementary-determining region

CD$_{xx}$ Common determinant *xx*
CEA Carcinoembryonic antigen
CETAF Corneal epithelial T cell activating factor
CF Cystic fibrosis
Cf Cationized ferritin
CFA Complete Freund's adjuvant
CFC Colony-forming cell
CFU Colony-forming unit
CFU-Mk Megakaryocyte progenitors
CFU-S Colony-forming unit, spleen
CGD Chronic granulomatous disease
cGMP Cyclic guanosine monophosphate *also known as* guanosine 3', 5'-phosphate
CGRP Calcitonin gene-related peptide
CH2 Hinge region of human immunoglobulin
CHO Chinese hamster ovary
CI Chemical ionization
CIBD Chronic inflammatory bowel disease
CK Creatine phosphokinase
CKMB The myocardial-specific isoenzyme of creatine phosphokinase
Cl *The chemical symbol for* chloride
CL Chemiluminescent
CLA Cutaneous lymphocyte antigen
CL18/6 Anti-ICAM-1 monoclonal antibody
CLC Charcot-Leyden crystal
CMC Critical micellar concentration
CMI Cell mediated immunity
CML Chronic myeloid leukaemia
CMV Cytomegalovirus
CNS Central nervous system
CO Cyclooxygenase
CoA Coenzyme A
CoA-IT Coenzyme A – independent transacylase
Con A Concanavalin A
COPD Chronic obstructive pulmonary disease
COS Fibroblast-like kidney cell line established from simian cells
CoVF Cobra venom
CP Creatine phosphate
Cp Caeruloplasmin
c.p.m. Counts per minute
CPJ Cartilage/pannus junction
Cr *The chemical symbol for* chromium
CR Complement receptor
CR1, CR2 & CR4 Complement receptor types 1, 2 and 4
CR3-α Complement receptor type 3-α
CRF Corticotrophin-releasing factor
CRH Corticotrophin-releasing hormone
CRI Cross-reactive idiotype
CRP C-reactive protein
CSA Cyclosporin A
CSF Colony-stimulating factor

CSS Churg-Strauss syndrome
CT Computed tomography
CTAP-III Connective tissue-activating peptide
CTD Connective tissue diseases
C terminus Carboxy terminus of peptide
CThp Cytotoxic T lymphocyte precursors
CTL Cytotoxic T lymphocyte
CTLA-4 *Known to be* co-expressed with CD20 on activated T cells
CTMC Connective tissue mast cell
CVF Cobra venom factor

2D Second derivative
Da Dalton (the unit of relative molecular mass)
DAF Decay-accelerating factor
DAG Diacylglycerol
DAO Diamine oxidase
D-Arg D-Arginine
DArg-[Hyp^3,DPhe^7]-BK A bradykinin B$_2$ receptor antagonist. Peptide derivative of bradykinin
DArg-[Hyp^3,Thi^5,DTic^7,Tic^8]-BK A bradykinin B$_2$ receptor antagonist. Peptide derivative of bradykinin
DBNBS 3,5-dibromo-4-nitroso-benzenesulphonate
DC Dendritic cell
DCF Oxidized DCFH
DCFH 2',7'-dichlorofluorescin
DEC Diethylcarbamazine
DEM Diethylmaleate
desArg^9-BK Carboxypeptidase N product of bradykinin
desArg^{10}KD Carboxypeptidase N product of kallidin
DETAPAC Diethylenetriaminepentaacetic acid
DFMO α-Difluoromethyl ornithine
DFP Diisopropyl fluorophosphate
DFX Desferrioxamine
DGLA Dihomo-γ-linolenic acid
DH Delayed hypersensitivity
DHA Docosahexaenoic acid
DHBA Dihydroxybenzoic acid
DHR Delayed hypersensitivity reaction
DIC Disseminated intravascular coagulation
DL-CFU Dendritic cell/Langerhans cell colony forming
DLE Discoid lupus erythematosus
DMARD Disease-modifying anti-rheumatic drug
DMF *N,N*-dimethylformamide
DMPO 5,5-dimethyl-l-pyrroline *N*-oxide
DMSO Dimethyl sulfoxide
DNA Deoxyribonucleic acid
D-NAME D-Nitroarginine methyl ester
DNase Deoxyribonuclease

DNCB Dinitrochlorobenzene
DNP Dinitrophenol
Dpt4 *Dermatophagoides pteronyssinus* allergen 4
DGW2, DR3, DR7 HLA phenotypes
DREG-56 (Antigen) L-selectin
DREG-200 A monoclonal antibody against L-selectin
ds Double-stranded
DSCG Disodium cromoglycate
DST Donor-specific transfusion
DTH Delayed-type hypersensitivity
DTPA Diethylenetriamine pentaacetate
DTT Dithiothreitol
dv/dt Rate of change of voltage within time

ε Molar absorption coefficient
EA Egg albumin
EACA Epsilon-amino-caproic acid
EAE Experimental autoimmune encephalomyelitis
EAF Eosinophil-activating factor
EAR Early phase asthmatic reaction
EAT Experimental autoimmune thyroiditis
EBV Epstein–Barr virus
EC Endothelial cell
ECD Electron capture detector
ECE Endothelin-converting enzyme
E-CEF Eosinophil cytotoxicity enhancing factor
ECF-A Eosinophil chemotactic factor of anaphylaxis
ECG Electrocardiogram
ECGF Endothelial cell growth factor
ECGS Endothelial cell growth supplement
E. coli *Escherichia coli*
ECP Eosinophil cationic protein
EC-SOD Extracellular superoxide dismutase
EC-SOD C Extracellular superoxide dismutase C
ED$_{35}$ Effective dose producing 35% maximum response
ED$_{50}$ Effective dose producing 50% maximum response
EDF Eosinophil differentiation factor
EDL Extensor digitorum longus
EDN Eosinophil-derived neurotoxin
EDRF Endothelium-derived relaxing factor
EDTA Ethylenediamine tetraacetic acid *also known as* etidronic acid
EE Eosinophilic eosinophils
EEG Electroencephalogram
EET Epoxyeicosatrienoic acid
EFA Essential fatty acid
EFS Electrical field stimulation
EG1 Monoclonal antibody specific for the cleaved form of eosinophil cationic peptide

EGF Epidermal growth factor

EGTA Ethylene glycol-bis(β-aminoethyl ether) N,N,N',N'-tetraacetic acid

EHNA Erythro-9-(2-hydroxy–3-nonyl)-adenine

EI Electron impact

EIB Exercise-induced bronchoconstriction

eIF-2 Subunit of protein synthesis initiation factor

ELAM-1 Endothelial leucocyte adhesion molecule-1

ELF Respiratory epithelium lung fluid

ELISA Enzyme-linked immunosorbent assay

EMS Eosinophilia-myalgia syndrome

ENS Enteric nervous system

EO Eosinophil

EO-CFC Eosinophil colony-forming cell

EOR Early onset reaction *also known as* EAR

EPA Eicosapentaenoic acid

EpDIF Epithelial-derived inhibitory factor *also known as* epithelium-derived relaxant factor

EPO Eosinophil peroxidase

EPOR Erythropoietin receptor

EPR Effector cell protease

EPX Eosinophil protein X

ER Endoplasmic reticulum

ERCP Endoscopic retrograde cholangiopancreatography

E-selectin Endothelial selectin *formerly known as* endothelial leucocyte adhesion molecule-1 (ELAM-1)

ESP Eosinophil stimulation promoter

ESR Erythrocyte sedimentation rate

e.s.r. Electron spin resonance

ET, ET-1 Endothelin, -1

ETYA Eicosatetraynoic acid

FA Fatty acid

FAB Fast-electron bombardment

Fab Antigen binding fragment

F(ab')2 Fragment of an immunoglobulin produced pepsin treatment

FACS Flow activated cell sorter

factor B Serine protease in the C3 converting enzyme of the alternative pathway

factor D Serine protease which cleaves factor B

factor H Plasma protein which acts as a co-factor to factor I

factor I Hydrolyses C3 converting enzymes with the help of factor H

FAD Flavine adenine dinucleotide

FapyAde 5-formamido-4,6-diamino-pyrimidine

FapyGua 2,6-diamino-4-hydroxy-5-formamidopyrimidine

FBR Fluorescence photobleaching recovery

Fc Crystallizable fraction of immunoglobulin molecule

Fcγ Receptor for Fc portion of IgG

FcγRI Ig Fc receptor I *also known as* CD64

FcγRII Ig Fc receptor II *also known as* CD32

FcγRIII Ig Fc receptor III *also known as* CD16

FcεRI High affinity receptor for IgE

FcεRII Low affinity receptor for IgE

FcR Receptor for Fc region of antibody

FCS Foetal calf (bovine) serum

FEV₁ Forced expiratory volume in 1 second

Fe-TPAA Fe(III)-tris [N-(2-pyridylmethyl)-2-aminoethyl] amine

Fe-TPEN Fe(II)-tetrakis-N,N,N',N'-(2-pyridyl methyl-2-aminoethyl)amine

FFA Free fatty acids

FGF Fibroblast growth factor

FID Flame ionization detector

FITC Fluorescein isothiocyanate

FKBP FK506-binding protein

FLAP 5-lipoxygenase-activating protein

FMLP N-Formyl-methionyl-leucyl-phenylalanine

FNLP Formyl-norleucyl-leucyl-phenylalanine

FOC Follicular dendritic cell

FPLC Fast protein liquid chromatography

FPR Formyl peptide receptor

FS cell Folliculo-stellate cell

FSG Focal sequential glomerulosclerosis

FSH Follicle stimulating hormone

FX Ferrioxamine

5-FU 5-fluorouracil

Ga G-protein

G6PD Glucose 6-phosphate dehydrogenase

GABA γ-Aminobutyric acid

GAG Glycosaminoglycan

GALT Gut-associated lymphoid tissue

GAP GTPase-activating protein

GBM Glomerular basement membrane

GC Guanylate cyclase

GC-MS Gas chromatography mass spectroscopy

G-CSF Granulocyte colony-stimulating factor

GDP Guanosine 5'-diphosphate

GEC Glomerular epithelial cell

GF-1 An insulin-like growth factor

GFR Glomerular filtration rate

GH Growth hormone

GH-RF Growth hormone-releasing factor

Gi Family of pertussis toxin sensitive G-proteins

GI Gastrointestinal

GIP Granulocyte inhibitory protein

GlyCam-1 Glycosylation-dependent cell adhesion molecule-1

GMC Gastric mast cell

GM-CFC Granulocyte-macrophage colony-forming cell

GM-CSF Granulocyte-macrophage colony-stimulating factor

GMP Guanosine monophosphate (guanosine 5'-phosphate)

Go Family of pertussis toxin sensitive G-proteins

GP Glycoprotein

gp45-70 Membrane co-factor protein

gp90ᴹᴱᴸ 90 kD glycoprotein recognized by monoclonal antibody MEL-14; *also known as* L-selectin

GPIIb-IIIa Glycoprotein IIb-IIIa *known to be* a platelet membrane antigen

GppCH₂P Guanyl-methylene diphosphanate *also known as* a stable GTP analogue

GppNHp Guanylyl-imidiodiphosphate *also known as* a stable GTP analogue

GRGDSP Glycine–arginine–glycine–aspartic acid serine–proline

Gro Growth-related oncogene

GRP Gastrin-related peptide

Gs Stimulatory G protein

GSH Glutathione (reduced)

GSHPx Glutathione peroxidase

GSSG Glutathione (oxidized)

GT Glanzmann Thrombasthenia

GTP Guanosine triphosphate

GTP-γ-S Guanarine 5'O-(3-thiotriphosphate)

GTPase Guanidine triphosphatase

GVHD Graft-versus-host-disease

GVHR Graft-versus-host-reaction

H Histamine

H₁, H₂, H₃ Histamine receptor types 1, 2 and 3

H₂O₂ *The chemical symbol for* hydrogen peroxide

Hag Haemagglutinin

Hag-1, Hag-2 Cleaved haemagglutinin subunits-1, -2

H & E Haematoxylin and eosin

hIL Human interleukin

Hb Haemoglobin

HBBS Hank's balanced salt solution

HCA Hypertonic citrate

H-CAM Hyaluronic acid cell adhesion molecule

HDC Histidine decarboxylase

HDL High-density lipoprotein

HEL Hen egg white lysozyme

HEPE Hydroxyeicosapentanoic acid

HEPES N-2-Hydroxylethylpiperazine-N'-2-ethane sulphonic acid

HES Hypereosinophilic syndrome

HETE 5,8,9,11,12 and 15 Hydroxyeicosatetraenoic acid

5(S)HETE A stereo isomer of 5-HETE

HETrE Hydroxyeicosatrienoic acid

HEV High endothelial venule

HFN Human fibronectin

HGF Hepatocyte growth factor

HHTrE 12(S)-Hydroxy-5,8,10-heptadecatrienoic acid

HIV Human immunodeficiency virus

HL60 Human promyelocytic leukaemia cell line

HLA Human leucocyte antigen

HLA-DR2 Human histocompatability antigen class II

HMG CoA Hydroxylmethylglutaryl coenzyme A

HMW High molecular weight

HMT Histidine methyltransferase

HMVEC Human microvascular endothelial cell

HNC Human neutrophil collagenase (MMP-8)

HNE Human neutrophil elastase

HNG Human neutrophil gelatinase (MMP-9)

HODE Hydroxyoctadecanoic acid

HO· Hydroxyl radical

HO$_2$· Perhydroxyl radical

HPETE, 5-HPETE & 15-HPETE 5 and 15 Hydroperoxyeicosatetraenoic acid

HPETrE Hydroperoxytrienoic acid

HPODE Hydroperoxyoctadecanoic acid

HPLC High-performance liquid chromatography

HRA Histamine-releasing activity

HRAN Neutrophil-derived histamine-releasing activity

HRf Homologous-restriction factor

HRF Histamine-releasing factor

HRP Horseradish peroxidase

HSA Human serum albumin

HSP Heat-shock protein

HS-PG Heparan sulphate proteoglycan

HSV, HSV-1 Herpes simplex virus, -1

^{3}HTdR Tritiated thymidine

5-HT 5-Hydroxytryptamine *also known as* Serotonin

HTLV-1 Human T-cell leukaemia virus-1

HUVEC Human umbilical vein endothelial cell

[Hyp^3]-BK Hydroxyproline derivative of bradykinin

[Hyp^4]-KD Hydroxyproline derivative of kallidin

^{111}In Indium^{111}

Ia Immune reaction-associated antigen

Ia+ Murine class II major histocompatibility complex antigen

IB4 Anti-CD18 monoclonal antibody

IBD Inflammatory bowel disease

IBMX 3-isobutyl-1-methylxanthine

IBS Inflammatory bowel syndrome

iC3 Inactivated C3

iC4 Inactivated C4

IC$_{50}$ Concentration producing 50% inhibition

ICAM Intercellular adhesion molecules

ICAM-1, ICAM-2, ICAM-3 Intercellular adhesion molecules-1, -2, -3

cICAM-1 Circulating form of ICAM-1

ICE IL-1β-converting enzyme

i.d. Intradermal

IDC Interdigitating cell

IDD Insulin-dependent (type 1) diabetes

IEL Intraepithelial leucocyte

IELym Intraepithelial lymphocytes

IFA Incomplete Freund's adjuvant

IFN Interferon

IFNα, IFNβ, IFNγ Interferons α, β, γ

Ig Immunoglobulin

IgA, IgE, IgG, IgM Immunoglobulins A, E, G, M

IgG1 Immunoglobulin G class 1

IgG$_{2a}$ Immunoglobulin G class 2a

IGF-1 Insulin-like growth factor

Ig-SF Immunoglobulin supergene family

IGSS Immuno-gold silver stain

IHC Immunohistochemistry

IHES Idiopathic hypereosinophilic syndrome

IκB NFκB inhibitor protein

IL Interleukin

IL-1, Il-2...IL-8 Interleukins-1, 2...-8

IL-1α, IL-1β Interleukin-1α, -1β

ILR Interleukin receptor

IL-1R, IL-2R; IL-3R–IL-6R Interleukin-1–6 receptors

IL-1Ra Interleukin-1 receptor antagonist

IL-2Rβ Interleukin-2 receptor β

IMF Integrin modulating factor

IMMC Intestinal mucosal mast cell

i.p. Intraperitoneally

IP$_1$ Inositol monophosphate

IP$_2$ Inositol biphosphate

IP$_3$ Inositol 1,4,5-trisphosphate

IP$_4$ Inositol tetrakisphosphate

IPF Idiopathic pulmonary fibrosis

IPO Intestinal peroxidase

IpOCOCq Isopropylidene OCOCq

I/R Ischaemia-reperfusion

IRAP IL-1 receptor antagonist protein

IRF-1 Interferon regulatory factor 1

I$_{sc}$ Short-circuit current

ISCOM Immune-stimulating complexes

ISGF3 Interferon-stimulated gene Factor 3

ISGF3α, ISGFγ α, γ subunits of ISGF3

IT Immunotherapy

ITP Idiopathic thrombocytopenic purpura

i.v. Intravenous

K *The chemical symbol for* potassium

K$_a$ Association constant

kb Kilobase

20KDHRF A homologous restriction factor; binds to C8

65KDHRF A homologous restriction factor, also known as C8 binding protein; interferes with cell membrane pore-formation by C5b-C8 complex

Kcat Catalytic constant; a measure of the catalytic potential of an enzyme

K$_d$ Equilibrium dissociation constant

kD Kilodalton

K$_D$ Dissociation constant

KD Kallidin

Ki Antagonist binding affinity

Ki67 Nuclear membrane antigen

KLH Keyhole limpet haemocyanin

K$_m$ Michaelis constant

KOS KOS strain of herpes simplex virus

λ_{max} Wavelength of maximum absorbance

LAD Leucocyte adhesion deficiency

LAK Lymphocyte-activated killer (cell)

LAM, LAM-1 Leucocyte adhesion molecule, -1

LAR Late-phase asthmatic reaction

L-Arg L-Arginine

LBP LPS binding protein

LC Langerhans cell

LCF Lymphocyte chemoattractant factor

LCR Locus control region

LDH Lactate dehydrogenase

LDL Low-density lipoprotein

LDV Laser Doppler velocimetry

Le^x(Lewis X) Leucocyte ligand for selectin

LFA Leucocyte function-associated antigen

LFA-1 Leucocyte function-associated antigen-1; *also known to be* a member of the β-2 integrin family of cell adhesion molecules

LG β-Lactoglobulin
LGL Large granular lymphocyte
LH Luteinizing hormone
LHRH Luteinizing hormone-releasing hormone
LI Labelling index
LIS Lateral intercellular spaces
LMP Low molecular mass polypeptide
LMW Low molecular weight
L-NOARG L-Nitroarginine
LO Lipoxygenase
5-LO, 12-LO, 15-LO 5-, 12-, 15-Lipoxygenases
LP(a) Lipoprotein(a)
LPS Lipopolysaccharide
L-selectin Leucocte selectin, *formerly known as* monoclonal antibody that recognizes murine L-selectin (MEL-14 antigen), leucocyte cell adhesion molecule-1 (LeuCAM-1), lectin cell adhesion molecule-1 (LeCAM-1 or LecCAM-1), leucocyte adhesion molecule-1 (LAM-1)
LT Leukotriene
LTA$_4$, LTB$_4$, LTC$_4$, LTD$_4$, LTE$_4$ Leukotrienes A$_4$, B$_4$, C$_4$, D$_4$ and E$_4$
L$_y$-1$^+$ (Cell line)
LX Lipoxin
LXA$_4$, LXB$_4$, LXC$_4$, LXD$_4$, LXE$_4$ Lipoxins A$_4$, B$_4$, C$_4$, D$_4$ and E$_4$

M Monocyte
M3 Receptor Muscarinic receptor subtype 3
M-540 Merocyanine-540
mAb Monoclonal antibody
mAb IB4, mAb PB1.3, mAb R 3.1, mAb R 3.3, mAb 6.5, mAb 60.3 Monoclonal antibodies IB4, PB1.3, R 3.1, R 3.3, 6.5, 60.3
MABP Mean arterial blood pressure
MAC Membrane attack molecule
Mac Macrophage (also abbreviated to MΦ)
Mac-1 Macrophage-1 antigen; a member of the β-2 integrin family of cell adhesion molecules (also abbreviated to MΦ1), *also known as* monocyte antigen-1 (M-1), complement receptor-3 (CR3), CD11b/CD18
MAF Macrophage-activating factor
MAO Monoamine oxidase
MAP Monophasic action potential
MAPTAM An intracellular Ca^{2+} chelator
MARCKS Myristolated, alanine-rich C kinase substrate; specific protein kinase C substrate
MBP Major basic protein
MBSA Methylated bovine serum albumin
MC Mesangial cells

MCAO Middle cerebral artery occlusion
M cell Microfold or membranous cell of Peyer's patch epithelium
MCP Membrane co-factor protein
MCP-1 Monocyte chemotactic protein-1
M-CSF Monocyte/macrophage colony-stimulating factor
MC$_T$ Tryptase-containing mast cell
MC$_{TC}$ Tryptase- and chymase-containing mast cell
MDA Malondialdehyde
MDGF Macrophage-derived growth factor
MDP Muramyl dipeptide
MEA Mast cell growth-enhancing activity
MEL Metabolic equivalent level
MEM Minimal essential medium
MG Myasthenia gravis
MGSA Melanoma-growth-stimulatory activity
MHC Major histocompatibility complex
MI Myocardial ischaemia
MIF Migration inhibition factor
mIL Mouse interleukin
MI/R Myocardial ischaemia/reperfusion
MIRL Membrane inhibitor of reactive lysis
mix-CFC Colony-forming cell mix
Mk Megakaryocyte
MLC Mixed lymphocyte culture
MLymR Mixed lymphocyte reaction
MLR Mixed leucocyte reaction
mmLDL Minimally modified low-density lipoprotein
MMC Mucosal mast cell
MMCP Mouse mast cell protease
MMP, MMP1 Matrix metalloproteinase, -1
MNA 6-Methoxy-2-napthylacetic acid
MNC Mononuclear cells
MΦ Macrophage (also abbreviated to Mac)
MPG N-(2-mercaptopropionyl)-glycine
MPO Myeloperoxidase
MPSS Methyl prednisolone
MPTP N-methyl-4-phenyl-1,2,3,6-tetrahydropyridine
MRI Magnetic resonance imaging
mRNA Messenger ribonucleic acid
MS Mass spectrometry
MSS Methylprednisolone sodium succinate
MT Malignant tumour
MW Molecular weight

Na *The chemical symbol for* sodium
NA Noradrenaline *also known as* norepinephrine

NAAb Natural autoantibody
NAb Natural antibody
NAC N-acetylcysteine
NADH Reduced nicotinamide adenine dinucleotide
NADP Nicotinamide adenine diphosphate
NADPH Reduced nicotinamide adenine dinucleotide phosphate
NAF Neutrophil activating factor
L-NAME L-Nitroarginine methyl ester
NANC Non-adrenergic, non-cholinergic
NAP Neutrophil-activating peptide
NAPQI N-acetyl-p-benzoquinone imine
NAP-1, NAP-2 Neutrophil-activating peptides -1 and -2
NBT Nitro-blue tetrazolium
NC1 Non-collagen 1
N-CAM Neural cell adhesion molecule
NCEH Neutral cholesteryl ester hydrolase
NCF Neutrophil chemotactic factor
NDGA Nordihydroguaretic acid
NDP Nucleoside diphosphate
Neca 5'-(N-ethyl carboxamido)-adenosine
NED Nedocromil sodium
NEP Neutral endopeptidase (EC 3.4.24.11)
NF-AT Nuclear factor of activated T lymphocytes
NF-$\varkappa$B Nuclear factor-$\varkappa$B
NgCAM Neural-glial cell adhesion molecule
NGF Nerve growth factor
NGPS Normal guinea-pig serum
NIH 3T3 (fibroblasts) National Institute of Health 3T3-Swiss albino mouse fibroblast
NIMA Non-inherited maternal antigens
NIRS Near infrared spectroscopy
Nk Neurokinin
NK Natural killer
Nk-1, Nk-2, NK-3 Neurokinin receptor subtypes 1,2 and 3
NkA Neurokinin A
NkB Neurokinin B
NLS Nuclear location sequence
NMDA N-methyl-D-aspartic acid
L-NMMA L-Nitromonomethyl arginine
NMR Nuclear magnetic resonance
NO *The chemical symbol for* nitric oxide
NOD Non-obese diabetic
NOS Nitric oxide synthase
c-NOS Ca^{2+}-dependent constitutive form of NOS
i-NOS Inducible form of NOS
NPK Neuropeptide K

NPY Neuropeptide Y
NRS Normal rabbit serum
NSAID Non-steroidal anti-inflammatory drug
NSE Nerve-specific enolase
NT Neurotensin
N terminus Amino terminus of peptide

$^1\Delta O_2$ Singlet Oxygen (Delta form)
$^1\Sigma O_2$ Singlet Oxygen (Sigma form)
$O_2^{\cdot-}$ *The chemical symbol for* the superoxide anion radical
OA Osteoarthritis
OAG Oleoyl acetyl glycerol
OD Optical density
ODC Ornithine decarboxylase
ODFR Oxygen-derived free radical
ODS Octadecylsilyl
OH$^-$ *The chemical symbol for* hydroxyl ion
$\cdot$OH *The chemical symbol for* hydroxyl radical
8-OH-Ade 8-Hydroxyadenine
6-OHDA 6-Hydroxyguanine
8-OH-dG 8-Hydroxydeoxyguanosine *also known as* 7,8-dihydro-8-oxo-2'-deoxyguanosine
8-OH-Gua 8-Hydroxyguanine
OHNE Hydroxynonenal
4-OHNE 4-Hydroxynonenal
OT Oxytocin
OVA Ovalbumin
ox-LDL Oxidized low-density lipoprotein

Ψa Apical membrane potential
P Probability
P Phosphate
P_aO_2 Arterial oxygen pressure
P_i Inorganic phosphate
p150,95 A member of the β-2-integrin family of cell adhesion molecules; *also known as* CD11c
PA Phosphatidic acid
pA$_2$ Negative logarithm of the antagonist dissociation constant
PAF Platelet-activating factor *also known as* APRL and AGEPC
PAGE Polyacrylamide gel electrophoresis
PAI Plasminogen activator inhibitor
PA-IgG Platelet associated immunoglobulin G
PAM Pulmonary alveolar macrophages
PAS Periodic acid–Schiff reagent
PBA Polyclonal B cell activators
PBC Primary biliary cirrhosis
PBL Peripheral blood lymphocytes
PBMC Peripheral blood mononuclear cells
PBN *N-tert*-butyl-α-phenylnitrone
PBS Phosphate-buffered saline
PC Phosphatidylcholine

PCA Passive cutaneous anaphylaxis
pCDM8 Eukaryotic expression vector
PCNA Proliferating cell nuclear antigen
PCR Polymerase chain reaction
PCT Porphyria cutanea tarda
p.d. Potential difference
PDBu 4α-Phorbol 12,13-dibutyrate
PDE Phosphodiesterase
PDGF Platelet-derived growth factor
PDGFR Platelet-derived growth factor receptor
PE Phosphatidylethanolamine
PECAM-1 Platelet endothelial cell adhesion molecule-1; *also known as* CD31
PEG Polyethylene glycol
PET Positron emission tomography
PEt Phosphatidylethanolamine
PF$_4$ Platelet factor 4
PG Prostaglandin
PGAS Polyglandular autoimmune syndrome
PGD$_2$ Prostaglandin D_2
PGE1, PGE$_2$, PGF$_2$, PGF$_{2\alpha}$, PGG$_2$, PGH$_2$ Prostaglandins E_1, E_2, F_2, $F_{2\alpha}$, F_2, H_2
PGF, PGH Prostaglandins F and H
PGI$_2$ Prostaglandin I_2 *also known as* prostacyclin
P_aO_2 Arterial oxygen pressure
PGP Protein gene-related peptide
Ph^1 Philadelphia (chromosome)
PHA Phytohaemagglutinin
PHD PHD[8(1-hydroxy-3-oxo-propyl)-9,12-dihydroxy-5,10 heptadecadienic acid]
PHI Peptide histidine isoleucine
PHM Peptide histidine methionine
P_i Inorganic phosphate
PI Phosphatidylinositol
PI-3,4-P2 Phosphatidylinositol 3,4-biphosphate
PI-3,4,5-P3 Phosphatidylinositol 3,4,5-trisphosphate
PI-3-kinase Phosphatidylinositol-3-kinase
PI-4-kinase Phosphatidylinositol-4-kinase
PI-3-P Phosphatidylinositol-3-phosphate
PI-4-P Phosphatidylinositol-4-phosphate
PI-4,5-P2 Phosphatidylinositol 4,5-biphosphate
PIP Phosphatidylinositol monophosphate
PIP$_2$ Phosphatidylinositol biphosphate
PK Protein kinase
PKA, PKC Protein kinases A and C
PKG cGMP-dependent protein kinase, protein kinase G
PL Phospholipase

PLA, PLA$_2$, PLC, PLD Phospholipases A, A_2, C and D
PLN Peripheral lymph node
PLNHEV Peripheral lymph node HEV
PLP Proteolipid protein
PLT Primed lymphocyte typing
PMA Phorbol myristate acetate
PMC Peritoneal mast cell
PMN Polymorphonuclear neutrophil
PMSF Phenylmethylsulphonyl fluoride
PNAd Peripheral lymph node vascular addressin
PNH Paroxysmal nocturnal hemoglobinuria
PNU Protein nitrogen unit
p.o. *Per os* (by mouth)
POBN α-4-Pyridyl-oxide-*N-t*-butyl nitrone
PPD Purified protein derivative
PPME Polymeric polysaccharide rich in mannose-6-phosphate moieties
PRA Percentage reactive activity
PRD, PRDII Positive regulatory domain, -II
PR3 Proteinase-3
PRBC Parasitized red blood cell
proET-1 Proendothelin-1
PRL Prolactin
PRP Platelet-rich plasma
PS Phosphatidylserine
P-selectin Platelet selectin *formerly known as* platelet activation-dependent granule external membrane protein (PADGEM), granule membrane protein of MW 140 kD (GMP–140)
PT Pertussis toxin
PTCA Percutaneous transluminal coronary angioplasty
PTCR Percutaneous transluminal coronary recanalization
Pte-H$_4$ Tetrahydropteridine
PUFA Polyunsaturated fatty acid
PUMP-1 Punctuated metalloproteinase *also known as* matrilysin
PWM Pokeweed mitogen
Pyran Divinylether maleic acid

q.i.d. *Quater in die* (four times a day)
QRS Segment of electrocardiogram

$\cdot$R Free radical
R15.7 Anti-CD18 monoclonal antibody
RA Rheumatoid arthritis
RANTES A member of the IL8 supergene family (*R*egulated on *a*ctivation, *n*ormal *T* expressed and *s*ecreted)
RAST Radioallergosorbent test
RBC Red blood cell

RBF Renal blood flow
RBL Rat basophilic leukaemia
RC Respiratory chain
RE RE strain of herpes simplex virus type 1
REA Reactive arthritis
REM Relative electrophoretic mobility
RER Rough endoplasmic reticulum
RF Rheumatoid factor
RFL-6 Rat foetal lung-6
RFLP Restriction fragment length polymorphism
RGD Arginine–glycine–asparagine
rh- Recombinant human – (prefix usually referring to peptides)
RIA Radioimmunoassay
RMCP, RMCPII Rat mast cell protease, -II
RNA Ribonucleic acid
RNase Ribonuclease
RNHCl *N*-Chloramine
RNL Regional lymph nodes
ROM Reactive oxygen metabolite
RO· *The chemical symbol for* alkoxyl radical
ROO· *The chemical symbol for* peroxy radical
ROP Retinopathy of prematurity
ROS Reactive oxygen species
R-PIA *R*-(1-methyl-1-phenyltheyl)-adenosine
RPMI 1640 Roswell Park Memorial Institute 1640 medium
RS Reiter's syndrome
RSV Rous sarcoma virus
RTE Rabbit tubular epithelium
RTE-a-5 Rat tubular epithelium antigen a-5
r-tPA Recombinant tissue-type plasminogen activator
RW Ragweed

S Svedberg (unit of sedimentation density)
SALT Skin-associated lymphoid tissue
SAZ Sulphasalazine
SC Secretory component
SCF Stem cell factor
SCFA Short-chain fatty acid
SCG Sodium cromoglycate *also known as* DSCG
SCID Severe combined immunodeficiency syndrome
sCR1 Soluble type–1 complement receptors
SCW Streptococcal cell wall
SD Standard deviation
SDS Sodium dodecyl sulphate
SDS-PAGE Sodium dodecyl sulphate-polyacrylamide gel electrophoresis
SEM Standard error of the mean
SGAW Specific airway conductance

SHR Spontaneously hypertensive rat
SIM Selected ion monitoring
SIRS Soluble immune response suppressor
SIV Simian immunodeficiency virus
SK Streptokinase
SLE Systemic lupus erythematosus
SLe^x Sialyl Lewis X antigen
SLO Streptolysin-O
SLPI Secretory leucocyte protease inhibitor
SM Sphingomyelin
SNAP *S*-Nitroso-*N*-acetylpenicillamine
SNP Sodium nitroprusside
SOD Superoxide dismutase
SOM Somatostatin *also known as* somatotrophin release-inhibiting factor
SOZ Serum-opsonized zymosan
SP Sulphapyridine
SR Systemic reaction
sr Sarcoplasmic reticulum
SRBC Sheep red blood cells
SRS Slow-reacting substance
SRS-A Slow-reacting substance of anaphylaxis
STZ Streptozotocin
Sub P Substance P

T Thymus-derived
α-TOC α-Tocopherol
t$_{1/2}$ Half-life
T84 Human intestinal epithelial cell line
TauNHCl Taurine monochloramine
TBA Thiobarbituric acid
TBAR Thiobarbituric acid-reactive product
TBM Tubular basement membrane
TBN di-*tert*-Butyl nitroxide
tBOOH *tert*-Butylhydroperoxide
TCA Trichloroacetic acid
T cell Thymus-derived lymphocyte
TCR T cell receptor α/β or γ/δ heterodimeric forms
TDI Toluene diisocyanate
TEC Tubular epithelial cell
TF Tissue factor
Tg Thyroglobulin
TGF Transforming growth factor
TGFα, TGFβ, TGFβ$_1$ Transforming growth factors α, β, and β_1
T$_H$ T helper cell
T$_H$0 T Helper o
T$_{Hp}$ T helper precursor
T$_H$0, T$_H$1, T$_H$2 Subsets of helper T cells
THP-1 Human monocytic leukaemia
Thy 1+ Murine T cell antigen
t.i.d. Ter in die (three times a day)
TIL Tumour-infiltrating lymphocytes
TIMP Tissue inhibitors of metalloproteinase

TIMP-1, TIMP-2 Tissue inhibitors of metalloproteinases 1 and 2
Tla Thymus leukaemia antigen
TLC Thin-layer chromatography
TLCK Tosyl-lysyl-CH$_2$Cl
TLP Tumour-like proliferation
Tm T memory
TNF, TNF-α Tumour necrosis factor, -α
tPA Tissue-type plasminogen activator
TPA 12-*O*-tetradeconylphorbol-13-acetate
TPCK Tosyl-phenyl-CH$_2$Cl
TPK Tyrosine protein kinases
TPP Transpulmonary pressure
TRAP Thrombospondin related anomalous protein
Tris Tris(hydroxymethyl)aminomethane
TSH Thyroid-stimulating hormone
TSP Thrombospondin
TTX Tetrodotoxin
TX Thromboxane
TXA$_2$, TXB$_2$ Thromboxane A$_2$, B$_2$
Tyk2 Tyrosine kinase

U937 (cells) Histiocytic lymphoma, human
UC Ulcerative colitis
UDP Uridine diphosphate
UPA Urokinase-type plasminogen activator
UTP Uridine triphosphate
UV Ultraviolet
UVA Ultraviolet A
UVB Ultraviolet B
UVR Ultraviolet irradiation
UW University of Wisconsin (preserving solution)

VAP Viral attachment protein
VC Veiled cells
VCAM, VCAM-1 Vascular cell adhesion molecule, -1, *also known as* inducible cell adhesion molecule MW 110 kD (INCAM–110)
VF Ventricular fibrillation
VIP Vasoactive intestinal peptide
VLA Very late activation antigen beta chain; *also known as* CD29
VLA α2 Very late activation antigen alpha 2 chain; *also known as* CD49b
VLA α4 Very late activation antigen alpha 4 chain; *also known as* CD49d
VLA α6 Very late activation antigen alpha 6 chain; *also known as* CD49f
VLDL Very low-density lipoprotein
V max Maximal velocity
V min Minimal velocity
VN Vitronectin
VO$_4^-$ *The chemical symbol for* vanadate
vp Viral protein
VP Vasopressin
VPB Ventricular premature beat

VT Ventricular tachycardia
vWF von Willebrand factor

W Murine dominant white spotting mutation
WBC White blood cell
WGA Wheat germ agglutinin

WI Warm ischaemia

XD Xanthine dehydrogenase
XO Xanthine oxidase

Y1/82A A monoclonal antibody detecting a cytoplasmic antigen in human macrophages

ZA Zonulae adherens
ZAS Zymosan-activated serum
zLYCK
Carboxybenzyl-Leu-Tyr-CH$_2$Cl
ZO Zonulae occludentes

Key to Illustrations

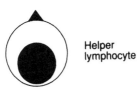 Helper
lymphocyte

 Suppressor
lymphocyte

 Killer
lymphocyte

 Plasma cell

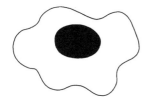 Bacterial or
Tumour cell

 Blood vessel
lumen

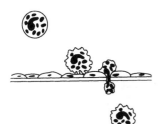 Eosinophil
passing through
vessel wall

 Neutrophil
passing through
vessel wall

Resting
neutrophil

Activated
neutrophil

Resting
eosinophil

Activated
eosinophil

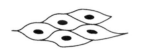
Smooth
muscle

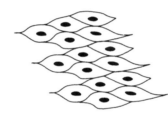
Smooth muscle
thickening

Smooth muscle
contraction

Normal blood
vessel

Endothelial cell
permeability

Resting
macrophage

Activated
macrophage

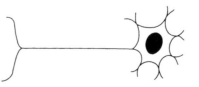
Nerve

 Intact epithelium

 Damaged epithelium

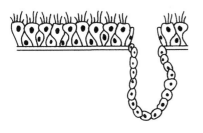 Intact epithelium with submucosal gland

 Normal submucosal gland

 Hypersecreting submucosal gland

 Normal airway

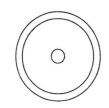 Oedema

 Bronchospasm

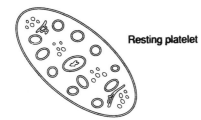 Resting platelet

 Activated platelet

 Airway hypersecreting mucus

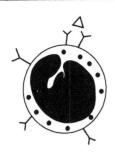

Resting
basophil

Activated
basophil

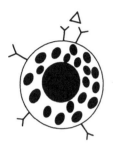

Resting
mast cell

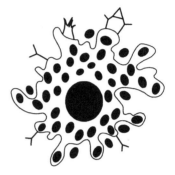

Activated
mast cell

Resting
chondrocyte

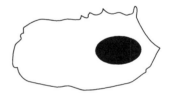

Activated
chondrocyte

Cartilage

Fibroblast

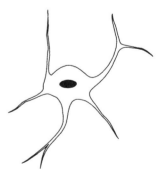

Dendritic cell/
Langerhans cell

Arteriole

Venule

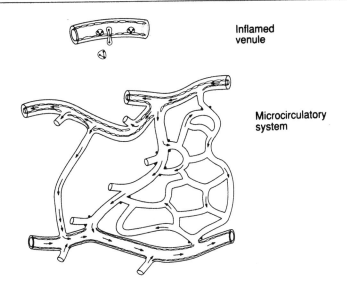

Inflamed
venule

Microcirculatory
system

Index